Aging and
Drug Therapy

Ettore Majorana International Science Series
Series Editor:
Antonino Zichichi
European Physical Society
Geneva, Switzerland

(LIFE SCIENCES)

Recent volumes in the series

A Continuation Order Plan is available for this series. A continuation order will bring delivery of
each new volume immediately upon publication. Volumes are billed only upon actual shipment.
For further information please contact the publisher.

Aging and Drug Therapy

Edited by

G. Barbagallo-Sangiorgi

Institute of Internal Medicine and Geriatrics
University of Palermo
Palermo, Italy

and

A. N. Exton-Smith

School of Medicine
University College London
London, England

Plenum Press • New York and London

Library of Congress Cataloging in Publication Data

International School of Physiopathology and Clinic of the Third Age (3rd: 1982: Erice, Italy)
    Aging and drug therapy.

    (Ettore Majorana international science series. Life sciences; v. 15)
    "Proceedings of the third course of the International School of Physiopathology and Clinic of the Third Age, held November 7–13, 1982, in Erice, Sicily, Italy"—T.p. verso.
    Includes bibliographies and index.
    1. Geriatric pharmacology—Congresses.
    I. Barbagallo-Sangiorgi, G. II. Exton-Smith, A. N. (Arthur Norman) III. Title. IV. Series: Ettore Majorana international science series. Life sciences; 15. [DNLM: 1. Drug Therapy—in old age—congresses. W1 ET712M v.15/WT 100 I617 1982a]
RC953.7.I58 1982   615.5′8′0880565   84-11772
ISBN-13:978-1-4612-9721-5   e-ISBN-13: 978-1-4613-2791-2
DOI: 10.1007/978-1-4613-2791-2

Proceedings of the Third Course of the International School of Physiopathology and Clinic of the Third Age, held November 7–13, 1982, in Erice, Sicily, Italy

©1984 Plenum Press, New York
A Division of Plenum Publishing Corporation
233 Spring Street, New York, N.Y. 10013
Softcover reprint of the hardcover 1st edition 1984

PREFACE

One of the greatest advances in Geriatric Medicine during the last quarter of a century has been the increased realization of the possibilities of treatment for elderly patients. Neglect has been replaced by a more optimistic therapeutic endeavour and countless old people have benefitted from this approach. But there is also a drawback, and this is the risk of hazardous side effects of medication which are often directly proportional to the biological potency of the drug and may be unpredictably increased due to changes in the senile organism.

In fact the anatomical and biological changes in old age alter both the kinetics of most drugs and the receptor response. On account of these changes the individual tolerance of aged patients to drug therapy may be quite different from that of younger subjects. Thus for a variety of reasons elderly patients receive more drugs, but they are at a higher risk of encountering adverse reactions, which often show atypical clinical features.

We can therefore speak of "geriatric iatrogenic disorders" and point out that some of these side effects are determined by a complex pathogenesis due to the particular pathophysiological condition in the elderly. It is important to encourage the habit of individually evaluating potential risk versus expected advantages of drugs in keeping with the same principles adopted in the evaluation of cost/benefit ratio.

It is necessary to study the complex problems posed by drug therapy in aged patients, emphasizing peculiarities concerning pharmacokinetics, pharmacological effects, indications for specific illnesses, the most convenient dose schedules, as well as the frequency and clinical features of possible side effects. The present Course is aimed at achieving an increasingly rational and conscious attitude towards drug treatment in the elderly, by maximum utilization of advantages and the reduction of drug side effects to a minimum.

The subjects proposed and the specific competence of our
lecturers make us hope that this goal may be achieved.

G. Barbagallo Sangiorgi
A. N. Exton-Smith

CONTENTS

CONTENTS

PHARMACOKINETICS IN ADVANCING AGE

I. H. Stevenson

Department of Pharmacology and Clinical Pharmacology
Ninewells Hospital and Medical School
Dundee, Scotland

INTRODUCTION

As has been extensively reviewed,[1,2,3] there is now a
considerable body of evidence to indicate that the response to many
drugs may change as people age. The three principal factors
contributing to such age-related alterations in drug effect are:

(i) altered pharmacokinetics
(ii) change in receptor sensitivity (or density)
(iii) impairment of normal homeostatic mechanisms

In general, but not inevitably, drugs produce exaggerated
effects in old people, particularly in the debilitated elderly, but
it is often difficult to identify clearly which of the three factors
is/are contributing. With a drug such as digoxin there is impaired
renal elimination, with benzodiazepines there may be an increase in
sensitivity of the elderly brain whereas with some antihypertensive
agents a decreased homeostatic compensation may be particularly
important. While the present paper focusses attention on only the
first of these - drug disposition or pharmacokinetics in the elderly
- it is important to accept that changes in this may often be
accompanied by alteration in the other two factors.

It is also important to realise that, as yet, it is not possible
to separate clearly age effects per se from those resulting from
ageing associated with the presence of disease. The elderly, while
referred to collectively, are a heterogeneous group rangeing from
"fit old folk" to the grossly debilitated. The presence of multiple
pathology and the consequent need in many cases to prescribe several
drugs concurrently adds a further complication.

Drug kinetics in the elderly has been the subject of a number of reviews[1,4,5,6] and may conveniently be considered under the headings of absorption, distribution, hepatic metabolism (including presystemic elimination) and renal excretion. The principal physiological changes occurring in old age are summarised in Table 1, together with their pharmacokinetic consequences.

Drug Absorption

There are a number of changes in the gastro-intestinal tract in the elderly which might be expected to alter drug absorption[7] e.g. increased gastric pH, decreased intestinal blood flow secondary to decreased cardiac output and alterations in gastric emptying time and gastro-intestinal motility. It is perhaps surprising therefore that, for a great many drugs, the available evidence[7] indicates that the rate and extent of drug absorption are largely unchanged in the elderly. Changes where they do occur are often small and unlikely to be of clinical significance particularly during chronic therapy. With digoxin and theophylline for example, the extent of absorption was similar in the two age groups and the sole difference was a delay in the absorption of digoxin in the elderly group of patients studied. As will be discussed in the section on hepatic drug metabolism and presystemic elimination, age-related changes in absorption are likely to be of greatest significance where there is a reduced first pass effect in the elderly. With a drug such as levodopa, both reduced first pass effect and more rapid gastric emptying with reduced gastric degradation are responsible for the greater area under the curve (AUC) in the elderly. After oral dosing, plasma drug concentration and AUC are often higher in elderly subjects than in young controls. However in almost all cases these changes are likely to result from impairment of metabolism (e.g. aspirin, chlormethiazole, metoprolol, propranolol and paracetamol), decreased renal excretion (propicillin, procaine) or a reduction in distribution volume (quinine, antipyrine) rather than from increased absorption. The only drug for which there is evidence of a substantial reduction in extent of absorption is prazosin, the bioavailability of which in the elderly is only some 60 per cent of that in younger subjects.[8] Since no alteration in extent of first-pass metabolism occurred, it was concluded that the gastro-intestinal absorption of prazosin decreases in old age.

Drug Distribution

The most important features of a drug's distribution are distribution in body fluids and the extent of binding to plasma proteins (usually to albumin but with some drugs to α_1, acid glycoprotein), to red cells and to body tissues, including the target organ. In old age there is a significant decrease in lean body mass and total body water, an increase in body fat (particularly in males) and a small but significant decrease in plasma albumin.

Table 1. Physiological changes occurring in old age and their pharmacokinetic consequences

PROCESS	PHYSIOLOGICAL CHANGES	PHARMACOKINETIC CHANGES
Drug absorption	↑ gastric pH	with many drugs, no alteration
	↓ gastrointestinal mobility	↓ rate with digoxin
	↓ intestinal blood flow	↓ extent with prazosin
	↓ mucosal cell absorbing area	
Drug distribution	↓ lean body mass	↓ volume of distribution of some hydrophilic drugs
	↓ total body water	↑ volume of distribution of some lipophilic drugs
	↑ body fat	extent of plasma protein binding of many drugs not significantly altered
	↓ plasma albumin	
Drug hepatic metabolism	↓ hepatic size	↓ metabolism of some but not all drugs
	↓ hepatic blood flow	↓ first pass effect
	↓ no. of functional cells	
	↓ microsomal enzyme activity	
Drug renal excretion	↓ renal blood flow	↓ elimination
	↓ glomerular filtration	
	↓ tubular excretion	

While it is difficult to generalise, the distribution of water-soluble drugs such as antipyrine and paracetamol may decrease in the elderly whereas lipid soluble drugs such as lignocaine, chlordiazepoxide and, in particular, diazepam appear to be more extensively distributed.   Overall, the elderly are smaller in body size than younger subjects and this may contribute in part, for example, to the higher blood levels of digoxin in the elderly following the same i.v. dose to young and elderly patients.

Plasma albumin concentrations are slightly lower in healthy old people than in their younger counterparts and may be markedly so in the poorly nourished or severely debilitated elderly and will result in an increase in the free pharmacologically active fraction of some drugs and in turn lead to a greater distribution and a more rapid elimination.   There is also some evidence to suggest that in the elderly, drugs may be more readily displaced from plasma protein binding sites by other drugs.[10]

The extent of plasma protein binding of phenytoin, carbenoxolone and temazepam is slightly reduced in the elderly.   On the other hand no detectable alteration occurs in the plasma protein binding of diazepam and with warfarin the evidence is contradictory. The practical significance of such changes in distribution volume and extent of protein binding is unclear but they are of major importance in the interpretation of other pharmacokinetic data, such as drug plasma half-life.   It is doubtful whether changes in protein binding are of clinical importance although the possibility of such changes influencing, or reflecting alteration in, a drug's penetration to its site of action cannot be excluded.

Drug Metabolism

The onset of drug effect is largely determined by rate of absorption and manner of distribution.   Duration of effect is much more influenced by rate of elimination, principally by metabolic degradation in the liver usually to more-polar, less-active metabolites or by renal excretion of parent drug or metabolites.   The hepatic clearance of antipyrine, a drug widely used as an index of liver microsomal oxidation, is reduced in the elderly, partly due to an age-related decrease in functional liver volume and partly to a reduced rate of hepatic metabolism.[11]   Several other drugs undergoing oxidation exhibit a similar reduction in clearance (e.g. chlordiazepoxide, theophylline) but for some other drugs (e.g. warfarin, diazepam) no age-related differences in clearance exist.   It is apparent therefore that there is no simple pattern of age-related change in drug metabolism.   Table 2 summarises some of the findings to date.   Changes, where they occur, are often small and may be less important than those brought about by environmental factors such as cigarette smoking.

Table 2. The influence of old age on the plasma elimination of extensively metabolised drugs

Drugs for which there is evidence of significantly reduced clearance	Drugs which seem to show no appreciable alteration in clearance
Acetanilide	Oxazepam
Antipyrine	Lorazepam
Chlormethiazole*	Temazepam
Labetalol*	Prazosin
Lignocaine*	Warfarin
Phenylbutazone	
Propranolol*	
Theophylline	

* Drugs undergoing extensive first pass elimination

A number cf drugs are so avidly extracted by the liver i.e.
by uptake into hepatic binding sites and by metabolism and that
their clearance depends on rate of delivery to the liver in the
blood.   In old age, a decrease in hepatic blood flow together
with possible reduction in the rate of hepatic metabolism is
responsible for the reduced elimination of such high-clearance
drugs as chlormethiazole, labetalol, lignocaine and possibly
propranolol (Table 2) although the evidence on the last-named
is conflicting.   With such highly-cleared drugs there is a marked
first-pass effect due to their extensive pre-systemic removal from
the blood on their first passage through the liver.   Their oral
bioavailability is therefore low but is increased in the elderly due
to a reduction in the first-pass extraction.

Drug metabolism ability may be enhanced by treatment with
enzyme inducing agents such as phenobarbitone or phenytoin or by
exposure to environmental factors such as cigarette smoking.   While
no conclusive data exist, there is some evidence to suggest that the
induction response may be reduced in the elderly.[12]   If this were
the case, the elderly as well as having a lower base-line ability to
metabolise some drugs would be less able to develop tolerance to
metabolised drugs.

Renal Excretion

The effects of age on renal function exert a profound influence
on the elimination of a number of drugs.   In many cases, drugs are
excreted by simple glomerular filtration and their rate of excretion
correlates with glomerular filtration rate (and hence with creatinine clearance) e.g. digoxin and the aminoglycoside antibiotics.   In
old age, renal function diminishes consequent on reduced renal blood
flow so that at age 65 there is a reduction of approximately 30 per
cent in glomerular filtration rate compared with young adults.
Tubular function also deteriorates with age and drugs such as
penicillin and procainamide which are actively secreted by renal
tubules show a marked reduction in clearance.   In addition to
physiological decline in glomerular and tubular filtration, the
elderly patient is particularly liable to renal impairment due to
dehydration, congestive heart failure, hypotension and urinary
retention or to intrinsic renal pathology e.g. diabetic nephropathy
or pyelonephritis which may further modify the renal handling of
drugs.

Where there is obvious renal disease, guidance for appropriate
dosage of renally-excreted drugs may be obtained from standard
tables. Even with normal blood urea or creatinine values, elderly
patients have a smaller renal reserve than do younger people and
therefore the dose of such drugs should always be chosen with this
in mind.   With some drugs this is particularly important because
of the serious effects of overdosage e.g. digoxin, lithium, amino-

glycoside antibiotics and chlorpropamide. In general, elderly patients are best treated with lower doses of renally-excreted drugs than are younger patients.

Much of the pharmacokinetic data on hepatic metabolism and renal excretion of drugs in the elderly has been obtained from single-dose studies and there is a paucity of data on age-related comparisons of steady-state drug levels with chronic dosing. With renally-excreted drugs such as digoxin, lithium, penicillin and streptomycin, adequate serum levels are obtained in the elderly with lower doses. With metabolised drugs it is again not possible to generalise. Plasma steady-state levels of propranolol and phenytoin increase with age as do those of some, but not all, tricyclic antidepressants.[13]

CONCLUSION

In the elderly, and particularly in the sick elderly, drug elimination is often significantly reduced. This occurs most consistently with drugs eliminated by renal mechanism where the extent of the reduction is such that without adjustment of dose excessive drug accumulation and toxicity may result. With drugs which are metabolised, the situation is less definite but the rate of metabolism of some drugs is appreciably slower in older age groups. The situation is complicated by dietary and smoking habits and by the presence of disease however. With drugs undergoing first pass elimination, there may be a marked increase in bioavailability of the first dose and exaggerated effects may be seen.

REFERENCES

1. J. Crooks and I. H. Stevenson, eds. "Drugs and the Elderly," MacMillan, London (1979).
2. R. E. Vestal, Drug use in the elderly: a review of problems and special considerations, Drugs 16:358 (1979).
3. J. Crooks and I. H. Stevenson, Drug response in the elderly - sensitivity and pharmacokinetic considerations, Age and Ageing 10:73 (1981).
4. E. J. Triggs, R. L. Nations, A. Long and J. J. Ashley, Pharmacokinetics in the elderly, Eur.J.Clin.Pharmacol. 8:55 (1975).
5. J. Crooks, F. O'Malley and I.H. Stevenson, Pharmacokinetics in the elderly, Clin.Pharmacokin. 1:280 (1976).
6. J. Koch-Weser, D. J. Greenblatt, E. M. Sellers and R. I. Shader, Drug disposition in old age, New Eng.J.Med. 306:1081 (1982).
7. I. H. Stevenson, S. A. M. Salem, K. O'Malley, B. Cusack and J. G. Kelly, Age and drug absorption in: "Drug Absorption," L. F. Prescott and W. S. Nimmo, eds., ADIS Press, New York (1980).

8. P. C. Rubin, P. J. W. Scott and J. L. Reid, Prazosin disposition
    in young and elderly subjects, Br.J.Clin.Pharmac.12:401 (1981).
9. E. Woodford-Williams, A. S. Alvarez, D. Webster, B. Landless
    and M. P. Dixon, Serum protein patterns in normal and
    pathological ageing, Gerontologia, 10:86 (1964).
10. S. Wallace, B. Whiting and J. Runcie, Factors affecting drug
    binding in plasma of elderly patients, Brit.J.Clin.Pharmac.
    3:327 (1976).
11. C. G. Swift, M. Homeida, M. Halliwell and C. J. C. Roberts,
    Antipyrine disposition and liver size, Eur.J.Clin.Pharmacol.
    14:149 (1978).
12. S. A. M. Salem, P. Rajjayabun, A. M. M. Shepherd and I. H.
    Stevenson, Reduced induction of drug metabolism in the
    elderly, Age and Ageing 7:68 (1978).
13. I. H. Stevenson, S. A. M. Salem and A. M. M. Shepherd,
    Studies on drug absorption and metabolism in the elderly in:
    "Drugs and the Elderly," J. Crooks and I. H. Stevenson,
    eds., MacMillan, London (1979).

BRAIN RECEPTORS AND AGING IN RELATION TO

DRUG EFFECT

S. Garattini and A. de Blasi

Institute of Pharmacology Research 'Mario Negri'
Milan, Italy

Receptors have always been considered by pharmacologists as the "magic sites" where drugs elicit their therapeutic or toxic effects. Only recently, however, have suitable methods been developed to measure receptors. Developments have flourished particularly on the central nervous system (CNS) where receptors are considered an essential step both for neurotransmission mediated by chemical transmitters and for effects induced by drugs.

Rather than even try to cover the field in this paper, an attempt will be made to summarize critically some of the most relevant findings concerning modifications observed in brain receptors during the process of aging. This is done as a logical introduction to understanding the major changes connected with brain aging and should help clarify some of the differences in the action of centrally acting drugs observed in elderly compared to adult subjects (see other chapters of this book).

NOTES ON BRAIN RECEPTOR MEASUREMENTS

It is beyond the scope of this article to discuss methodological questions related to the determination of brain receptors. Briefly, such determinations are based on detection of the total binding of labelled compounds (neurotransmitters or drugs) to brain structures. In the example reported in Figure 1, ^{3}H-spiroperidol a neuroleptic agent which binds to dopaminergic sites,[1] has been utilized (top curve). The additions of an excess of unlabelled(+)-butaclamol displaces the labelled compound from the specific saturable sites but not from the aspecific, unsaturable sites (lower curve in the graph). The specific binding sites (receptors) are determined

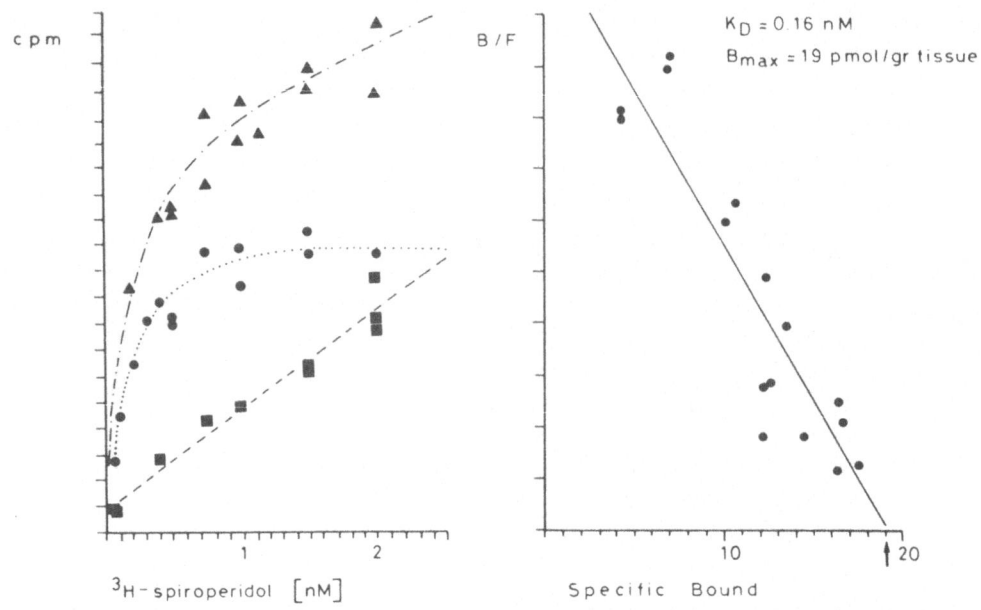

Fig. 1. ^{3}H-spiroperidol binding in rat striatum.
B/F = Bound/Free

▲ total bound; ■ aspecific bound;
● specific bound

by the difference (middle curve). Mathematical elaborations of the
data according to the Scatchard plot[2] results on the curve reported
in the right-hand graph of Figure 1 from which the affinity of the
ligand (^{3}H-spiroperidol) for the receptors (Kd) and the number of
receptors (B_{max}) per unit of weight or proteins can be calculated.

It is important to understand that such methods, though they
have gained popularity for their simplicity and ease of performance,
present a number of limitations such as the following:

i. The specific binding is obtained by difference; scient-
    ifically it would be sounder to determine it by a direct
    measurement;
ii. The binding sites do not necessarily represent receptor
    sites because high affinity binding sites have been found
    even with inert materials (e.g. plastic). Recently it
    was suggested to use the term "acceptors" for binding
    sites which are not receptors in the strict sense;
iii. The biological material (e.g. mambranes) is manipulated
    before binding is studied; thus acceptors may be unmasked
    which may not be physiologically present;
iv. Binding sites available in vitro are not necessarily available
    to the ligand injected in vivo.

Data on receptors and receptor-induced changes should therefore be regarded critically and in the context of knowledge available about pharmacokinetics, pharmacodynamics and mechanisms of action of drugs.

SIGNIFICANCE OF RECEPTORS FOR DRUG ACTION

Several studies indicate that the action of some psychoactive drugs may be related to their interaction with brain receptors. A correlation has been reported for instance between the affinity for dopamine and benzodiazepine receptors and the clinically effective doses of neuroleptics and anxiolytics respectively.[3] Furthermore the frequency of extrapyramidal side effects of neuroleptics was found to be inversely related to the in vitro affinity for muscarinic receptors while the sedative-hypotensive actions of the same drugs appear to be correlated with their binding to α_1 noradrenergic receptors.[3] In addition the peripheral cholinergic side effects of tricyclic depressant agents depend on their affinity for central cholinergic receptors.[3]

However these correlations are only an approximation because studies of the affinity of drugs for receptors in vitro suffer several limitations. In vivo drugs are absorbed, excreted and metabolized in a different way; they may form active metabolites and they may be differently affected by the blood/brain barrier.

Methods are therefore being developed to study drug-receptor interactions in vivo. It has been found for instance that several benzodiazepines occupy about 50 per cent of brain receptors in vivo at doses equally blocking metrazol convulsions.[4] However this relation is not without exceptions since a new benzodiazepine, estazolam, increases the number of benzodiazepine receptors at anticonvulsant doses.[4]

RECEPTOR ADAPTATION AND MODULATION

Receptor number and affinity may change in relation to a number of factors including continuous treatment with drugs affecting the action of neurotransmitters. Increased or decreased neurotransmission results in compensatory mechanisms which respectively reduce or raise the number of specific receptor sites. This receptor adaptation may in some cases explain the onset of tolerance or supersensitivity to certain drugs. In the case of tricyclic antidepressant agents it is the change in receptors which is believed today to explain their therapeutic effect in depressed patients.[5]

A recent study reported that prolonged treatment with cis-flupentixol - a potent neuroleptic agent - induced an increase in

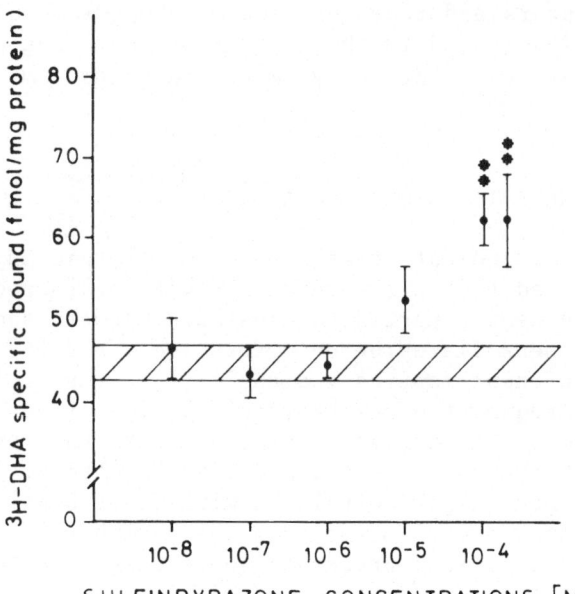

Fig. 2.   Effect of sulfinpyrazone on human
          intact lymphocyte β-adrenergic
          receptors. Each bar is the M ± S.E.
          of 5 subjects. The shaded area
          shows the control values. ^{3}H-DHA
          concentration = 2 nM
          ** p <0.01 Dunnett's test.

in striatal dopamine receptors (measured by the binding H-spiro-
peridol) which paralleled an increased stereotyped response to
apomorphine (dopaminergic supersensitivity) and a decreased
cataleptic response to trifluoperazine (neuroleptic tolerance).

    In another study the number of serotonin receptors in the brain
was either raised or lowered by repeated treatment with metergoline
(a serotoninergic antagonist) or fenfluramine (an indirect serotoninergic agonist).   In the presence of these changes, a direct
serotoninergic agonist (m-chlorophenylpiperazine) modified its
pharmacological effect according to the changes in serotonin
receptors.

    Changes in receptors (affinity or numbers) also occur acutely
(modulation) after treatment with drugs both in vitro or in vivo.
For instance topfizopam, a benzodiazepine which has no anticonvulsant activity, does increase the number of brain benzodiazepine
receptors in vivo.   During this condition diazepam becomes more
anticonvulsant against metrazol without any change in the levels of
brain diazepam.[4]

β-adrenergic receptors present in intact human lymphocytes can be
modulated by Sulfinpyrazone.   The addition in vitro of this compound
at concentrations similar to those present in plasma at therapeutic
doses raises the number of β-adrenergic receptors, (see Figure 2),
the effect being on B_{max} without any change in K_D (data not shown).

CHANGES OF BRAIN RECEPTORS WITH AGING

In addition to the difficulties discussed above the interpretation of receptor changes related to age is still hazardous.
Different experimental models have been utilized but their compara-
bility has not been assessed.   The natural history of the various
strains may perhaps be more important than age itself.   Tissues,
including the brain, undergo changes in composition with aging and
therefore it is difficult to decide whether changes in receptors
are primary events or depend upon other changes.   Aging is
characterized by an increase of glial cells in relation to neuronal
cells[8] and therefore receptors located on neurons may appear diluted.
Table 1 shows for instance that ^{3}H-Gaba binding is reduced in
relation to age in rat brain when values are expressed per unit of
weight but they appeared unchanged when expressed per mg of proteins.

A number of studies have now been made of receptor changes
induced by age although there are occasional major discrepancies in
the results.   Examples of these discrepancies are reported in
Table 2.

In our experimental studies on male Sprague-Dawley CD-COBS rats
(Charles River) aged 21-23 months we have reached the following
conclusions, in agreement with other authors:

i.   In the brain of aged animals changes in receptors are not the
      results of non-specific, widespread degeneration of the neurons
      because different brain areas may show different changes for
      the same acceptor.   For instance ^{3}H-spiroperidol binding - a
      marker of neuroleptic sites - is very much reduced in the
      striatum, moderately reduced in the limbic area and unchanged
      in the cortex as reported in Figure 3 (see also ref. 10);
ii.   In a given brain area the receptor changes are not non-
      specific but they may be selective.   For instance in the
      cerebral cortex there is a reduction of β-adrenergic receptors
      (ligand: ^{3}H-dihydroalprenolol) without any change of α_1-
      adrenergic (ligand: ^{3}H-WB4101) or serotoninergic receptors
      (see Figure 4,A);
iii. Even in the same class of receptors there may be selective
      changes in some subclasses and not in others as a result of
      aging.   For instance in the striatum the dopaminergic
      receptors are decreased when the antagonist subclass is
      measured (^{3}H-spiroperidol binding) but unchanged when the
      agonist subclass is determined (^{3}H-ADTN binding)(see Figure 4B).

Table 1. ^{3}H-GABA receptor binding in brain regions of young and old rats

Brain Region	n	Age (months)	Specific binding fmol/mg protein	fmol/mg tissue
Hippocampus	5	3	483 + 34	3.9 + 0.3
		21-23	473 + 70	2.6 + 0.3*
Cerebellum	6	3	1236 + 114	9.3 + 1.2
		21-23	1131 + 132	5.9 + 0.6*

Values are means + S.E. of n separate experiments

* p <0.05 two tailed Student's t test.

Up to now there is agreement amongst different laboratories about a decrease of dopaminergic receptor in the nigrostriatal pathway. This finding seems to be independent of the method used and of the experimental model, as it recurs in several strains and species.[9,10,11,12,13,14,15,16,17,18,19,20] Possibly this lowered number of dopaminergic receptors also accounts for the impaired nigrostriatal functions characteristic of aging.[21]

β-adrenergic receptors are another class which clearly decrease with age. A reduction has been found in several brain areas[9,17,22,23,24] but not in myocardial tissue (results not shown) or in lymphocytes.[25,26] Although these differences require additional confirmation it is important to note that a change in a given class of brain receptors does not necessarily reflect on the same class of receptors in peripheral organs.

RECEPTOR ADAPTATION WITH AGING

Although many investigations have been made on age-dependent changes in receptor number or affinity, little is known about whether the responsiveness of receptor adaptative mechanisms is altered with aging. It would be of considerable interest to know this because the ability to adjust receptor sensitivity is one important mechanism for maintaining homeostasis. Adaptative responses at least in some cases, are probably the first alteration occurring in aging subjects.

For example, the adaptive mechanisms of β-adrenergic receptors become impaired in old age. Weiss and his co-workers have[24]

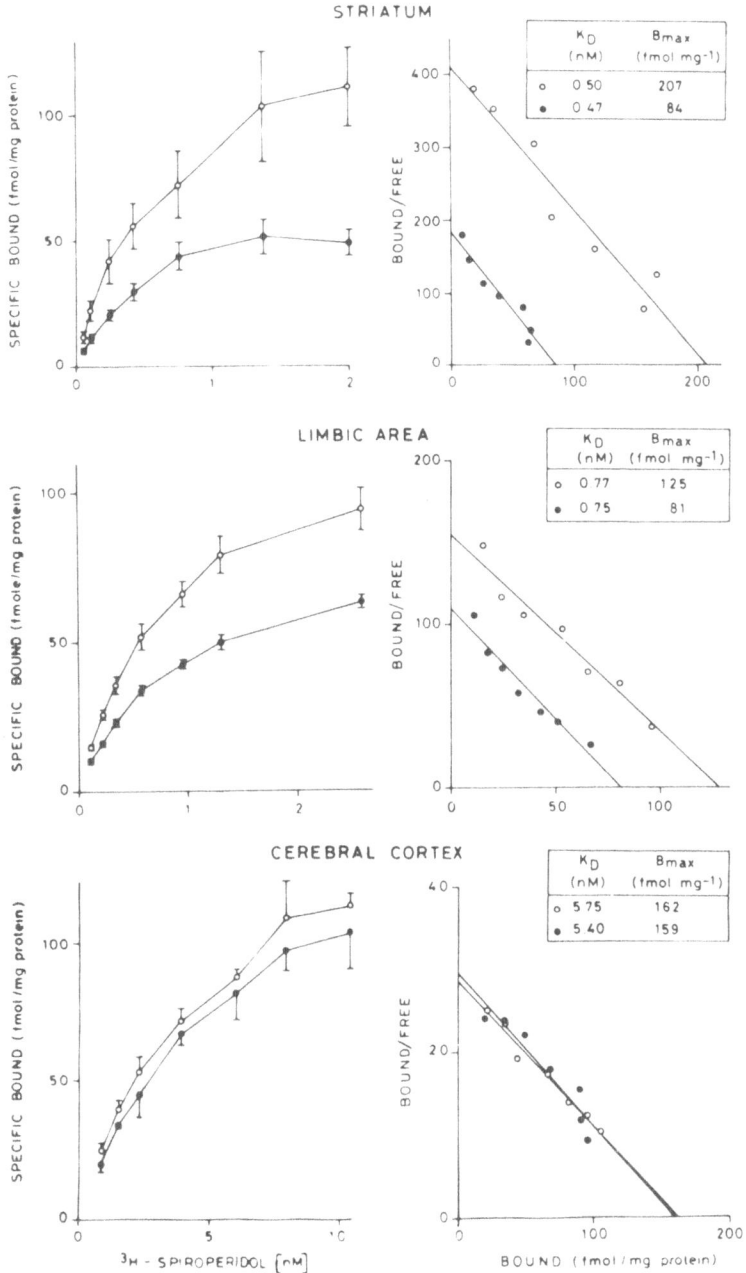

Fig. 3. Saturation curves and Scatchard analysis of ^{3}H-spiroperidol specific binding in brain regions of young and old rats. On the left, saturation curves are given as mean ± S.E.M. of four separate experiments. On the right, Scatchard plots of individual experiments representative of their group.
o 3 months   ● 21–23 months

Table 2. Age-related changes in different receptor binding in the CNS

Receptor binding	^{3}H-ligand	change (reference)	
	^{3}H-ADTN	↓ (11,14,19,20	= (9,10)
DOPAMINE	^{3}H-spiroperidol	↓ (9,10,13,15,16,17,18,19,20)	
	^{3}H-haloperidol	↓ (12)	↑ (35)
α_1-ADRENERGIC	^{3}H-WB4101	↓ (17)	= (9)
β-ADRENERGIC	^{3}H-DHA	↓ (9,17,22,23,24)	
	^{3}H-Diazepam	= (36,37,38)	↑ (37)
BENZODIAZEPINE	^{3}H-Flunitrazepam	↓ (9)	= (39,40)
GABA	^{3}H-GABA	↓ (41)	= (9)

extensively investigated the ability of β-adrenergic receptors to compensate for altered adrenergic input in rat brain. Repeated administration of desmethylimipramine, a noradrenaline uptake blocker, reduced the number of β-adrenergic receptors in cerebral cortex and pineal gland of young and old rats suggesting that aged animals still maintain the ability to elicit subsensitive responses following increased adrenergic input. This is not the case for the supersensitive response. In fact chronic treatment with reserpine, which causes long-term depletion of catecholamines, resulted in an increased number of β-adrenergic receptors in cerebral cortex and cerebellum of young rats but had no such effect in aged rats.[22,24] Moreover in young animals light exposure, decreasing adrenergic input to the pineal gland, increased the density of pineal β-adrenergic receptor binding sites. By contrast, there was no significant light-induced increase in 3H-DHA binding in the pineals of 24-month-old rats.[22,24]

An altered β-adrenergic receptor response was also evident from studies of short-term receptor adaptation. As shown in Figure 5, exposure of young rats to stress induced a β-adrenoceptor decrease in intact lymphocytes, with maximal effect after 30 minutes; between 90 and 180 minutes receptor number returned

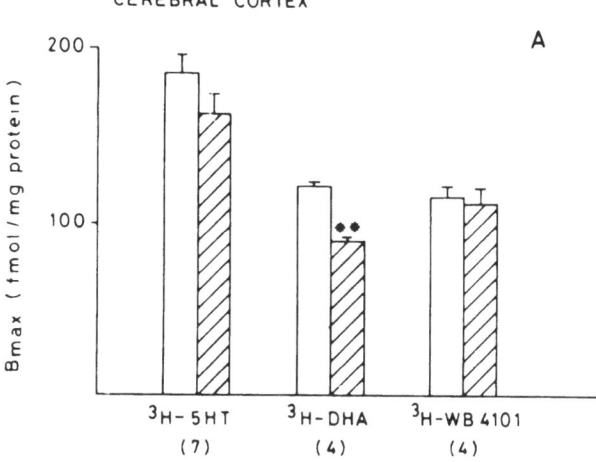

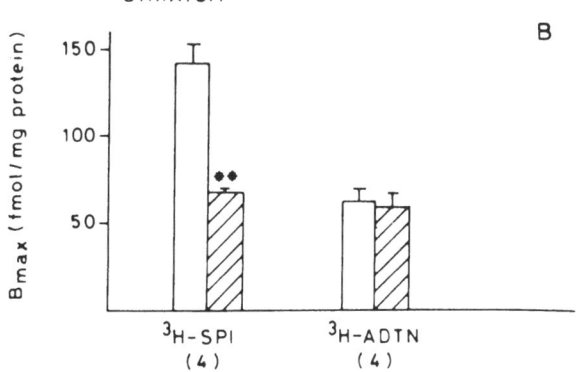

Fig. 4. Different receptor binding in 3-month
(open columns) and 21-23-month old
(shaded columns) rats. K_D values
were similar in young and old rats.
** p <0.01 students' t test

to normal. This β-adrenergic receptor desensitization, which may
represent a means of short-term adaptation to stress, is altered
in aged animals, being absent after 30 minutes stress, and a
significant decrease in β-adrenergic receptors is only seen after
180 minutes of stress.

The dopamine receptor system seems to behave differently.
Randall et al[27] studied the effect of chronic treatment with haloperidol (a dopamine antagonist) or bromocriptine (a dopamine agonist)
on 3H-spiroperidol striatal binding and apomorphine-induced stero-
typed behaviour in 5,12 and 24-26 month-old C57B/6NNia and C57Bl/
6J mice. Chronic bromocriptine treatment induced desensitization

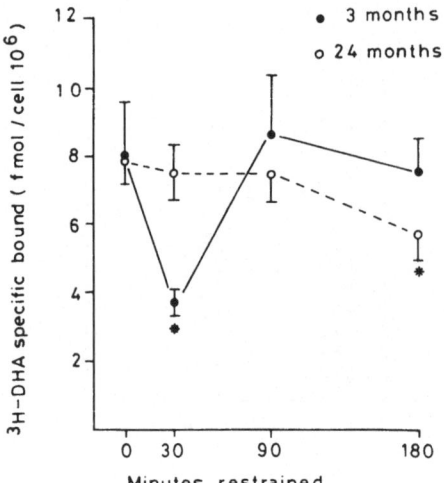

Fig. 5.   Effect of stress on β-
          adrenergic receptors of
          intact lymphocytes of
          young and old rats.
          ³H-DHA concentration =
          2 nM
        * P <0.05 Dunnett's test.

at all ages, chronic haloperidol induced supersensitivity in younger
mice (5 and 12 months) while old mice were unaffected.   Several
groups have studied dopamine receptor sensitivity in aged rats of
different strains (Sprague-Dawley, Fisher 344, Wistar), investigating
the effect of chronic neuroleptic treatment (haloperidol, fluphen-
azine) or specific nigrostriatal lesion on dopamine striatal
receptors (3H-spiroperidol or 3H-ADTN binding), dopamine-stimulated
adenylate cyclase activity, apomorphine induced sterotypy and rotat-
ional behaviour.   Despite small differences, all these authors
concluded that in old rats the ability to regulate dopamine receptors
remains virtually intact.[11,13,18,28]

CONCLUDING REMARKS

     Possible changes of brain receptors must be considered among
other factors when interpreting differences in the response to
psychoactive drugs in aged subjects.   Although those studies must
still be considered in their infancy they have already established
that loss of receptors in the aged brain should not be considered a
uniform or non specific effect related to gliosis and loss of
neuronal cells.   Certain specific changes occur in given brain
areas and given receptors which however may only be observed in
given species or strains.   While these studies are useful to

increase our knowledge, it seems important - as an ultimate goal -
to understand changes that occur in man's brain as it ages.   In
this respect it is particularly interesting to consider those
changes which are present across animal species because they may
have particular significance.   Dopaminergic and β-adrenergic
receptors appear to decrease in aged animals.   The changes of
dopaminergic receptors are accompanied by a decrease in other
dopaminergic parameters such as dopamine levels, uptake, turnover,
[29,30,31] tyrosine hydroxylase[32] dopamine stimulated adenylcyclase
[33,34] and by a functional decrease.[21]   However, the remaining
receptors appear to be capable of adaptation to various stimuli,
a finding which suggests that the loss of dopaminergic neurons may
be at the basis of the reduced number of dopaminergic receptors.

Changes in β-adrenergic receptors are instead accompanied by
a relative reduction in adaptative capacity.[22,24]

A decrease of receptors during aging may therefore have
different significance depending on the receptor capacity to maintain
their adaptive properties.   Extensive studies on brain receptors,
from various aspects - number, affinity, adaptation and modulation
- are a necessary step to obtaining better knowledge of the functions
of the aged brain and may pave the way to more rational therapeutic
approaches.

REFERENCES

1. J. E. Leysen, W. Gommeren and P. M. Laduron, Spiperone: A
    ligand of choice for neuroleptic receptors. 1. Kinetics and
    characteristics of in vitro binding, Biochem.Pharmacol.
    27:307 (1978).
2. G. Scatchard, The attractions of proteins for small molecules
    and ions, Ann.N.Y.Acad.Sci. 51:660 (1949).
3. I. Creese, Receptor binding as a primary drug screening device,
    in:"Neurotransmitter Receptor Binding," H. I. Yamamura,
    S. J. Enna and M. J. Kuhar, eds., Raven Press, New York (1978).
4. T. Mennini and S. Garattini, Benzodiazepine receptors':
    Correlation with pharmacological responses in living animals,
    Life Sci. 31:2025 (1982).
5. S. J. Peroutka, and S. H. Snyder, Long-term antidepressant
    treatment decreases spiroperidol-labeled serotonin receptor
    binding, Science 210:88 (1980).
6. K. Murugauah, A. Theodorou, S. Mann, A. Clow, P. Jenner and
    C. D. Marsden, Chronic continuous administration of neuro-
    leptic drugs alters cerebral dopamine receptors and increases
    spontaneous dopaminergic action in the striatum, Nature
    296:570 (1982).

7. T. Mennini, E. Poggesi, S. Caccia, C. Bendotti, F. Borsini and
    R. Samanin, Adaptive changes in central serotonin receptors
    after long-term drug treatment in rats, in: "Neurotransmitters
    and their Receptors," U. Z. Littauer, Y. Dudai, I, Silman,
    V. I. Teichberg and Z. Vogel, eds., Wiley & Sons, Chichester
    (1980).

8. P. W. Landfield, R. K. Baskin, and T. A. Pitler, Brain aging
    correlates: Retardation by hormonal-pharmacological treat-
    ments, Science 214·581 (1981).

9. A. De Blasi, S. Cotecchia, and T. Mennini, Selective changes of
    receptor binding in brain regions of aged rats, Life Sci.
    31:335 (1982).

10. A. De Blasi, and T. Mennini, Selective reduction of one class
    of dopamine receptor binding sites in the corpus striatum of
    aged rats, Brain Res. 242:361 (1982).

11. I. D. Hirschhorn, M. H. Makman, and N. S. Sharpless, Dopamine
    receptor sensitivity following nigrostriatal lesion in aged
    rat, Brain Res. 234:357 (1982).

12. J. A. Joseph, R. E. Berger, B. T. Engel, and G. S. Roth, Age-
    related changes in the nigrostriatum: A behavioural and
    biochemical analysis, J. Gerontol. 33:643 (1978).

13. J. A. Joseph, C. R. Filburn, and G. S. Roth, Development of
    dopamine receptor denervation supersensitivity in the neo-
    striatrum of the senescent rat, Life Sci. 29:575 (1981).

14. P. Levin, J. K. Janda, J. S. Joseph, D. K. Ingram, and G. S.
    Roth, Dietary restriction retards the age-associated loss
    of rat striatal dopaminergic receptors, Science 214:561 (1981).

15. M. H. Makman, H. S. Ahn, L. J. Thal, N. S. Sharpless, B. Dvorkin,
    S. G. Horowitz, and M. Rosenfeld, Aging and monoamine
    receptors in brain, Fed.Proc. 38:1922 (1979).

16. M. Memo, L. Lucchi, P. F. Spano, and M. Trabucchi, Aging process
    affects a single class of dopamine receptors, Brain Res.
    202:488 (1980).

17. C. H. Misra, H. S. Shelat, and R. C. Smith, Effect of age on
    adrenergic and dopaminergic receptor binding in rat brain,
    Life Sci. 27:521 (1980).

18. C. H. Misra, H. Shelat, and R. C. Smith, Influence of age on
    the effects of chronic fluphenazine on receptor binding in
    the rat brain, Eur.J.Pharmacol. 76:317 (1981).

19. J. A. Severson, and C. E. Finch, Reduced dopaminergic binding
    during aging in the rodent striatum, Brain Res. 192:147 (1980).

20. L. J. Thal, S. G. Horowitz, B. Dvorkin, and M. H. Makman,
    Evidence for loss of brain (^{3}H) spiroperidol and (^{3}H)ADTN
    binding sites in rabbit brain with aging, Brain Res. 192:185
    (1980).

21. J. F. Marshall, and N. Berrios, Movement disorders of aged rats:
    Reversal by dopamine receptor stimulation, Science 206:477
    (1979).

22. L. H. Greenberg, and B. Weiss, β-adrenergic receptors in aged
    rat brain: reduced number and capacity of pineal gland to
    develop supersensitivity, Science 201:61 (1978).

23. A. Maggi, M. J. Schmidt, B. Ghetti, and S. J. Enna, Effect of
     aging on neurotransmitter receptor binding in rat and human
     brain, Life Sci. 24: 367 (1979).
24. B. Weiss, L. Greenberg, and E. Cantor, Age-related alterations
     in the development of adrenergic denervation supersensit-
     ivity, Fed.Proc. 38:1915 (1979).
25. I. B. Abrass, and P. J. Scarpace, Human lymphocyte beta-
     adrenergic receptors are unaltered with age, J.Gerontol.
     36:298 (1981).
26. R. Landmann, H. Bittiger, and F. R. Bühler, High affinity
     beta-2-adrenergic receptors in mononuclear leucocytes:
     Similar density in young and old normal subjects, Life Sci.
     29:1761 (1981).
27. P. K. Randall, J. A. Severson, and C. E. Finch, Aging and the
     regulation of striatal dopaminergic mechanisms in mice,
     J. Pharmacol.Exp.Ther.219:695 (1981).
28. M. Roffman, F. Cordasco, and A. Kling, Apomorphine-induced
     stereotypy in mature and senescent rats following cessation
     of chronic haloperidol treatment, Commun.Psychopharmacol.
     4:283 (1980).
29. S. Algeri, G. Calderini, G. Lomuscio, G. Toffano, and F. Ponzio,
     Catecholamines and adaptive mechanisms in senescent rats,
     in:"Apomorphine and other Dopaminomimetics,Vol.2 Clinical
     Pharmacology," G. U. Corsini and G. L. Gessa, eds.,
     Raven Press, New York (1981).
30. V. Jonec, and C. F. Finch, Ageing and dopamine uptake by sub-
     cellular fractions of the C57BL/6J male mouse brain,
     Brain Res. 91:197 (1975).
31. S. N. Pradhan, Central neurotransmitters and aging, Life Sci.
     26:1643 (1980).
32. E. G. McGeer, H. C. Fibiger, P. L. McGeer, and V. Wickson,
     Aging and brain enzymes, Exp. Gerontol.6:391 (1971).
33. S. Govoni, P. Loddo, P. F. Spano, and M. Trabucchi, Dopamine
     receptor sensitivity in brain and retina of rats during
     aging, Brain Res. 138:565 (1977).
34. M. H. Makman, H. S. Ahn, L. J. Thal, N. S. Sharpless, B. Dvorkin,
     S. G. Horowitz, and M. Rosenfeld, Evidence for selective
     loss of brain dopamine-and histamine-stimulated adenylate
     cyclase activities in rabbits with aging, Brain Res.192:177
     (1980).
35. J. K. Marquis, A-S. Lippa, and R. W. Pelham, Dopamine receptor
     alterations with aging in mouse and rat corpus striatum,
     Biochem.Pharmacol. 30:1876 (1981).
36. J. E. Heusner, and H. B. Bosmann, GABA stimulation of ^{3}H-
     diazepam binding in aged mice, Life Sci. 29:971 (1981).
37. M. Memo, P. F. Spano, and M. Trabucchi, Brain benzodiazepine
     receptor changes during ageing, J. Pharm.Pharmacol. 33:64
     (1981).
38. C. C. Tsang, K. V. Speeg Jr., and C.R. Wilkinson, Aging and
     benzodiazepine binding in the rat cerebral cortex, Life Sci.
     30:343 (1982).

39. N. W. Pedigo, J. N. McDougal, T. F. Burks, and H. I. Yamamura,
     Benzodiazepine receptor binding in frontal cortex and
     cerebellum from aged rat, Fed.Proc.40:311 (1981).
40. N. W. Pedigo, H. Schoemaker, M. Morelli, J. N. McDougal,
     J. B. Malick, T. F. Burks, and H. I. Yamamura, Benzodiaze-
     pine receptor binding in young, mature and senescent rat
     brain and kidney, Neurobiol.Aging 2:83 (1981).

CLINICAL TRIALS IN OLD SUBJECTS

S. M. Chierichetti

Institute of Psychology
University of Rome, Italy

It is undeniable that in recent years there has been a progressive increase in the number of articles, symposia and books on the medical, therapeutical and social problems of old people, but it is equally true that this upsurge of interest is relatively recent and the studies carried out so far have raised more questions than they have offered solutions, especially in the field of clinical pharmacology.

Up to 1975 there were practically no papers published on geriatric clinical pharmacology in the three major journals in this field (the European and British Journals of Clinical Pharmacology, and Clinical Pharmacology and Therapeutics), and it is only since 1976 that interest in geriatrics has risen so that 11 articles were published in 1977 (1.9% of the total), and 22 (2.9%) in 1980.[1]

This outburst of interest in the study of old age and senescence can be attributed to the fact that, in the industrialized nations at least, the over-60s are the fastest-growing sector of the population, and the sector which creates problems faster than any other, so much so that the White House Conference on Aging held in 1981 drew the attention of the American public to the importance of the elderly, politically, as a voting force.

Nonetheless, despite this new interest, it is certainly too early to talk about gerontopharmacology as an independent discipline, although it is generally agreed that the principles of pharmacology have to be substantially modified when applied to the elderly; however, we have only just started to decide in which way.[2] It is not so much that experimental designs for clinical trials have to be changed or new designs invented, but that they must be adapted

23

to these newly recognized needs, and they must take into account
that a different and still rather unknown "species" is being
investigated.   Just as data on experimental animals, even though
obtained with sophisticated methodologies, only serve to show that
a new molecule could be studied in man and in which possible
indications, so pharmacokinetic, pharmacodynamic, therapeutic and
toxicologic data obtained in the adult cannot necessarily be applied
to treatment of the elderly without prior careful verification.
This is not easy since, as I have said, we still know too little
about old subjects and the problems connected with aging.   In fact,
old age is that stage of life when many changes occur in the
individual, not only physiological, psychological and intellectual
changes, but also existential changes which, at least not yet, are
not measurable in scientific terms.

In many areas of pharmacology it would be very difficult to
make a critical assessment of the present state of research.
According to Friedel,[1] in geriatric pharmacology, unfortunately, it
is less difficult since a review of the literature shows clearly
that at least three indisputable conclusions can be drawn:

1)   Many older patients are overmedicated
2)   The elderly are at increased risk for adverse drug reactions
3)   The elderly patient is often difficult to treat.

MANY OLDER PATIENTS ARE OVERMEDICATED

This depends, fundamentally, on two factors:

a)   Multiple pathologies: Multiple pathology is in fact the rule
     rather than the exception in the elderly patient.[3]   One must
     consider that an average of five to six different diseases may
     be found in a group of hospitalized elderly patients, and one
     survey showed that of every 100 persons over 65, there were
     33 with arthritis, 17 with heart conditions, 16 with hyper-
     tension, 12 with gastrointestinal diseases, 10 with mental
     disorders, 5 with diabetes, 4 with asthma, and 3 with chronic
     bronchitis.[4]
b)   Polypharmacy is an almost inevitable conseqence of multiple
     pathology.   In the U.K., the elderly, who constitute 12 per
     cent of the population, are responsible for about 30 per cent
     of the expenditure on prescription drugs.[5]
     The situation is about the same in the U.S.: 10 per cent of the
     population is over 65, and they account for more than 25 per
     cent of all prescription drugs.   This means that the elderly
     take more drugs than any other age group.

THE ELDERLY ARE AT INCREASED RISK FOR ADVERSE DRUG REACTIONS

There are two main reasons also for this:

a) The first is that the overmedication of old people just
    referred to results in the probability of adverse drug
    reactions being greater in this age group.
b) The second and more important reason is that the elderly are
    more susceptible to adverse drug reactions because of the
    multiple age-dependent changes which influence the drugs'
    ADME and the sensitivity of the receptors.

THE ELDERLY PATIENT IS OFTEN DIFFICULT TO DIAGNOSE

This aspect is certainly underestimated, whereas in reality
it is extremely important. The geriatricians have shown us how
easy it is for an old person to be hospitalized with a diagnosis
of dementia, even severe dementia, and then it is discovered that
the patient was only suffering from dehydration or a chronic
urinary infection, and that it is enough to treat these conditions
correctly to obtain a dramatic recovery of the patient's mental
faculties.   Furthermore, with difficulties in diagnosis there
are obviously difficulties in making decisions on how to treat the
patient.

What I have said so far should be taken into consideration both
by the physician prescribing treatment for an old patient and by
the clinical pharmacologist who must plan clinical trials in the
elderly.   But there are also other considerations.

To plan a clinical study in the elderly, and to conduct it
correctly, account must be taken not only of the multiple pathologies
always present, of the chronic treatments being given which may be
standardized but not eliminated before starting the study, and of
the diagnostic difficulties which cause incredible complications
in admission and exclusion criteria of the subjects for the study,
but there are other no less important factors to be considered.
I propose to deal with each group separately.

PHYSIOLOGICAL FACTORS

With aging, there are physiological changes that may alter
pharmacokinetics as well as target organ sensitivity. These include:

Changes in Absorption

Reduced gastrointestinal blood flow, changes in pH and in
motility, and reduced numbers of absorbing cells and intestinal

surface area can affect the active transport system and alter the
absorption rate of certain drugs.   Moreover, we lack knowledge of
the effect of bedrest (very usual in the elderly) on pharmacokinetics.

Changes in Distribution

Some important changes occur with increasing age: decrease in
weight (total body weight and weight of individual organs), and an
overall reduction in the percentage of body water, with a relative
increase in body fat from 15 per cent (in young adult men) to 36
per cent at 76 years (from 33 to 45% in women).

In consequence, the water-soluble drugs reach higher peak
serum levels (digoxin, gentamicin, propranolol, etc.), and fat
soluble drugs (sedatives, for instance) may accumulate and their
action be prolonged due to the increased volume of distribution.

Changes in Drug Binding

The albumin: total protein ration in the elderly is decreased
because of the lower (20%) plasma albumin level.   This means an
increase in the ratio of free to albumin-bound drugs, and a decrease
of binding sites, which make drug interaction more likely in the
case of multiple drug therapy.

Changes in Metabolism and Excretion

Impairment of liver and renal function are obvious factors
influencing the activity and/or toxicity of drugs.   This is why
it is obviously necessary to re-study pharmacokinetics in the
elderly not only after a single dose, but also, especially, after
repeated administrations, to establish the steady-state and chronic
dosage pharmacokinetics.

Moreover, while it is clear that the metabolism of many drugs
is impaired in the elderly, the studies performed up to now have
been almost exclusively based on the disappearance rate of unchanged
drug.   There is, therefore, a need to study the precise patterns
of metabolism in the elderly as well as the minor metabolic
pathways.

Also renal function is impaired in the elderly, resulting in
a reduction of the renal excretion of various drugs: this makes it
difficult to adjust the dose to keep the plasma levels of a drug
within the therapeutic range, also because it is not clear whether
this range varies with age.

Changes in Target Organ Sensitivity

Different responses to some drugs occur in the elderly that cannot be fully explained by the above physiological changes. This is particularly true for altered response to CNS drugs due to depletion or alteration of certain neurotransmitters and probably to the altered integrity of the brain-blood barrier. But it is also true for other organs, due to the deterioration of enzyme systems, and the reduction and/or alteration of receptors.

Assessment of drug sensitivity constitutes a major problem, as there is a lack of reliable methods to determine the target organ response.

In fact, homeostatic mechanisms are impaired in thermoregulation, glucose homeostasis, coordination of motor function, and postural hypotension. All these variables complicate the investigation of drug effect, and need to be constantly taken into account.

Intercurrent illnesses, so common in the elderly, are also important in modifying drug response.

The measurement of the distribution, density and "in vitro" activity of receptors is now possible for several drugs. The demonstration of a depletion of certain neurotransmitters in the elderly may lead to a more rational use of some therapies, i.e. cholinergic drugs in certain confusional states, and monoaminergic drugs in Parkinsonism, chronic senile cerebral insufficiency and possibly in Alzheimer's disease.

These last two points, changes in pharmacokinetics and impaired capacity of adaptation to adverse reactions, as well as the fact that we do not really know how the patient's chronological age compares with his physiological age, should make us re-evaluate the true "therapeutic ratio" for the elderly.

PSYCHOLOGICAL FACTORS

Undoubtedly the elderly show slight deficits in cognitive functioning and a slowing in processing which often presents as a slower response speed and caution not expected in younger subjects. On the other hand, physiological and psychological processes are probably more related to the health status of the subject than to the subject's age. Moreover, the results of the interaction between psychological and physiological factors and the reaction to external stimuli are much more complex in the elderly than in the adult "normals".

The compensatory model[6] demonstrates this altered or complex response to "normal stimuli". The physiological changes

evidenced in the elderly population which cause the behavioural
changes often observed both in experimental settings and in more
naturalistic settings can be seen as the result of the natural
process of aging.   Such changes or deterioration may be perceived
by the elderly individual, and may be attributed to some disease
process or significant limiting process, thus resulting in
behavioural adaptation.   This "compensation" against changes often
takes the form of negative adaptation unless "people strategies"
are utilized.   Negative adaptation can be at least partially
reduced by modifying the traditional approach to the patient.

PRACTICAL PROBLEMS

Patient Comliance in the Elderly

     The psychological concept of compliance is that the patient
expects the treatment to have a beneficial effect and acts accord-
ingly, i.e. he takes the prescribed treatment.   For this to happen,
these positive expectations have to be experienced by the patient.

     There is no scientific literature which demonstrates a reduction
with increasing age, in compliance or non-adherence as it is now
being termed.   On the other hand, as already seen, 10 per cent of
the population of over 65 take more than 25 per cent of all
prescription drugs.   If for no other reason, old persons would
seem to be at exceptional risk for prescription non-adherence.

     A recent study[7] which focused on non-adherence or non-compliance
of the elderly produced some interesting and surprising results.
Non-adherence to prescribed drugs was high; of these cases there was
a drug overuse in 90 per cent and underuse in only 10 per cent.   A
total of 43 per cent of the patients studied were non-adherent to
one or more prescriptions.   Most non-adherence was intentional,
and within this category the most common reason given was that the
patient did not believe in the drug (52%), while unintentional
non-adherence occurred in a relatively smaller proportion (23%).

     However, there are also other factors which play an important
role in non-compliance of the elderly:
- reduced memory capacity
- lack of assistance by family, friends, nursing personnel, etc.
- impaired hearing or reduced intellectual capacity, which means
  that the patient may have problems in understanding the doctor's
  or nurse's instructions; however, this depends very much on the
  patient's educational background, social position and environment.

     We must always also keep in mind the general "rigidity" of the
elderly.   Rigidity means that the elderly have reduced adaptive
resources at all levels, including physical and psychological.

This rigidity often makes it difficult to change from one therapy taken for a long time to a new therapy. We can be sure that a high percentage of patients will take the new drug but will continue with the daily old "pill" too.

Physician's Compliance

Too often we complain about the lack of compliance of the patient, forgetting the phenomenon of the "non-compliant doctor". Very often, in fact, the therapeutic regimen recommended by hospital doctors at the discharge from hospital of an old subject are changed by the general practitioner.

A similar phenomenon occurs in the trialists who do not follow the guidelines of a trial protocol.

There are other frequent causes of non-compliance:

- Pharmacological forms: the size of the tablets and capsules. Many subjects already have difficulties in swallowing particularly large tablets or capsules and when these are prescribed, the patient's compliance may be poor.

- Taste: it is unreasonable to expect a patient to take a nasty-tasting medicine regularly for a long time (for example, potassium supplement tablets).

- Side effects: although these may be mild and transient, they could lead to poor compliance also because the patient expects benefits, not side effects.

- Long-term treatment: compliance tends to decrease during long-term treatment, and careful follow-up by the doctor and/or nurses and family is needed.

CLINICAL STUDIES IN OLD SUBJECTS

Provided all the possible sources of variation in the elderly compared with the adult subject are taken into account, it is reasonable to suppose that a clinical study of the elderly may be performed correctly, provided also that the aims of the study are clear and that some additional points are borne well in mind:

- Firstly, the great variability of the population under study, varying from the prematurely aged to the oldest "normals", and from the institutionalized patients to the elderly living at home.

- Secondly, the drug should be assessed in a sample of patients who represent the population which is likely to be treated by the drug.

- Thirdly, efficacy should be assessed in terms of relative therapeutic objectives.

- Fourthly, when interpreting the results, the famous 5 per cent "p" values of significance should be regarded from the clinical point of view rather than from the statistician's point of view. In other words, a statistically significant change must represent a clinical benefit.

- Last but not least, we have to answer the question: could this active drug be useful.   This consideration is, of course, always valid, but it is particularly important for the elderly because they have a diminished life expectancy, and precisely for this reason the quality of their life may be more important than the quantity.

REFERENCES

1. R. O. Friedel, Introduction to the satellite symposium to the
      XII International Congress of Gerontology, Bremen, July
      10-11 1981: Influence of old age on the effect of drugs,
      Gerontoly 28:5 (1982).
2. N. B. Köln, Influence of old age on the effect of drugs: opening
      remarks, Gerontology 28 (Suppl.1) S. Karger, Basel (1982).
3. M. C. Hamdy, Drug therapy in the elderly, J.Pharmacother. 3:69
      (1980).
4. J. Crooks, J. Shepherd, I. Stevenson, Health Bulletin, 33:222
      (1975).
5. K. O'Malley, T. Judge, J. Crooks, Drug treatment, in: "Principles
      and Practice of Clinical Pharmacology and Therapeutics,"
      G. S. Avery, ed., Adis Press, Sydney (1976).
6. R. A. Hussain, "Geriatric Psychology: a Behavioral Perspective,"
      Van Nostrand Reinhold Co. New York (1981).
7. J. K. Cooper, D. W. Love, P. R. Raffael, International
      prescription non adherence (non compliance) by the elderly,
      J.Am.Geriat.Soc. 30:329 (1982)

GENERAL SYMPTOMATOLOGY OF PATHOGENIC DRUG USAGE

IN OLD PATIENTS

G. Barbagallo Sangiorgi

Department of Medical Pathology
University of Palermo, Italy

PRELIMINARY CONSIDERATIONS

Drug administration involves the risk of unwanted side effects in patients of any age. This risk is considerably higher in older patients,[5] as is shown by several statistics on this subject.[8,16,18,21] Higher risk of unwanted side effects in aged subjects is due to the concomitance of a number of factors[1,5,17] i.e.

a) aged patients often suffer from more than one ailment at a time and are therefore treated with various drugs characterized by different mechanisms of action;
b) reduced patient's compliance with the physician's prescriptions involving mistakes both by defect and by excess;[11]
c) self-prescription, particularly of laxatives, analgesics and often also of diuretics;
d) senile alteration in pharmacokinetics[1,11,13,19] concerning both liver metabolism (due to alterations of microsome enzyme activity) ana plasma transport (due to decreased serum albumin concentration) and excretion function (due to decreased renal clearance); the influence exerted by environmental factors (by smoking, that favours senile alterations in pharmacokinetics) and by coexisting pathologic conditions has been demonstrated for some drugs such as antipyrin and propranolol;
e) senile alterations of receptor response to drugs determine a different sensitivity of several target organs to drug action. This explains why drugs reaching the same plasma concentration and hence the same bioavailability may bring about a marked potentiation effect in some subjects.

A correct interpretation of the pathogenesis of adverse drug

31

effects in the elderly involves consideration of the following
factors:

a) decreased margin of functional reserve and hence of homeostasis,
   which is a typical feature associated with senile involution of
   the different organs;
b) biochemical alterations typical of aging, changes of water and
   salt balance, of enzyme, metabolic and neurotransmitter pools:
   all elements on or through which many drugs exert their
   biological action;
c) anatomical changes in the parenchyma, in connective tissues,
   in the vascular system and in microcirculation, which on one
   hand are responsible for reduced homeostasis and for biochemical
   changes but can also determine an altered response to certain
   drugs, e.g. the lack of haemodynamic adjustment under the
   action of antihypertensives.

Although the fact that aged patients are increasingly exposed
to risks due to drug adverse effects has been well documented,
it is also quite true that it may be very difficult to recognize
this condition in clinical practice, and for several reasons:[5]

a) symptoms are frequently, at least partly, non specific,
b) symptoms may be ascribed to disease or to aging,
c) it may be difficult to ascertain which drugs the patient is
   actually taking or has been taking and in what doses.

The latter circumstance is rendered even more difficult by the fact
that aged patients often believe they should take some drugs for
their lifetime. Patients usually do not volunteer information
during anamnesis unless the physician's questions are particularly
aimed at obtaining such information. Enquiry should be therefore
completed by the physician's actual search for any residual drugs.
This sort of enquiry - which we refer to as "drug anamnesis' is
quite indispensable to discover the iatrogenic nature of certain
clinical symptoms in aged subjects.[3]

d) drug adverse effects may be caused by drug treatment which had
   been started quite a long time ago and had been well tolerated
   for a long time.

This point is particularly important since in most cases the close
relationship between the beginning of treatment and the onset of
adverse effect(s) is the most useful and valid guideline for
recognizing the iatrogenic origin of some symptoms. In many
cases in fact iatrogenic symptoms appear immediately after the
beginning of treatment and may be therefore easily identified and
traced back to the drug(s) taken. In other cases, however, and
particularly in aged patients, side effects may appear even
several months after the beginning of treatment and it is quite

difficult to ascribe them to drugs under these circumstances.
Delayed onset of side effects may be due to various causes:

a) some time must elapse before adverse effects become manifest
   (e.g. the potassium depleting effect of diuretics;[4]
b) adverse effects may be due to interaction with drugs taken
   successively;[2,4]
c) a forthcoming illness may trigger clinical manifestations of
   subclinical damage caused by drugs taken previously (e.g.
   heart diseases appearing during diuretic treatment);
d) progressive aging of target organs increases their sensitivity
   to drug action and causes alteration in pharmacokinetics.

GENERAL SYMPTOMATOLOGY

   It may be therefore particularly difficult, in some individual
cases, to recognize the iatrogenic nature of certain symptoms,
and it is also quite probable that official statistics do not
reflect, at least as far as aged patients are concerned, the actual
frequency of drug side effects or of any kind of disorders which
may be responsible for a worsening of the aged patients' quality of
life, for a reduction or loss of self-sufficiency involving further
individual, social problems.

   The importance of this problem and the responsibility of the
physician in taking decisions should be particularly stressed.
In fact in most cases these problems concern drugs whose purpose is
to treat ailments or diseases which may determine loss of self-
sufficiency.   The decision to discontinue such treatment may
therefore involve greater risks than the decision to continue them.

   The purpose of the present lecture is to try to determine
which symptoms may be suspected as due to iatrogenic effects and
which guidelines should be followed to identify such pathological
conditions.   I will not deal with some symptoms which are easily
identified and do not show essential differences when compared to
those occurring in other ages of life, such as allergic syndromes
and cutaneous symptoms, whose clinical features are similar in all
ages of life, as well as disorders of the bladder or of the eyes
due to administration of anticholinergic drugs.   It may be quite
difficult to explain cases of jaundice resulting from drugs; in
fact, due to the chemical structure of the drug, cholestasis may
occur and the patient may be mistakenly considered a candidate for
surgical treatment.   This very important and specific aspect of
the problem will not be dealt with in this lecture. Our purpose is
to draw attention to the fact that in older patients many symptoms
usually ascribed to aging or to vascular diseases, are often caused
by drugs administered for various purposes.

It should also be pointed out that drug side effects often show peculiar clinical features in aged patients in that they are often non-specific and may be completely or partially different from those observed in younger subjects.

General symptoms of iatrogenic damage in aged subjects are reported in Table 1. These symptoms are such as to suggest the need to discontinue the administration of the drug.[5]   It should also be stressed that the list of drugs responsible for each group of symptoms is far from complete.   On the other hand, our purpose is not that of exhaustively treating this problem but only of outlining its most important aspects in everyday clinical practice.   The onset of such symptoms in aged patients should prompt the physician to question first (and not by exclusion) whether they may be due to adverse iatrogenic effects and to search for the cause.

Some of the symptoms reported in Table 1 are common to several drugs, while others are more directly correlated with specific pharmacological effects.   Among the most common symptoms we must mention asthenia and muscle weakness, which are often associated with the use of diuretics, sedatives and tranquillisers, antihypertensives, cardioactive drugs, beta-blocking agents and muscle relaxants.

The second most common symptom is anorexia: refusal of food is often the first clinical sign of excess drug intake or of drug intolerance when the dosage is correct.   Anorexia too is a non-specific symptom and is often associated with nausea and vomiting: these symptoms can follow the administration of various drugs, but are particularly common with cardioactive drugs.   Anorexia in aged patients is the most sensitive symptom of digitalis intoxication, and appears even earlier than electrocardiographic signs. Anorexia may appear even after parenteral administration of low doses of digitalis or strophantin.   Refusal of food without apparent reason on the part of aged patients should therefore be ascribed to pharmacological effects.   In fact in at least 50 per cent of cases persistent anorexia may be ascribed to iatrogenic effects and it regresses only slowly on discontinuation of treatment. Antidepressants, anti-Parkinsonian agents, tranquillisers and anti-arrhythmic drugs such as quinidine and procainamide, should be mentioned among the other drugs responsible for anorexia.   Anorexia is often associated with nausea and vomiting with meals; these symptoms are due to the same causes mentioned for anorexia, which usually is the first to appear.

Alterations in intestinal function and particularly obstinate constipation are also among the most common iatrogenic effects and are usually caused by psychoactive drugs, anticholinergic agents, antihypertensives, antiarrhythmics (disopyramide), antibiotics and diuretics.   The latter drugs exert this action through

Table 1.   The most common symptoms of iatrogenic damage in aged subjects:

- Asthenia, muscle weakness

- Anorexia, nausea, vomiting

- Changes in intestinal function (generally
  constipation, but also diarrhoea)

- Secondary anaemia and bleeding into the
  digestive tract

- Hypersomnia

- Apathy, depression

- Disorders of memory

- Mental confusion, disorientation as
  to time and space

- Vertigo

- Parkinsonism and other extrapyramidal
  syndromes

- Orthostatic hypotension

- Disorders of heart rate

potassium depletion,[7] which aggravates the natural tendency to
hypokinesia, typical of old age.   In many cases a vicious circle
sets in: diuretics → potassium depletion → constipation → laxative
intake → further potassium depletion.   Sometimes disagreeable
consequences may ensue, such as an intestinal obstructive syndrome
due to faecal impaction.   Diarrhoea too may be caused by adverse
drug effects following treatment with antibiotics, digitalis, and
with some antiarrhythmic drugs such as quinidine and procainamide.

   Sideropenic anemia should also be mentioned among the symptoms
of iatrogenic damage and is generally caused by both steroidal and
non-steroidal anti-inflammatory agents.   These drugs may in fact
bring about not only acute haemorrhagic gastritis, which is quite
rare and easily detectable, but also chronic bleeding, which often
escapes the physician's attention.   These anaemic syndromes
(due to iron loss) may be ascribed to neoplastic formations, since
the age of these patients exposes them to a high risk of tumors.
Diuretics too are among the drugs that may cause acute or chronic
blood loss into the digestive tract, although these drugs are rarely
suspected of causing this particular adverse effect.

   Neuropsychiatric symptoms[12] are equally important for their
frequency and clinical relevance: such symptoms are prevailing -

though not exclusively - as side effects of psychoactive and antidepressant drugs and include hypersomnia often associated with
apathy, detachment from the environment and psychical depression.
These side effects, which may also appear singly, follow treatment
with various drugs and particularly with psychoactive compounds.
Hypersomnia can follow repeated administration of even low doses of
sleep inducing drugs. This excess effect in aged subjects is due
to the longer half-life of the drug, exceeding 24 - 30 hours.
Hence single doses of a benzodiazepine compound taken every night
may bring about a summation effect lasting all the following day.
Among other drugs responsible for hypersomnia we should mention
not only barbiturates - whose administration is very risky in aged
patients - but also tranquillisers, psychoactive drugs, antihypertensives and diuretics.   Cinnarizine and flunarizine too, may
cause unpleasant and persistent drowsiness and apathy in certain
subjects.   Aged subjects are particularly prone to psychic
depression due to senile changes in brain catecholamine levels, but
this symptom appears quite often as a side-effect of such drugs
as antihypertensives, anti-Parkinsonian agents, corticosteroids,
some antibiotics and antituberculosis agents (cycloserine) and
some antiblastic compounds (e.g. Vincristine).

We should now like to draw your attention to another group of
symptoms, shown in Table 1, namely mental confusion, disorders of
memory, [15] vertigo and latent or manifest Parkinsonism.   These are
of clinical importance since they may be interpreted as symptoms
of brain insufficiency, rather than as adverse effects of drugs.
The list of drugs responsible for such symptoms is quite a long
one.   Mental confusion, or disturbances of memory may be due to
treatment with antihypertensives, anticholinergic agents, anti-
Parkinsonian drugs, tranquillisers, sedatives, hypnotics and
even to antidepressants, particularly (but not only) if they also
exert an anticholinergic action.   Mental confusion and disturbances
of memory may also be caused by diuretics through a complex
mechanism involving potassium depletion on the one hand, and haemodynamic changes due to a sudden lowering of blood pressure on the
other.   Among further possible causes of mental confusion let us
also recall cimetidine, hypoglycemic drugs, corticosteroids and
digitalis compounds.

Vertigo in aged patients may result from treatment with
various drugs including antihypertensives, sedatives, antidepressants,
anti-Parkinsonian agents, diuretics, anti-arrhythmic drugs such as
quinidine and procainamide.   In most cases vertigo caused by these
drugs occurs when patients pass from the reclining to the standing
position and are related to postural variations in arterial pressure.

Another important, possibly iatrogenic neuropsychiatric symptom
in aged patients is Parkinsonism, which may be manifest but is
generally of the latent type, with vague, indefinite symptoms

(slight muscle hypertonus with small step gait, reduced mimicry, torpid ideation, indifference to the environment) often without tremor or only with sporadic ones.   Such Parkinsonism may be due to antipsychotic drugs (as phenothiazines and their derivatives), antihypertensives (particularly reserpine), to antidepressants and even to some drugs used for their action upon the digestive tract (metoclopramide).   Extrapyramidal syndromes such as akatisia, akinesia etc. may also be ascribed to iatrogenic causes.

An important group of adverse side effects involves the cardiovascular system: the most frequent one is orthostatic hypotension, which is also potentially dangerous since it involves the risk of falling down.   Orthostatic hypotension may be caused by psychosedatives, antihypertensives, diuretics, anti-Parkinsonian and antipsychotic drugs.   Disorders of heart rhythm are also among the symptoms of iatrogenic damage and include both hyperkinetic (sinus tachycardia, various kinds of extrasystoles tachycardia and atrial fibrillation) and hypokinetic ones (disorders of a.v. conduction).   The drugs most frequently responsible for these symptoms[4] are diuretics (through potassium depletion) and cardioactive drugs, and particularly the association of these two groups of drugs, as is often used in the treatment of congestive heart failure.   Other drugs, however, may be responsible for these side effects, i.e. antipsychotics, anticholinergic agents and antidepressants.   Myocardial failure may also be included among iatrogenic syndromes due to the incongruous use of beta-blocking agents or of tachycardia-inducing vasodilators.

FINAL CONSIDERATIONS

As a conclusion I would like to stress once more that symptoms of iatrogenic damage are in practice non-specific, since almost all drugs may cause one or the other of these symptoms according to the individual organ dysfunction induced by aging.   I would also like to summarize the subject and suggest some general rules[10,11,17,20] as listed in Table 2.   A responsible and comprehensive approach to the patient and his symptoms and ailments is particularly necessary in geriatric practice.

This attitude derives from the firm conviction that diseases of aged patients deserve to be studied with the same accuracy and precision required for patients of any other age in order to clarify the whole of the aged patient's pathology - so complex it may be - and to establish which condition should be treated first. The choice of the appropriate drug, or of drug combinations if necessary, should also depend on the knowledge of the pharmacokinetic features of the compounds selected and of the possible interactions with other drugs.   Another rule to be borne in mind is to avoid drug administration before having considered its useful effects

Table 2.   General criteria for correct drug administration

- Pathologic conditions in aged patients should be
  evaluated as precisely as possible and as a whole,
  establishing which ailment should be treated first

- Drug associations should be avoided as far as possible-
  and in any case limited to those drugs whose kinetics,
  dynamics and reciprocal interferences are well known.

- Expected advantages and potential risks caused by
  administration of the drug(s) chosen should be carefully
  evaluated and compared before starting treatment

- Caution in prescribing doses should be exerted, based on
  serum drug assay, when feasible, or on creatininaemia,
  or - even better - on creatinine clearance values

- Therapeutic effectiveness of the drugs chosen should be
  carefully monitored

- Enquiries about former drug treatments should always be
  made

- In aged patients any new symptoms should first be
  ascribed to drug treatment rather than to illnesses;
  drug administration should therefore be discontinued.

versus potential risks.   Drugs should be administered only when
expected benefits are greater than potential risks.

Aged patients should be treated with minimum effective dose
of drug and this may be achieved by first administering only 1/2
or 1/3 of the dose usually administered to younger subjects and by
gradually increasing doses until the therapeutic effect is
achieved.   It should also be borne in mind that the margin between
useful and toxic effect is very narrow in aged subjects.   For some
drugs it may be useful to adjust doses according to serum creatinin
levels or - even better - to creatinin clearance, although it
should be recalled that delayed renal excretion is but one of the
pharmacokinetic phases undergoing senile alterations and that the
evaluation of other factors such as liver metabolism and plasma
protein binding is more difficult.

Assay of the plasma concentration of some drugs may contribute
to the adjustment of dosage and to the avoidance of some adverse
effects, as is the case with digoxin, lithium, quinidine, salicylates etc.   Careful monitoring of therapeutic effect is also
necessary in order to evaluate the effectiveness of treatment, to

adjust the dosage, to detect any decrease in activity and the onset of any adverse effects. It should be finally borne in mind that symptoms of iatrogenic damage are often non-specific in aged patients and may be ascribed either to aging itself or to the disease which is just being treated.

Two conclusions may be drawn:

a) when approaching a new aged patient a careful anamnesis should be taken as to drugs taken hitherto,
b) any new symptom arising in an aged patient under treatment should be attributed to drugs rather than to disease - a wise and realistic rule.

Only by keeping in mind these rules can we be sure that treatment will not be more dangerous than disease in aged patients.

SUMMARY

Drug administration to aged patients may frequently determine adverse effects appearing as disorders of caenesthesia and with symptoms affecting various organs and apparatuses. The most common symptoms and the drugs causing them are considered, as well as the greater frequency of adverse side effects in the elderly and the difficulties encountered in identifying such effects in medical practice. Symptoms are in fact non-specific and may be well ascribed to aging itself or to the diseases affecting aged patients.

REFERENCES

1. G. Barbagallo Sangiorgi, Principi di farmacologia geriatrica, Rec.Progr.Med. 58:317 (1975).
2. G. Barbagallo Sangiorgi, Le interazioni tra farmaci con particolare riguardo al problema in geriatria, Giorn.Geront. 25:769 (1977).
3. G. Barbagallo Sangiorgi, L' "Anamnesi" in geriatria, Giorn.Geront. 30:325 (1982).
4. G. Barbagallo Sangiorgi, A. Di Sciacca, F. Durante, G. Frada Jr. G. Costanza, G. Cupidi, La terapia diuretica e la terapia antiipertensiva nell'anziano, Giorn.Geront. 30:749 (1982).
5. G. Barbagallo Sangiorgi, A. Di Sciacca, G. Frada Jr, G. Cupidi, M. Barbagallo, Il danno jatrogenico da farmaci, Nuovo Boll. Farmacol.Clin. 3:157 (1981).
6. G. Barbagallo Sangiorgi, A. Di Sciacca, G. Frada Jr, R. Malta, C. Botindari, Principi farmacologici della terapia del dolore nell'anziano, Giorn.Geront. 29:391 (1981).
7. G. Barbagallo Sangiorgi, A. Di Sciacca, A. Pardo, Potassium depletion in aged patients; an evaluation through red-blood-cell potassium determination, Age & Ageing 8:190 (1979).

8. G. J. Caranasos, R. B. Stewart, L. E. Cluff, Drug-induced
    illness leading to hospitalization, J.Amer.Med.Ass.
    228:713 (1974).
9. C. M. Castleden, C. F. George, The effect of aging on the
    hepatic clearance of propranolol, Brit.J.Clin.Pharmacol.
    7:49 (1979).
10. K. A. Conrad, R. Bressler, eds.,"Drug Therapy for the Edlerly,"
    C. V. Mosby Co., St. Louis (1982).
11. J. Crooks, I. H. Stevenson, eds., "Drugs and the Elderly,"
    MacMillan Press Ltd. London (1979).
12. W. Davison, Iatrogenic brain failure, in: "The Aging Brain:
    Neurological and Mental Disturbances," G. Barbagallo
    Sangiorgi and A. N. Exton-Smith, eds., Plenum Press, New York
    and London (1980).
13. S. Garattini, Prospettive farmacologiche per i problemi
    biomedical posti dall'invecchiamento, Atti 19° Congr.Naz.Soc.
    It.Geront.Geriatr., Parma 7-9 Ottobre, Suppl. 48:8 (1972).
14. S. Garattini, P. L. e Morselli, Interazioni tra farmaci,
    Ferro Edizioni, Milano (1975).
15. H. M. Hodkinson, Confusional states in the elderly, in: The
    Aging Brain, Neurological and Mental Disturbances,"
    G. Barbagallo Sangiorgi and A. N. Exton-Smith, eds.,
    Plenum Press, New York and London (1980).
16. N. Hurwitz, Admission to hospital due to drugs, Brit.Med.J.
    1:539 (1969).
17. L. F. Jarvik, D. J. Greenblatt, D. Harman, eds.,"Clinical
    Pharmacology and the Aged Patient," (Aging, Vol.16),
    Raven Press, New York (1981).
18. J. W. Smith, L. G. Seidl, L. E. Cluff, Studies on the epidem-
    iology of adverse drug reactions. V. Clinical factors
    influencing susceptibility, Ann.Int.Med. 65:629 (1966).
19. R. E. Vestal, Drug use in the elderly: a review of problems
    and special considerations, Drugs 16:358 (1978).
20. O. L. Wade, Drug therapy in the elderly, Age & Ageing 1:65 (1978)
21. J. Williamson, Adverse reactions to prescribed drugs in the
    elderly, in: "Drugs and the Elderly," J. Crooks and I. M.
    Stevenson, eds., MacMillan Press Ltd., London (1979).

EPIDEMIOLOGICAL CONSIDERATION OF ADVERSE REACTIONS TO DRUGS

J. Williamson

University Department of Geriatric Medicine
Edinburgh

There must be few doctors or informed members of the public
who remain unaware of the worldwide rising tide of drug consumption.
More and more drugs are being produced and promoted for a widening
range of disease conditions but a rather more disturbing phenomenon
is that the man in the street (and to an even greater extent the
woman in the street) increasingly seems to expect his doctor to
provide medication which will have the effect of making life's
problems go away. Trethowan (1975)[1] in a thoughtful article
pointed out that drugs are now being demanded for "those who dislike
their jobs, fear redundancy, cannot get along with their wives"
etc. Nor is it fair for doctors to blame the public and patients
for this phenomenon since it has been the medical profession and
the pharmaceutical industry which have created these expectations.
Not only is irrational prescribing of this sort medically wrong and
wasteful but since it distracts attention from the underlying
stresses and causative factors it militates against effective
relief by merely patching over the cracks in the patient's life.

In the forefront of all studies of drug consumption and adverse
reactions are the elderly for a variety of reasons:
(a) they comprise the most rapidly growing part of the population;
(b) they suffer from multiple diseases;
(c) they are prone to experience multiple social and emotional
    problems such as low income, poor housing, isolation and
    widowhood. Some of these social stresses manifest themselves
    in a somatic fashion with headache, insomnia, giddiness,
    feelings of breathlessness, etc. Thereafter patients may be
    given drugs, usually powerful psychotropics.

In addition some of the disturbances of function associated

41

with age change may be mistakenly attributed to disease. This is
especially liable to occur in extreme old age (now the most rapidly
increasing group). For example, diminished postural control may be
presented as giddiness or "lightheadedness" and the patient then
assaulted with a phenothiazine.   While this may be highly effective
in controlling the rotatory vertigo of labyrinthitis it can do
nothing to help the above old person, but on the contrary renders
her liable to Parkinsonism and/or postural hypotension.

Other papers in this course will emphasise the enhanced dangers
to the elderly from reduced metabolism and excretion of drugs and
increased target organ sensitivity.

WHAT DRUGS ARE THE ELDERLY CONSUMING?

It is difficult to find out the drug-taking habits of the
general population and most studies are concerned with patients
attending hospitals or general practitioners.   I am, however, able
to give some hitherto unpublished data on the drugs being taken by
a random sample of old people in Edinburgh.   This sample has been
fully described elsewhere (Milne et al).[2]

Table 1 summarises the findings and indicates the widespread
use of drugs by the general elderly population.

It will be seen that certain drugs are used significantly more
frequently by females, especially analgesics, psychotropics and
haematinics.   It is also noteworthy that almost 3 per cent of
females were receiving thyroxine replacement therapy.

This information may be contrasted with the findings of
another survey carried out by myself under the aegis of the British
Geriatrics Society.   This was based upon a multicentre study of
patients at the time of admission to Geriatric Medicine Departments
in Britain in 1975 to 1976.   The method was described elsewhere
(Williamson and Chopin 1980).[3]

Table 2 provides a summary of the main findings.

It will be seen from Table 2 that more than four out of five
patients were receiving prescribed drugs with a mean of close to
three drugs per patient.

Table 3 shows the most commonly prescribed drugs in the
patients admitted to hospital.   There are differences from the
findings in the general elderly population.

ADVERSE REACTIONS

Participants in the study were all senior, experienced

Table 1. Drugs used by randomly sampled old people
          Age 62-92   215 males   272 females

Drug Group	Males %	Females %	
Analgesics	28.4	39.7	($p < 0.01$)
Laxatives	38.1	49.3	($p < 0.05$)
Hypnotics	10.7	18.8	($p < 0.05$)
Psychotropics	6.0	15.1	($p < 0.01$)
Antacids	20.9	23.9	
Tonic/Vitamins	7.0	10.7	
Haematinics	3.3	15.8	($p < 0.01$)
"Cardiac"	11.2	12.1	
Hypotensives	1.4	5.5	($p < 0.05$)
Diuretics	5.1	9.6	
Thyroxine	0.3	2.9	
Other	18.6	14.0	

Consultants in Geriatric Medicine and they were asked to record adverse reactions and to specify the drug thought to be responsible. Figure 1 shows the most commonly offending drugs.   It has already been shown in Table 2 that adverse reactions were recorded in 248 patients or 15.3 per cent of those who were taking drugs. A single drug was responsible in 179 patients, two drugs in 59 patients and in five patients three drugs were believed to be contributing to the adverse reaction picture.

It will be seen from Figure 1 that the greatest number of reactions was due to diuretics with psychotropics next in order of frequency.   The lower histogram is more revealing since it portrays the relative risk associated with each drug.   Thus, although diuretics caused the largest number of reactions, this was mainly due to the fact that they were the most commonly prescribed of all drugs.   The lower histogram indicates that the most dangerous substances were the anti-Parkinsonians, the hypotensives and the psychotropics, all with about 13 per cent of adverse reactions. The only drug in which a significant sex difference in liability to reaction was recorded was digitalis.   In males, 16.1 per cent on digitalis preparations had reactions compared to only 9.4 per cent of females ($p < 0.05$).

Table 2.   Patients in sample, numbers receiving prescribed drugs
           and with adverse reactions

	Males	Females	Total
No. in sample	677	1321	1998
No. taking prescribed drugs	547 (80.8%)	1078 (81.6%)	1625 (81.3%)
No. with adverse reactions	82 (12.1%)	166 (12.6%)	248 (12.4%)
Proportion of drug takers with adverse reactions	15.0%	15.4%	15.3%
Total number of drugs causing adverse reactions	101	211	312
Total number of drugs taken	1528	3148	4676
Average number of drugs per drug taker	2.79	2.92	2.88

Age and Adverse Reactions

Comparing the prevalence of adverse reactions in the patients aged less than 75 with those aged 75+, there was no significant difference.   This contrasts with previous studies in which higher rates were found in older groups (Hurwitz 1969).[4]   I interpret this as meaning that for patients admitted to geriatric units mere chronological age is much less significant than the presence of multiple diseases.

Adverse Reactions related to Numbers of Prescribed Drugs

There was a highly significant stepwise increase (see Figure 2) in adverse reactions with increasing number of drugs being taken, rising from 10.8 per cent where one drug was being taken to 27.0 per cent in those receiving six drugs (p <0.001).

The Contribution of Adverse Reactions to the Need for Hospital Admission

Participating physicians were asked to indicate whether they considered an adverse reaction had contributed to the patient's need for admission.   They were further asked to indicate whether the reaction was a "sole" cause or a "contributory" cause.

Table 3. Patients admitted to Geriatric Units.
Most commonly prescribed drugs.

Drug Group	Males	Females	Total
Diuretics	243 (35.9)	504 (38.2)	747 (37.4)
Analgesics and Antipyretics	157 (23.2)	391 (29.6)	548 (27.4)
Antidepressives, tranquil- lisers and psycho- mimetics	144 (21.3)	329 (24.9)	473 (23.7)
Hypnotics, sedatives and anticonvulsants	153 (22.6)	291 (22.0)	444 (22.2)
Digitalis preparations	124 (18.3)	277 (21.0)	401 (20.1)
Potassium supplements	102 (15.1)	211 (16.0)	313 (15.7)
Haematinics	60 (8.9)	191 (14.5)	251 (12.6)
Antibiotics and other antibacterials	97 (14.3)	132 (10.0)	229 (11.5)
Laxatives	47 (6.9)	89 (6.7)	136 (6.8)
Vasodilators	51 (7.5)	57 (4.3)	108 (5.4)
Hypotensives	32 (4.7)	75 (5.7)	107 (5.4)
Bronchodilators	58 (8.6)	43 (3.3)	101 (5.1)
Anti-Parkinsonians	39 (5.8)	61 (4.6)	100 (5.0)
Miscellaneous nervous system drugs	22 (3.2)	73 (5.5)	95 (4.8)
Vitamins	31 (4.6)	63 (4.8)	94 (4.7)
Insulin and Hypo- glycaemics	27 (4.0)	60 (4.5)	87 (4.4)
Thyroid preparations	3 (0.4)	46 (3.5)	49 (2.5)

The figures in brackets represent the percentages of the total in the sample.

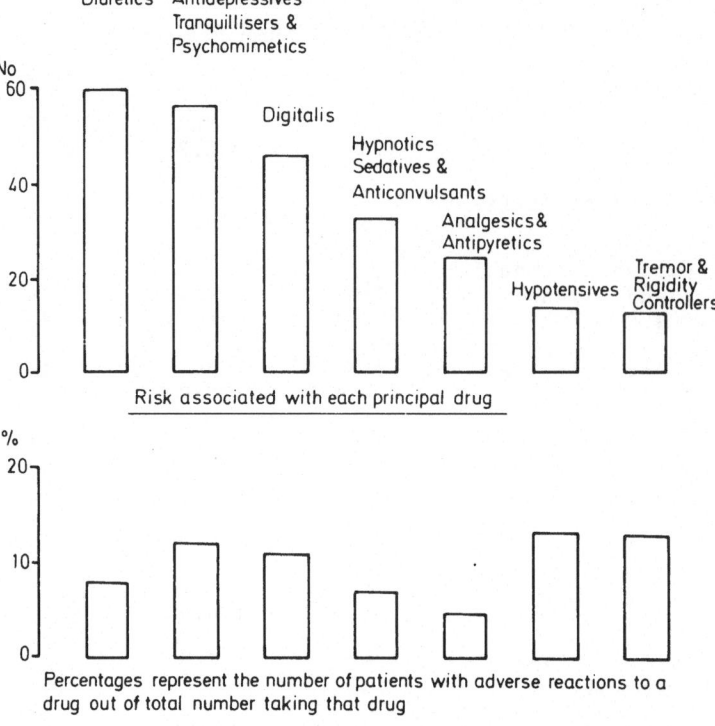

Fig. 1.   Principal Drugs Associated with Adverse
          Reactions. The top histogram represents
          the total numbers of reactions to each
          drug while the lower shows the number of
          patients with adverse reactions as a
          percentage of all patients taking each
          drug group (By permission of "The Pract-
          itioner").

    Figure 3 shows the role of each principal drug group in this
context.

    Figure 3 shows that, although the psychotropics formed the
group responsible for the largest total number of reactions
related to admission, adverse reactions to hypotensives were
proportionately most likely to contribute to need for admission.
Of adverse reactions most likely to be the sole cause for admission,
those due to anti-Parkinsonian drugs carried the highest risk.

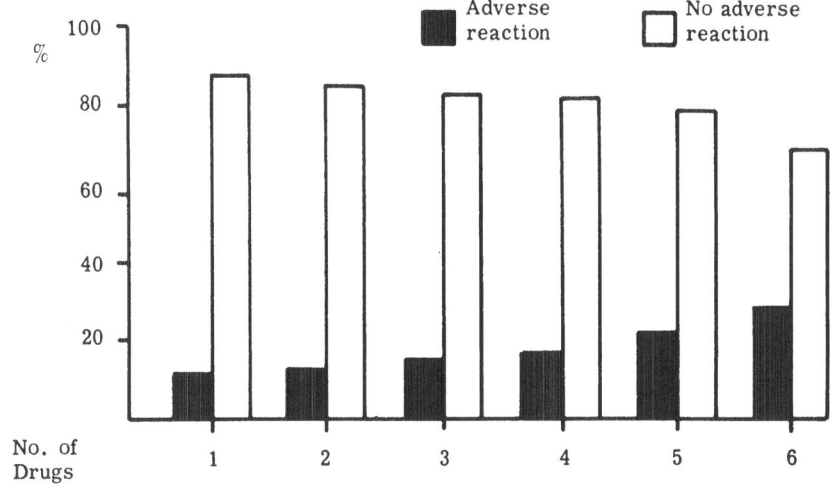

Fig. 2. Prevalence of Adverse Reactions related to number
         of prescribed drugs (By permission of "The Practitioner").

Outcome of Adverse Reactions

        Participants were asked to record whether full or partial
recovery occurred after observed adverse reactions or sequelae.
Full recovery occurred in 169 patients (or 68.7 per cent of all with
reactions). The chance of recovery varied with different drug
groups:

	Full Recovery
Digitalis Group	80.4%
Diuretics	73.3%
Hypotensives	71.4%
Antidepressants	61.3%
Hypnotics	54.4%
Anti-Parkinsonians	46.2%

        It is significant that less than half the patients with adverse
reactions to drugs used for Parkinsonism managed to recover fully
from the reaction (or its sequelae).

DISCUSSION

        Interest and concern is widely expressed about so-called
iatrogenic disease of which adverse drug reactions are by far the
most common and serious form.   In February 1982 in my own department
we were horrified to discover that at one time six out of 40

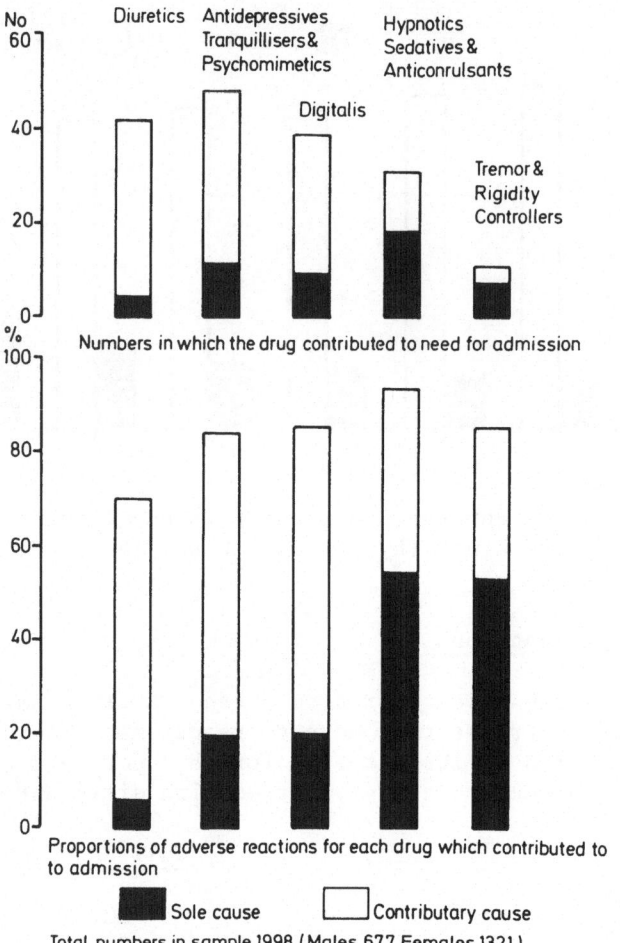

Fig. 3. Contribution of Drug Groups to Need for
         Admission. Top histogram shows the
         numbers of patients with adverse
         reactions to each drug in which the
         reactions contributed to need for
         admission. Bottom histogram shows the
         relative chance of each drug contrib-
         ution in this way.
         (By permission of "Age and Ageing")

patients in the admission/assessment ward were there entirely
because of adverse reactions to prescribed drugs and another two
patients had been admitted partly because of adverse reactions.

The findings reported here indicate that the dangerous groups are psychotropics, hypotensives, anti-Parkinsonians, hypnotics and digitalis.   During the period of the multicentre study about 140,000 patients were being admitted annually to Geriatric Medicine Departments in Great Britain.   It is likely that about 17,000 would have adverse reactions at time of admission.

In 55 (2.8%) patients an adverse reaction was judged to be the sole cause of admission (with anti-Parkinsonians carrying the highest risk) and in 154 (7.7%) a contributory cause.   Extrapolating these findings would indicate that in Britain in 1975-76 about 4000 patients were admitted to geriatric units solely because of adverse reactions and in another 11,000 such reactions were a contributory cause.

Even more alarming is the finding that substantial numbers of patients with adverse reactions never made a full recovery.   Once again the anti-Parkinsonian drugs emerge with the blackest record since less than half of the patients with adverse reactions to this group made a full recovery.

There is no doubt that a very large proportion of adverse reactions could be avoided if prescribers took more care.   Clinical Pharmacologists tell us that what is needed is more prominence to be given to their subject in medical education.   While this is undoubtedly true I believe that it is equally important that proper importance should be attached to the teaching of Gerontology and Geriatric Medicine.   This should form a significant part of every undergraduate curriculum and all "generalists" should have postgraduate instruction in this field.   The most urgent need is for general practitioners to be adequately educated in the needs of elderly patients since the great bulk of medical care of this group rests in their hands.   It is clear that, at present, this is not happening.

REFERENCES

1. W. H. Trethowan, Pills for personal problems, Brit.med.J.
       3:749 (1975).
2. J. S. Milne, M. M. Maule and J. Williamson, Method of sampling
       in a study of older people with a comparison of respondents
       and non-respondents, Br.J.Prev.Soc.Med. 25:37 (1971).
3. J. Williamson and J. M. Chopin, Adverse reactions to prescribed
       drugs in the elderly: A multicentre investigation,
       Age and Ageing 9:73 (1980).
4. N. Hurwitz, Predisposing factors in adverse reactions to drugs,
       Brit.med.J. 1:536 (1969).

DRUG COMPLIANCE

M. J. Denham

Geriatric Department
Northwick Park Hospital and Clinical Research Centre
Harrow, Middlesex

Taking drugs has become part of our way of life and indeed
William Osler considered that man has an inborn craving for medicine.
The enormous expenditure on drugs in the British National Health
Service testifies to his opinion.   However, it is also evident that
many normal, sensible people default or make mistakes in their
treatment.   Such actions not only waste money, but can be a potential danger to both the patient and the community, as might occur if
diabetic, tuberculous or epileptic patients fail to take their
medication in the prescribed manner.

MEASURING COMPLIANCE

Compliance may be measured directly or indirectly. Direct
methods involve measuring the blood or urine concentration of the
drug, its metabolites or a marker.   Unfortunately, this is not
applicable to all drugs and usually only digoxin, anti-epileptic
drugs, some antibiotics and hypotensive agents are estimated.   The
technique is further limited by the rate of drug elimination.
Those medicines with a long half-life give a steady state plasma
concentration, which can reflect compliance over the previous weeks.
However, the measurement of the plasma concentration of a drug with
a short half-life yields less information and can only indicate
whether or not the last dose has been taken.   The drug concentration
in urine fails to distinguish between under and over dosages, and
gives little insight into the overall level of drug consumption.

Because of the limitations of direct drug estimations, indirect
methods are often used instead.   One popular technique is to count
the number of tablets in the patient's possession, compare it with

the number prescribed and so construct a percentage:

$$\text{Compliance \%} = \frac{\text{Number of tablets taken by the patient}}{\text{Number of tablets prescribed to the patient}} \times 100$$

Non-compliance may therefore be defined as a deviation of more than, say, 10% from absolute adherence to the prescription. However, the method assumes that the tablets are actually taken regularly by the patient, are not shared by others, and that the more astute patient will not dispose of an appropriate number of tablets in order to make compliance appear better than it is. Porter[1] found that a number of his patients had correct tablet counts, but had negative urine tests. Another limitation of tablet counts is that they may not demonstrate altered patterns of drug defaulting. The omission of one tablet per day over a period of a few months may have less clinical impact than if the same number were omitted during one week. A second method involves interviewing the patient, which does have the advantage that non-compliance can be identified and corrected, but it does rely on truthful answers. Willcox et al[2] reported that about one third of their psychiatric patients said they were taking their tablets, but urine tests were negative. Consequently, it is not surprising that there may be discrepancies between tablet counts and the verbal report obtained at the interview. In order to improve the reliability of interviews, questions should be phrased in a way which avoids guilt. A third method assesses the patient's demand for new supplies of drugs, but this may be a cloak for hoarding. A 71 year old man admitted following an attempted suicide had in his possession 10,685 tablets and capsules, which had been prescribed over the previous 17 months.[3] A fourth method examines the clinical outcome due to drug treatment, but this may not be easy to establish.

With all these difficulties, it is not surprising that there is no universally agreed definition for measuring compliance.

THE EXTENT OF NON-COMPLIANCE

The lack of any agreed definition makes it difficult not only to assess the extent of non-compliance, but also to compare the numerous studies on the subject. However, it does seem that the elderly, the mentally ill, and those on long term medication are likely to default. Hospital organised studies of Schwartz et al[4] and MacDonald et al[5] for example, who used interviews and drug histories, found that as many as three quarters of their elderly patients made drug medication errors, and as many as one quarter made potentially serious errors, mostly of under-dosage. Parkes et al[6] using interviews, reported that 25 per cent of their schizo-phrenic patients were defaulting in their medication one year after

discharge from hospital.   Moulding et al[7] and Addington[8] using a
medication monitor or supervision, discovered that as many as one
third of their tuberculous patients were not taking their tablets
correctly, while Gibberd et al[9] using blood concentrations, found
that many epileptic patients were clearly not taking their tablets in
the manner prescribed.

Community organised studies show similar variability in results.
While Drury et al[10] found that non-compliance in general practice was
only 14 per cent, Porter[1] discovered that one third of his patients
failed to complete courses of antibiotics, while Gatley[11] reported
that approximately two thirds of his patients made medication errors.
Waters et al[12] found that 27 per cent of prescription forms for
analgesics were not even presented to the pharmacy for dispensing.

THE CAUSES OF IMPAIRED DRUG COMPLIANCE

The factors concerned with drug compliance can be classified in
several ways.   One of the more satisfactory relates the problem to
the patient, the illness, the physician, the treatment and the
pharmacist, noting however, that more than one factor may apply in a
particular situation.   Once again, unfortunately, the lack of an
agreed definition makes it difficult to draw consistent conclusions
and so identify a typical 'at risk' patients.

The Patient

Many features have been studied, but few consistent results have
emerged.

Age and sex. Impaired compliance can occur at all ages. The
majority of studies have found no evidence that the elderly comply
less well than younger people.[1,10,13,14,15,16]   However, Schwartz
et al[3] and Boyd et al[17] considered drug defaulting was worse in the
elderly.   Compliance does not correlate significantly with sex,
although Hulka et al[14] found trends suggesting that women complied
better than men, while Wandless et al[16] found the opposite.

Social factors. Hulka et al[14] and Parkin et al[15] found no
correlation between compliance and social class, education occupation,
income or religion.   Porter,[1] Willcox et al,[2] Schwartz et al[4] and
Gibson and O'Hare[18] found that those living alone were more likely to
make medication errors than those living with others.   Schwartz et
al[4] also found that those who coped with difficulty at home made more
errors in drug taking than those who coped rather better.   Compliance
may be reduced where patients have to pay the cost of the drugs.[19]

Mental state. Patients may not take the drugs correctly due to
mental impairment, loss of memory or confusion. MacDonald et al[5]

found that patients with low mental test scores had poor levels of
compliance, even when counselled, compared with more mentally alert
patients.   However, Wandless et al[16] found no correlation between
mental test scores and compliance, but their patients had had more
time to become familiar with their medication routine.

The Illness

The patient's perception of his illness can be an important
factor in compliance and may result in a positive decision to stop
taking the medication.   Factors concerned include lack of symptoms,
little faith in, or lack of response to the treatment and early
response to therapy.   Sackett et al[20] found limited compliance in
nearly half of their asymptomatic hypertensive men, which did not
respond to education.   Hemminki and Heikkila[19] discovered that non-
compliant elderly patients did not take their tablets because they
felt they had no further need for the medication, or because the
drugs did not appear to work.   Bergman et al[21] showed that 71 per
cent of children ceased taking their medication by the sixth day of
a ten day course of treatment, in spite of their parents being told
of the need to complete the course.   Waters et al[12] reported that
7 per cent of all prescription forms were not even presented to the
chemist for dispensing.   This phenomenon occurred mostly in younger
people and may have been due to the need to obtain a sickness note
rather than the patient considering the medication of little worth
or value.

The Physician

The relationship between the patient and the doctor can have a
considerable impact on compliance.   The factors concerned are the
level of expectation and satisfaction with the prescribed treatment,
and the rapport between patient and doctor which relates to the
personalities involved.   However, even more important is the
explanation of the need for the drug, together with instructions on
how and when to take it.[10] [22] [23]   MacDonald et al[5] demonstrated
that half of a group of fairly mentally alert, but uncounselled
patients made mistakes in their medication, the proportion increasing
to three quarters when less mentally alert patients were included.
Parkin et al[15] found that patients who failed to comply sometimes
did so because they were unsure of the correct dose and therefore
tended to revert to dosages they had before entering hospital.
Schwartz et al[4] showed that 21 per cent of those who made serious
compliance errors did so because of inaccurate knowledge of their
drugs.   Wootton[24] discovered that nearly half of a group of patients
given suppositories failed to remove the wrapper before insertion.
However, Law and Chalmers[25] found that 75 per cent of their patients
who were on regular medication knew details of their drug therapy
without having to read the labels on the bottles.

The Treatment

The nature of the drug can influence compliance. Those which act on the central nervous system, antibiotics and analgesics are more likely to be omitted than cardiac drugs.[10],[14],[15],[19]   It is often considered that drug formulation is important, with poor compliance being associated with drugs which are difficult to swallow, suppositories and greasy skin preparations, but Parkin et al.[15] found that compliance did not relate to the difficulty or unpleasant- ness of administration.   Side effects of drugs may induce some patients to stop their medication - postural hypotension due to anti- hypertensive drugs, or mouth ulceration due to Cetiprin are examples - but others may put up with the symptoms.[19]

The number and timing of drugs is important. The more drugs given per day, the worse the compliance.[1],[11],[13],[14],[15],[18],[26],[27],[28] Drugs which need to be taken more than 3 - 4 times daily are diffi- cult to fit into a daytime schedule and patients may be concerned whether the tablets should or should not be taken with food.

The Pharmacist

The labelling on the drug container is most important. Lack of precise instructions or the use of terms such as 'to be taken as directed' mean that drugs may be taken incorrectly.[16],[29],[30]   In addition, patients may have considerable difficulty in reading badly written labels.   Wandless et al,[16] Das[27] and Jenkins[31] found that over half of their patients experienced difficulty.   On the other hand, Schwartz et al[4] found that of five severely visually handi- capped patients, only one made potentially serious errors in reading the instructions.

Impaired eyesight and manual dexterity in the elderly can prevent patients opening drug containers.   Pathy[32] for example, found that many elderly people were unable to open the manufacturers' drug containers.   Rheumatoid patients have similar problems.[33]   The situation has been compounded following the introduction of child resistant containers, which aim to reduce the risk of accidental overdose in children.   Typical examples are the click-lock, pop-lock, and snap-safe bottles.   Several studies have demonstrated the inability of old people to open these new containers, particularly the click-lock.[26],[32],[34]   Consequently, it was suggested that the elderly might find blister, bubble or strip pack containers easier to manage, but once again studies have failed to demonstrate this.[27],[35],[36],[37]

THE DANGERS OF NON-COMPLIANCE

Poor compliance can have serious effects on the patient and his illness, sometimes with repercussions on the community.   Those

patients most at risk are those on drugs with a narrow therapeutic
range, and those in whom loss of control of the disease is serious.
It is clearly most important for the patient and the community for
diabetic, epileptic and tuberculous patients to take their medication
correctly.   Regression analysis has shown that blood pressure only
begins to fall significantly when patients take more than 80 per
cent of their treatment.[20]   Streptococcal infections become twice
as common when compliance with penicillin prophylaxis falls below one
third.[38]   However, there appears to be little evidence of patients
coming to much harm if they fail to complete a 10-day course of
antibiotics for a chest infection.

Poor compliance, or mis-prescribing can result in hoarding,
which can be very extensive.   Women appear to hoard more than men.[25]
A study of 500 households in Hartlepool[39] resulted in the return of
43,000 unwanted pills or capsules, while a recent campaign in
Glasgow produced $2\frac{1}{4}$ tons of medicines.   A similar campaign in
Birmingham produced over 1/3 million tablets/capsules, but this was
calculated to be only 3 per cent of the potential total.[40]   The
majority of the returned drugs were diuretics, respiratory medicines,
analgesics and antibiotics.   There was strong evidence that only
half of the courses of antibiotics were taken.   Over 70 per cent of
the drugs were over one year old, and some showed definite signs of
deterioration.

There are three potentially serious consequences of drug hoarding.
Firstly, the patient may confuse bottles of newly prescribed drugs
with hoarded ones, leading to inappropriate dose/drug schedules.
A second problem is that some drugs deteriorate with age.   Most
commercially available drugs have a shelf life of several years, but
some compounds such as glyceryl trinitrate are inherently unstable,
and will evaporate on repeated exposure to air, thus reducing their
potency.[41]   If tetracycline tablets are allowed to become wet and
then dry out, unusual degradation products may be formed, which can
cause the Fanconi syndrome.[42]   Evaporation of the alcohol base of
the opiate Nepenthe can lead to a dangerous increase in the concentration of the active constituent.[43]   Thirdly, patients may use
hoarded drugs inappropriately.   A man kept a bottle of steroid eye-
drops which had been prescribed for iritis.   He used the drops again
when he subsequently developed more pain in the eye.   Unfortunately,
his new symptoms were due to a dendritic ulcer with the result that
he perforated his cornea and required an emergency graft.

IDENTIFYING THE PATIENT WITH LOW COMPLIANCE

From what has been said earlier, there is clearly no easy way
of detecting the patient who does not comply fully with his medication.   However, clues can be obtained by watching for failure to
attend out-patients or the general practitioner's surgery, monitoring

the therapeutic goal and questioning the patient about his ability
to take the tablets.   If necessary, blood samples can be taken to
measure the drug concentration, while in severe cases patients may
need to be admitted to hospital for monitoring and treatment.

COMPLIANCE TRIALS

     Many studies have attempted to improve compliance. Sackett[44]
has emphasised that such trials or studies should be subjected to
the same rigorous testing as any new drug or operation. He has
suggested six criteria by which compliance trials should be judged:

1. Was compliance a legitimate issue?
2. Did the subjects in the trial form a credible patient sample?
3. Did the random selection of patients really work?
4. Was there a fair evaluation of the methods used to improve
    compliance?
5. Were the goals of compliance and the achievement of treatment
    measured?
6. Was there a search for side effects and toxicity?

Most studies attempt to improve compliance but few assess the clinical benefit.

MEASURES TO IMPROVE COMPLIANCE

     Several methods are used to improve compliance,by making it
mentally simpler for the patient to understand the drug regime and
physically easier to take the tablets. The results vary. However,
it may well be easier to treat non-compliant patients who deliberately refuse medicine, as opposed to those who do not comprehend.

Medication Regime

     It is important to rationalise the drug regimen. Where
possible, not more than 3 - 4 different drugs should be given each
day, with doses being related to meal times or other regular
activities. Some drugs with long half-lives, such as L-thyroxine,
which are sometimes given in split doses, can be given once a day
without any harm to the patient. Drugs with few side effects are
to be preferred to those with many. Although the combination of
drugs into one tablet has many well known disadvantages, it may be
helpful in this situation.

Patient Instruction and Counselling

     Ideally, it has been argued, patients should be given the
following information about their medicines:

a) The name, purpose, timing and route of administration of the drug.
b) Information to indicate whether the drug is/is not working.
c) What should be done if a dose is missed.
d) A list of the most important drug interactions and side effects.
e) The expiry date of the drug.

Such a comprehensive list requires a marked increase in the time
spent with the patient, as well as collaboration between the doctor
and the pharmacy department.   This helps to explain why only
relatively few details are actually given to the patient.

However, the majority of compliance studies have shown that
limited simple, clear instructions, especially if repeated over a
period of time, as well as an explanation of the need for therapy
and enquiries about problems with taking tablets, will generally
improve compliance. [5,9,26,45,46,47,48,49,50]   On the other hand,
Sackett et al [20] was unable to improve compliance in asymptomatic
hypertensive men by giving a full explanation of the need for treatment.   Furthermore, Wilber and Barrow [51] and McKenny et al [52] found
that compliance deteriorated once instructions/counselling stopped.

Written Information

Written instructions coupled with drug record cards can usefully
support verbal instructions.[26,45,53]   However, written instructions
were considered by Sackett et al[20] to be less effective than
strategies aimed at altering or modifying behaviour, while Malahy
found them to have little effect.

Containers

In the light of experience with child resistant containers,
bliser and bubble packs, it has been suggested that solid medicines
should be dispensed in opaque, amber or tinted containers.   However,
Atkinson et al, [26] Davidson [35] and Gibson [37] found that clear coloured
glass, screw or snap-top containers were best for older people,
since they are easier to open and enable the medicine to be identified by its colour.

Labelling

Clear, explicit labels are required, which should be printed
in large letters or type-written.   Vague instructions, such as
'take as directed' should not be used.   It is better to indicate
that the drug is 'to be taken four times daily, every six hours
before meals, until the tablets are all used up' rather than just
'take four times a day'.   Braille labels are available for the blind.

Memory Aids

A number of different aids have been tried with a varying degree

of success.   Dosette boxes are widely used, but do require someone
to fill them.   The elderly may find them inconvenient and difficult
to open or to manipulate, resulting in spilling of the contents.
Some patients do not find them child-proof.   Rehder et al[50] found
the boxes aided compliance, but Das[27] and Crome et al[49] did not.
MacDonald et al[5] and Wandless and Davie[45] have used pill wheels to
help compliance, but found the devices increased errors, due to
difficulty in extracting the tablets from them.   However, they did
find tear-off calendars helpful.   Fogden et al[36] found blister packs
to be of limited value.   Calendar packs can be helpful provided
flexibility of dose is not required.

Supervision

    Many elderly people are responsible for taking their own drugs.
Consequently, good communication is required between all caring
staff regarding drug dosages.   Follow-up in the patient's home by
the doctor, nurse or health visitor can be particularly helpful for
those elderly who are confused and/or living alone without close
relatives or neighbour support.   Patients should be asked to return
all their tablets at clinic visits and show that they understand
how to take them.

SUMMARY

    The errors the elderly make in their medication can be reduced
if the cause(s) can be identified and corrected.   Compliance will
be improved if the doctors responsible for prescribing for the
elderly keep the drug regime as simple as possible; prescribe the
patient just a few tablets each day, using drugs which are as free
from side effects as possible; give clear re-inforced explanations
supported by written instructions, use precise, well labelled, clear
coloured containers with judicious use of special packaging as
necessary, and arrange for community supervision.   Improved follow-
up should reduce the dangers of hoarding.

REFERENCES

1.   A. M. Porter, Drug defaulting in general practice, Brit.med.J.
       1:218 (1969).
2.   D. R. G. Willcox, R. Gillan, E. H. Hare, Do psychiatric out-
       patients take their drugs? Brit.med.J.   2: 790 (1965).
3.   S. E. Smith, K. C. Stead, Non-compliance or mis-prescribing,
       Lancet 1:937 (1974).
4.   D. Schwartz, M. Wang, L. Zeitz, and M.E.W. Goss, Medication
       errors made by elderly chronically ill patients, Am.J.Public
       Health 52:2018 (1962).

5. E. T. MacDonald, J. B. MacDonald, and M. Phoenix, Improving drug compliance after hospital discharge, Brit.med.J. 2:618 (1977).

6. C. M. Parkes, G. Brown, and E. Monck, The general practitioner and the schizophrenic patient, Brit.med.J.1:972 (1962).

7. T. S. Moulding, G. D. Onstad and J. A. Sbarbaro, Supervision of out-patient drug therapy with medication monitor, Ann.Int.Med. 73:559 (1970).

8. W. W. Addington, Patient compliance: The most serious remaining problem in the control of tuberculosis in the United States, Chest, 6:741 (1979).

9. F. B. Gibberd, J. F. Dunne, A. J. Handley,and B. L. Hazleman, Supervision of epileptic patients taking phenytoin, Brit.med.J. 1:147 (1970).

10. V. S. M. Drury, O. L. Wade, and E. Woolf, Following advice in general practice, J.R.Coll.Gen.Pract.26:712 (1976).

11. M. Gatley, To be taken as directed, J.R.Coll.Gen.Pract. 16:39 (1968).

12. W. H. R. Waters, N. V. Gould and J. E. Lunn, Undispensed prescriptions in a mining general practice, Brit.med.J. 1:1062 (1976).

13. B. Malahy, The effect of instruction and labelling on the number of medication errors made by patients at home. Am.J.Pharm. 23:283 (1966).

14. B. S. Hulka, L. L. Kupper, J. C. Cassell and R. L. Efird, Medication use and misuse: physician-patient discrepancies, J.Chron.Dis. 28:7 (1975).

15. D M. Parkin, C. R. Henney, J. Quirk, and J. Crooks, Deviation from prescribed drug treatment after discharge from hospital, Brit.med.J. 2:686 (1976).

16. I. Wandless, J. C. Mucklow, A. Smith, and D. Prudham, Compliance with prescribed medicines: a study of elderly patients in the community, J.R.Coll.Gen.Pract.29:391 (1979).

17. J. R. Boyd, T. R. Covington, W. F. Stanaszek, and T. T. Coussons, Drug defaulting Part II. Analysis of non-compliance patterns, Am.J.Hosp.Pharm.31:485 (1974).

18. I. I. J. M. Gibson, and M. M. O'Hare, Prescription of drugs for old people at home, Geront.clin. 10:271 (1968).

19. E. Hemminki, and J. Heikkila, Edlerly people's compliance with prescriptions and quality of medication, Scand.J.Soc.Med.

20. D. L. Sackett, R. B. Haynes, E. S. Gibson, B C. Hackett, D. W. Taylor, R. S. Roberts, and A. L. Johnson, Randomised clinical trial of strategies for improving medication compliance in primary hypertension, Lancet 1:1205 (1975).

21. A. B. Bergman, and R. J. Werner, Failure of children to receive penicillin by mouth, New Engl.J.Med. 268:1334 (1963).

22. J. R. Boyd, T. R. Covington, W. F. Stanaszek and T. R. Coussons, Drug defaulting Part I. Determinants of compliance, Am.J.Hosp.Pharm.31:362 (1974).

23. D. V. Lundin, Medication taking behaviour of the elderly.Drug. Intell.Clin.Pharm. 12:518 (1978).

24. J. Wootton, Prescription for error, Nursing Times, 71:884 (1975).
25. R. Law, and C. Chalmers, Medicines and elderly people: a general practice survey, Brit.med.J. 1:565 (1976).
26. L. Atkinson, I. Gibson, and J. Andrews, An investigation into the ability of elderly patients continuing to take prescribed drugs after discharge from hospital and recommendations concerning improving the situation, Gerontol. 24:225 (1978).
27. B. C. Das, Drug taking is a hazardous business for the old, Mod.Geriat.7:22 (1977).
28. A. J. Taggart, G. D. Johnstone, and D. G. McDevitt, Does the frequency of daily dosage influence compliance with digoxin therapy? Br.J.Clin.Pharmac. 1:31 (1981).
29. J. R. Powell, T. J. Cali, and J. A. Linkewich, Inadequately written prescriptions, J.Amer.Med.Assoc. 226:999 (1973).
30. M. Bliss, Prescribing in general practice, Lancet 2:248 (1976).
31. G. H. C. Jenkins, Drug compliance and the elderly patient, Brit.med.J. 1:124 (1979).
32. M. S. Pathy, Drug administration in the elderly, Unpublished (1977).
33. D. I. R. Mason, P. M. Brooks, and M. E. Mavrikakis, Study of opening medicine bottles in patients with rheumatoid arthritis, J.Clin.Pharm.1:171 (1976).
34. K. A. Bellamy, and M. I. Barnett, A study of child resistance and geriatric acceptability of a range of dispensing containers, Brit.J.Pharm.Pract. 2:4 (1981).
35. J. R. Davidson, Presentation and packaging of drugs for the elderly, J.Hosp.Pharm.31:180 (1973).
36. J. Fogden, P. Hopkins, and V. Wright, The packaging of antirheumatic drugs, Occup.Therap.39:32 (1976).
37. I. I. J. M. Gibson, Are drugs in the right container? Mod. Geriat.8:38 (1978).
38. M. Markowitz, Eradication of rheumatic fever, Circulation 41:1077 (1970).
39. D. M. Dunlop, Drug control and the British Health Service, Ann.Int.Med. 71:237 (1969).
40. D. W. Harris, D. S. Karandikar, M. G. Spencer, R. H. Leach, A. C. Bower, and G. A. Mander, Returned medwcines campaign in Birmingham 1977, Lancet 1:599 (1979).
41. M. O'Hanrahan, K. McGarry, J. G. Kelly, J. Horgan, and K. O'Malley, Diminished activity of glyceryl trinitrate, Brit.med.J. 284:1183 (1982).
42. J. Montoliu, M. Carrera, A. Darnell, and L. Revert, Lactic acidosis and Fanconi's syndrome due to degraded tetracycline, Brit.med.J. 283:1576 (1981).
43. Pharmaceutical Journal, A Report, 224:213 (1980).
44. D. L. Sackett, Compliance trials and the clinician, Arch.Int. Med. 138:23 (1978).
45. I. Wandless, and J. W. Davie, Can drug compliance in the elderly be improved? Brit.med.J. 1:359 (1977).
46. C. Blaxendale, M. Gourlay, I. I. J. M. Gibson, A self medication retraining programme, Brit.med.J. 2:1278 (1978).

47. R. Spector, P. McGrath, N. Uretsky, R. Newman, and P. Cohen,
     Does intervention by a nurse improve medication compliance?
     Arch.int.Med. 138:36 (1978).
48. S. E. Norell, Improving medication compliance; a randomised
     clinical trial, Brit.med.J. 2:1031 (1979).
49. P. Crome, M. Akehurst, and J. Keet, Drug compliance in elderly
     hospital inpatients. Trial of Dosette box. Practitioner
     224:782 (1980).
50. T. T. L. Rehder, L. K. McCoy, B. Blackwell, W. Whitehead, and
     A. Robinson, Improving medication compliance by counselling
     and special prescription container. Am.J.Hosp.Pharm.
     37:379 (1980).
51, J. A. Wilber, and J. G. Barrow, Reducing elevated blood pressure.
     Experience found in a community. Minnesota Med. 52:1303
     (1969).
52. J. M. McKenney, J. M. Slining, D. Devine, M. Barr, and H. R.
     Henderson, The effect of clinical pharmacy services on
     patients with essential hypertension, Circulation 48:1104
     (1973).
53. M. Klaber, Aid to drug compliance, Brit.med.J. 1:302 (1978).

THE TREATMENT OF SLEEP DISORDERS

Peter J. Cook

Honorary Lecturer in Clinical Pharmacology
Department of Medicine
Royal Free Hospital, London

Insomnia or sleep difficulty is primarily a symptom rather than an illness. The prevalence of the symptom rises with age so that about 30 per cent of people over the age of 65 feel they are sleeping poorly with a higher incidence in females (for a review, see Mendelson, 1980)[1]. Surprisingly, many people who complain of insomnia, perhaps the majority, do not get substantially less sleep than those who feel their sleep is adequate.[2] The complaint is often associated with other conditions of which both the subject and the physician may be unaware (Table 1). These include primary psychiatric disorders, such as depression and schizophrenia, sleep induced ventilatory impairment (the sleep apnoea syndromes), nocturnal myoclonus and the restless legs syndrome. Drug induced mental excitement due to concurrent use of CNS stimulants, such as sympathomimetic drugs or caffeine, and rebound hyperexcitability following withdrawal of regular sedation, including alcohol, may also be missed. Hypnotic treatment for some of these disorders can be inappropriate and, in the case of the sleep apnoea syndrome, actually harmful. This brief review will assume that insomnia due to pre-existing medical conditions or CNS stimulation has been excluded. The aim of therapy for the remaining patients is an improvement in the subjective quality of sleep since most subjects have sleep latency and sleep times that are within normal limits.

MEASUREMENT OF HYPNOTIC EFFICACY

Hypnotic efficacy can be measured in one of 4 ways. The most objective measures are provided by the sleep EEG. EEG studies give accurate information about the time of onset and duration of sleep. Unfortunately there is no clear relation between objective measures

Table 1. Conditions associated with insomnia

Physical discomfort
   e.g. pain, dyspnoea, nocturia

Distress secondary to a change in circumstances
   e.g. admission to hosptal, bereavement

Primary psychiatric disorders
   e.g. depression, schizophrenia

Disruption of 24 hour sleep cycle
   e.g. day time napping, acute 'phase shift'

Sleep apnoea syndromes

Nocturnal myoclonus

Restless legs syndrome

CNS stimulants
   Sympathomimetic drugs
   e.g. adrenergic bronchodilators, cold remedies

   Xanthine derivatives
   e.g. aminophylline, caffeine

   Sedative withdrawal
   e.g. hypnotics, antipsychotics, alcohol

such as sleep latency, number of awakenings and total sleep time,
and the complaint of insomnia. EEG studies are also expensive and
have usually been performed with small numbers of highly selected
young and middle-aged insomniac subjects which limits their
relevance to everyday practice. No EEG sleep studies appear to
have been carried out with elderly subjects.

Patient questionnaires provide a more subjective measure.
They have the advantage of being directed at the principal complaint
which is the subjective quality of sleep. However, they are
unreliable as measure of sleep parameters. Untreated insomniacs
tend to overestimate the time they require to get to sleep and
underestimate their duration of sleep.[3] Most hypnotic drugs cause
amnesia and alter the perception of time which makes these estimates
even less reliable.   It can of course be argued that the duration
of sleep is irrelevant as long as the patient feels that the quality
of sleep has been improved.   Another disadvantage is that the
questionnaire is usually administered the morning after the hypnotic
has been given when the patient is still under the influence of the

drug. This causes patients to underestimate any impairment of psychomotor performance and rate themselves as fully alert when objective
testing shows quite clearly that they are not.

An alternative method of assessment is to arrange for independent
observers, usually nursing staff, to observe the subject during the
night. Unfortunately, the results of these observations correlate
poorly with EEG measures.[1]

The non-benzodiazepine drugs currently available for use as
hypnotics in the U.K. and their elimination half lives in young
subjects are shown in Table 2. Little is known about the pharmacokinetics of most of these drugs in the elderly. Amylobarbitone has
been reported to be metabolised more slowly in the elderly[13]
whereas chlormethiazole pharmacokinetics are unchanged.[10] The
metabolism of chloral and its cogeners, triclofos and dichloral-
phenazone is probably also unchanged. The active metabolite,
trichlorethanol, is eliminated in the same way as ethanol whose
pharmacokinetics have been shown to be unchanged with aging.[14]

The benzodiazepine drugs which are licensed for use as hypnotics
are shown in Table 3. These drugs are now used more frequently than
non-benzodiazepine drugs and their use will be discussed in more
detail. The drugs all share the same basic pharmacological
properties but there are considerable differences in their pharmacokinetics (Table 3). The magnitude of these differences are clearly
illustrated in Figure 1. There are also considerable variations
between individuals, particularly in the elderly, and there is some
evidence that the elimination is more prolonged in frail elderly
inpatients when compared with healthy elderly controls.[20]
Flurazepam and temazepam show sex-linked differences in pharmacokinetics. The mean elimination half life of temazepam in young men
has been found to vary between 8 - 12 hours in different studies
compared with 12 - 16 hours in young females and the mean elimination
half life in elderly males has been found to vary between 10 - 17
hours compared with 17 - 19 hours in elderly females.[22,23,24,25,26]

HYPNOTIC EFFICACY - SHORT TERM USE

All hypnotic drugs are effective for the first few days of use,
provided the dose is large enough. This has been clearly demonstrated
in EEG studies with young and middle-aged insomniacs (Table 4). The
drugs reduce the time required to fall asleep (sleep latency) and/or
reduce the number of awakenings. A report by the Institute of
Medicine in the USA concluded that the beneficial effect was generally
to reduce the time needed to fall asleep by 10 - 20 minutes and to
lengthen the total sleep time by 20 - 40 minutes.[41] The significance
of these changes is unclear since patient questionnaires indicate
that patients often derive satisfaction even when the sleep para-

Table 2.   Elimination half lives of non benzodiazepine hypnotics
           available in the U.K.

	Mean (hrs)	Range (hrs)	Reference
Barbiturates			
Hexobarbitone	4	3-5	Breimer[4]
Heptabarbitone	8	6-11	Breimer[5]
Cyclobarbitone	12	8-17	Breimer[6]
Amylobarbitone	24	8-40	Inaba[7]
Quinalbarbitone (Secobarbital)	25	19-34	Breimer[8]
Pentobarbitone	27	18-48	Reidenberg[9]
Butobarbitone	38	34-42	Briemer[8]
Non Barbiturates			
Chlormethiazole	5	4-6	Nation[10]
Chloral Hydrate	9	8-11	Sellers[11]
Dichloral-phenazone	9	8-11	Sellers[11]
Glutethimide	12	5-22	Curry[12]
Methyprylone	-	-	-
Chlormezanonone	-	-	-

meters, as measured by EEG, are unchanged. This satisfaction appears
to depend on some effect of the drug on the quality of sleep which
is as yet undefined.

     The effect of a drug on sleep latency depends on the rate of
absorption and the rate of cerebral uptake as well as the dose.
Thus, the benzodiazepines oxazepam and lorazepam, which are both
absorbed and taken up into the brain relatively slowly, are not
generally used as hypnotics.   They have a delayed onset action and
are likely to be less effective in patients who specifically complain
of difficulty in getting off to sleep.

     Chloral hydrate is probably the least potent of the hypnotics
shown, in Tables 2 and 3. One small study found no significant
effects on the sleep EEG of normal volunteers given 0.5G, which is
the lowest recommended dose.[1]   A study in elderly insomniacs which
used patient questionnaires also found that this dose was ineffect-

Table 3. Pharmacokinetics of benzodiazepine hypnotics in healthy volunteers

Drug	Sex	Elimination half life (hrs)		Time to Steady State▽ (days)	Drug Accumulation△ (doses)	References
		Young	Elderly	Elderly	Elderly	
Long Acting						
Flurazepam**	M	74	160	27	X10	Greenblatt[15]
	F	90	120	20	X7.7	
Nitrazepam	M/F	29–33	33 (40)	6(7)	X2.5(2.9)	Castleden[16] Kangas[17]
Flunitrazepam	M	(31)	(24)	(4)	X2.0	Kanto[18]
Short Acting						
Lormetazepam	M/F	10	14.5	2	X1.5	Humpel[19]
Texazepam*	M/F	8	11(18)	2(3)	X1.3(1.7)	Huggett[20]
Triazolam*	M/F	2.5	2.6	1	0	Jochemsen[21] Upjohn[22]

▽ Time to SS = $4 \times t_{\frac{1}{2}}$

△ Accumulation = $\dfrac{1}{1-e^{\,0.693/t_{\frac{1}{2}} \cdot t}}$, where t = dosing interval in hrs

( ) = patients   * one   ** or more, active metabolites

Table 4.   Results of sleep studies of hypnotics in young subjects

	Reduction in time awake during first week	Evidence of tolerance after first week	Rebound Insomnia	References
Barbiturates and Non Barbiturates				
Quinalbarbitone 100 mg	Yes	Yes	Yes(mild)	Kales[27]
Pentobarbitone 100 mg	Yes	Yes	No	Kales[28]
Chloral Hydrate 1.0 g	Yes	Yes	Yes	Kales[27]
Glutethimide 0.5 g	Yes	Yes	No	Kales[27]
Benzodiazepines				
Flurazepam 30 mg	Yes	No(up to [4]/52)	No	Kales[29]; Vogel[30] Oswald[31]
Nitrazepam 5-10 mg	Yes	No(up to [4]/52)	Yes	Adam[33] Kales[34]
Flunitrazepam 1-2 mg	Yes[1]	Yes	Yes	Bixler[35]
Lormetazepam 1-2.5 mg	Yes	No(up to [3]/52)	Yes	Oswald[31,32]
Temazepam 15-30 mg	Yes[2]	Yes	Yes	Bixler[36] Mitler[37]
Triazolam 0.5-3 mg	Yes	Yes[3]	Yes	Vogel[30,38] Roth[39] Kales[40]

[1] 1 mg dose weakly effective   [2] little effect on sleep latency   [3] Roth 1976, [39] found no evidence of tolerance

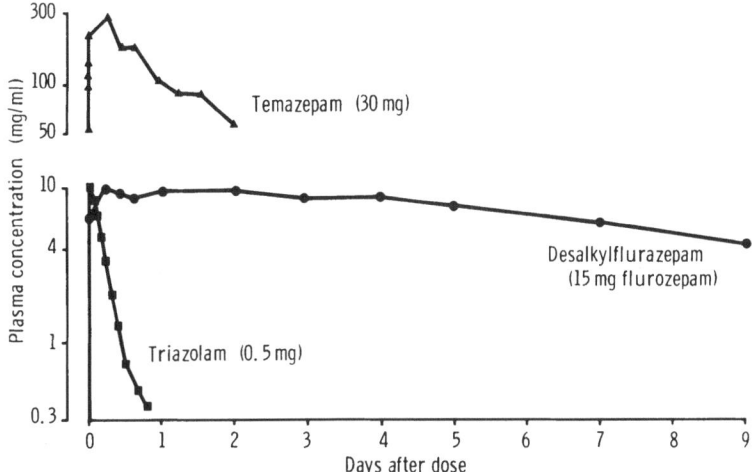

Fig. 1.   Plasma concentrations of desalkylflurazepam,
temazepam and triazolam in a healthy elderly
volunteer who participated in three pharmacokinetic studies (taken from Greenblatt et al[71]
1982, with permission from the editors)

ive.[42]   A dose of 1 g administered to 4 insomniacs was found to
produce a 40 per cent reduction in sleep latency but no change
occurred in the sleep time after first falling asleep.[27]   Glutethimide also appears to be relatively ineffective.[27,43]   Temazepam
30 mg has little effect on sleep latency when it is administered in
hard gelatin capsules (Figure 2).   This may have been due to slow
absorption.   The drug is now administered in soft gelatin capsules
which give rise to higher peak plasma concentrations.[24]   However,
the rate of absorption is still very variable and no EEG sleep
studies, which have used this formulation, have been reported.

HYPNOTIC EFFICACY - INTERMEDIATE AND LONG TERM USE

Patients are often prescribed hypnotic drugs for months or years,
yet most studies of hypnotic efficacy have only been continued for
two weeks or less.[27]   Nonetheless, several studies have shown that
tolerance occurs to many of the shorter acting drugs during the first
week of use (Table 4, Figure 2 and 3).   On the other hand, the
longer acting drug flurazepam (30 mg) and nitrazepam (5 mg) have
been shown to remain effective for up to 4 and 10 weeks of use
respectively.[28,33,45]   A 5 week EEG and questionnaire study of the
efficacy of temaze am (30 mg in hard capsules) found that the effect
on sleep latency was lost during the second week but the drug

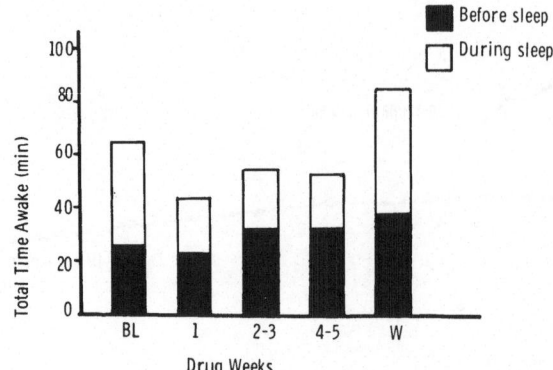

Fig. 2. Average time awake over a 3 night
period measured by continuous EEG
recording during sleep of 14
insomniac subjects, before (BL),
during and after (W), regular
treatment with temazepam 30 mg nocte,
given in hard gelatin capsules.
Combined results from 2 studies.[36,37]

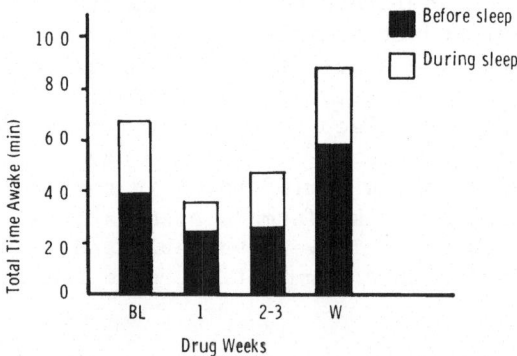

Fig. 3. Average time awake over a 2-4 night
period measured by continuous EEG
recording during sleep of 36
insomniac subjects, before (BL),
during and after (W), regular
treatment with triazolam 0.5 mg nocte.
Combined results from 4 studies.
[30,38,39,40]

continued to prolong the total sleep time until the end of the study. This loss of effect was not seen in the subjective scores since the subjects continued to rate their sleep latency as substantially improved.   More recently, a long term placebo controlled study which used patient questionnaires, found that both nitrazepam 5 mg, and lormatazepam 2 mg continued to improve the subjective quality of sleep with no loss of effect for up to 6 months. Most of the effect of nitrazepam on sleep latency was lost after the first two weeks.[32]   These results emphasize that the sedative properties of hypnotic drugs, as measured by their effect upon the total sleep time, may be unrelated to their effect on the perceived quality of sleep.

TOLERANCE, DEPENDENCE AND THE HYPNOTIC WITHDRAWAL SYNDROME

Tolerance to the sedative effects of hypnotics occurs during regular dosing and leads to a change in the effects of the drugs on sleep time and day time alertness, yet this subject has received almost no attention.   For instance, most young patients show no clinical evidence of day time sedation during regular dosing with flurazepam, in spite of the considerable degree (approximately 6 fold) of drug accumulation that occurs in the body before steady state plasma concentrations are reached.[45]   If a six fold dose were given acutely it would cause considerable sedation. This change in sedative effect has been demonstrated in an experiment carried out at the Royal Free Hospital (Figure 4).[46]   The dose of diazepam required to achieve a defined level of sedation in patients about to undergo endoscopy was up to three fold higher in patients who had been taking small doses of sedatives, mainly benzodiazepines, on a regular basis.

Many patients become dependent on benzodiazepine hypnotics but, unlike most drugs of addiction, they do not usually increase their dosage. This is also true when drugs like diazepam are taken for anxiety.[47]   This suggests that tolerance to the anxiolytic effects, which are mediated by a different class of benzodiazipine receptors from the sedative effects, does not occur. It is possible that the anxiolytic effect is more important that the sedative effect for improving the quality of sleep in insomniac subjects.

A series of studies conducted in General Practice, have demonstrated that about a fifth of patients commenced on hypnotics for insomnia are still taking their hypnotics four years later. About 20 per cent of the patients began their long term hypnotic prescriptions in hospital and patients appeared to be just as liable to become dependent on nitrazepam as on amylobarbitone.[48,49]

Barbiturates are well recognised as drugs of addition.[50] They produce tolerance with regular use and abrupt withdrawal of large

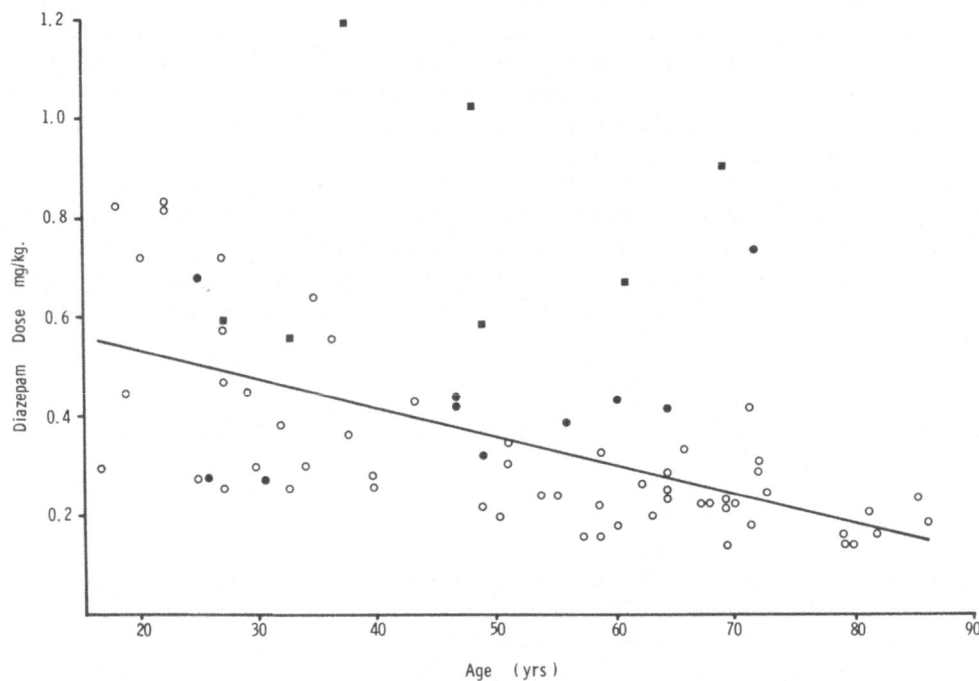

Fig 4.   Relation between intravenous dose of diazepam required to
          achieve a predetermined level of sedation and age in
          patients about to undergo endoscopy. ● = patients who were
          receiving 40 G alcohol/day. ■   = patients who were
          receiving treatment with sedatives

doses leads to a well defined withdrawal syndrome, characterised by
rebound hyperexcitability of the central nervous system.   The
syndrome can include delirium insomnia, coma, anorexia and convulsions.   The onset and intensity of this syndrome depends upon both
the dose and the rate of elimination of the drugs from the brain.
There is an inverse relation between the elimination half life of
the drug and the intensity of the withdrawal symptoms.   This
relationship is also seen with opiate drugs.   Withdrawal of drugs
with a short half life such as pentobarbitone or morphine leads to
more severe symptoms than that which occurs after withdrawal of
drugs with a long half life such as phenobarbitone or methadone.
During the last few years, it has become apparent that a similar,
though milder, withdrawal syndrome also occurs when benzodiazepines
which have been taken regularly, are suddenly discontinued.[50,51]
The syndrome is characterised by insomnia and day time anxiety.

     Some degree of insomnia occurs after stopping most hypnotic
drugs (Table 4).   Rebound insomnia has been demonstrated with all

of the benzodiazepine hypnotics shown, with the exception of flurazepam.[52]   Day time anxiety has also been observed during regular dosing with benzodiazepine hypnotics which have very short half lives such as fosazepam[53] and triazolam.[54]   These observations indicate that there is also an inverse relationship between the elimination half life of benzodiazepines and the onset and intensity of the withdrawal syndrome.   CNS hyperexcitability may become manifest during the following day if elimination is rapid enough. Presumably the withdrawal syndrome plays an important part in the development of hypnotic dependence.   Unfortunately, no studies of hypnotic withdrawal appear to have been carried out in the elderly and the incidence of these adverse effects is unknown.

DAY TIME SEDATION

    Open epidemiological studies of the incidence of adverse effects in in-patients receiving regular benzodiazepine therapy have shown that the frequency of day time sedation increases with age in patients receiving diazepam, chlordiazepoxide, nitrazepam or flurazepam.   The highest incidence of sedation occurs in patients receiving either nitrazepam or flurazepam.[45,55,56] About 30 - 40 per cent of patients over 70 who are receiving regular doses of 30 mg of flurazepam at night, suffer adverse affects, chiefly unwanted sedative and ataxia during the day.[45,57] Unwanted sedation usually occurs within the first few days of commencing therapy and is more common in patients who have been in hospital for some time, and in patients with hypoalbuminaemia or chronic renal failure. Our experience at the Royal Free Hospital suggests that elderly patients who are chronically ill and those with intellectual impairment are also more sensitive to the sedative effects of the drugs.[58]

    The brain becomes more sensitive to the sedative effects of benzodiazepines with aging.   Thus the dose and plasma concentration of diazepam required to produce sedation falls with age (Figure 4). [59,60]   Similar changes in response can be shown in aged rats. This change in sensitivity is not due to changes in cerebral concentration or altered receptor binding.[46]   An increased cerebral response in the elderly has also been demonstrated with nitrazepam[16] and temazepam.[61]   The increase in the frequency of unwanted sedation that occurs during aging may be due more to this change in cerebral response (pharmacodynamics) than to changes in pharmacokinetics.

    Comparative studies of the residual effects of single doses of hypnotics in young and middle aged subjects have shown an overall relationship between both the dose and the elimination half life, and the duration of effect. Those benzodiazepines with a long elimination half life such as flurazepam or nitrazepam impair psychomotor performance for a longer period of time than benzodiazepines with a relatively short half life such as temazepam or triazolam.   Similarly barbiturates with a long elimination half life such as pentobarbitone or quinalbarbitone produce more residual impairment than heptabarbitone.[62,63,64,65]   The principal exception to this rule is diazepam which has an elimination half life of 20 - 40 hours in young volunteers but appears to be almost free of residual effects at 10 hours after night time doses of up to 10 mg.[62]

There is much less information about residual effects in the elderly.   Although epidemiological evidence suggests that flurazepam produces day time sedation in the elderly more commonly than benzodiazepines with shorter half lives, the residual effects of single doses have not been adequately investigated.   There is no doubt that 10 mg of nitrazepam produces marked residual sedation in elderly subjects and this dose should be avoided[16,66,67] On the other hand, single doses of chlormethiazole 384 mg and temazepam 20 mg appear to be free of residual effects in healthy elderly subjects.[44,68]

Regular dosing with most benzodiazepine hypnotics leads to drug accumulation in the elderly (Table 3) but few studies of residual effects have been carried out under these conditions.   A weeks dosing with either nitrazepam 5 mg or flurazepam 15 mg produces more residual effects than shorter acting drugs such as fosazepam 60 mg or placebo.[69]   Two studies which used patient questionnaires have found that regular dosing with flurazepam 15 - 30 mg produces increasing residual effects with time.[15,70]   In one of these studies the increase in day time sedation was shown to parallel the rise in plasma concentrations.[15]   Another group have compared a week of treatment with nitrazepam 5 mg, and chlormethiazole 384 mg in a double blind crossover study.[68]   Neither drug produced a significant impairment of the reaction time the morning after dosing or an increase in adverse effects when compared with placebo.

We have recently completed a controlled trial of the hypnotic and residual sedative effects of seven regular night time doses of either nitrazepam 5 mg (N), temazepam 20 mg (T) or placebo (P) in 58 elderly inpatients.[58]   Plasma temazepam and nitrazepam concentrations rose by 50 per cent and 113 per cent between the morning of day 1 (Day 1 am) and day 7 (Day 7 am) (Figure 5).   Patients reported sleeping well more often on the morning after the first dose of both hypnotics (Day 1) ($p < 0.05$) but there was no difference between the drugs and placebo after the seventh dose (Day 7) (Figure 6).   Reaction time was unchanged on the morning after the first dose (1 am) but was significantly prolonged on the morning after the seventh dose (7 am) of both hypnotics (Figure 7). The E test completion time tended to be prolonged after the first dose of both drugs (1 am) and was further prolonged on the morning after the seventh dose of nitrazepam (7 am) (Figure 8).   Both the reaction time and the letter E test time tended to return to

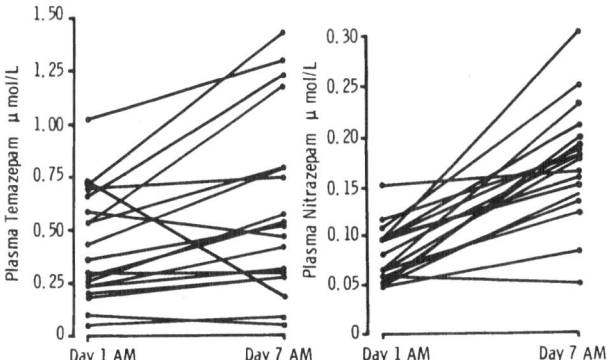

Fig. 5. Plasma concentrations of temazepam
         and nitrazepam on the morning after
         the 1st and 7th doses.

normal during the afternoon after the 7th dose (7 pm). Thus, plasma
accumulation was associated with a deterioration in day time
performance.   There was an inverse correlation between IQ and
impairment of performance which was significant for the E test time
though not the reaction time.   These findings support previous
clinical observations that it is patients with some degree of brain
failure who are particularly at risk when taking these drugs.

CHOICE OF DRUGS

     Barbiturates are probably best avoided as they are no more
effective than benzodiazepines and are dangerous when taken in over
dose. They also produce hepatic enzyme induction which can lead to
drug interactions and osteomalacia.   Benzodiazepine drugs with a
long elimination half life, such as flurazepam, nitrazepam or
flunitrazepam are effective but they carry a high risk of day time
sedation.   However, this risk is dose dependent and the drugs may
sometimes be indicated in healthy elderly patients with severe
insomnia provided small doses are used.   Unfortunately, the 5 mg
dose of flurazepam, which is available in the USA, is not available
in the UK.   A 5 mg dose of nitrazepam should probably be the
maximum dose for use in elderly subjects.   Drugs with a very short
elimination half life such as triazolam or fosazepam are also
effective hypnotics but they can produce day time anxiety in young
subjects and may be more likely to cause rebound insomnia. However,
these adverse effects have not yet been demonstrated in the elderly.
On the other hand short acting drugs carry a lower risk of day time
sedation and may be indicated for short term use.   Drugs with an
intermediate half life such as lormetazepam, temazepam and other
benzodiazepines with a half life of between 6 and 24 hours represent
a compromise between these two extremes.   The regular dosing trial

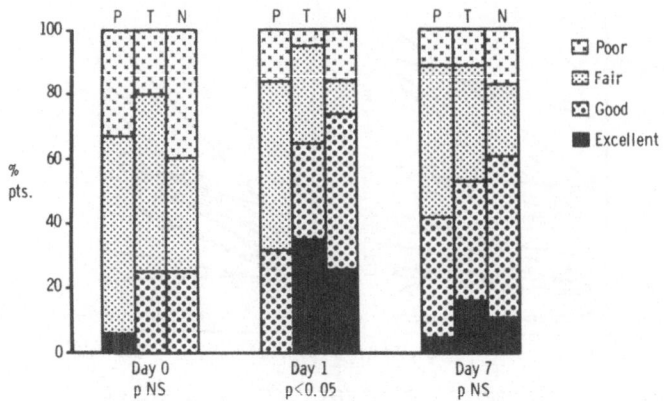

Fig. 6. Answers to the question "How did you
         sleep last night?" on the morning before
         and after the 1st dose and the morning
         after the 7th dose of either temazepam
         20 mg (T), nitrazepam 5 mg (N) or placebo.

conducted at the Royal Free Hospital showed that a drug such as
temazepam which has an intermediate half life can still produce
appreciable residual sedation in susceptible patients.   Chronic
inpatients, those with a low plasma albumin or chronic renal
failure and those with some degree of intellectual impairment are
at the greatest risk. Benzodiazepine drugs are best avoided
altogether in these types of patients.

     Chloral hydrate is rather a weak drug but has the advantage of
being safe.   It can also be taken with alcohol. The principal
adverse effects are gastric irritation, flatulence and mild chronic
acidosis due to the metabolite trichloracetic acid.   Chlormethiazole is a non benzodiazepine hypnotic which has given promising
results in the elderly.   It is more rapidly absorbed than most
benzodiazepines and has a short half life. Its effect on sleep
time has not yet been evaluated in sleep EEG studies.   The
principal side effect is nasal irritation but this only appears to
be a problem if the patient is awake during the first few hours
after dosing.

     Hypnotics are often given to elderly patients in hospital for
temporary sleep disturbances due to either acute illness, changes in
the environment, or a sleep pattern that does not fit in with the
ward routine.   They should only be given for very short periods
for this purpose, and should always be stopped before the patient
is discharged in order to avoid life long dependence   For those
patients who are already dependent on their drugs, changing to low
doses of flurazepam may make it easier to discontinue therapy.

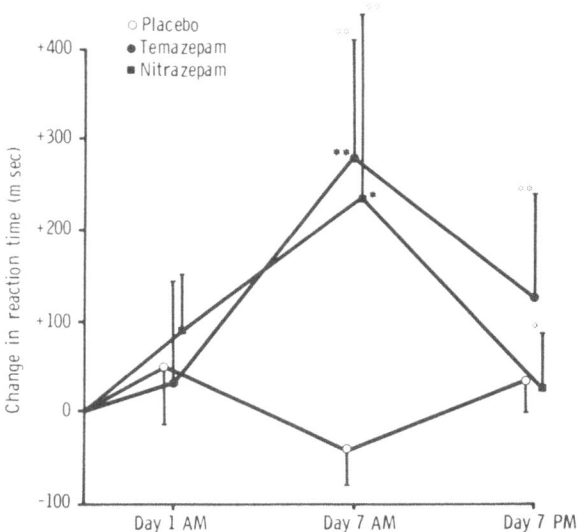

Fig. 7.   Change in choice reaction time.

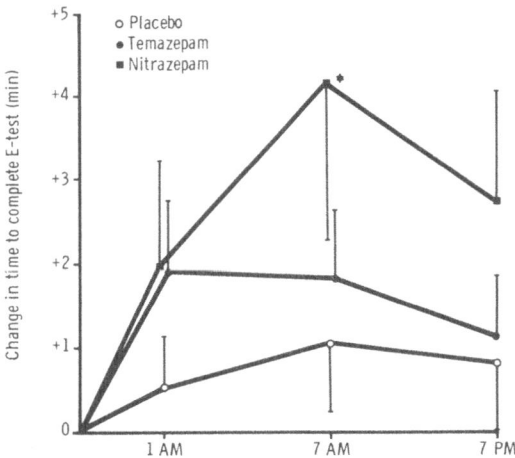

Fig. 8.   Change in time required to complete
a letter E deletion test.

Plasma concentrations persist for several weeks after the drug is discontinued and rebound insomnia does not occur.

REFERENCES

1. W. B. Mendelson, "The Use and Misuse of Sleeping Pills," Plenum, New York (1980).
2. M. A. Carskadon, W. C. Dement, M. M. Mitler, C. Guilleminault, V. P. Zarcone, and R. Spiegel, Self-reports versus sleep laboratory findings in 122 drug-free subjects with complaints of chronic insomnia, Am.J.Psychiat. 133:1382 (1976).
3. B. L. Frankel, R. D. Coursey, R. Buchbinder, and R. Snyder, Recorded and reported sleep in chronic primary insomnia, Arch.Ger.Psychiat. 33:615 (1976).
4. D. D. Breimer, and J. M. Van Rossum, Pharmacokinetics of (+) and (-) and - hexobarbitone in man after oral administration, J.Pharm.Pharmacol. 25:762 (1973).
5. D. D. Breimer, and A. G. De Boer, Pharmacokinetics and relative bioavailability of hepatabarbital and heptobarbital sodium in man after oral administration, Eur.J.Clin.Pharmacol. 9:169 (1975).
6. D. D. Breimer, and M. A. C. M. Winten, Pharmacokinetics and relative bioavailability of cyclobarbital calcium in man after oral administration, Eur.J.Clin.Pharmacol. 9:443 (1976).
7. T. Inaba, B. K. Tang, L. Endrenyl, and W. Kalow, Amobarbital - a probe of hepatic drug oxidation in man, Clin.Pharmacol. Ther. 20:439 (1976).
8. D. D. Breimer, Pharmacokinetics and biopharmaceutical aspects of hypnotic drug therapy. in: "Clinical Pharmacy and Clinical Pharmacology," W. A. Gouveia, G. Tognoni, and E. Van der Kleijn, eds., New York: North Holland (1976).
9. M. M. Reidenberg, D. T. Lowenthal, W. Briggs, and M. Gasparo, Pentobarbital elimination in patients with poor renal function, Clin.Pharmacol.Ther. 20:67 (1976).
10. R. L. Nation, J. Vine, E. J. Triggs, and B. Learoyd, Plasma levels of chlormethiazole and two metabolites after oral administration to young and aged human subjects, Eur.J.Clin. Pharmacol. 12:137 (1977).
11. E M. Sellers, M. Lang-Sellers, and J. Koch-Weser, Comparative metabolism of chloral hydrate and triclofos, J.Clin.Pharm. 18:457 (1978),
12. S. H. Curry, D. Riddall, J. G. Gordon, P. Simpson, T. B. Binns, R. K. Rondel, and C. McMartin, Disposition of glutethimide in man, Clin.Pharmacol.Ther.12:849 (1971).
13. R. E. Irvine, J. Grove, P. A. Toseland, and J. R. Trounce, The effect of age on the hydroxylation of amylobarbitone sodium in man, Br.J.Clin.Pharmacol. 1:41 (1974).

14. R. E. Vestel, E. A. McGuire, J. D. Tobin, R. Andreas, M. S. Norris, E. Mezey, Ageing and ethanol metabolism, Clin.Pharmacol.Ther. 21:343 (1977).

15. D. J. Greenblatt, M. Divoll, J. S. Harmatz, D. S. MacLaughlin, R. I. Shader, Kinetics and clinical effects of flurazepam in young and elderly non-insomniacs, Clin.Pharmacol.Ther. 30:475 (1981).

16. C. M. Castleden, C. F. George, D. Marcer, and C. Hallett, Increased sensitivity to nitrazepam in old age, Brit.med.J. 1:10 (1977).

17. L. Kangas, E. Iisalo, J. Kanto, V. Lehtinen, S. Pynnonen, I. Ruikka, J. Salminen, M. Sillanpaa, and E. Syvalahti, Human pharmacokinetics of nitrazepam : effect of age and diseases, Europ.J.Clin.Pharmacol. 15:163 (1979).

18. J. Kanto, L. Kangas, L. Aaltonen, and H. Hikle, Effect of age on the pharmacokinetics and sedative effect of flunitrazepam, Int.J.Clin.Pharmacol.Ther.Toxicol. 19:400 (1981).

19. M. Humpel, B. Neiuweboer, W. Milius, H. Hanke, H. Wendt, Kinetics and biotransformations of lormetazepam. II. Radio-immunologic determinations in plasma and urine of young and elderly subjects : first pass effect. Clin.Pharmacol.Ther. 28:673 (1980).

20. A. Huggett, R. J. Flanagan, P. Cook, P. Crome, D. Corless, Chlormethiazole and temazepam, Brit.med.J. 281:475 (1981).

22. Upjohn Ltd., Technical Report no. 3129-81-9122-029. Effects of age and sex on the pharmacokinetics of triazolam and temazepam in healthy volunteers. December (1981).

23. P. Bittencoutt, A. Richens, P. A. Toseland, J. F. C. Wicks, and A. N. Latham, Pharmacokinetics of the hypnotic benzo-diazepine, temazepam, Br.J.Clin.Pharmacol. 8:37 (1979).

24. L. M. Fuccella, Bioavailability of temazepam in soft gelatin capsules, Brit.J.Clin.Pharmacol. 8:31 (1979).

25. M. Divoll, D. J. Greenblatt, J. S. Harmatz, R. I. Shader, Effect of age and gender on disposition of temazepam, J.Pharm.Sci. 70:1104 (1981).

26. S. Salem, C. Kinney, and D. McDevitt, Pharmacokinetics and psychomotor effects of nitrazepam and temazepam in healthy elderly males and females, Brit.J.Clin.Pharmacol.13:601 (1982).

27. A. Kales, E. O. Bixler, J. D. Kales, and M. B. Scharf, Comparative effectiveness of nine hypnotic drugs : sleep laboratory studies, J.Clin.Pharmacol. 17:207 (1977).

28. A. Kales, J. D. Kales, E. O. Bixler, and M. B. Scharf, Effectiveness of hypnotic drugs with prolonged use: flurazepam and pentobarbitol, Clin.Pharmacol.Ther. 18:356 (1975).

29. A. Kales, E. O. Bixler, M. Scharf, and J. D. Kales, Sleep laboratory studies of flurazepam : a model for evaluating hypnotic drugs, Clin.Pharmacol.Ther. 19:576 (1976).

30. G. W. Vogel, K. Barder, P. Gibbons and A. Thurmond, A
     comparison of the effects of flurazepam 30 mg and triazolam
     0.5 mg on the sleep of insomniacs, Psychopharmacology
     47:81 (1976).
31. I. Oswald, K. Adam, S. Burrow, and C. Idzikowski, The effect
     of two hypnotics on sleep. Subjective findings and
     skilled performance, Advances in the Biosciences 12:51
     (1978).
32. I. Oswald, C. French, K. Adam, and J. Gilham, Benzodiazepine
     hypnotics remain effective for 24 weeks, Brit.med.J.
     284:860 (1982).
33. K. Adam, L. Adamson, V. Brezinova, and W. M. Hunter,
     Nitrazepam : lastingly effective but trouble on withdrawal,
     Brit.med.J. 1:1558 (1976).
34. A. Kales, M. B. Scharf, and J. D. Kales, Rebound insomnia :
     a new clinical syndrome, Science 201:1039 (1978).
35. E. O. Bixler, A. Kales, C. R. Soldatos, and J. D. Kales,
     Flunitrazepam, an investigational hypnotic drug : sleep
     laboratory evaluations, J.Clin.Pharmacol. 17:569 (1977).
36. E. D. Bixler, A. Kales, C. R. Soldatos, M. B. Scharf, and J. D.
     Kales, Effectiveness of temazepam with short, intermediate
     and long term use : sleep laboratory evaluation, J.Clin.
     Pharmacol. 110-18 (1978).
37. M. M. Mitler, M. A Carskadon, R. L. Phillips, W. R. Sterling,
     V. P. Zarcone, R. Spiegel, C. Guilleminault, and W. C.
     Dement, Hypnotic efficacy of temazepam : a long-term sleep
     laboratory evaluation, Br.J.Clin.Pharmacol. 8:63 (1978).
38. G. Vogel, A. Thurmond, P. Gibbons, K. Edward, K. Sloan, and
     K. Sexton, The effect of triazolam on the sleep of
     insomniacs, Psychopharmacologia (Berlin) 41:65 (1975).
39. T. Roth, M. Kramer, and T. Lutz, Intermediate use of triazolam:
     a sleep laboratory study, J.Int.Med.Res.4: 59 (1976).
40. A. Kales, J. D. Kales, E. O. Bixler, M. B. Scharf, and E.
     Russek, Hypnotic efficacy of triazolam : sleep laboratory
     evaluation of intermediate term effectiveness, J.Clin.
     Pharmacol. 16:399 (1976).
41. F. Solomon, C. White, D. Parron, and W. Mendelson, Sleeping
     pills, insomnia and medical practice, New Engl.J.Med.
     300:803 (1979).
42. P. Piccione, F. Zorick, T. Lutz, T. Grisson, M. Kramer and
     T. Roth, The efficacy of triazolam and chloral hydrate in
     geriatric insomniacs, J.Int.Med.Res. 8:361 (1980).
43. L. Goldstein, J. Graedon, D. Willard, F. Goldstein, and R. R.
     Smith, A comparative study of the effects of methaqualone
     and glutethimide on sleep in male chronic insomniacs,
     J.Clin.Pharmacol. 110;258 (1970).
44. R. S. Briggs, C. M. Castleden, C. A. Kraft, Improved hypnotic
     treatment using chlormethiazol, and temazepam, Brit.med.J.
     1:601 (1980).

45. D. J. Greenblatt, M. D. Allen, R. I. Shader, Toxicity of high
    dose flurazepam in the elderly, Clin.Pharm.Ther. 21:355
    (1977).
46. P. J. Cook, Diazepam pharmacodynamics and ageing. MD thesis
    Bristol University,(1982) Submitted to examiners.
47. J. Marks,"The Benzodiazepines. Use, overuse, misuse, abuse,"
    MTP Press Ltd., England (1978).
48. L. Johnson, and A. D. Clift, Dependence of hypnotic drugs in
    general practice, Brit.med.J. 4:613 (1968).
49. A. D. Clift, Sleep disturbance in general practice, in:
    "Sleep disturbance and hypnotic drug dependence," A. D.
    Clift, ed., Excerpta Medica (New York) (1975).
50. J. H. Jaffe, Drug addiction and drug abuse, in: "The
    Pharmacological basis of Therapeutics," 6th Edition,
    A. Gilman, and L. S. Goodman, eds., Macmillan, New York
    (1980).
51. H. Petursson, M. H. Lader, Withdrawal from long-term benzo-
    diazepine treatment, Brit.med.J. 283:643 (1981).
52. A. Kales, M. B. Scharf, J. D. Kales, R. Soldatos, Rebound
    insomnia. A potential hazard following withdrawal of
    certain benzodiazepines, J.Amer.Med.Assoc. 241:1692 (1979).
53. S. Allen, and I. Oswald, Anxiety and sleep after fosazepam,
    Br.J.Clin.Pharmac. 3:165 (1976).
54. K. Morgan, and I. Oswald, Anxiety caused by a short life
    hypnotic, Brit.med.J. 284:942 (1982).
55. Boston Collaborative Drug Surveillance Program, Clinical
    depression of the CNS due to diazepam and chlordiazepoxide
    in relation to cigarette smoking and age, New Engl.J.Med.
    288:277 (1973).
56. D. J. Greenblatt, and M. D. Allen, Toxicity of nitrazepam in
    the elderly: a report from the Boston Collaborative Drug
    Surveillance Program, Brit.J.Clin.Pharm.
57. J. K. Martilla, R. J. Hammel, B. Alexander, and R. Zushiak,
    Potential untoward effects of long term use of flurazepam
    in geriatric patients, J.Amer.Pharm.Assn. 17:692 (1977).
58. P. J. Cook, A. Huggett, R. Graham-Pole, I. T. Savage, and
    I. M. James, Hypnotics accumulation and hangover in elderly
    in-patients: a controlled double blind study of temazepam
    and nitrazepam, Brit.med.J. 286:100 (1983).
59. M. M. Reidenberg, M. Levy, H. Warner, C. B. Coutinho, M. A.
    Schwartz, G. Yu, J. Cheripko, Relationship between diazepam
    dose, plasma level, age and CNS depression, Clin.Pharmacol.
    Ther. 23:371 (1978).
60. H. G. Giles, G. M. MacLeod, J. R. Wright, E. M. Sellers,
    Influence of age and previous use on diazepam dosage
    required for endoscopy, J.Can.Med,Assn. 118:513 (1978).
61. C. G. Swift, J. M. Haythorne, P. Clarke, I. H. Stevenson,
    The effect of ageing on measured responses to single doses
    of oral temazepam, Brit.J.Clin.Pharm. 11:413 (1981).

62. A. N. Nicholson, Performance studies with diazepam and its
     hydroxylated metabolites, Br.J.Clin.Pharm. 8:39 (1979).
63. T. Roth, P. Piccione, P. Salis, M. Kramer, and M. Kaffeman,
     Effects of temazepam, flurazepam and quinalbarbitone on
     sleep : psychomotor and cognitive function, Br.J.Clin.Pharm.
     8:47 (1979).
64. T. Roth, F. Zorick, J. Sicklesteel, and E. Stepanski, Effects
     of benzodiazepines on sleep and wakefulness, Br.J.Clin.
     Pharmacol. 11:31 (1981).
65. I. Hindmarch, Effects of hypnotic and sleep-inducing drugs on
     objective assessments of human psychomotor performance and
     subjective appraisals of sleep and early morning behaviour,
     Br.J.Clin.Pharmacol. 8:43 (1979).
66. A. Harenko, A comparison between chlormethiazole and nitrazepam
     as hypnotics in psycho-geriatric in-patients, Curr.Med.Res.
     Opn. 10:657 (1975).
67. M. Linnoila and M. Viukari, Efficacy and side effects of
     nitrazepam and thioridazine as sleeping aids in psycho-
     geriatric in patients, Br.J.Psychiat.128:566 (1978).
68. A. N. Exton-Smith and D. J. Witts, A comparison of chlormeth-
     iazole and nitrazepam as hypnotics in elderly subjects, with
     a note on the pharmacokinetics of chlormethiazole, in:
     "Current Trends in Therapeutics in the Elderly," A. N.
     Exton-Smith, ed., Medical Education Services Ltd. Oxford,
     (1980).
69. M. Viukari, M. Linnoila and V. Aalto, Efficacy and side effects
     of flurazepam, fosazepam and nitrazepam as sleeping aids
     in psychogeriatric patients, Acta.Psychiat.Scand. 57:27
     (1978).
70. J. M. Fillingim, Double blind evaluation of temazepam,
     flurazepam, and placebo in geriatric insomniacs, Clin.Ther.
     4:369 (1982).
71. D. J. Greenblatt, M. Divoll, D. R. Abernethy and R. I. Shader,
     Benzodiazepine hypnotics : kinetic and therapeutic options,
     Sleep 5:S23 (1982).

ASSESSMENT AND MANAGEMENT OF CONFUSED OLD PEOPLE

IN GENERAL HOSPITALS

R. Jones

Senior Lecturer and Consultant Psychiatrist
Department of Health Care of the Elderly
University of Nottingham

The confused animal soon attracts malevolent attention from its fellows. Unusual activity or bizarre behaviour alert others to its vulnerability to attack. Nature gives rapid and ruthless short shrift to such animals. When the highest level of the nervous system cannot perceive and respond to the world effectively survival is in peril. Modern day human society hopefully deals with such problems in a more compassionate and understanding manner - though it is clear that in times gone by some confused elderly people received very brutal treatment indeed. Nowadays it is increasingly appreciated that confused elderly people need the closest scrutiny by health care professionals, and most authorities agree that the facilities of a general hospital are necessary for this.

The acute states of confusion are of course as old as mankind, but it is probably in the towering works of Hippocrates, around 400 years B.C., that we find the first written reference to such matters. Hippocrates talked about "phrenitis" and realised that the seat of the disorder was the brain. The accurate clarity of his observations echoes down the centuries and one is struck also by the colour of the descriptive phrases for some features such as the "wandering of the wits." Describing the groping, picking movements of some profoundly ill people he referred to "carpholgia" (meaning straw collecting) and also Areteus referred to "crocydismos" (wool tuft picking). Celsus in the first century A.D. was probably the first to use the term delirium, and over the intervening years there have been many other great names - too many to recall now - who have contributed to the field. In the 20th century the contributions of Bonhoeffer,[1] Wolff and Curran,[2] and Engel and Romano[3] have probably been the most influential. From

their work a consensus has emerged that mostly there was "no
evidence ..... (of) ..... any specific relationship between a
particular noxious agent and the form and content of the associated
psycho-biological disturbances"; "variability was the dominant
feature."   They felt that what was seen in delirium (or "exogenous
psychoses," Bonhoeffer) was closely related to the nature, intensity,
rate of development and reversibility of the underlying 'noxious
agent.'

   However, despite my confident tone and talk of concensus there
is, of course, considerable confusion as to precisely what we are
talking about here.   Many different syndrome names have been used
and continue to be used.   It is perhaps wise to define our terms
a little more precisely.   Both the Oxford English Dictionary and
Lishman[4] agree that the term "confusion" is most appropriately
used to refer to lack of clarity of thinking.   Thus the term
"confused old people" refers to those elderly who are unable to
think with their "customary clarity and coherence."   Sometimes
this may be a reflection simply of fatigue or other factors, and
certainly this sort of problem is sometimes seen in functional
psychiatric disorder.   In using the term however most people are
probably thinking of the organic mental disturbances, and of course
it occurs as a feature both in dementing states (chronic organic
reactions) and in acute confusional states.   It is really the
assessment and management of acute confusional states which I am
considering.

   Some people do not like the term 'acute confusional state' but
it is probably the one most commonly used by British psychiatrists.
It may be defined as "an organic brain syndrome of acute onset and
transient duration, characterised by global cognitive impairment,
and due to widespread disturbance of cerebral metabolism."   This is
the definition used by Lipowski for the clinical state - though
actually he prefers to term this "delirium."[5]

   Most people would link such states to the concept of "clouding
of consciousness" but here again there are some problems of
definition and often problems of detection.   Rather like an
elephant it is quite obvious when it is there though difficult to
sum up in a brief sentence.   Basically clouding of consciousness
is a state of diminished alertness, diminished awareness, and
often it is associated with diminished wakefulness, with drowsiness.
It seems sensible to think of there being a continuum with coma at
one extreme end and at the other the hypersensitive concentrated
keen expectancy of the Olympic sprinter under the starting gun.
Clouding of consciousness is a recognisable point, though not
necessarily an easily objectifiable point, on this continuum and
is when mild impairment starts to supervene.   Clouding is often
thought of as the hallmark of acute confusional states, but mild
degrees may be difficult to detect with confidence.   Additionally

experience with the elderly suggests it is a much less marked
feature in elderly patients with acute confusional states than would
be useful.   It seems likely that the intensity and abruptness of
onset of underlying physical 'noxious agents' are important in
determining the prominence of clouding.   With a number of elderly
people a multiplicity of factors may give rise to their state,
perhaps insidiously over a period.   They therefore may evidence
clouding much less.   This has led us to a discussion of some of the
general features of acute confusional states.   They should be
familiar and there is not time to discuss them further now but
Table 1 summarises those mainly found.

ASSESSMENT

An acute confusional state is a syndrome indicative of many
possible pathologies, not an entity in its own right.   There is
generally an important physical cause, especially when it is severe,
but often a multiplicity of factors are contributing.   It would be
fruitless to try to list all the possible physical causes as such a
list should actually be so extensive as to more or less include all
the disease states to which man is heir.   It is important, however,
to have a framework in mind of the possible categories of cause and
to be able to mentally 'fill in' causes under each category.
Table 2 illustrates such a framework.

Particular causes could be picked out and some seem to be much
more important with the elderly than others.   These would include
urinary and chest infections, heart failure, and cerebro-vascular
disease.   Epilepsy is one frequently overlooked but the importance
of unwanted effects of drugs, drug intoxication and withdrawal
states cannot be too strongly stressed.

The point with diagnosis is that 'confusion of unknown origin'
should be equally unsatisfactory as a diagnosis as 'pyrexia of
unknown origin'.   However our general approach to the patient and
to the problem are equally important and two particular concepts
are worth elaborating in this connection.

A diagram which most of our students see (Figure 1, derived
from Wattis[6]) illustrates the first concept.

A frail old lady balances precariously on a three-legged stool.
To aid her balance she uses a balancing pole, metaphorically
representing her various personality strengths and weaknesses.   The
three legs of the stool represent respectively her physical health,
her social "health" and her psychiatric health.   The message is
the interdependence of all these different aspects. Catastrophe
may result from major problems with any one of them or from smaller
problems with two or more aspects.   Also, if the structure comes

Table 1. Features of an acute confusional state

Short history

Clouding of consciousness

↓ Wakefulness

Temporo-spatial disorientation

Variability of features

Worse at night

↑ Motor activity − Restlessness
                           − Plucking, Picking

↓ Attention   ↓ Concentration

↓ Memory   (↓Registration, ↓Recall)

Anxiety, Suspicion, Agitation

Misinterpretations, Illusions, Hallucinations

Thinking disorganised

Delusions (usually transitory, primitive)

Speech abnormalities

tumbling down, from whatever cause, the fall itself may lead to further damage.   Every aspect needs to be given attention and repair is likely to be necessary in a mumber of ways to restore stable balance.   A broad perspective is essential for successfully dealing with the problem of the elderly and certainly for successfully dealing with an acute confusional state.

The second concept is summed up by another diagram (Figure 2, derived from Jolley[7]) which our students have sometimes unkindly dubbed a "weather forecast map."

It shows diagrammatically important structures and relationships involved in the maintenance of normal consciousness.   Three particular structures are shown.   Firstly there is the cerebral cortex, and this relates secondly to the arousal centres in the reticular activating system of the brain stem, and thirdly to the sleep-waking centre.   All three of these centres, it is suggested, smoothly interact with each other in the maintenance of normal consciousness (and indeed in the production and maintenance of normal sleep).   Additionally all three receive a constant inflow of sensory data from lower nervous levels, predominantly from the external world, but also from the internal world of the individual. Experimental evidence has shown that the normal conscious mental state becomes disrupted in individuals deprived of this external

Table 2. Framework of causes

INTRACRANIAL   Space occupying lesions

                      Trauma

                      Infections

                      Cerebrovascular accidents

                      Epilepsy

EXTRACRANIAL   Infections

                      Metabolic

                      Anoxic

                      Vascular

                      Endocrine

                      Vitamin deficiency

                      Intoxication/withdrawal

                      Physical agents

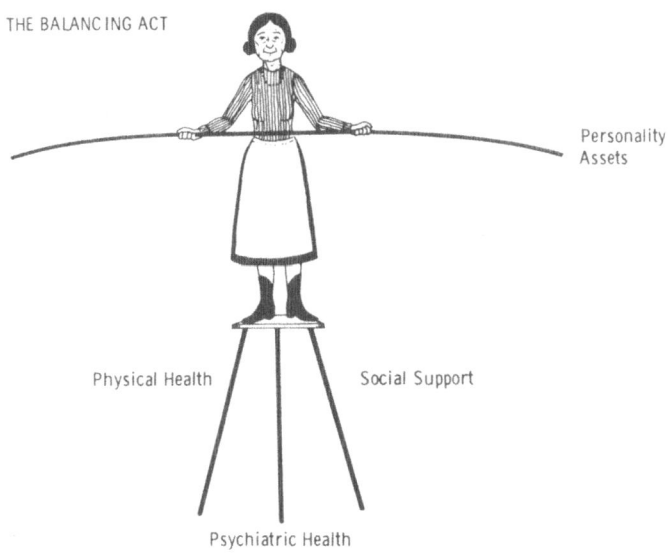

Fig. 1. 'The Balancing Act'

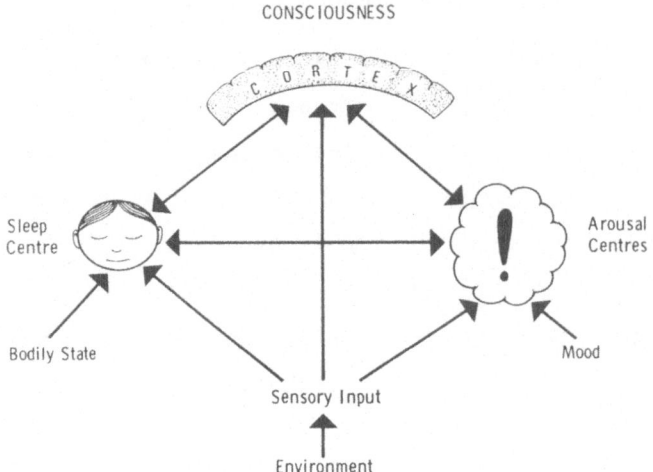

CONSCIOUSNESS

Fig. 2.   'A Possible Model of Consciousness'

sensory input.[8] So effective in fact is sensory deprivation that
it has become used as a means of 'torture' for breaking down
individuals.   Clearly sensory deprivation is extremely relevant to
the elderly with their high levels of visual impairment and deafness.
Such sensory impairment increases vulnerability to acute confusional
states.   Experience also suggests that "over-loading" with excessive sensory stimulation, for example in the form of pain, may be
disruptive and contribute to an acute confusional state.   Thus a
painful fracture may provoke such problems or sometimes simple but
painful constipation may be important.

On this diagram we also indicate a contribution from the state
of emotional arousal and from the mood state of the individual.   It
is a matter of common observation that over-anxiety, over-arousal
can disrupt the efficient cognitive performance of the individual.
Probably we have all experienced disruption of our train of thought
when in states of great apprehension such as in oral examinations.
Similarly observation suggests that mood can play an important part
in vulnerability to confusion, presumably through its effect on
arousal, and there is evidence of an appreciable problem of acute
confusional states occurring in the setting of depression in the
elderly.

This triangular system of cortex, arousal centres and sleep-
waking centre is more vulnerable to disruption if some part of it
such as say, the cortex, has already been significantly impaired
by for instance a dementing process.

Clearly, these are additional influences to the influence of the prevailing bodily state with all the various disease conditions and bodily abnormalities which may arise.

The point is that disease states in the elderly, not least acute confusional states, are likely to be caused by a number of different factors working together. Often only one of these factors is recognisably medical in the classical sense, and it may appear relatively minor in degree. This then makes the occurrence of the acute confusional state difficult to understand unless these broader perspectives are employed. This unravelling of all the various strands combing together to produce the acute confusional state is the essential task of the assessment of such patients. Only when we thoroughly understand how the acute confusional state has come about are we in a position to treat it correctly, to encourage recovery and to aim for a reasonable, stable return to normality.

With this perspective it can be seen that the clinical history of the development of the state is of prime importance. The gathering of information on how matters have developed is the first and perhaps the most essential tool in the assessment process. Its importance cannot be too strongly emphasised. Sometimes in the busy rather anonymous general hospital ward setting an elderly confused patient seems marooned without any information being available, giving every appearance of having arrived without trace from outer space. Always however there are relatives, there are neighbours or there are friends and their vital information must be sought at the earliest opportunity before perhaps they disappear anonymously back into the community.

This full history will inform the clinical examination which is the next most important aspect of our assessment. The physical disorders giving rise to the state will usually be substantially detectable on careful physical examination. This must be meticulous and patient. Skimping will bear no returns. Frequently the examination will have to be completed over a period of time according to the patient's co-operativeness, and in order not to produce higher levels of arousal than are necessary.

The results of the full history and examination will determine investigations. It would be useless here to try to cover the full range of these as they could reasonably be expected to encompass practically the full range of medicine and psychiatry. We may go anywhere from a simple blood count to the highly sophisticated CAT scan. We must establish the cause but it is worth reflecting at this point that it is certainly possible to expensively over-investigate. Our aim must be an understanding of the individual and what has happened rather than simply a predetermined battery of highly technological investigations.

MANAGEMENT

The principal lines of management of the acutely confused
patient can be summarised under the following sections:-

1. Treatment of the main causative factors.
2. Nursing measures and rehabilitation.
3. Promotion of normality and good health.
4. Use of drugs.
5. Follow-up and after care.

Let us examine them in more detail.

1. The process of assessment should elucidate the main
causative factors and our aim will be to reverse with treatment those
which are readily reversible and to ameliorate those which are
relatively immoveable. This section obviously includes the
straightforward treatment of infections with antibiotics as appropriate, or perhaps correction of cardiac arrhythmias; but it also
includes the less glamorous though sometimes even more essential
tasks such as the simple removal of ear wax giving rise to auditory
impairment. Quite often there will be problems such as those of
irreversible visual impairment which we cannot substantially change;
but it may well be that there is much that we can do to modify the
environment or perhaps people's behaviour such as to help overcome
the problem. Some physical problems are not easily overcome but
generally this section of treatment is relatively straightforward
provided we have taken the care and patience necessary to elucidate
all the important factors in the first place.

2. Good nursing is essential to the proper care and recovery
of the confused patient. The observations which the nurses make
and the care which they provide will be vital. Certain important
themes in this good nursing care can be defined. Firstly it is
important to try and make things as clear as possible for the patient.
Thus the lighting needs to be clear, either well lit or, if necessary, properly dark. Misinterpretations, illusions and hallucinations are most likely to arise, in a setting of apprehension, where
the lighting is such as to be uncertain and where threatening
shadows cluster.

As far as possible the nurses themselves should be few, familar
and regular in their appearance. Often this is very difficult on
a busy general ward, but the closer one can come to this ideal
standard of as few different nurses being involved as possible the
greater the dividends for the running of the ward with a less
disruptive confused patient. The nurses themselves must be
patient, tolerant and clear in their approach. The confused mind
needs time to understand what is going on and frequent explanation
of what people are trying to do. The patient will eventually learn

where he is and what is going on, but many repetitions will
generally be necessary.   Slow, clear patient repetition using the
patient's name and reiterating the circumstances will promote a
gradually returning grasp and contact with reality.   Reassurance
and calmness are also important elements.

Whilst the nurses apply these general principles they will also
frequently need to use more specific nursing skills for particular
medical problems and for simple but vital functions such as the
maintenance of adequate hydration.   Ensuring that the confused
patient has an adequate fluid intake is a difficult and skilled task
in its own right.   It needs to be linked to proper estimation or
if possible actual measurement of fluid output as well.

Observation in the nursing management is similarly of great
importance.   Obviously there will often be specific clinical
observations which the nurses are charting but frequently in addition nurses will observe other clinical material which could be
highly important in the management of the patient.   A particularly
treatable background to confusion in the elderly, for instance, is
depression and very often the first or perhaps the only clue that
depression is present is the good nurse's observation of some
utterances or other reflections of mood.

3.   The promotion of normality and good health are also very
much within the sphere of good nursing care.   This means the nurse
aiding the patient to return as far as possible and as quickly as
possible - without excessive stress - to the normal pattern and
rhythm of healthy life.   This means, for instance, a pattern of
activity during the day with rest and sleep at night time.   Aided
mobility maybe necessary at first.   The nurses may help the patient
to physically become upright and start to walk initially.   Resumption
of normal motor activity probably plays an important part in helping
the patient to reassert his grip on normality.

Attention to fluid balance has been mentioned but it is also
important to attend to nutrition.   Particularly with elderly people
it is highly important to attend to simple matters such as bowel
habit and bladder control.

The nurses can help to restore normality to the patient's
confused perception by ensuring that items are brought in from
home which are recognisable, familiar, and which have a power of
significance, helping to draw the disorientated mind back into the
realms of reality.   Useful in this sort of way may be small
ornaments, photographs or simple pictures.   Sometimes the switching on of a regularly followed radio programme may aid in this
process of restoration.   The nurses have every opportunity to
exploit their imagination and ingenuity to the full here.   However
we are all involved in the 'nursing care' of the confused patient

and whilst it is important for the nurses to apply these principles it is important for all of us also to try to follow them to the full in our approach.

4.   The use of drugs is deliberately placed fourth on my list as it is so frequently made automatically the first line.   These other aspects need to be stressed.   Nevertheless the use of drugs should not be shirked and frequently they will be needed.   The very restless and over-aroused patient may well need some calming medication to enable him to benefit from other approaches, and to help overcome possibly life threatening resistance to attempts at hydration, etc.   An important point is that the right dose is the minimum necessary to produce a reasonable response.   This must be individually tailored to suit the patient and often the regime will need to be tailored as well to a particular patient's pattern of disturbances through the 24 hours.   Regimes should be individual- ly set up in this way and frequently reviewed as gradual build-up of medication is a common danger.   Ultimately we must bear in mind that all tranquillising drugs may, in injudicious dosages, cause confusion.   There is a thin line between health and hindrance with their use.

Everybody has their own favourite, familiar drugs.   It is wise to employ a small repertoire of drugs whose repetitious use ensures thorough grasp of their limitations and problems.   Thioridazine is a rather more powerful member of the family but can similarly be satisfactorily used with care.   It tends to have a greater propensity for sedation and I therefore prefer Haloperidol from the butyrophenones.   My experience is that this tends to be very effective in reducing arousal whilst avoiding problems of diminished consciousness.   There is a greater risk of extrapyramidal side- effects with Haloperidol but by starting with small doses such as 0.5 twice or thrice daily, and by aiming for a gradual calming rather than a dramatic sedative effect, this is usually not a great problem.   The benzodiazepines tend to linger so long in the system, particularly with the elderly,[9] that I have avoided their use with the confused elderly.   This attitude has been encouraged by the not infrequent finding that they have been an important contribution - perhaps in the form of sleeping tablets - to the problem of confusion in the first place.   If night time sedation is necessary chlormethiazole seems a trustworthy, quickly excreted and safe agent.[10] Sometimes combination may be necessary with a small dose of pheno- thiazine for lasting effectiveness throughout the whole night.

5.   With these measures and, hopefully the skills of the physiotherapist and occupational therapist, our patient will soon achieve normality and discharge.   It is vital that once so much effort has been invested in elucidating the causes that all these lessons are not just simply forgotten on discharge.   The essence of

what has been done and what has emerged should be communicated to
all those concerned outside hospital.   This obviously means the
family doctor, usually the family and patient themselves, and often
other services such as Social Services and Nursing Services who are
providing help to the home.   These people need to know what has
happened and what can helpfully be done to avoid problems in the
future.   Essentially it is a negligent waste of resources if all
our effort in hospital is not followed by some continuing surveillance after return home.   An important aspect of this effective
follow-up and after care will be effective communication and co-
operation between the various services and people involved with the
old person.

PREVENTION

     Prevention is often a rather difficult task and it is clearly
not likely that we will be able to prevent all the various diseases
to which the elderly are heir.   However we can identify some
people as more vulnerable than others to acute confusional states.
Thus, those elderly with marked special sense impairment, or others
who are perhaps socially isolated and depressed, are particularly
at risk.   The family doctor may see ways to ameliorate important
factors in those at risk; or he may provide early intervention when
illness threatens to usher in confusion.   Surveillance and attempts
to maintain the good health of the elderly would be helpful, but
also certain at-risk groups could be given special attention in an
attempt to minimise problems.

     Another aspect of prevention, of course, is limiting the
problem caused by an illness once it has already started, limiting
the resultant disabilities.   Very often it may be disadvantageous
for elderly persons to be uprooted from their home and 'carted'
into hospital away from all that they know and all that is reassuring.   It may mean that their task of recovery and rehabilitation
is made more difficult.   One way of avoiding this sort of problem
is to have a service which aims to respond rapidly to problems in
the community and which aims to try to begin assessment whilst the
patient is still contained at home.   Often it will prove possible
to conduct this assessment adequately in the home setting and to
arrange for adequate care from home nursing using the family or
domiciliary services.   Providing the illness is not too serious
or the treatment required is not too specialised the elderly patient
will very often have a better prospect of recovery in the familiar
and reassuring environment of his own friendly home.   It is our
view that a great deal of time and money for both patients, families
and services can be saved by rapid and thorough initial assessment
of elderly people with confusion in their own homes through services
based on general hospitals going out into the community.

REFERENCES

1. K. Bonhoeffer,"Exogenous Psychoses" translation of "Zur Frage
    der Exogenen Psychosen," Zentbl, Nervenheilk, 1909, 32,
    499-505, translated by H. Marshall, in: "Themes and
    Variations in European Psychiatry," S. Hirsch and M. Shepherd,
    eds., John Wright and Sons, Bristol (1974).
2. H. G. Wolff and D. Curran, Nature of delirium and allied states,
    Arch. of Neur. and Psychiatry 33:1175 (1935).
3. G. L. Engel and J. Romano, Delirium, a syndrome of cerebral
    insufficiency, J.Chron.Dis. 9:260 (1959).
4. A. Lishman, "Organic Psychiatry," Blackwell Scientific
    Publications (1978).
5. Z. L. Lipowski, "Delirium," Charles C. Thomas, Springfield (1980).
6. J. Wattis, unpublished lecture, Nottingham University (1980).
7. D. Jolley, Acute confusional states in the elderly, in:
    "Acute Geriatric Medicine," D, Coakley, ed., Croom Helm,
    London (1981).
8. D. O. Hebb, "Textbook of Psychology," Saunders, Philadelphia
    (1972).
9. C. M. Castleden, C. F. George, C. Hallett and D. Marcer,
    Increased sensitivity to nitrazepam in old age, Brit.med.J.
    1:10 (1977).
10. S. Nayal, C. M. Castleden, C. F. George and D. Marcer,
    Age and Ageing 7, Supplement, 50 (1978).

TREATMENT OF DEPRESSION IN THE ELDERLY

R. M. Fraser

Consultant Psychiatrist in Geriatric Medicine
St. Pancras Hospital
London, England

It is still inadequately recognised that the term 'depression',
like so many others in psychiatry, takes in its medical context a
meaning somewhat different from that of its normal lay usage,
and from this failure of recognition arise many of the difficulties
that are experienced in the diagnosis and treatment of the
syndrome.   Clinical depression is more than a mere alteration in
mood; it is a true 'depression' of a range of vital functions.
There is depression of activity, of appetite, communicative
ability, memory, gastro-intestinal function and sleep, there is
alteration of endocrine function and there is distorted and pathological thinking.   Illness is defined as <u>loss of function</u>,
and thus clinical depression can be regarded as a true illness,
many of its functional impairments being life-threatening, especially
in elderly people.   The incidence of suicide in the elderly is high[1,2]
but death and chronic morbidity are just as likely, indeed more
likely, to arise from self-neglect and its consequences.   Malnutrition, accidents, pressure sores, and deep venous thrombosis are just
some of the dismal results of depressive illness, and perhaps of
even greater concern is the insidious loss of independence that
overtakes old people in hospital when their condition has not been
speedily recognised and treated.

In old people in particular, then, depression will not be
manifest only, or primarily, as a state of sadness or pessimism, but
as impairments in other areas as well; so in the diagnosis of a
depressed, or possibly depressed patient, one must begin from a
careful assessment that takes in not only mood change or pathological
thinking, but also the range of 'vegetative' functions -- sleep,
appetite, diurnal variation, bowel habit, weight activity and
interests.   It is a good discipline to use one of the diagnostic

scales that lists the changes that commonly occur in depression
indeed where they can be scored, such as the Hamilton Scale.
This does sensitise the clinician to the breadth of the depressive
spectrum and the life-threatening impairments that may occur, and
also is a useful check on progress by way of fairly objective
measures.

CLASSIFICATION

If we decide that a patient is depressed, how do we define or
classify his depression? Is it necessary, before commencing
treatment, to decide what sub-type of depression ne is suffering
from? What sub-types of depression are said to exist? Two
systems of classification at present compete for prominence. First,
there is the traditional sub-division of depressive illness into
'endogenous' or 'neurotic' forms. The description of these sub-
types must be familiar: the endogenously depressed patient is said
to exhibit guilt, retardation, sleep and appetite disturbance and
diurnal variation as prominent features, and he is said to respond
well to ECT or tricyclic antidepressants. The 'neurotic' depressive
is said to have evolved his symptoms in more direct response to
environmental stress; he is said to show a mixture of depressive,
anxiety and 'hysterical' elements, and to respond poorly to tricyclic
antidepressants and ECT. Whatever value this subdivision may have
in the treatment of younger patients (and this is doubtful) it is
unhelpful and indeed positively misleading where elderly patients
are concerned. These descriptive terms not only confuse clinical
presentation with aetiology but they fail in relation to the elderly
in that in the older age group the spectrum of depressive symptomatology is much wider, with considerable overlap between sub-types.[3]
For example, elderly patients with anxiety and agitation may
respond very well to antidepressants and ECT,[4] so that assignment of
such patients to the 'neurotic' category can lead to these patients
being under-treated, and indeed does so.

Depression is best described in purely clinical terms, such as
'retarded depression' or 'agitated depression'; in fact, one can
scarcely improve on the old-style diagnostic formulation in
psychiatry where the presenting features are briefly described
without any attempt to fit them into an artificial system - for
example, one might say, 'the patient shows severe retarded
depression with loss of weight and appetite, preoccupation with bowel
function and suicidal wishes.' Aubrey Lewis[5] has said, however,
that the principal dimension which is clinical interest in depression is that of severity, and this is particularly true when we
consider the elderly and the degree of risk attaching to the range
of disorders within the syndrome.

A newer means of classifying depressive illness relies

principally on biochemical differences; in this system depression is either of Type A or Type B.[6] Type A approximates clinically to 'endogenous' or 'retarded' depression, is related to deficiency in the activity of central catecholamines (principally noradrenaline and dopamine) and there is reduced methoxyhydroxyphenylglycol (MHPG, a noradrenaline metabolite) in the urine. Type B corresponds to 'neurotic' depression and is related to deficiency in the serotonergic system; urinary MHPG is normal. These observations, neat though they appear, have not yet proved their value as clinical predictors - and indeed matters cannot be as simple as this, since there are several glaring anomalies. For example, some new drugs that have highly selective effects on amine re-uptake such as zimelidine (on serotonin) and nomifensine (on dopamine and nora-drenaline) are not particularly selective in the types of depression for which they are clinically useful. The same disappointing returns come from another relatively new diagnostic test - the dexamethasone suppression test. This relies on the observation that in major depression there is hypersecretion of cortisol and resistance of the pituitary-adrenal axis to suppression with dexa-methasone, a synthetic steroid which in normals suppresses the synthesis of ACTH by affecting feedback receptors.[7] Thus following administration of dexamethasone patients with 'melancholia' show a level of serum cortisol that remains elevated in comparison with normals, and this test is highly specific in that it is positive only in 5 per cent of normal subjects. However, it is a relatively insensitive test in that only about 50 per cent of depressed patients actually yield positive DMTs. For this reason, and also because it has yet to prove its predictive value, this test is still far from being a substitute for clinical observation, and indeed adds to it little or nothing. Tests like these do, sadly, come and go; I wonder how many people now recall Funkenstein's test, for example, which was meant to predict a good outcome with ECT on the basis of a rise in blood pressure following injection of noradrenaline. The DMT and also MHPG may or may not go the same way, but at least it is safe to say at present that, once physical causes for a patient's illness have been excluded, and depression has been diagnosed on the basis of both affective and vegetive alterations, there are no further tests likely to be helpful in deciding how to treat.

MANAGEMENT

The first necessity in management is the protection of the patient against the consequence of his illness, and for this reason an elderly patient with major depression is best treated in hospital, at least initially, and in particular if the patient lives alone. The only safe alternative to this would be regular attendance at a day hospital, where at least medication and food intake could be carefully controlled.

Psychotherapy is not indicated at this stage. A supposition does
exist that depression can be treated by detecting the cause of the
illness and by some way eradicating this cause, but matters in
practice are hardly ever so simple. There is a certain failure
within psychiatry to distinguish between prevention and treatment;
in my view treatment must come first, and prevention measures can
follow later. Treatment, further, is best commenced without delay.
It is often said that depression is a 'self-limiting condition', and
this may be true in theory, but it is also true that while this
self-limiting is in progress many life-threatening and irreversible
complications can occur, to which, again, the elderly are particularly
prone. I am not referring just to the physical consequences of
self-neglect, but what is sometimes even more difficult to deal
with, the insidious loss of independence of a patient who has been
over-long in hospital and who, even though free from distinct
clinical depression, has lost the self-esteem, the resolution, and
the habits of self-care that will enable him to return to normal
life outside an institution.

CHOICE OF ANTIDEPRESSANT

    How, then, is an antidepressant to be chosen? The range of
choice is not in fact particularly wide. In the first place, it is
doubtful whether the tricyclic antidepressants are now to be
considered the drugs of first choice in the treatment of the elderly.
It is still by no means clear how these compounds exert their antidepressant action - though, of course, uncertainty about the mode of
action of a treatment has never been a reason for not using it.
Current thinking is that the tricyclic antidepressants down-grade the
sensitivity of post-synaptic monoamine receptors and thus, by way of
feedback mechanisms, stimulate the release of excess monoamine into
the synaptic cleft.[8] This model answers as well as those which it
replaces - in other words not very well, as there remains some
theoretical objections - but the tricyclics are nevertheless potent
antidepressants. The main difficulty in practice arises from the
non-specificity of their action; the first-generation tricyclics,
imipramine and amitryptyline, for example, interfere in at least
five different neurotransmitter systems. Structurally the tricyclics closely resemble acetylcholine, and their most troublesome
side-effects result from blockade of acetylcholine receptors. The
peripheral effects of cholinergic blockade are normally listed as
dry mouth, visual disturbance, postural hypotension and mild tachycardia, most of which sound relatively inocuous, but all of which
can be exaggerated and even dangerous in old people. A postural
drop of more than 20 mm Hg in systolic pressure is not, in my
experience, unusual and falls are common; I have also seen severe
glaucoma on a couple of occasions recently. Further, the central
anticholinergic effects of tricyclics are only now being recognised.
Imipramine and amitriptyline have a more potent anticholinergic

action within the central nervous system than any other antidepressants, amitryptyline being by far the more potent acetylcholine
blocker of the two.   This would account for the confusion that
sometimes results from treatment with tricyclics and, since much
more is now known about the relationship between senile dementia
and acetylcholine depletion, one might now reasonably regard
amitryptyline in particular as contraindicated in the elderly
The drug is said to have a sedative action valuable in agitated
depression, but this is not proven:[9] in the elderly, indeed, no one
can say exactly where sedation leaves off and confusion and ataxia
begin, and one would rather not have to decide.   It does say
something, in fact, about the side-effects of amitryptyline that the
drug is the almost invariably used reference compound in studies
designed to throw up the relative non-toxicity of the newer antidepressants.   Imipramine is not a good choice in the elderly either,
but it can sometimes be quite effective in the treatment of agitated
depression if combined with a small dose of tranquilliser, for
example thioridazine.   I have found that the best starting dose is
not more than 10 mg of each drug three times daily.   This combination makes for early relief of anxiety and thus helps with
compliance; there is also some evidence that, as a result of
competitive inhibition with the phenothiazine at breakdown sites,
the antidepressant activity of the tricyclic is enhanced;[9]
the evidence is not, however, very strong.   The tricyclic with the
weakest anticholinergic action is prothiaden, and this may be the
preferred drug of the group, but it still has to be remembered that
all the tricyclics are extremely toxic in overdose; and that about
10 per cent of suicides annually are by tricyclic overdose.[10]

Two of the more recently introduced drugs now compete strongly
for pride of place as the antidepressant of choice for elderly
patients.   First mianserin - a tetracyclic with a somewhat
uncertain mode of action, being a rather weak inhibitor of both
serotonin and noradrenaline re-uptake, but one shown in many
clinical studies to be an antidepressant having equal potency to
imipramine and amitryptyline but with only weak and peripheral anticholinergic effects;[11] specifically there is no effect on the ECG.
Patients have survived heavy overdosage[12] and, although blood
dyscrasias are reported as a rare side-effect,[13] I would suggest
that mianserin is at present the drug of first choice for elderly
depressed patients.   It may well, however, be displaced by
zimelidine, which is at present the subject of several clinical
trials in the elderly, including one by Professor Exton-Smith and
myself.   Zimelidine was synthesised as a selective blocker of
serotonin re-uptake; it is a two-ring compound which has already
been shown to have equivalent antidepressant effect to the tricyclics.
In our study, although its antidepressant effect has not yet been
evaluated, it does show remarkable freedom from unwanted side-
effects in elderly patients.   It is not very sedating, which may
often be a disadvantage in view of the frequent occurrence of

mixed states in the elderly, and a small dose of diazepam concurrently may well be helpful; diazepam is probably the tranquilliser of choice, since zimelidine has been shown to have no effect
on benzodiazepine receptors.[14]

To conclude the survey of drugs - there remain two antidepressants that may be effective as a second line of treatment
where others have failed.   Lithium may be valuable in the rapid
control of bipolar affective illness, but severe tremor and ataxia
are common side-effects in old people and, of course, renal
function needs to be watched closely.   Lithium tremor can sometimes
be controlled with propranolol, but the role of lithium in the
elderly depressive is probably limited to long-term prophylaxis
in patients who have previously shown a good response to the drug.
Monoamine oxidase inhibitors may also be hazardous in use.   Although
these drugs are effective blockers of serotonin and noradrenaline
degradation, problems as with tricyclics arise from their non-
selectivity in that they block the degradation of pressor amines in
other drugs and in some foods (principally those containing tyramine)
and dangerous interactions can occur.   These effects are not
particularly rare; I have seen two definite instances and one
probable instance of thalamic stroke in the past six months.   At
the same time, I have seen patients respond to MAOIs who have failed
to respond to tricyclics and ECT.   They are usually effective in
depression of 'mixed' type with a major personality of phobic
component.   They are slow to act, and benefit may not be evident
until after three or four weeks of treatment.   Advances towards
safer treatment with MAOIs are being made with the identification
of drugs that will selectively inhibit Type A MAO, the substrates
of which are said to be principally noradrenaline and serotonin,
but the results of investigations are conflicting.[8]   However, the
MAOIs should remain available as a second-line treatment for some
elderly depressed patients.

To sum up in the area of antidepressant drugs - and leaving
aside lithium and the MAOIs - there appears to be remarkably little
to choose in terms of clinical efficacy between the commonly
available antidepressants.   The newer drugs do not have greater
antidepressant potency than the tricyclics, and they do not act
more rapidly.   The choice for an elderly patient will, therefore,
tend to turn on relative freedom from side-effects, and we do now
seem to have antidepressants that are remarkably safe and well
tolerated in this age-group.

ELECTROCONVULSIVE THERAPY

No paper on the management of depression would be complete
without reference to ECT.   Electroconvulsive therapy has been
recognised as highly effective and, indeed, quickly effective, in

severe depression for over forty years now, but the hazards of ECT
do multiply with advancing age, the main risks being of cardiac
arrhythmia, peaking of systolic blood pressure leading to stroke,
and memory impairment.   The majority of patients who suffer major
morbidity following ECT are aged over 60,[15] so we must ask whether
the use of ECT is ever justified in old people and, if so, in what
circumstances?   I will conclude by describing the outcome of a
series of studies by a colleague and myself designed to answer just
these questions.

Now, the clinical paradox in this area is that, even given this
susceptibility to clinical hazards, the elderly have always been
considered to respond very well to ECT, even better than younger
people.   This is, however, largely on impression reproduced by one
text from another without much experimental data.   What, now,
does the evidence show? Our first study was a minor one intended
to investigate the pattern of confusion to be expected immediately
after each treatment.[16]   In a double-blind trial we found that full
orientation was regained within about an hour, although elderly
patients took on average five times as long to recover as did
patients under 65.   But a dramatic shortening of recovery time
resulted from the use of current applied unilaterally to the non-
dominant hemisphere; recovery time was, in some patients, decreased
twenty-fold - in other words, the difference was between fifteen
minutes and five hours.   Incidentally, unilateral and bilateral ECT
were given alternately to each patient, so controlling for individual
variation.

But is unilateral ECT as effective as bilateral ECT?   In a
second study, this time with 33 depressed patients aged from 65 - 86,
we found no difference in efficacy between the two techniques; both
groups showed an equivalent rate of recovery.[4]   All patients in
this part of the study had been receiving tricyclic antidepressants
for at least four weeks without benefit, but after five ECTs there
was significant relief from depression, and all patients but one
had shown complete recovery at the end of three weeks.

What clinical features made for the best outcome with ECT?
Pathological guilt, impairment of work interests, and agitation were
the features most strongly distinguishing the best responders,
followed by depressed mood (oddly, not first in rank order) and
psychic anxiety.

What were the effects on memory? Most scores on the test
battery we employed were impaired before treatment, but during ECT
all of these improved until, by three weeks after the end of
treatment, all had reached normal values.   No new areas of impairment were evident. No other side-effects, either subjective or
objective, were recorded, apart from some complaints of drowsiness
and ataxia in the immediate post-treatment period.

Now, these results do illustrate the remarkable safety and
efficacy of ECT in elderly depressed patients, especially when the
current is applied unilaterally, but two of the findings that were
rather less expected are worth highlighting. First, the observation that there was extensive cognitive impairment in most of the
depressed patients, to the extent that at least two were considered
to be dementing. This illustrates the fact that it is most unsafe
to diagnose dementia on the basis of memory loss alone, and that
the full range of cerebral functions must be assessed before a
diagnosis of dementia is considered.[17,18,19] Secondly, we do see
again here the wide spectrum of depressive illness that responded
to treatment; if one had not taken account of the totality of
symptoms in these patients, some of them might have been dismissed
as 'neurotic' or 'hysterical'; within the range of depressive
functional impairment depressed mood was _not_ necessarily the most
prominent feature.

Is ECT, then, the treatment of first choice in the severely
depressed elderly? Sir Martin Roth[20] has recently said that it is
and I would not in general dissent, but would also say quite
frankly that part of the answer to the question will depend on the
local ECT facilities. In the UK, at any rate, this factor is a
variable one.[21] But given ideal conditions, ECT should be
considered the first line of treatment where the patient is in
severe distress or there are life-threatening aspects of the
depression, and where there is, of course, absence of significant
cardiovascular or respiratory disease. In less severe depression,
while ECT might still in theory be the most rapidly effective
treatment, one does have to bear in mind the fear, rational or not,
that patients and their relatives do have of ECT and the many
practical difficulties that can arise over consent and with organisation of the treatment. It is, therefore, best to begin with one
of the less toxic oral antidepressants and, other things being
equal, to continue with the same drug for at least three weeks and
preferably four weeks before considering a change.

REFERENCES

1. P. Sainsbury, Suicide in later life, Geront.Clin.4:161 (1962).
2. K. Schulman, Suicide and parasuicide in old age: a review,
        Age and Ageing 7:201 (1978).
3. M. D. Blumenthal, Heterogeneity and research on depressive
        disorders, Arch.Gen.Psych. 24:524 (1971).
4. R. M. Fraser, and I. B. Glass, Unilateral and bilateral ECT
        in elderly patients, Acta.Psychiat.Scand. 62:13 (1980).
5. A. Lewis, States of depression: their clinical and aetiological
        differentiation, Brit.med.J. ii:875 (1938).
6. J. J. Schildkraut, and P. J. Orsulak, Recent studies in the
        role of catecholamines in the pathophysiology and

classification of depressive disorders, in:"Neuroregulatory and Psychiatric Disorders," E. Usdin, D. A. Hamburg and J. D. Barchas, eds., Oxford University Press, Oxford (1977).

7. G. M. Asnis, E. J. Sachar, U. Halbreich, S. Nathan, L. Ostrow, and F. S. Halpern, Cortisol secretion and dexamethasone response in depression, Am.J.Psychiatry 138:9 (1981).

8. A. R. Green, and D. W. Costain, "Pharmacology and Biochemistry of Psychiatric Disorders," John Wiley & Sons, Chichester (1981).

9. L. F. Gram, and K. F. Overø, Drug interaction: inhibitory effect of neuroleptics on metabolism of tricyclic antidepressants in man. Brit.med.J. i:463 (1972).

10. R. H. Girdwood, Deaths from tricyclic overdosage, Brit.med.J. i:501 (1974).

11. R. H. S. Mindham, Tricyclic antidepressants and amine precursors, in: "Psychopharmacology of Affective Disorders," E. S. Paykel, and A. Coppen, eds., Oxford University Press, Oxford (1979).

12. P. K. Bridges, and T. R. E. Barnes, New antidepressant drugs, J. Pharmacother. Spring. 12 (1978).

13. C. E. Page, Mianserin-induced agranulocytosis, Brit.med.J. 284:1912 (1982).

14. A. Carlsson, Some current problems relating to the mode of action of antidepressant drugs, in: "Recent Advances in the Treatment of Depression," A. Carlsson, C-G. Gottfries, G. Holmberg, K. Modigh, S. Torgny, and S-O. Ögren, eds., Acta.Psychiat. Scand. Suppl. 290, Munksgaard, Copenhagen (1981).

15. J. A. Abramczuk, and N. M. Rose, Pre-anaesthetic assessment and prevention of post-ECT mortality, Br.J.Psychiatry 134:582 (1979).

16. R. M. Fraser, and I. B. Glass, Recovery from ECT in elderly patients, Br.J.Psychiatry 133:524 (1978).

17. M. Hare, Clinical check-list for diagnosis of dementia, Brit.med.J. ii:266 (1978).

18. P. N. Nott, and J. J. Fleminger, Presenile dementia: the difficulties of early diagnosis, Acta Psychiat.Scand. 51:210 (1975).

19. M. A. Ron, B. K. Toone, M. E. Garralda, and W. A. Lishman, Diagnostic accuracy in presenile dementia, Br.J.Psychiatry 134:161 (1979).

20. M. Roth, Treatment of depression in the elderly, in: "Recent Advances in the Treatment of Depression," A. Carlsson, C-G. Gottfries, G. Holmberg, K. Modigh, S. Torgny, and S-O. Ögren, eds., Acta.Psychiat.Scand. Suppl. 290, Munksgaard, Copenhagen (1981).

21. J. Pippard, and L. Ellam, "Electroconvulsive Therapy in Great Britain, 1980: A Report to the Royal College of Psychiatrists" Gaskell, London (1981).

TREATMENT OF CHRONIC BRAIN FAILURE

A. Agnoli, N. Martucci and V. Manna

Neurological Clinic, University of Rome and
Neurological Clinic, University of L'Aquila
L'Aquila, Italy

The treatment of any pathological condition can be
- preventive;
- aetiopathogenetic or causal;
- symptomatic.

Preventive therapy entails a correct picture of the risk factors or predisposing factors. A correct diagnosis of the disease and the underlying aetiopathogenetic mechanisms is necessary to apply a causal therapy. In the case of symptomatic treatment, knowledge of the pathogenic mechanisms and the altered function from which the symptoms arise is necessary.

From these premises it can be expected that any therapeutic approach to "chronic brain failure" requires an analysis of what this term means, considering that it is a rather generic one. Indeed, with this term can "chronic brain failure" be identified with the "psycho-organic syndrome" mentioned by the English language authors?

This syndrome was described by Mayer-Gross et al[1] in 1963 as a syndrome with an "arteriosclerotic" pathogenesis, characterized by a set of neurological and psychic symptoms as in Table 1.

Already in 1966 Noyes and Kolb[2] reviewed this concept and, from a strictly vascular standpoint, they moved to looking at it as an altered neuronal function   Indeed, as they did for other organs, they introduced the concept of "general cerebral failure" characterized by a reduction in the metabolic cerebral processes which contribute to a decreased brain function which is reversible if it is acute and which, when it is chronic, can lead to damage

Table 1. Symptoms of psycho-organic syndrome
          (From W. Mayer-Gross, E. Slater, M. Roth)

A. Psychiatric Symptoms

    - mood depression
    - anxiety
    - hypochondrial delusions
    - episodes of confusion and/or excitement
    - emotional instability
    - unchanged personality
    - Korsakov-psychosis
    - sleep disturbances

B. Neurological Symptoms

    - focal signs and/or symptoms
    - pseudobulbar syndrome
    - extrapyramidal disorders
    - memory disturbances (mainly short term)
    - epileptiform attacks

C. Somatic Symptoms

    - headache
    - dizziness
    - tinnitus
    - fatigability

to the tissues with a loss of neurons and consequent irreversibility.
Therefore, according to these authors, the distinction can be made
between "acute brain syndrome" and "chronic brain syndrome" characterized by a cerebral damage which is persistent and irreversible

Table 2. Symptoms of psycho-organic syndrome
          (From A.P. Noyes and L.C. Kolb)

    - episodes of confusion and excitement
    - clouding of consciousness
    - incoherence
    - mental fatigability
    - lessening of initiative
    - emotional instability
    - impairment of attention
    - tendency to depression
    - focal neurological disturbances
    - epileptiform attacks
    - memory disturbances

and which can be symptomatologically characterized by an alteration
of the higher cerebral functions with the symptoms listed in Table 2.
As can be seen by comparing the two tables, the symptomatological
pictures present a high degree of similarity although the pathogenetic concepts are different.   All these symptoms cannot be
attributed to the Alzheimer-type Dementia (see Table 3), since they
differ profoundly from a clinical standpoint.   On the other hand
they cannot be due to the Multi-Infarct Dementia, although they
have common aspects (see Table 4), because in MID, due to the very
definition of dementia, there is a prevalence of symptoms of a
severe decay of the higher functions of the central nervous system.

In 1981, Plum,[3] in an article on chronic vascular brain disease,
proposed the following classification:
1. Neurasthenic-depressive reaction to aging
2. Cerebrovascular transient ischaemic attacks
3. Alzheimer senile dementia
4. Post-stroke syndromes
5. Multi-infarct dementia (Table 5)

The psycho-organic syndrome in this new classification must be
identified in the "neurasthenic-depressive reaction of aging" and,
in this connection, the same author points out that in some case
series in the literature, the decrease in regional CBF was of such
a degree as to justify the presence in these cases of a pathology
having the vascular factor as the predominant one, and proposes
"chronic cerebrovascular disease" as its definition.   Whereas in
other cases there is probably a circulatory disorder, especially
in the microcirculation, the dominant pathogenesis being, however,
a changed function of the neurotransmitter systems and cerebral
energy metabolism.

It must be pointed out that the definition of chronic cerebrovascular failure appears in some works.   This definition must be
attributed only to those cases in which later examinations with
positron emission tomography (P.E.T.) will make it possible to

Table 3.   Clinical characteristics of Alzheimer-
           type dementia (From F. Plum)

        - insidious amnesic dementia
        - bright, alert with preserved social amenities
        - fluctuation with environmental stress
        - "impure" defects in language function
        - absence of sensorimotor defects
        - apraxia frequent
        - gyral atrophy by C.T. scan

Table 4. Clinical characteristics of multi-infarct
          dementia (From F. Plum)

- Abrupt onset with successive multifocal progression

- Multifocal or diffuse pyramidal, extrapyramidal,
  neurological signs

- Slovenly appearance, often speech or motor defects

- Dull, slow behaviour

- History of hypertension, smoking, diabetes,
  hyperlipaemia

establish a discrepancy between O_2 demand, O_2 consumption, and glucose reserves with the correct identification of a mechanism of "chronic cerebrovascular failure."

From the above it can be concluded that with the definition of chronic brain failure we can identify a clinical syndrome which is very similar to the psycho-organic syndrome or to the neurasthenic-depressive reaction of aging.

From a pathogenetic standpoint, chronic brain failure has several factors with great variation between cases.

PATHOGENETIC ASPECTS

Brain blood flow

A syndrome of such complexity has an important start in the vascular factor despite being the expression of a decrease not only in the brain blood flow and consequent decrease in oxygen and glucose supply to the neurons, but is also the expression of a primitive reduction in the neuronal metabolico-enzymatic and neuro-transmitter activity.

The decrease in the brain blood flow and oxygen consumption connected with the pathological cerebral aging is a well known phenomenon.

Table 5. Neurological diseases in the elderly
          (From F. Plum)

1. Neurasthenic-depressive reaction of aging
2. Cerebrovascular transient ischaemic attacks
3. Alzheimer-senile dementia
4. Post-stroke syndrome
5. Multi-infarct dementia

The evaluation of the blood flow is important in the study of the pathogenetic factors. However, it has not provided any relevant data when it was studied on a whole hemisphere. Significant and interesting are instead the circulatory variations when the brain blood flow is studied in various brain regions through an evaluation of both the regional blood flow and the oxidative tissue metabolism by means of the measurement of the oxygen consumption.

The reduction in these parameters is significant especially in older patients with factors of vascular risks, hypertension and signs of peripheral arteriosclerosis.

However, a significant reduction in regional blood flow was shown also in patients with a dementia syndrome of cellular genesis with a good correlation between the decrease in the blood flow, cerebral atrophy and clinical localization of the neuro-psychological deficit.

At the present time it is possible to have information on both brain flow and glucose and oxygen metabolism with a tridimensional presentation of the picture achieved with P.E.T. which has a good resolution power also to visualize the subcortical structures.

The importance of this technique in the field of chronic brain failure must be pointed out and we must also consider that various pathogenetic mechanisms can produce similar clinical pictures. With this technique it is possible to place an objective differential diagnosis useful to define the nosological and pathogenetic picture in chronic brain failure and consequently the therapeutic treatment.

The fall in the brain blood flow can be the cause of chronic brain hypoxia in the arteriosclerotic patients in particular, with consequent damage to the nervous tissue and a secondary reduction in the neuronal energy metabolism.

In the subjects with a primary neuronal degeneration and a decrease in the energy and neurotransmitter metabolism it is possible to find a decrease in the regional blood flow as an adaptation to the decrease in the energy requirements.

Only with the study of the bidemensional and tridemensional regional flow is it possible to distinguish in the group of subjects presenting with a psycho-organic syndrome those which can be classified as presenting with chronic cerebrovascular disorders or chronic cerebrovascular insufficiency.

Neuronal Metabolism

The literature does not provide any data on the enzyme changes
in the neuronal energy metabolism which can be correlated to the
psycho-organic syndrome, in that this is a clinical model which
cannot be reproduced in the laboratory animal. There are, however,
data on these changes in cerebral aging in experimental animals
such as senescent rats as well as rats with acutely induced
cerebral ischaemia and hypertensive rats with experimentally induced
cranial vascular damage.

Taking into account what has previously been stated, it is
likely that the changes in neuronal metabolism in the psycho-
organic syndrome are very similar to those found in the senescent
brain and in models of chronic ischaemia.

A number of data published in the literature have confirmed
that the changes in the steric conformation and in the activity of
various enzymes of neuronal energy metabolism are correlated to the
aging of the neurons and to a decrease in enzyme-protein synthesis
(Benzi et al,[4]).

The enzymes studied under this standpoint were mainly lactic
dehydrogenase for the anaerobic glycolysis, citratosynthetase and
malatodehydrogenase for the Krebs circle, total NADH-cytochrome C
reductase and cytochrome oxidase for the respiratory chain.

Biochemical studies of this kind carried out on the cerebral
tissue in toto have not resulted in the localization of the
various distribution of enzyme activity in the cell and neuronal
tissue which is functionally and anatomically heterogeneous.

It must also be pointed out that in the cultures of fibro-
blasts of patients with SDAT a marked drop in the phosphofructo-
kinase activities was found.

The data found in the literature also confirm that in neuronal
aging there is a progressive decrease in the enzyme activity of a
number of substances involved in energy metabolism. This can be
explained either as a decreased protein transcription of the
neuronal DNA or as a decreased "faithfulness of transcription" of
the genetic code with the production of inactive enzymatic or
receptorial proteins, or as changes in the protein synthesis
mechanism, or, more likely, as due to information errors present
in the genic unit.

It is therefore likely that pathogenic insults acting on
either DNA or protein synthesis are at the root of the physiological
processes of cerebral senescence.

In this connection we must bear in mind the "slow" virus hypothesis of dementia syndromes based on the observations of this pathology in some populations such as the Chamarros, the Anyu and the Jaquai of New Guinea which is sometimes associated with degenerative neurological pathologies of the motor neurons and the extrapyramidal system. On the other hand the very appearance of these syndromes in restricted areas and populations cannot but suggest a participation by the genetic heritage in their pathogenesis.

Closely connected with the observations on energy metabolism and neuronal protein synthesis are the structural biochemical considerations resulting from studies carried out on senescent mammals. Various authors have demonstrated that, unlike what occurs in other tissues, in the brain there is an increase in the molar ratio between spermidine and spermine following a more rapid decrease in the spermine content. These polyamines appear to interact directly with the nucleic acids, in particular with RNA, at a neuronal level, thus stimulating protein synthesis in particular conditions.

A decrease in the correct protein synthesis can also explain the number and quality variations in the neuronal membrane receptors and enzymes involved in the synthesis of the neurotransmitters which have a non-secondary role in the pathogenesis of the clinical disturbances which appear in the dementia syndromes. (Table 6)

Various authors have shown a decrease in the total lipid content in human brains during aging as well as a change in the percentage of lipid composition with non-homogeneous changes in the various lipid fractions.

This, together with the findings published in the literature which have shown a decrease in lipid synthesis and/or catabolism, suggest that there is a total hypofunction of the neuron membrane which, as we know, has a fundamental role in the neuronal cell physiology.

Neurotransmitter Aspects

The findings presented in this paragraph refer essentially to what has been shown in the large group of dementias. Since many neurological and psychic symptoms are common, the only difference being in their severity, they must be taken into account in view of a therapeutic approach.

Research in the neurochemical field in dementia is shifting rapidly and progressively from a "static and compositional"

approach to a molecular biology approach.   This change was
synthesis derive?   It was recently suggested that these disturbances result from a neuron degeneration (therefore with a neuronal
loss) in structures projecting towards the cortex.   In particular
the mega-neurons of the Meynert's "nucleus basalis megacellularis"
and the nuclei of the septum which project to the hippocampus.
This datum was demonstrated very recently in brains with SDAT.
At least half of the CAT activity is lost in these nuclei as well
as half of that of Broca's band.

Besides the proven impairment of the subcortical cholinergic
neurons protruding into the cortex, it was also demonstrated that
in some SDAT cases other pathways functioning with different
neurotransmitters are impaired.

It must be pointed out that during physiological and pathological aging of the central nervous system there occur a number
of significant changes in the various enzymes involved in the
neurotransmitter synthesis.

McGeer and McGeer[5] in 1975 have demonstrated a decrease in
tyrosine hydroxylase activity in the human.   The areas provided
with a greater tyrosine hydroxylase activity are the black substance
compacta, the putamen, the caudate, the nucleus accumbens, the
hypothalamus and the globus pallidus.

The highest level of dopadecarboxylase is found in the preoptic
area and therefore in the putamen, the substantia nigra, the
caudate nucleus, the olfactory tubercle, the hypothalamus, and
the n. accumbens.   Dopadecarboxylase and tyrosine hydroxylase
activity undergo a substantial decrease with age.

The enzyme which is responsible for the GABA synthesis, that
is GAD or decarboxylase of the glutamic acid, is present in a
higher concentration in the areas connected with the extrapyramidal
system and, above all, with the substantia nigra.   With aging
there is a decrease in the activity of this enzyme comparable to
that observed in CAT, predominantly in the thalamus, the cortex,
the diencephalon and the cerebellum.   It is, however, less
evident in the basal ganglia.

Recent studies have also pointed out a decrease in GABA binding
with the specific receptor in the substantia nigra and not in the
striatum.   This could suggest a functional dependence of the GABA
ergic system on the dopaminergic system as if the latter had a
"marker" role of the GABA ergic system activity.

The oxidase activity increases with brain age as well as with
that of the peripheral organs innervated by sympathetic fibres.

A decrease in the serotoninergic tonus induced by the admini-

Table 7. Neurotransmitters in aging and SDAT

	AGING	ALZHEIMER-TYPE DEMENTIA
DA	↓	↓
DOPA DEC.	↓	⇄
HVA	↓	↓
NA	↓	↓
T.H.	↓	⇄
MAO	↑	↑
MHPG	↓	↓
SHT	↑ ⇄	⇄
SHIAA	↑	↕
CAT	⇄	↓
AchE	↓	↓
GABA	↓	↓

stration of diets poor in tryptophan or rich in parachlorophenyl-alanine appears to interfere positively with some signs of somatic and cerebral aging. Recent acquisitions on the role played in the central nervous system by various polypeptides acting as neuro-modulators have prompted a search for their possible role in the pathogenesis of both dementia and aging. A correlation between memory deficit and neuroptides has been recently postulated. There is also a marked loss in the locus ceruleus neurons, already evident in aging, but even more marked in SDAT. It was pointed out that this loss can be as high as 80 per cent. Consequently there are severe disturbances in the noradrenergic innervation to the cerebellum, the spinal cord and the brain cortex.

In concluding the discussion of this aspect, if there are evident neurotransmitter disturbances, the cholinergic one in particular, a problem arises in connection with the pathogenesis of this selective damage in subcortical structures with a cortical damage which is only secondary.

The hypothesis of neuronal damage are multiple; that is, genetic, viral, autoimmune, trophic. An alternative hypothesis is possible, that the damage is primarily in the pre- and post-synaptic terminals with a secondary neuronal reaction (Table 7).

In consideration of what has been discussed in both the clinical and pathogenetic sections, we consider the classification in Table 8 more up to date.

Table 6. Receptor modifications in the aging brain

AGING BRAIN

↓

ALTERED RECEPTORIAL PROTEIN SYNTHESIS

↓

MODIFIED STRUCTURE OF RECEPTORS

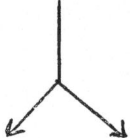

HYPOFUNCTION   HYPERSENSIBILITY FOR
OF RECEPTORS   SYNTHESIS REDUCTION
BY ALTERED   OF MACROMOLECULAR
PLASTICITY   INHIBITOR SYSTEMS

caused in the last few years by the general progress made in
analytical technologies, that is, by the possibility of carrying
out subcellular fractionings of the specific brain regions,
separate the neurons from the glia and even carry out nervous
tissue cultures.

The background of the study of the involvement of various
neurotransmitters is provided by studies indicating an involvement
of the pre-synaptic terminals in dementia.

In particular, considering that acetylcholine has an important
role in memory processes, this neurotransmitter was studied in
particular, this resulting in the identification of the activity of
the two enzymes, namely, choline acetyl transferase (CAT) and
acetylcholinesterase (AchE). These are highly concentrated in the
caudate nucleus and the putamen besides the n. accumbens, the
amygdala, the cerebral cortex, especially in the hippocampus. In
SDAT, a significant decrease in the function of the pre-synaptic
cholinergic system in the frontal hippocampus area has been
demonstrated. CAT and AchE decrease in parallel with cerebral
aging. The decrease in CAT activity is more evident in the
putamen and the caudate.

These enzyme changes are restricted to the pre-synaptic
terminals, the post-synaptic receptors remaining unchanged. This
observation is of a vital importance in view of possible therapy.

The problem arising now is the following: from what do these
disturbances in the enzyme responsible for the acetylcholine

Table 8. Clinical characteristics of the aging brain

Aging Brain:

Psycho-organic syndrome (or neuro-
depressive reaction of aging):

- memory disturbances
- extrapyramidal disorders
- pseudobulbar syndrome

Senile dementia Alzheimer type (S.D.A.T.)
Multi infarct dementia (M.I.D.)
Chronic cerebrovascular disorders
(C.C.V.D.)

"Chronic brain failure" is identified in this classification
with the "psycho-organic syndrome," whereas the definition of
"chronic cerebrovascular disorders (CCVD) can apply to the
condition in subjects who do not only answer the five directives
proposed by the "Ad hoc Committee, Paris, 1980 (Table 9) but who
through the procedures for the study of the bidimensional district
flow show a vascular pathology in the districts from which
the symptoms derive functionally.

THERAPY OF CHRONIC BRAIN FAILURE

The present therapy of chronic brain failure can be divided
into two aspects:

1. The first has the purpose of correcting the vascular risk
   factors and preventing all possible further acute vascular
   accidents.
   This kind of treatment must be adopted in all patients following an accurate clinical investigation in order to decrease
   the risk of a shift towards a more severe and disabling
   pathology such as MID.
2. If we take into account what we have claimed in connection with
   pathogenesis, the theoretical pharmacological approach should
   include an intervention on the following mechanisms:
   - macrocirculation
   - cerebral circulation
   - microcirculation
   - neurotransmission
   - neuronal energy metabolism
   - integrative functions of the CNS

However, from the previous discussion it appears that an
aetiopathogenetic treatment is not possible and therefore it

Table 9.   Chronic cerebrovascular disorders
          (According to the "Ad Hoc Committee,
          Paris, 1980")

1. One or more focal C.V. disorders (including
    T.I.A.)

2. Focal or diffuse neurological and/or
    psychiatric signs (e.g. depression)

3. Presence of risk factors (mainly hypertension)

4. Positivity of complementary exams
    (E.E.G., CT scan)

5. Mild compromising of superior functions of
    C.N.S.

becomes clear that our present capacities are aimed at the
correction of some psychic or neurological symptoms.   This requires
careful clinical monitoring of the patient, possibly combined with
neuropsychological and psychometric investigations, an evaluation
of the metabolites of the main neurotransmitters, a neuroendocrino-
logical investigation, a study of regional CBF through a non-invasive
procedure, or, where it is applicable, with PET scan and a computerized EEC examination.   All the findings obtained with these methods
will help to determine the neurological and psychic symptoms
requiring a priority approach.

In the following tables we examine the actual possibilities of
a therapeutic approach to each symptom.   In Table 10 we have listed
the most recent treatments of the psychic symptoms of the "psycho-
organic syndrome".   The choice of antidepressants results from the
consideration that in the aging brain and in involutional depression
there is an increase in the MAO activity and a prevalent involvement
of the dopaminergic and noradrenergic systems.   At the present time
the antidepressants listed are the most selective ones on these
systems.

For sleep disturbances, the barbiturates are the drugs of choice
as they neither interfere with the neurotransmitters nor cause paradoxical phenomena as do the benzodiazepines.   The use of 5-hydroxi-
triptophan is rational; however, its clinical outcome is not always
very satisfactory.

In Table 11 we have listed the drugs which are active on the
neurological symptoms.   As for the antispastic drugs, it must be
pointed out that it is justified to attempt a first approach with
diazepam, as it induces fewer side effects.   Baclofen has a contraindication in the cases presenting with epileptiform attacks, in
that it lowers the convulsive threshold.

Table 10. Symptomatic treatment of psycho-organic syndrome

Psychiatric Symptoms	Treatment
Depressive Mood	- IMAO
	- IMAO b(?) Inhibitors
	- Reuptake of DA and NE
	. Amineptine
	. Nomifensine
	. Nortriptiline)
	. Desimipramine) (low dosage)
	. Imipramine)
Anxiety and Hypochondriacal Delusions	- Benzodiazepines (low dosage)
Episodes of Confusion and Korsakov Psychosis	- Phenothiazines
	- Thioxanthenes
	- Butyrophenones
Sleep Disturbances	- Barbiturates
	- 5 HTP
	- Hypnotic Benzodiazepines

For the control of the extrapyramidal symptoms (plastic hyper-tonia, bradykinesia) the first choice drugs must be the receptorial agonists in that we know that the long term treatment with L-Dopa often induces a further drug-induced pathology. In addition to this, it is now possible to control the peripheral effects of the DA agonists through the use of a DA blocker which does not pass the B.E.E. as does domperidone.

The memory disorders have been of great interest for researchers in the last few years. We have now two possibilities:
- to increase the functional tonus of the cholinergic transmission in the cortex with some Ach precursors or with drugs reducing the acetylcholinesterase activity.
- to increase the functional tonus of the mesocortical dopaminergic systems projecting to the cortex with an activating function through DA receptor agonists.

Preventive therapy of the epileptic seizures is the typical therapy of all the other epilepsy forms. In the table we have listed the barbiturates and sodium valproate, both being the drugs which in the older patient cause fewer side-effects on the CNS.

Table 11. Symptomatic treatment of psycho-organic syndrome

Neurological Symptoms   Treatment

Pyramidal Signs   − FKT
(mainly spasticity)   − Antispastics:

                                     . Baclofen
                                     . Dantrolene
                                     . Diazepam

Extrapyramidal Disorders   − DA Receptor Agonists:

                                     . Bromocriptine
                                     . Lisurid
                                     . Dihydroergocristine

                                 − DA Precursors:

                                     . L-dopa   ) (low
                                     . L-dopa + IDD) (dosage)

Memory Disturbances   − Drugs acting upon DA System

                                   Receptor) . Bromocriptine
                                   Agonist   ) . Lisurid
) . Dihydroergo-
                                                 cristine

                                 − Drugs acting upon ACH System

                                   Precursors) . Choline
) . Lecithine
) . Phosfatidil-
)   serine ?

                                 − ACHE   ) . Physostigmine
                                   Inhibitors) . Pyridostigmine

Epileptiform Attacks   − Barbiturates

                                 − Sodium Valproate

     In Table 12 we have listed some of the drugs which are active
on somatic symptoms. However, it must be pointed out that in
vertigo the drugs of election remain the first three in the table,
both because they cause fewer side-effects and because some of them
have an action also on other symptoms. With nicergoline, for
instance, because it is also an alphalytic, it is possible to
control moderate hypertension.

Table 12. Symptomatic treatment of psycho-organic symptoms

Somatic Symptoms   Treatment

Headache   1. - Dihydroergocristine

                                    2. - Dihydroergotamine

                                       - 1. or 2. + Trazodone

Dizziness and Tinnitus   - Cinnarizine

                                    - Flunarizine

                                    - Nicergoline

                                    - Antihistamines

Tinnitus   - Bametane

Fatigability   - Nootropic drugs:

or/and   Piracetam

Apathy   - Psychostimulants:

or/and   Methylphenidate
                                     Dextroamphetamine
Memory Disturbances   Prolintane

                                    - Eumetabolic:

                                     Pyritinol
                                     Naftidrofuryl
                                     Cidicholine

When properly used the psychostimulants have proved to be able
to act positively both on fatigability and apathy. However, they
are not commercially available in Italy.

Less astonishing, however at times satisfactory are the results
which can be attained with the "nootropic and eumetabolic" drugs.
In connection with the nootropic drugs, there are controlled
studies published in the literature proving their action on
disorders of memory, especially when they are associated with Ach
precursors.

In conclusion, it must be stressed that, when there is a
syndrome with multiple symptoms, it is utopian and sometimes
negative, due to the pharmacological interferences, to act on each
individual symptom. Therefore a careful choice of the objectives
must be made according to the principle that the life of persons

who already suffer the condition of "being old' must be made less distressing and painful.

REFERENCES

1. W. Mayer-Gross, E. Slater, and M. Roth, Psichiatria Clinica.
    Sansoni Edizioni Scientifiche, Firenze, (1963).
2. A. P. Noyes, L. C. Kolb, "Modern Clinical Psychiatry," Sixth
    edition, W. B. Saunders Company, Philadelphia, London (1966).
3. F. Plum, What is chronic cerebral vascular disease? Clinical
    considerations, in:Proc. Int. Cerebrovascular Diseases, SIR,
    Drugs and Methods INCUD, Pergamon Press, Paris (1980).
4. G. Benzi, E. Arrigoni, F. Dagani, F. Marzatico, D. Curti,
    M. Polgatti, and R. F. Villa, Drug interference on the age
    dependent modification of the cerebral enzymatic activity
    related to energy transduction, in: "Aging, Vol. 13,"
    L. Amaducci, A. N. Davison and P. Antuoni, eds., Raven Press,
    New York (1980).
5. E. G. McGeer and P. L. McGeer, Age changes in the human for
    some enzymes associated with metabolism of catecholamines,
    GABA and acetylcholine, in: "Neurobiology of Aging,"
    J. M. Ordy and K. R. Brizzee, eds., Plenum Press, New York
    (1975).

PERSPECTIVE IN THE TREATMENT OF DEMENTIA

L. Bracco, D. Insitari and L. Amaducci

Department of Neurology
University of Florence
Florence, Italy

After long years of search for new and truly effective pharmacological treatment for aged and demented humans suffering cognitive
deterioration, results are far from impressive. Nevertheless, it
could be worth emphasizing some bodies of evidence which may
represent valuable tools for prevention and treatment of degenerative dementia, namely of senile dementia of the Alzheimer type (SDAT).
In the course of normal aging, in fact, some phenomena occur that
may play a role in mental deterioration; opportune clinical management of them, therefore, could reduce the possibility of a dementia
syndrome. Blood-brain barrier permeability increases with aging
(Link and Tibbling;[1] Amaducci et al[2]) and a defect of this system
could permit the accumulation within the brain parenchyma of potentially toxic environmental compounds, including trace metals. It is
well known, besides, that hypothyroidism and diabetes, both causing
dementing illness, are characterized by an elevation of CSF proteins
(Tourtellotte,[3]). Moreover, in Binswanger's disease, as well as in
other subcortical areteriosclerotic encephalopathies associated with
mental deterioration there is a marked hypodensity of periventricular
white matter; this finding has been attributed to chronic oedema of
the centrum semiovale, possibly following blood-brain barrier
alteration (Loizou et al[4]). A recent exemination of 20 of these
cases showed that 30 per cent of them had diabetes. So it is
possible that blood vessel damage, hypertension and diabetic micro-
angiopathy together produce this clinical syndrome. On the other
hand, these data agree with the report that elevation of serum-glucose
concentration accentuates brain injury due to hypoxia/anoxia in
laboratory animals (de Courten-Myers et al[5]), and with the observation of Plum[6] that prognosis for TIA in diabetic patients is worse
in respect to non diabetics. Probably, diffuse demyelinating lesions
in the white matter follow cerebral oedema, the development of which

is influenced by other factors besides hypoxia; one such factor may be acidosis (Bakey and Lee[7]; Feigin et al[8]).

In this context it is worth remembering that astrocytes associated with senile plaques in normal aging and SDAT brain have been demonstrated to have marked evidence of glial fibrillary acidic protein (GFAP), indicating that the development of senile plaques is accompanied by an increased formation or retention of this protein (Duffy et al[9]).   A similar increase of GFAP density is found following tissue acidosis suggesting a possible influence of pH in pathogenesis (Amaducci et al[10]).   Thus acidosis and diabetes, much more frequent later in life, could underlie some pathological mechanism playing a role in CNS degenerative diseases. such as senile dementia, and maximal care has to be taken in order to prevent them.

From a neurochemical point of view some evidence suggests that one major feature in SDAT is a disruption of the cholinergic neurotransmitter system (for a review see Bartus et al[11]), while the severe loss of neurons in the nucleus basalis of Meynert (Whitehouse et al[12]), thought to provide the primary cholinergic input to the cortical mantle, offers the possibility that the decrease of cholineacetyltransferase (CAT) activity in Alzheimer's patients may reflect a specific loss of cholinergic input to the cortex.   In addition the cholinergic system is involved also in non-transmitter functions such as the control of the permeability of the blood-brain barrier, epithelium, exocrine glands and of the red blood cells (rbc) and structurization and development of the cells including neurons and their membranes (Karczmar[13]; Karczmar et al[14]).   Hanin et al[15] established a correlation between plasma and rbc cholinergic parameters and the clinical state in a variety of neuropsychiatric patients: rbc choline is significantly higher in a sub-category of depressed patients and in individuals with Gilles de la Tourette disease.   Lithium, in addition, elevates rbc choline levels and inhibits choline transport in human erythrocytes; this leads to a large increase in intracellular choline presumably because choline derived from the breakdown of membrane phospholipids is no longer able to leave the cells; this suggests that in erythrocytes the choline transport system normally mediates a net efflux of choline.

Specifically, the exchange of choline plasma and rbc in vivo may be an important factor determining the availability of free choline for transport across the blood-brain barrier, and thus its availability at cholinergic nerve terminals in the brain for subsequent synthesis of acetylcholine (Ach).   Thus red blood cell cholinergic phenomena could provide an index of generalized cholinergic function in vivo and it is conceivable that certain specific neurological and psychiatric illnesses may be associated with a generalized defect in CNS nerve membrane function which, conveniently is also reflected in the rbc membrane (Hanin et al[15]).

In spite of the fact that the "cholinergic hypothesis" of dementia is supported by a large body of evidence, efforts attempting to enhance the synthesis and release of Ach by providing abundant amounts of the precursor substances choline or lecithin are disappointing. Nevertheless, drugs having anticholinergic effects seem to produce deficits most closely mimicking the natural age-related memory impairments, while marked improvement on memory tasks is achieved with anticholinesterase substances (Bartus[16]; Christie et al[17]), and muscarinic agonist arecoline (Sitaram et al[18]). These reports suggest that other mechanisms besides Ach deficit could play a role in senile dementia. It is possible that the aged brain is unable to incorporate or utilize abundant precursor substances, as several parameters that reflect energy production are decreased in the aged brain (Leong et al[19]; Sims et al[20]) and recent evidence shows a reduction of the activity of certain cholinergic pathways in elderly (Sherman et al[21]). Ach, moreover, is one of the neurotransmitters whose metabolism is closely related to oxidative metabolism: in fact, it appears to be an important neurotransmitter in the pathophysiology of hypoxic syndromes (Gibson and Blass[22]), while the decrease of PaO_2 induces a substantial decline in the Ach synthesis (Gibson and Duffy[23]).

Nevertheless, drugs activating cerebral metabolism (for example, Piracetam), although they facilitate transcallosal evoked potentials in cats (Giurgea and Mayersoons[24]),reduce ATP and DNA recovery time after hypoxia in rats (Gobert[25]), improve performances in oxygen-deprived (Sara and Lefevre[26]) or aged animals (Burnotte et al[27]) and accelerate septo-hippocampal release of Ach (Wurtman et al[28]) have failed to demonstrate a real efficacy in respect to placebo in clinical trials in elderly and demented people (Mindus et al[29]; Lloyd et al[30]). On the other hand the possibility exists that simultaneous administration of piracetam, a pharmacological agent enhancing oxidative metabolism, and choline enables the CNS to function more effectively, as shown in aged rats (Bartus et al[11]) and in humans (Friedman et al[31]).

A new trend of pharmacological studies, moreover, could start by recent evidence suggesting the possibility that choline can be synthesized de novo in neurons by the precursors serine and methionine (Blusztajn et al[32]). Alternatively, Gibson and Peterson[33] claim that 3,4-diaminopyridine (3,4-DAP), a potassium channel blocker than enhances the influx of calcium into nerve terminals, ameliorates some of the effects of low oxygen on the calcium-dependent release of Ach impaired by hypoxia, so that release only declined to 80 per cent (2.5% O) or 71 per cent (0% O_2) when compared to 100 per cent in the^{2}absence of the drug. 3,4-DAP, moreover, stimulated the in vivo synthesis of Ach in the hippocampus and striatum and improved performance on a standardized behavioural test. Therefore, at present, the rationale for an effective management of dementing people is represented by the employment of drugs affecting both oxidative and neurotransmitter mechanisms. In this respect it is possible that the good results with Hydergine observed by some authors in patients affected by mental deterioration (for review see Loew and Weil[34]) are due to several pharmacological effects: dihidroergotoxine probably normalizes PO_2 distribution (Meier Ruge et al[35]) and exerts serotonine and dopamine effects on biochemical and functional tests in animals (Vigouret et al[36]).

A dopaminergic activity is claimed also for phosphatidilserine; this agent, in fact, increases the Ach output from the cerebral cortex in the rat (Casamenti et al[37]) and antagonizes scopolamine-induced EEG alterations stimulating subcortical dopaminergic pathways (Mantovani et al[38]).

It should be remembered that CAT asymmetry has been demonstrated at the level of the posterior temporal region corresponding to Wernicke's area (Amaducci et al[39]). An analogous left-bias of CAT in the globus pallidus is significantly reduced in older compared to younger patients (Glick et al[40]). On the other hand norepinephrine concentration appears higher in the right side of the human thalamus (Oke et al[41]). Thus it is possible that the cognitive disorders characteristic of SDAT are related to a modification of the normal asymmetry following the CAT activity decrease in the left hemisphere; nevertheless it is well known that symptoms such as minimization and indifference reactions, quite frequent in aging and demented people, are present in patients suffering from a lesion of the right hemisphere (Gainotti[42]). Indeed it should be kept in mind that pharamacological agents may have different effects in relation to the site of the lesion: that is cholinomimetics drugs could be more effective on the left hemisphere, while catecholaminergic activating drugs could stimulate the right one. Nevertheless it is possible that the presence in human aging and dementia brain of a peculiar aspect (i.e. neurofibrillary tangle and senile plaques) absent in most experimental models, makes difficult every attempt fitting experimental data to clinical practice.

The acetylcholinesterase (AChE) staining of processes in senile plaques, in fact, suggests that tangle formation is related to cholinoceptive neurons (Perry and Perry[43]). This hypothesis is confirmed by Struble et al[44] whose observations indicate that the neurites in senile plaques consist, in part, of presynaptic cholinergic axons, many of which, probably, originate from neurons in the basal forebrain. Thus changes in cortical cholinergic innervation have an important role in the pathogenesis of the neuritic plaques; the structural lesions, besides, could trap synaptic connections and it is likely that, when morphological features are established, functional damage is also irreversible. So, whatever the therapeutic trial it could be effective only in very early stages of the disease.

REFERENCES

1. H. Link and G. Tibbling, Principles of albumin and IgG, analyses in neurological disorders. II, Relation of the concentration of the proteins in serum and CSF. Scan.J.Clin.Lab.Invest. 37:391 (1977).
2. L. Amaducci, R. Capparelli, and E. Galli, 1982 Studio longitudinale delle IgG e di altre proteine liquorali nelle malattie infiammatorie del SNC, in: "International Workshop on: New techniques of cerebrospinal fluid analysis," Piccin, ed., Padova, (in press).
3. W. W. Tourtellotte, On cerebral fluid IgG quotients in multiple sclerosis and other diseases. A review and a new formula to estimate the amount of IgG sinthesized per day by the CNS, J.Neurol.Sci. 10:279 (1970).
4. L. A. Loizou, B. E. Kendall, and J. Marshall, Subcortical arteriosclerotic encephalopathy: a clinical and radiological investigation, J.of Neurol.Neurosurg.Psychiat. 44:294 (1982).
5. G. M. de Courten-Myers, S. I. Yamaguchi and R. E. Myers, Accentuation of hypoxic/anoxic brain injury by post exposure infusion of glucose solution,Neurology 32:110 (1982).
6. F. Plum, What makes the brain vulnerable to ischemia, Neurology 32:106 (1982).
7. L. Bakey and J. C. Lee, The effect of acute hypoxia and hypercapnia on the ultrastructure of the CNS, Brain 91:697 (1968).
8. I. Feigin, G. Budzilovich, S. Weinberg, and J. Ocatu, Degeneration of the white matter in hypoxia acidosis and edema, Journ.Exper.Neurol. 32:125 (1973).
9. P. E. Duffy, M. Rapport, and L. Graf, Glial Fibrillar Acidic Protein and Alzheimer-type Senile Dementia, Neurology 30:778 (1980).
10. L. Amaducci, K.I. Forno, and L. F. Eng, Glial fibrillar acidic protein in cryogenic lesions of the rat brain, Neurosc.lett. 21:27 (1981).
11. R. T. Bartus, R. L. Dean, B. Beer and A. S. Lippa, The cholinergic hypothesis of geriatric memory dysfunction, Science 217:408 (1982).
12. P. J. Whitehouse, D. C. Price, R. G. Struble, A. W. Clark, J. T. Coyle,and M. R. De Long, Alzheimer's disease and Senile Dementia: loss of neurons in the Basal Forebrain. Science 215:1237 (1982).
13. A. G. Karczmar, Overview: cholinergic drugs and behaviour. What effects may be expected from a "cholinergic" diet? in: "Choline and Lecithin in Brain Disorders. Nutrition and the Brain", A. Barbeau, J.H. Growdon and R.J. Wurtman, eds., Raven Press, New York (1979).
14. A. G. Karczmar, S. Nishi, S. Minota, and G. Kindel, Electrophysiology, acetylcholine and acetylcholinesterase of immature spinale ganglia of the rabbit. An experimental study and a review. Gen.Pharmacol. 11:127 (1980).

15. I. Hanin, D. G. Spiker, A. G. Mallinger, U. Kopp, J. M.
     Himmelcoch, J. F. Neil and D. J. Kuffer, Blood choline and
     its meaning in psychiatric and neurologic disease states,
     in: "Cholinergic Mechanisms: Phylogenetic Aspects, and
     Clinical Significance," G. Pepeu and H. Ladinsky, eds.,
     Plenum Press, New York (1980).

16. R. T. Bartus, Physostigmine and recent memory: effects in
     young and aged non human primates, Science 206:1087 (1979).

17. J. E. Christie, A. Shering, J. Ferguson, A. Z. M. Glen,
     Physostigmine and arecholine effects of intravenous infusion
     in Alzheimer presenile dementia, Brit.J.Psych. 138:46 (1981).

18. N. Sitaram, H. Weingartner, J. C. Gillin, Human serial learning:
     enhancement with arecholine and choline, impairment with
     scopolamine, Science 201:274 (1978).

19. S. F. Leong, J. C. K. Lai, L. Lim, J. B. Clark, Energy-metabol-
     ising enzymes in brain regions of adult and aging rats,
     J. Neurochem. 37:1548 (1981).

20. N. R. Sims, C. C. T. Smith, A. N. Davison, D. M. Bowen, R. H. A.
     Flack, J. S. Snowden, D. Neary, Glucose metabolism and
     acetylcholine synthesis in relation to neuronal activity
     in Alzheimer's disease, Lancet 1:333 (1980).

21. K. Sherman, J. E. Kuster, R. L. Dean, R. T. Bartus, E. Friedman,
     Presynaptic cholinergic mechanism in brain of aged rats
     with memory impairments, Neurobiol.Ag. 2:99 (1981).

22. G. E. Gibson, J. P. Blass, Impaired synthesis of acetylcholine
     in brain accompanying mild hypoxia and hypoglycemia,
     J.Neurol. 27:37 (1976).

23. G. E. Gibson, T. E. Duffy, Impaired synthesis of acetylcholine
     by mild hypoxic hypoxia of nitrous oxide, J.Neurochem.
     36:28 (1981).

24. C. Giurgea, F. Mayersons, Differential pharmacological activity
     of three types of cortical evoked potentials, Arch.Int.
     Pharmacodyn. 188:401 (1970).

25. J. C. Gobert, Genese d'un medicament: le piracetam. Metabol-
     ization et recherchem J.Pharma.Belg. 27:281 (1972).

26. S. J. Sara, D. Lefevre, Hypoxia induced amnesia in one-trial
     learning and pharmacological protection by piracetam,
     Psychopharmacologica 25:32 (1972).

27. R. E. Burnotte, J. C. Govert, J. F. Temmerman, Piracetam
     (2-pyrrolidinone acetamide) induced modifications of the
     brain polyribosome pattern in ageing rats, Biochem.Pharmacol.
     22:811 (1973).

28. R. J. Wurtman, S. G. Magil, D. K. Reinstein, Piracetam
     diminishes hippocampal acetylcholine levels in rats,
     Life Sci. 28:1091 (1980).

29. P. B. Mindus, B. Cronholm, and S. E. Lavander, Does piracetam
     counteract the ECT-induced memory dysfunction in depressed
     patients? Acta.Psych.Scand. 51:319 (1975).

30. S. Lloyd-Evans, J. C. Brocklehurst, and M. K. Palmer, Piracetam in chronic brain failure, Curr.Med.Res.Opin. 6:351 (1979).

31. E. Friedman, K. A. Sherman, S. H. Ferris, B. Reisberg, R. T. Bartus and M. K. Schneck, Clinical response to choline plus piracetam in senile dementia: relation to red-cell choline levels, New Eng.J.Med. 304:1490 (1981).

32. J. K. Blusztajn, S. H. Zeisel, and R. F. Wurtman, Phospholipids methylation and cholinergic neurons, in: "Transmethylation," R. Bochardt, C. Creveling, E. Usdin, eds., Elsevier, Amsterdam (in press).

33. G. E. Gibson, C. Peterson, Decrease in the release of acetylcholine in vitro with low oxygen, Biochem.Pharmacol. 31:11 (1982).

34. D. M. Low, and C. Weil, Hydergine in senile mental impairment, Gerontology 28:54 (1982).

35. W. Meier-Ruge, H. Emmenegger, A. Enz, P. Gygax, P. Iwangoff, and N.Wiernspergez, Pharmacological aspects of dydydrogenated ergot alkaloids in experimental brain research, Pharmacology (Supp.1) 16:45 (1978).

36. J. M. Vigouret, H. R. Bürki, A. L. Jeton, P. E. Züger, D. M. Loewe, Neurochemical and neuropharmacological investigations with four ergot derivates: bromocriptine, dihydroergotoxine, CF 25-397 and CM 29-712, Pharmacology (Supp,1) 16:156 (1978).

37. F. Casamenti, P. Mantovani, L. Amaducci, and G. Pepeu, Effect of phosphatidylserine on acetylcholine output from the cerebral cortex of the rat, J.Neurochem. 32:529 (1979).

38. P. Mantovani, A. C. Bonetti, and G. Pepeu, Effects of phosphatidylserine on brain cholinergic mechanism, in: "Phospholipids Metabolism in the Nervous System," G. B. Ansell, L. A. Horrocks, and G. Porcellati, eds., Raven Press, New York (in press)

39. L. Amaducci, S. Sorbi, A. Albanese, and G. Gainotti, Cholineacetyltransferase (ChaT) activity differs in right and left human temporal lobes, Neurology 31:799 (1981).

40. S. D. Glick, D. A. Ross, L. B. Hough, Lateral asymmetry of neurotransmitter in human brain, Brain Res. (in press).

41. A. M. Oke, R. Keller, I. Mefford and R. N Adams, Lateralization in norepinephrine in human thalamus, Science 200:1411 (1978).

42. G. Gainotti, Emotional behaviour and hemispheric side of the lesion, Cortex 8:41 (1972).

43. R. H. Perry, and E. K. Perry, Histochemistry of brain cholinesterase in normal old age and Alzheimer's disease, in: "Aging of the Brain and Dementia," L. Amaducci, A. N. Davison, P. Antuoni, eds., Raven Press, New York (1980).

44. R. G. Struble, L. C. Cork, P. J. Whitehouse, D. J. Price, Cholinergic innervation in neuritic plaques, Science 216:413 (1982).

PHYSIOPATHOLOGY AND TREATMENT OF TRANSIENT ISCHAEMIC ATTACKS

C. Loeb

Department of Neurology
University of Genova, Italy

This presentation will be divided in three sections:

1. Clinical definition of TIA
2. Some comments on physiopathology
3. Medical and surgical treatment: an overview

CLINICAL DEFINITION OF TIAs

As is well known the term transient ischemic attack (TIA) denotes a cerebral dysfunction of ischemic nature lasting less than 24 hours, with a tendency to recur (Millikan).[1] Over the last twenty years, the recognition of such a clinical entity and of its relevance as a warning of impending stroke offered the unique opportunity to plan preventive medical and surgical strategies to ward off its redoubtable evolution (see Loeb[2] for references).

It can therefore be stated that the recognition of TIAs represents the most important clinical advance of the last two decades in the field of cerebrovascular diseases (Marshall[3]) to the extent that on their basis, studies aimed at preventing stroke became "mandatory" (Barnett[4]). If, however, we take into proper consideration the studies of a number of authors, including a few personal ones, the need for a reassessment of TIAs clearly emerges.

In fact the features essential to the definition of TIA, besides the ischemic pathogenesis are the following:
(1) 24 hours duration and (2) complete reversibility.
- The 24 hours limit was chosen arbitrarily; quite a few authors (Loeb[2]), think that usually the duration of these attacks is about 1 hour or even some minutes.

Moreover, ischemic episodes clearing completely after 24 hours from the onset, labelled as Strokes with full recovery (SFR) (Loeb and Priano[5]), Reversible ischemic neurological deficit (RIND) (Millikan[1]) or Protracted reversible ischemic neurological deficit (PRIND) (Gratzl and Schiemedek) and, more recently, protracted transient ischemic attacks (PTIA) (Loeb) have been described.
Therefore the comprehensive label of reversible ischemic attacks (RIA) was introduced to cover both TIAs and PTIAs (Loeb).
Quite recently, Humphreys and Marshall[7] pointed out that TIAs lasting less than five minutes had a higher incidence of completed stroke than TIAs lasting longer.
- Complete reversibility, on the other hand, has been disregarded as well. Some ischemic attacks leaving slight neurological deficits have been considered as TIAs by the great majority of authors (see Barnett[4]); or as TIA-IR (TIA with incomplete recovery) by Fieschi et al,[8] a label that includes patients showing persistence of minimal signs at the neurological examination but subjectively unnoticed. Computed tomography (CT) investigation of cases with veritable TIAs and PTIAs showed in about 30 per cent of them low density areas in sites compatible with their previous clinical picture (Oltenau-Nerbe;[9] Perrone et al;[10] Ladurner et al,[11]) thus raising obvious questions about the concept of reversibility. In fact, if the requirement of complete regression were extended to include CT findings, these cases should be classified as minor stroke or partial non prgressing stroke (Barnett[4]) (PNS) instead of RIA.
- According to some vascular surgeons (Austin et al;[12] Gratzl and Schemiedek;[6] Heilbrun et al[13]) TIAs, PTIAs (or RIND) as well as cases with residual slight neurological signs definable as partial non progressing stoke (PNS) can be treated as a single group with respect to prospective surgical treatment.

The point is that the term TIA now has too many meanings - as Tyler[14] pointed out - and the literature is hopelessly confused, since TIA, PTIA and PNS are, though not explicitly, often lumped together. Actually the question arises whether TIA, PTIA and PNS significantly differ from one another from the clinical, prognostic and therapeutic standpoints or, on the contrary, any attempt to differentiate them on these grounds should be considered a futile one (Loeb[15]).

From the comparative evaluation of 85 cases with PNS and 121 cases with TIAs or PTIAs, with a follow-up of about 6 years, emerged that clear clinical and CT differences existed in these three groups. At the present time it seems therefore reasonable, on clinical grounds to maintain the distinction between TIA, PTIA and PNS.

In fact TIA differs from PTIA in some clinical features, e.g.

duration and gradual onset and, on the other hand, RIA differs from PNS owing to positive CT findings, higher recurrence of RIA and CS as well as lower survival.   Obviously TIA and PTIA should be characterized by a complete clinical regression and normal CT, as pointed out, quite recently, also by Biller et al.[16] When clinical regression is partial and/or CT shows low attenuation areas, the case should be labelled as one with PNS.

In summary, in this study, TIA denotes a sudden episode of cerebral focal dysfunction of ischemic nature, characterized by a complete clinical regression within 24 hours and a normal CT scan.

SOME COMMENTS ON PHYSIOPATHOLOGY

No attempt will be made here to discuss all the theories amply reported in several fairly recent contributions (Paulson;[17] Hutchinson and Acheson;[18] Russell[19]).

According to the majority of the authors the main pathophysiological mechanism responsible for TIA is embolism or thromboembolism. A TIA, therefore, is the outcome of a critical fall of cerebral PaO_2 brought about by a reduction of cerebral blood flow (CBF) in its turn determined by a transient thromboembolism of the involved artery.

The critical CBF values are influenced by: (a) modality (i.e. sudden or gradual) of onset; (b) efficiency of the collateral circulation, contributed to, among other factors by the adequacy of the cardiac function and systemic pressure; (c) arterial blood O_2 saturation and (d) efficiency of the autoregulation mechanism and of the vasodilatory response to hypoxia.

According to the embolic theory, in the intervals between TIAs the autoregulation should be normal in the involved areas; in fact, this was found to be the case by Skinhøj et al.[20] The reverse is held to be true by the hemodynamic hypothesis: in other words, the sutoregulation is chronically impaired in those cortical areas perfused below the average normal value, namely, with marginal blood supply.   In fact, quite recently loss of autoregulation has been observed in up to 90 per cent of cases with TIA (Austin et al;[12] Renou et al;[21] Heilbrun et al[13]).

On the other hand these data do not appear to entirely solve the pathogenetic problem, first of all because the value of CBF determinations as a reliable index of neuronal function is debatable (Waltz[22]); moreover, the topographical correlation between low CBF values and the size of the lesion is often unsatisfactory (Yamaguchi et al[23]).

Therefore, the pathophysiology of TIA still remains somewhat obscure, even those authors - constituting the majority - who favour the embolic or thromboembolic pathogenesis do not seem to discard, at least for some cases, the "hemodynamic theory" (Loeb[2]).   As is well known, this last one postulated the superposition of transient reductions of perfusion pressure brought about by falls of systemic blood pressure and/or cardiac output upon pre-existing chronic focal reductions in CBF due to lesions of small or great arteries.

The embolic theory of TIA is supported by the following data (van der Drift and Kok;[24] Russell[19]): (a) the attacks, always sudden, do not usually bear any relationship to postural variations, hypotension or cardiac abnoramilties; (b) the episodes do not occur within the same district; according to Waltz,[22] however, a stereo-types behaviour can be accounted for even by embolization; (c) embolic sources can usually be identified; (d) emboli have been visualized during their progression within retinal vessels (Fisher;[25] Russell;[26,27,28] Ashby et al:[29] Gunning et al[30]), and (e) no attacks can be brought about by even considerably lowering the blood pressure in patients who were currently experiencing TIAs (Loeb et al;[31] Kendell and Marshall;[32] Drake and Drake[33]).

It is a reasonable conclusion that the great majority of the cases of TIA are due to embolism or thromboembolism and only a smaller number of cases are probably due to hemodynamic mechanisms. The claimed differences in physiopathology between carotid and vertebrobasilar attacks is far from being accepted and only few authors agree on this point (Barnett[4]).

Nevertheless, in our opinion, probably in some 15-20 per cent of the cases the pathophysiology still remains somewhat obscure. It is true that at least some of these cases are due to some other pathophysiological mechanisms:

a) subclavian arterial steal cover only few cases, being, however, an hemodynamic mechanism
b) coagulation abnormalities, specially fue to contraceptive pills and other drugs, accounts for some cases
c) non atherosclerotic vasculopathies, such as granulomatous angiitis, polyarteritis nodosa, disseminated lupus erythematosus, homocistinuria, fibromuscular hyperplasia may be responsible for some cases.

TREATMENT

Over the last twenty years anticoagulants, cervical endoarterectomy and more recently antiplatelet drugs and vascular microsurgery have been employed in cases with TIAs.   The value of these approaches has been difficult to assess mainly due to the huge

variability of the natural history of TIAs.   It seemed therefore
useful to refer to some recent reports which represent the
experience of various groups of researchers or community studies.

Anticoagulant treatment seems beneficial in reducing the
incidence of TIAs and/or completed strokes mainly in the first
3 - 6 months (Whisnant[34]).   After 6 months there is no significant
difference in morbidity and mortality between the treated and
control groups (Whisnant[34]).   The need for wide randomized
investigations is stressed by Toole et al.[35]

If carefully monitored by regular controls of prothrombin time,
anticoagulant treatment turns out to be relatively safe (Whisnant
et al[36]).   Aspirin seems to reduce the risk of subsequent completed
strokes in males (Barnett[37]), but the results of its therapeutic
trial, although promising, cannot be considered as conclusive
(Fields et al[38]).

Endoarterectomy is indicated in patients with TIAs exhibiting
ulcerated plaques and/or accessible stenoses of one or both carotids
or sometimes vertebral arteries in the neck (Toole et al[35]).   In
cases with carotid involvement surgery should not be unduly delayed
(Millikan and McDowell[39]).   Associated diseases such as hypertension
or myocardial infarction constitute, however, additional and
unacceptable risks (Sundt et al[40]).

The establishment of microsurgical anastomoses between the
superficial temporal and cortical middle cerebral arteries seems
promising, at least in cases not amenable to conventional surgical
techniques (Yasargil and Yonekawa;[41] Austin et al[12]).   Along with
Millikan and McDowell,[39] however, it is reasonable to conclude that
the indications and limits of the whole procedure await further
assessment (Reichman[42]).   If, on the other hand, one leaves aside
for a moment the results of the long-term investigations and joint
studies and tries to cope with the individual patient (Millikan[43])
quite often a pervading feeling of helplessness may creep in.

To this account the reader might find it expedient to refer
to the recent "Guidelines for management of TIAs" (Sandok et al[44])
with minor modifications, as follows:

Subject with Carotid TIAs

In the absence of relevant medical risk factors (severe hypertension
and heart or lung diseases) and if an exhaustive angiographic study
has disclosed a lesion relevant to the clinical picture without
signs of diffuse or muliple involvement, surgery can be undertaken
by a skilled operator.   At any rate the angiographic risk, even
when the femoral route is employed, cannot be overlooked and has to
be carefully estimated (Faught et al[45]).   Digital angiography can
probably overcome this difficulty.

Anticoagulant treatment can be begun in cases with TIAs of less than 2 months duration not amenable to surgery, and should be pursued for at least 3 - 6 months thereafter switching to aspirin. The complications of anticoagulant treatment should not be under-estimated: the risk of intracranial hemorrhage in patients with TIAs of 55 - 74 years of age is about eight times greater in treated than in untreated cases.

Treatment with aspirin is begun in patients unfit for surgery, either following the initial course of anticoagulants as described above or when the last episode occurred more than 2 months previously, being subsequently pursued at least for a 2 year period free from further attacks.

Subjects with Vertebrobasilar TIAs

Indications for surgical treatment are here even more selective and the efficacy of surgery far less established than in cartoid TIAs (Blaisdell[46]).   Anticoagulant and antiplatelet therapies constitute, therefore, the mainstay of treatment which should be carried out along the lines described above.

Some new data should modify our approach:

a) Medical therapy. The previously used dosage of Aspirin (ASA) was probably too large.   Some quite recent trials (Boysen et al;[47] Loeb et al[48]) showed that large doses of ASA (equal to 1 g/day) acts as an inhibitor of cyclo-oxygenase in platelets and in the vessel wall resulting in platelet aggregation. 25 - 50 mg ASA daily acts probably as an inhibitor of the plate-let aggregation since such a small amount may not affect the cyclo-oxygenase in the vessel wall so that prostacyclin inhibiting the platelet aggregation may act.   Further studies, however, are necessary to clarify this point.

b) Surgical treatment. The indication of by-pass surgery for cere-brovascular ischemic disease remain controversial.   The demon-stration of increase in blood flow and metabolism is a questionable point since the natural evolution of the ischemic lesion is toward improvement.   The increase in CBF and metabolism must be greater than expected in natural evolution and the demonstration that CBF determination is a reliable index of neuronal function is still debatable (Waltz[22]).

As Hachinski[49] pointed out "it becomes imperative to demon-strate that the benefits of the operation outweigh its risks and are better than the natural history and medical therapy."

REFERENCES

1. C. H. Millikan, Ad hoc Committee on cerebrovascular disease.
    A classification and outline of cerebrovascular diseases,
    Part II. Stroke 6:565 (1975).
2. C. Loeb, Protracted transient ischemic attacks, Eur.Neurol.
    19:1 (1980).
3. J. Marshall, The natural history of cerebrovascular disease,
    in: "Modern Concepts of Cerebrovascular Disease,"
    J. S. Meyer, ed., Spectrum, New York (1975).
4. H. J. M. Barnett, The pathophysiology of transient ischemic
    attacks, Med.Clin.North Am. 63:649 (1979).
5. C. Loeb and A. Priano, Strokes with full recovery. A reapp-
    raisal, in: "Cerebral Vascular Disease," Meyer, Lechner,
    Reivich and Eichhorn, eds., Thieme, Stuttgart (1973).
6. O. Gratzl, P. Schmieder, Microneurosurgical arterial anastomo-
    ses in patients with prolonged reversible ischemic neuro-
    logical deficits (PRINDI), in: "Microvascular anastomoses
    for cerebral ischemia," Fein and Reichman, eds., Springer,
    New York (1978).
7. P. R. D. Humphreys and J. Marshall, Transient ischemic attacks
    and strokes with recovery prognosis and investigations,
    Stroke 12:765 (1981).
8. C. Fieschi, S. Bernardi, Epidemiologia e storia naturale
    dell'ischemia cerebrale transitoria, L'Ospedale Maggiore
    74:229 (1979)
9. V. Oltenau-Nerbe, P. Schmiedek, E. Kazner, Comparison of
    regional blood flow and computerized tomography in patients
    with cerebrovascular disease and brain tumors, in:
    "Cranial Computerized Tomography," Lanksch and Kazner, eds.,
    Springer, Berlin (1976).
10. P. Perrone, R. Candelisi, G. Scotti, Evaluation in patients
    with transient ischemic attack. Correlation between
    clinical and angiographic findings, Eur.Neurol.18:217 (1979).
11. G. Ladurner, W. D. Sager, L. D. Iliff, and H. Lechner,
    A correlation of clinical findings and CT in ischemic cerebro-
    vascular disease, Eur.Neurol. 18:281 (1979).
12. G. Austin, G. Haugen, and W. Schuler, Transient ischemic
    attacks and metabolic aspects of their relief by microneuro-
    surgical anastomosis, in: "Microvascular Anastomoses for
    Cerebral Ischemia," Fein and Reichman, eds., Springer,
    New York (1978).
13. M. P. Heilbrun, O. H. Reichman, R. E. Anderson and T. S. Roberts,
    Regional cerebral blood flow studies following superficial
    temporal-middle cerebral artery anastomosis, in:"Micro-
    vascular anastomoses for cerebral ischemia," Fein and
    Reichman, eds., Springer, New York (1978).
14. H. R. Tyler, Cerebrovascular disease update, in: "Current
    Neurology," H. R. Tyler and D. M. Dawson, eds., Houghton
    Mifflin, Med. Div., Boston (1979).

15. C. Loeb, Transient ischemic attack protracted transient
    ischemic attack and completed stroke. XVI Venetian
    Symposium, May 29-30 (1982).

16. J. Biller, D. W. Laster, G. Howard, Cranial computerized
    tomography in carotid artery transient ischemic attacks,
    Eur.Neurol. 21:98 (1982).

17. O. B. Paulson, Cerebral apoplexy (stroke). Pathogenesis,
    pathophysiology and therapy as illustrated by regional
    blood flow measurement in the brain, Stroke 2:327 (1971).

18. E. C. Hutchinson, E. J. Acheson,"Strokes. Natural history,
    pathology and surgical treatment," Saunders, London (1975).

19. R. W. R. Russell, "Cerebral Arterial Disease," Churchill
    Livingstone, Edinburgh-London (1976).

20. E. Skinhøj, K. Hoedt-Rasmussen, O. B. Paulson, and N. A. Lassen,
    Regional cerebral blood flow and its autoregulation in
    patients with transient focal cerebral ischemic attacks,
    Neurology, Minneap. 20:485 (1970).

21. A. M. Renou, J. M. Caille, P. Constant, Vasoreactivite a
    l'alfatesine en pathologie cerebrale, J. Neuroradiol.
    5:257 (1978).

22. A. C. Waltz Anatomy and physiology pertinent to strokes
    in: "Microvascular anastomoses for cerebral ischemia,"
    Fein and Reichman, eds., Springer, New York (1978).

23. T. Yamaguchi, A. G. Waltz, and H. Okazaki, Hyperemia and
    ischemia in experimental infarction. Correlation of histo-
    pathology and regional blood flow, Neurology 21:565 (1971).

24. J. H. Van der Drift and N. K. D. Kok, Transient ischaemic
    attacks, in: "Assessment in cerebrovascular insufficiency,"
    Stöckler, Kuhn, Hall, Becker and Van der Veen, eds.,
    Thieme, Stuttgart (1971).

25. C. M. Fisher, Intermittent ischemia, in: "Cerebral vascular
    Diseases,"Wright and Millikan, eds., Grune and Stratton,
    New York (1958).

26. R. W. R. Russell, Observation on the retinal blood-vessels in
    monocular blindness, Lancet ii:1422 (1961).

27. R. W. R. Russell, Atheromatous retinal embolism, Lancet
    ii:1354 (1963).

28. R. W. R. Russell, The source of retinal emboli, Lancet ii:789
    (1968).

29. M. Ashby, N. Oakley, I. Lorenzi, Recurrent transient monocular
    blindness, Brit.med.J. ii:894 (1963)

30. A. J. Gunning, G. W. Pickering, A. H. T. Robb-Smith, Mural
    thrombosis of the internal carotid artery and subsequent
    embolism, Q.Jl.Med. 33:155 (1964).

31. C. Loeb, L. Garello, P. Pastorino, Aspetti clinici dell'
    insufficienza cerebrovascolare, Sist.Nerv. 13:334 (1961).

32. R. E. Kendell and J. Marshall, Role of hypotension in the
    genesis of transient focal cerebral ischaemic attacks,
    Brit.med.J. 10:344 (1963).

33. W. H. Drake and M. A. L. Drake, Clinical and angiographic
     correlates of cerebrovascular insufficiency, Am.J.Med.
     45:253 (1968).

34. J. P. Whisnant, Transient ischaemic attacks, Clin.Exp.Neurol.
     Proc.Aust.Ass.Neurol. 14:1 (1977).

35. J. F. Toole, R. Janeway, K. Choi, Transient ischemic attacks
     due to atherosclerosis. A prospective study of 160 patients,
     Archs Neurol. 32:5 (1975).

36. J. P. Whisnant, E. F. Niall, M. B. Cartlidge and L. R. Elveback,
     Carotid and vertebral-basilar transient ischemic attacks:
     effect of anticoagulants, hypertension, and cardiac
     disorders on survival and stroke occurrence. A population
     study, Ann.Neurol. 3:107 (1978).

37. H. J. M. Barnett, A randomized trial of aspirin and sulfin-
     pyrazone in threatened stroke.   The Canadian Cooperative
     Study Group, New Eng.J Med. 289:53 (1978).

38. W. S. Fields, N. A. Lemak, R. F. Frankowski and R. J. Hardy,
     Controlled trial of aspirin in cerebral ischemia. II.
     Surgical group. Stroke 9:309 (1978).

39. C. H. Millikan, and F. H. McDowell, Treatment of transient
     ischemic attacks, Stroke 9:299 (1978).

40. T. M. Sundt, B. A. Sandok, and J. P. Whisnant, Carotid end-
     arterectomy. Complications and pre-operative assessment of
     risk. Mayo Clin.Proc. 50:301 (1975).

41. M. G. Yasargil and Y. Yonekawa, Experience with the STA-
     cortical MCA anastomosis in 46 cases, in:"Microvascular
     anastomoses for cerebral ischemia," Fein and Reichman,
     eds., Springer, New York (1978).

42. O. H. Reichman, Neurosurgical microsurgical anastomosis for
     cerebral ischemia. Five years experience, in: "Cerebro-
     vascular Diseases, 10th Conference," Scheinberg, eds.,
     Raven Press, New York (1976).

43. C. H. Millikan, Clinical investigation of things or people,
     Neurology Minneap. 28:744 (1978).

44. B. A. Sandok, A. J. Furlan, J. P. Whisnant and T. M. Sundt,
     Guidelines for the management of transient ischemic
     attacks, Mayo Clin.Proc. 53:665 (1978).

45. E. Faught, S. D. Trader and G. R. Hanna, Cerebral complications
     of angiography for transient ischemia and stroke,
     Prediction of risk, Neurology, Minneap. 29:4 (1979).

46. W. Blaisdell, Extracranial arterial surgery in the treatment
     of stroke. in: "Cerebral Vascular Diseases," Trans.
     8th Princeton Conf. McDowell and Brennan, eds.,
     Grune & Stratton, New York (1973).

47. G. Boysen, J. Bottcher, J. S. Olsen, Minimal dosage of ASA
     needed for platelet inhibition in patients with cerebro-
     vascular disease. 11th Salzburg Conference on Cerebral
     Vascular Disease, Sept. 23-25 (1982).

48. C. Loeb, U. Armani, C. Gandolfo, Effects of low vs high
        dosages of aspirin on platelet aggregation in patients with
        cerebrovascular disease.   A follow up cooperative study.
        11th Salzburg Conference on Cerebral Vascular Disease,
        Sept. 23-25 (1982).
49. V. Hachinski, Extracranial/intracranial bypass surgery:
        technique or therapy? 11th Salzburg Conference on
        Cerebral Vascular Disease, Sept. 23-25 (1982).

PSYCHODYNAMIC TREATMENT OF THE ELDERLY

I. Simeone

Psychiatrist, Assistant Director of
Geneva Institute of Geriatry
Depts. of Psychiatry and Medicine
University of Geneva, Switzerland

Among diagnostic and treatment methods which are becoming more
numerous every year in geriatric clinics, the use of psychodynamic
know-how, i.e. psychoanalysis, in the investigation and therapy of
mental troubles remains unusual.   This is in contrast with the
almost exclusive use that one makes of it in adult and child
psychiatry.

Fortunately, psychogeriatrics, which is a new branch of medicine,
has begun in many countries to develop a more precise and clear
identify.   New generations of doctors and geriatricians are becoming
increasingly aware of old age psychiatry whilst the psychogeriatric
specialist himself remains a consultant or trainer.   It goes
without saying that the approach to the psychic make-up of the aging
man and the understanding of his psychological troubles is essentially of psychoanalytical origin as in other fields of psychiatry.

The contribution and the requirements of psychogeriatrics are
becoming increasingly large.   Until a few years ago there was no
similar progress in the psychological field of aging corresponding
to the rapid progress in geriatrics.   Slowly but surely, the
psychodynamic models of senescence are beginning to replace the old
neurological theories based on pathological deficit and anatomy.
At the same time, the concept of evolution is gaining on that of
involution; new light is being shed on the theories of communication,
the significance of the "crisis" family therapy, leaving behind the
ideas of the old psychiatric phenomenology.

PSYCHOPATHOLOGY

With every old person, we find a constant psychological suffering from the clinical point of view. It may be a former neurotic
trouble, but more often we see a new condition which expresses
itself with a psychological problem appearing for the first time in
senescence.

The psychic make-up may be the result or the cause of the
general clinical state. In addition, we always speak of psychosomatic medicine, never of somato-psychic medicine which could be
an original contribution on the part of geriatrics to medical
knowledge. We can sometimes define this particular "regression"
with a very large number of patients as the manifestation of a
"diffuse neuroticisation of the senescence".

This neuroticisation may show itself in the poor adaptation to
the passing of time, signify fear of madness, be a sign of existential worry about the signs of death. In a couple, this may
represent conjugal conflicts of a new type or the obsessional fear
of the death of the spouse or the tensions which originate in the
daily face to face interaction after retirement or which surface
when the children leave an emotional vacuum behind them.

This neuroticisation may finally be an over-compensation for
all these worries with an increase of authority, of possession, of
power or re-activation of other causes which remained hidden until
this moment. If the regression goes further and the anxiety
attacks the whole personality, then the clinical reality will be
that of a psychosis. Under the pulsionary impact, the archaic
defences and "psychotisation" occurs which we can often see in
senescence. Delirious manifestations and convictions of being
persecuted are the most usual clinical aspects. But over and above
these exact causes, of which a neurotic or psychotic state has the
above-mentioned causes, there is always the anxiety expressed with
worry about their own future in the face of change ("am I still
normal or not?") and the fear of the final disappearance provoked by
the promonitory signs each day.

BODY AND HALLUCINATIONS

This existential questioning, which goes hand in hand with the
process of aging, is rarely expressed orally between the doctor and
patient. Often this feeling does not even rise up to the conscious
level and remains anchored in the depths of the unconscious of the
patient who is the last to recognise it. This does not, however,
prevent the internal manifestations of the unconscious fear from
being active and producing the psychological problem. The symptoms
that "camouflage" this anxiety express themselves most frequently
through the bodily reality of the elderly subject.

This silent body remaining in the background for a whole life suddenly becomes a protagonist and active. The psychodynamic explanation of this phenomenon still has to be correctly formulated. But it does not prevent many questions occurring to us. Does the patient speak in a somatic language because the somatic symptom is the only one that everyone can understand easily? And especially if it is a doctor who is the listener?

With the help of recorded meetings, Haynal and Garrone[1] tell us that the "audio-visual analysis of the first meeting clearly shows that in addition to information relating to the poor functioning of his body, the patient transmits messages which go far beyond this field." As for the doctor – the listener – he may receive these messages, select them or eliminate and discourage everything that does not directly concern the "mechanical" perturbations. It is always the first meeting that orientates and dictates the continuation of the relationship as we are reminded by Balint.[2]

It is undeniable that a bodily symptom is more socially acceptable than a psychic symptom. It is easier to express oneself with somatic illness rather than with psychological symptomatology. We are sure that extensive research in this area will produce its fruit in the future and enlighten us on this subject.

Let us be content, for the moment, to note that the bodily reality of the aging subject is infiltrated with the psychic experience and emotions felt. The forgotten memories of the past surface strongly into a fragile present. The "hallucination", this basic component of psychism, no longer retains its autonomy as it has become "body"; black thoughts or relational exchanges with the environment are given flesh, the feeling of inferior is drawn into a hypochondriacal experience. It is not necessary to list the digestive problems, the rebellious head pains and other pain symptomatologies and handicaps without foundation. In time, these complaints increase and the subject can only pose the eternal, false question: "is it the beginning that will lead me to death?"

THE PSYCHOLOGICAL RESPONSE OF THE DOCTOR

A correct psychotherapeutic attitude or psychotherapy to be carried out without delay are the only adequate medical responses to these acute or chronic psychological conflicts whatever they are but which express themselves in such a way. We know that the psychotherapy of the practising doctor is not the same as that of the psychiatrist or the psychoanalyst but it is a myth that psychotherapy should belong exclusively to the specialist. Quite the contrary since the psychodynamic understanding of the problem should be a part of the knowledge of the doctor and one of his tools for treatment. This is why we can say that the psychodynamic knowledge

and the psychological therapies extend horizontally through
medicine and are not to be limited to the single specialist.[3]

Paradoxically, it is the elderly patient who forces us to
discover the real and basic meaning of medicine as well as the
role of the doctor responsible. Whilst medical training has
always been directed (but for how much longer?) towards the study
of organic pathology, the practice of geriatrics forces us to
consider first of all the man from the global aging point of view
at the same time as the other aspects of his sickness. This
separation between the somatic aspects of a given sickness and the
psychic reactions is usual (although illusory) with adults but does
not exist with the child. Neither is it to be found with the
elderly person. Body and psyche form a unity and this truth
(particularly evident with the advance of age) is difficult to
accept at the same time as we are confronted with the symptoms that
the patient shows.

WHAT IS A PSYCHOTHERAPEUTIC ATTITUDE?

It is the managing of the doctor – patient relationship in a
way other than that suggested by common sense. In current practice,
this means that personal feelings or projections, suggestive or
hasty intervention, ambiguous or paternal attitudes should be avoided
as being useless, even dangerous. Just as it may be partly the
case of a placebo (with rare exceptions), it is useless and
dangerous as its prescription only dispels ignorance and prejudice.
One of the key ideas which motivates the doctor – patient relationship according to correct psychotherapeutic principles is that the
patient is not a "broken mechanical part" that it is necessary to
repair but that his sickness is to be situated in the whole of his
personality. Each symptom has its significance and its symbolism
that we must not separate from the rest of the personality.

On the contrary, if we try to reintegrate the symptom in the
whole of the individual, his psychic experience, his affective
resonance and his relational world will give us the response to our
diagnostic and treatment investigations. The interrogation must,
in fact, allow the possibility of a dialogue. Any diagnosis, no
matter how benign or dramatic it may be, must be severed from the
experience that the patient has of his sickness in his vision of the
world and that we must respect. This understanding must, of
necessity, extend to the spouse, to the family, to the children.
Often, in family therapy, if there is one designated patient, there
are in reality two or three patients.

Another of the difficulties of an exact psychotherapeutic
attitude in geriatrics is to find the "exact distance" in that
relational space that is infiltrated by death, inadequacy of the

therapy, social distress, rejection.   In addition, our personal
convictions will be only too present in the relational distance that
one can have with an old alcoholic that social pressure forces us to
treat, doggedly attached to his liberty, unworried about his future;
in fact, this distance may be absent in the relationship with an old
lady who does not dare express her desires but remains stubborn in
her apparently puerile convictions and which may have a serious
effect on her health. One of the dangers which awaits a psychologic-
ally correct attitude is first the illusion of being able to apply
adult standards : to want to treat the old man as an adult is a
source of error in geriatric practice.

Another source of confusion may be the following : if one says
that there is not an organic foundation and that there is a psychic
problem in a particular sickness, often it is the doctor who puts
himself into the position of the patient to resolve it.   With his
advice and his instructions, he of course acts with his personality;
which is unfortunately quite different from that of the patient.
We can substitute ourselves for the patient, in an exemplary and
admirable way, but always with the result of creating a dependance
of the patient on us.   It should not be forgotten that the elderly
patient is looking in the majority of cases for the "magic"
protection coming from the doctor.   The psychotherapeutic aim is
exactly in the opposite direction, namely, to help the patient become
autonomous and independent.

PSYCHOTHERAPY

Whilst the term "psychotherapy" designates all the curative
methods which use psychological means, psychotherapeutic interventions
divide technically into two major categories : "expressive" and
"re-covering" psychotherapies.   In the former, one conventionally
says that there is a baring of the unconscious conflicts and the two
types of treatment used are::

1) conventional psychoanalytic therapy
2) psychotherapy based on psychoanalytical suggestion

In the latter, the transfer is used in a more careful way, by taking
into account the psychological possibilities of the patient, of the
force of his ego, with a more limited therapeutic aim fixed in advance.
These treatments are usually reserved for the specialist.

Among the "re-covering" psychotherapies, we find the so-called
"supporting psychotherapies", i.e. the support of the ego and rein-
forcement of the ego.   This second class represents the greatest
number of psychotherapies carried out among which we can include the
psychotherapies carried out by non-specialist doctors.   These

therapies allow the subject to be helped effectively by showing the
presence of conflicts without baring the deep mechanisms which are
related to them.

In this vast field of psychotherapies, a separate place is
occupied by group therapy which is used a great deal in institutions
where the institutional dynamic mixes with the effective dynamic
of the patients.   We should also remember the fruitful use that we
are beginning to see in family therapy in the manner that one uses
it in psychiatry of the adult or child.   And finally we should not
forget the non-verbal psychotherapy or so-called "bodily meditation"
which is among the most useful and rewarding in the elderly, from
the self-training of Schultz to the psychotonic re-education of
Ajuriaguerra.   This covers treatments used with success by general
practitioners after a suitable training.   The good results that
are obtained with the elderly subjects can be explained by the
involvement and place of treatment being the body itself.

The aim of all psychotherapy is cure and this can only be
obtained by the disappearance of the symptoms and/or the modification
of the personality.   All these treatments try, directly or
indirectly and without exception, to restore the esteem the patients
have for themselves and to convince them that they can control their
feelings and their behaviour.[4]

PSYCHOTHERAPY AT AN ADVANCED AGE

In connection with the advance in age, Freud thought, and the
idea persists in the psychoanalytical field, that from 50 years old
and on, the psychic system has become so rigid that a psychoanalytical treatment could not bring about any meaningful modification.
About the same time, K. Abraham[5] rectified this severe conclusion
by carrying out successful psychoanalyses of elderly subjects and
stated that the neurotic age of the patient is more important that
the actual age of the patients.

Since then, there has been a long silence in psychoanalysis
until the 50's when geriatric psychotherapy saw a new beginning in
the U.S.A.   Today in Europe, psychotherapeutic interest has begun
to extend and timidly started to adopt a certain shape thanks to
the contribution of psychoanalysts to clinical gerontology.   For
example, at the end of October 1982 the Second Congress of the
International Association of Psychoanalytical Gerontology was held
in Paris.[6,7,8]

This reticence on the part of psychoanalysis to deal with
senescence has been criticised by certain psychoanalysts: "the
attitude of psychoanalysis towards aging has been paradoxical:
it has emphasized the maternal and paternal image in the dynamic

of the psyche as practised, and refused to consider elderly subjects either from the point of view of systemic research or from the point of view of a possible therapy.   Does this state of affairs perhaps hide the fear of examining more closely the lacks in the parental images?[9]

From the psychodynamic point of view according to Grotjahn,[10] there remain three paths open to the old man: 1) integrate and accept life as it has been lived, 2) react with rigidity of the ego which tends to hold itself in a defensive posture following patterns of internal adaptation which are more or less neurotic, 3) have a neurotic, and often psychotic, regression.   In the two latter possibilities, we may be called upon to intervene.

The major problem that we have to face is the diminution of the capacities and the retirement from the previous areas of activity which bring about a readjustment of the equilibrium of the psyche and thus a questioning of old conflicts which were resolved poorly or not at all throughout the life.   It should be remembered that success in adult life may have been achieved in a neurotic state that old age is going to reveal.   But these old conflicts will find themselves having to live along with the fear of death, reactivated in its term by the narcissistic traumatism of becoming old.   Some authors[11] believe the fear of death is a symbolic representation of the old fear of castration.

The defences which come into play are multiple and specific: regression increases (and we meet it at every step from obsessional ideas to autistic withdrawal and transitory fits of madness); negation and denial become abnormally large; troubles with the memory may be synonymous with repression; somatic reactions are of an infinite scope.

TRANSFER AND COUNTER-TRANSFER

But the most important thing we can learn from the practice of psychodynamics lies mainly in the clarification of transfer and counter-transfer.   "Transfer" is the unconscious behaviour which pushes a man to reproduce in those that surround him, the feelings which joined him to his parents, to the other children or to his images of substitution.   Thus those who surround us remind us, without our realising it, of the first persons of our lives and because of this, we "transfer" love, hate, jealousy and aggressivity to these different persons in a more or less transparent manner.   Our patient "transfers" to us.   The elderly patient easily sees in the therapist, his son or a representative of the younger or future generation against which he feels all the jealousy and aggressivity (inverted Oedipus).   But the transfer may also be the opportunity for the aging person to clarify these unconscious relations with respect to a true or imaginary son.

In practice, the old patient immediately establishes a relationship that Goldfarb calls "parentified" in living the therapy as a powerful but ambivalent parental figure capable of doing good or bad, that must be feared and to whom one must submit. [12,13,16]

"Counter-transfer" is, on the other hand, all the reactions of the therapist himself with respect to the patient. "As in all psychotherapy, the more the therapist masters his counter-transfer, the less he needs to fear that the situation or the transfer may be fully exploited".[14]   For the treatment of elderly people, this is of major importance due to the limited time and of the potential fragility of the subjects, especially from the somatic point of view.

In the psychotherapy of an elderly person, it is supposed that the therapist has resolved his ambivalence and his unconscious guilt with respect to his parents. The fragility of the therapist, thereafter, can only manifest itself when confronted with a patient who is complaining and aggressive, depressed or lacking in all resources, impregnated with death. The weakening of his ego causes the outpouring of emotions and lack of inhibition that cannot leave one indifferent. In addition, the narcissism of the therapist is tested by the easier expression on the part of the elderly patient of all that the adult pushes back with his younger force (but which will not always be so young). From this there is a certain confused guilt which affects the therapist confronted with the jealousy of the patient towards himself.

Certain rules have to be observed in order to establish an adequate counter-transfer: not to protect the patient too much by the bias of interpretations which are too benign with the false idea that elderly patients are naturally fragile and that one must not disturb them too much; but they must not be subjected to aggression with raw interpretations; do not leave neutrality without good reason and, at the same time, do not fear to assume the role of a protector which may last for some time.

Basically, in order not to go on forever, as Dedieu-Anglade[14] says "two kinds of risk should be recognised in transfer: not to do something which should be done (authentic analysis of the transfer), or to do something which should not be done (useless gratification)".

CONCLUSIONS

We can thus summarise some special remarks concerning psycho-therapy as it relates to old age.

Although there exist some contrary examples in the literature,[15] at the present state of our knowledge we cannot carry out a real psychoanalysis as "the basic conflict is always pushed back

dynamically, the structures of the senile ego remaining inaccessible
by conventional methods".[11] However, the field of the psychotherap-
ists still remains a field to be crossed and, no doubt, is rich in
satisfaction.   It is necessary first of all to carry out an
accurate evaluation of the request in order not to undertake a
treatment which may later reserve some disagreeable surprises.
We hold reservations about psychotherapeutic treatment of psychiatric
troubles such as: serious neurotic state, delirious or hallucinatory
psychotic syndromes etc.   These forms benefit, anyhow, from various
pharmacological treatments and a psychotherapeutic attitude, possibly
enlarged, of the environment.

    Finally, when the indications have been well localised, it is
necessary to limit oneself to a brief therapy, spaced out into two
sessions per week each of a maximum length of 30 minutes which may
later be continued in a very sporadic manner, once a month or every
two months and possibly being spread over some years.[13]   The
important conditions for these treatments are that one takes into
account the bodily and social reality since this kind of psychotherapy may be at times quite pragmatic and cover an aspect which
does not always correspond to the initial form.

REFERENCES

1. A. Haynal, G. Garrone, "L'entretien medecin-malade," Folia
        Psychopractica, 4 (1978).
2. M. Balint, "Techniques psychotherapeutiques en medicine,"
        ed. Payot, Paris (1966).
3. J. Willi, "Qu'est-ce que la medecine psychosociale?"
        Schweizerische Arztezeitung/Bulletin des medecins suisses
        Band 63, Heft 31 (1982).
4. J. D. Frank, Therapeutic factors in psychotherapy, Psycho-
        therapy Amer.J. 25:350 (1971).
5. K. Abraham, "Oeuvres completes" ed. Payot, Paris, T.II,
        92 (1966).
6. G. Abraham, and I. Simeone, La therapie breve dans l'age
        avance, Psychologie Medicale 12:597 (1980).
7. K. Fortini, A.-L. Vouga, La psychotherapie breve chez la
        personne agee hospitalisee, in: "Introduction a la Psycho-
        geriatrie, I. Simeone and G. Abraham, eds., SIMEP, Lyon
8. J. Wertheimer, L'intervention psycho-therapeutique en geria-
        trie, Praticiens et 3eme age 1:171 (1982).
9. G. Abraham, Ph. Kocher, and G. Goda, Psychanalyse et vieillisse-
        ment, Psychotherapies 4:229 (1981).
10. M. Grotjahn, Analytic psychotherapy with the elderly,
        Psychoanal.Rev. 42:419 (1955).

11. M. Burner, "L'abord psychotherapique du malade age"
     La psychiatrie de la vieillesse, cours de perfectionement
     de la Societe Suisse de Psychiatrie 61-75 (1970).
12. G. Dedieu-Anglade, Psychotherapie au cours du troisieme age,
     Confrontations psychiatriques 5:167 (1970).
13. J. A. M. Merloo, Assistance mentale et psychotherapie du
     vieillard, Medecine et Hygiene 1059 (1973).
14. G. Dedieu-Anglade, "Psychotherapie de troubles psychonevrotiques
     au cours du vieillissement," E.M.C., 37541, Alo (1981).
15. H. Segal, Fear of death: notes on the analysis of an old man,
     Internat.J.Psychoanal.39:178 (1958).
16. A. I. Goldfarb, Psychotherapy of aged persons, Psychoanal.
     Rev.42:180 (1955).

NUTRITION OF ELDERLY PATIENTS IN LONG CARE WARDS

Peter H. Millard, Eleanor Peel and Sue Thomas

St. George's Hospital Medical School
London, England

In this paper I first make some general comments on nutrition in the elderly and, then, look at the special problems of long term care and report the results of a nutritional study carried out on 14 elderly long stay patients in a large psychiatric hospital in south London.

Abraham Maslow[1] separated our needs into first, physiological, the need for food, water and warmth; secondly, the need for defendable space; thirdly, the need to love and belong; fourthly, the need for esteem and fifthly, the need for self-fulfilment. Food is one of man's most basic needs and pleasures, but even in the 1980's there are groups of people who may be at risk from malnutrition. The elderly are one such group, where increasing physical and mental disability, disease and drugs are just a few of the factors which may contribute to their poor nutrition. In addition food fads and fancies can affect their nutritional intake.

The basic drives of hunger and satiety are controlled by centres in the hypothalamus: the ventromedian nucleus controlling satiety and the lateral nucleus controlling hunger. These centres in the short term are influenced by the sight and smell of food, temperature, blood levels of glucose, insulin and amino acids, gastrointestinal stimuli and the effects of digestive hormones.

Longer term controls are one's food intake, energy reserves and energy expenditure. Appetite can be increased by stimulating the appetite centre and decreased by stimulating the satiety centre. Few drugs stimulate the appetite centre but, phenothiazines and pizotifen stimulate appetite, especially for carbohydrates.

Reduction of appetite in the elderly may occur without precipitating pathology and lead to a vicious circle with modification of the threshold for hunger. Initially reduction in appetite may be biological due to a reduction in energy expenditure and cellular metabolism, a normal accompaniment of aging, or it may be secondary to loss of teeth and ill-fitting dentures causing poor mastication or, loss of sense of smell, decrease in taste perception and changes in the gastro-intestinal secretion and motility. There also may be traditional myths such that it is normal for an old person to eat little, which may lead to the provision of insufficient food, both in quantity and quality.

Chronic subnutrition may lead to a vicious circle with weakness, anorexia, acceleration of senescence and dependency and concomitant increased risk of infection, ulceration, depression and confusion and lowered resistance to illness, thus allowing the elderly person readily to slide into the bed-bound state. In long term care wards, whatever the quality of catering, loss of appetite may be associated with apathy, or as a side effect of medications that had been given to the patients, or associated with poor presentation of the food. Appetite loss must be differentiated from diminished intake, for energy requirements decrease with increasing age due to a diminution in mental and physical exercise. Taste changes and the type of food eaten alters in association with the change in taste. Taste perception for sweet, salty and acid food may decrease but the bitter taste remains. An adequate intake of calories, proteins, vitamins and trace elements is essential.

To find out whether a person has appetite loss systematic enquiry should be made of the patient, relatives and, or care attendants with direct questioning with regard to the normal pattern of eating, the present modification and precipitating circumstances. Factors that ameliorate appetite should be identified and the severity of anorexia assessed by checking for weight loss. It must be remembered that anorexia can be secondary to somatic or psychiatric pathology. The individual case requires diagnosis by history, examination and special investigations. Occult infections, for example tuberculosis, can cause anorexia, as can drugs, e.g. alcohol, digoxin, biguanides or indomethacin. Occult neoplasia as well as constipation can also cause anorexia.

In patients with symptoms, anorexia can be associated with a loss of teeth, mouth ulcers, candidiasis secondary to antibiotic therapy, reflux oesophagitis, stricture, gastric ulcer, etc. In the elderly loss of co-ordination of the reflexes of swallowing (presbyoesophagus) can occur. Without other symptoms loss of appetite, especially in elderly women, can be due to an interaction of psychological and social factors.

In the long term hospital the most important factors to concentrate upon are the presentation of food, its content and the environment in which it is eaten.   The social ritual of eating, as a stimulant to appetite should not be forgotten.   The sight and smell of food stimulates hunger but poor presentation can cause anorexia.   At meal times elderly patients should be up and dressed in their own clothes, and sitting with others, preferably at round tables.   The table should be covered with tablecloths and flowers displayed.   A patient with a severe eating disability may prefer to eat on his own, but for the others group eating in an attractive setting is a bonding activity.   The food itself should be presented as normally as possible.   Far too many institutions present pureed diet constantly to the elderly when even those with difficulty in chewing can appreciate a soft diet.   A soft diet maintains the quality of food whilst the pureed diet loses the aesthetic qualities. Some consideration should be given to the service of food 'en pension', whereby the patients are encouraged to serve themselves and others from a serving dish in the middle of the table.

Concentration on the social aspects of the delivery of food, ways of tempting the sick to eat, of overcoming the problems of eating in bed, examining the repetitive menu, being aware of the loss of vitamins in bulk cooking and trying work with the elderly via the medium of patient committees to examine the quality and quantity of the food will greatly enhance the nutritional status of long term patients.

That this is necessary was aptly demonstrated in a nutritional study of 14 long stay patients in a psychiatric hospital undertaken by Sue Thomas[2] who is Senior Dietitian in the Department of Geriatric Medicine at St. George's Hospital.   A six day weighed food survey was carried out on a random sample of 14 patients, seven men and seven women, selected from two adjacent long stay wards.   The mean age of the patients was 77 years.   The average length of hospital stay for the whole group was 18 years and six months, with a range of six months to 62 years.   The general health of the group was reported to be good at the outset of the survey, with the exception of one man who was taking antibiotics for cellulitis of the arm.   The psychiatric diagnoses were chiefly that of schizophrenia, senile dementia and depression. Drugs taken were mainly the major tranquilisers, e.g. thioridazine, though three patients were taking phenobarbitone.   Six of the 14 had neither their own teeth nor dentures.   Six of the subjects took regular trips into the hospital grounds, five of these were men, the remaining eight seldom, if ever, went out.

During the study period each item of food or drink eaten between 7.30 a.m. and 7.00 p.m. was weighed by a dietitian. During the night the nursing staff recorded any extra food or drink taken, estimating the quantity consumed.

The daily pattern of meals was 8.00 a.m. breakfast, 1.00 p.m. lunch, 3.00 p.m. tea and 5.00 p.m. supper.   In the evening a drink was given only if requested.   Seven (50%) were given pureed diets. Computer analysis of the nutritional data was carried out using standard food tables.

Samples of hospital vegetables were collected immediately after cooking and at the time of serving and analysed for vitamin C content.   Blood samples were obtained from 13 of the 14 patients on the first day for biochemical and haematological analysis. Serum albumin, alkaline phosphatase, alanine transaminase, gamma glutamyl transferase, calcium, phosphate, ascorbic acid and vitamin D were estimated in the Chemical Pathology Department at St. George's Hospital.   Haemoglobin, serum and red cell folate, serum iron and total iron binding capacity were estimated in the Department of Haematology.

The results showed that the nutritional content of the diet was adequate except with regard to vitamin C, D and folate.   Serum ascorbic acid levels below the standard value of greater than 11.36 μmol/litre were found in five patients, four of whom were women and in all the vitamin C intake was correspondingly low.   Analysis of the food taken demonstrated that the vitamin C content of the vegetables was one quarter of that estimated in the food tables, the only patients who were found to have normal vitamin C levels were those patients who were drinking a daily glass of fruit juice. No clinical examinations were carried out and it should not be assumed that the deficient intake of vitamin C together with a low serum ascorbic acid level meant that the patients had scurvy, but rather that they were at risk from it.   That there is vitamin C deficiency in long term care wards is well known but its significance without clinical scurvy was highlighted by Schorah et al[3]   on long stay patients in a department of geriatric medicine who showed slight but significant improvement in appetite, daily living activities and interest when those patients who were known to have low ascorbic acid levels were given a one month course of this vitamin.   In our study those patients drinking a daily fruit juice had normal levels of vitamin C and following the survey we recommended that a daily fruit juice was made a standard part of the hospital diet.   This we consider is the best way of providing this vitamin.

The mean vitamin D intake of the groups was 2μg.   Few foods contain natural vitamin D and even with the inclusion of fortified foods in the hospital diet it would be difficult to achieve an adequate intake.   The mean serum vitamin D levels of the whole group was below the lower limit of normal for summertime values.

Six, two men and four women, had serum vitamin D levels of 11 nmol/litre or less, but their serum alkaline phosphatase, calcium and phosphate levels were then all within normal limits.   Two of these were on phenobarbitone and their low serum vitamin D levels were accompanied by alkaline phosphatase levels at the upper limit of normal and abnormally raised gamma glutamyl transferase levels probably associated with the effect of phenobarbitone on the liver. Higher serum vitamin D levels were found in the six patients who took regular trips into the grounds.   Five of these were men who were on average, younger and more mobile than the women.   The choices in ensuring adequate vitamin D levels lie between daily supplementation, change in ultraviolet light irradiation within the wards, or supplementation on a longer term basis.

All the patients had serum folate levels below the lower limit of normal.   However, none of the red cell folate levels showed serious signs of depletion.   The significance of these findings is difficult to interpret, but the results were in keeping with the dietary folate intakes which were all below the recommended daily intake.   It is probably, therefore, that the folate deficiency was not sufficient to cause clinical symptoms.

In conclusion, two centuries after the value of citrus fruits in preventing scurvy was realised by mariners who were sailing in boats across the ocean, we still find vitamin C deficiency in institutions.   We consider that the simple recommendation of a daily glass of orange juice will not only improve the vitamin C status of the patients but also help aesthetically to improve the presentation of the food.   Unlike the mariners, our patients are usually confined to the wards and we are now evaluating the effects of a bi-annual dosage of 2.5 mg of calciferol on the vitamin D status.

REFERENCES

1. A. H. Maslow,"Motivation and Personality," ed. 2. Harper and Row, New York (1970).
2. S. Thomas, A study of the nutritional status of 14 long-stay patients in Springfield Hospital, London. Geriatric Teaching and Research Unit, St. George's Hospital, (1981).
3. C. J. Schorah, D. Scott, A. Newill, D. B. Morgan, Clinical effects of vitamin C in elderly patients with low blood – Vitamin C levels, Lancet 1:403 (1979).

DISORDERS OF FLUID AND ELECTROLYTE METABOLISM IN THE ELDERLY

A. Borghetti

Institute of Semeiotica Medica
University of Parma
Parma, Italy

In this paper no attempt will be made to treat all the basic principles of water and electrolyte metabolism, or to make a complete exposition. For this purpose, reference must be made to available textbooks.[1] We will, instead, focus the attention on some aspects of the matter which are less frequently emphasized and will indicate some rules of thumb that should be helpful for clinicians.

A proper approach to the management of fluid and electrolyte disorders should begin with a correct estimate of the abnormalities present. The extracellular concentration of ions and other blood parameters are commonly used for diagnostic purposes. However, in most conditions, extracellular markers cannot be considered reliable indicators and calculations based on them are only approximate: for example normal serum sodium and potassium levels can coexist with mild, moderate and even profound deviation from a normal balance (Table 1). As a general rule, the clinician's assessment will be improved by detailed monitoring of external balance from day to day (Figure 1). For water, this criterion meets with the difficulty of evaluating the insensible losses through skin and lungs. For a rough estimate of the state of hydration one can resort to serum osmolarity, which varies inversely with the changes in water balance. As water is lost from the ECF, serum osmolarity is increased, while water accumulation is signalled by decreasing plasma tonicity. The following is a representative formula:

$$\text{Plasma osmolarity} = 2.0(Na+K) + \frac{BUN}{2.8} + \frac{Blood\ glucose}{18}$$

(where: Na-K mEq/l, BUN and Blood glucose mg/dl)

Table 1. Discrepancies between the plasma value and
external balance of Na and K in 13 elderly
subjects.   They were in a state of coma
caused by acute cerebrovascular accidents and
received total parenteral nutrition (TPN).
The plasma concentration of Na and K was either
increased, stationary or decreased during the
period of treatment, but these changes were
markedly unrelated to the external balance.

Plasma Na (mEq/l)	Na balance (mEq/l)	Plasma K (mEq/l)	K balance (mEq/l
140-141	+ 1030	4,3-4,5	- 435
140-140	- 445	4,1-4,3	- 245
138-139	+ 380	4,2-4,0	- 0
138-141	- 330	4,2-4,1	+ 30
140-140	- 200	4,5-3,5	- 210
139-142	- 900	4,0-4,0	- 100
142-136	0	3,8-4,7	+ 80
140-166	- 75	3,8-4,5	- 250
138-151	- 400	3,9-4,7	0
139-151	+ 250	4,4-3,9	- 50
141-166	+ 840	4,5-4,1	- 60
145-125	0	4,5-5,5	- 400
143-126	+ 125	4,7-3,0	+ 30

This formula indicates that, in addition to water changes, cations
(here the factor 2 takes into account the accompanying anions),
glucose and urea can contribute to the fixing of the value of
plasma osmolarity. If renal function is preserved and ADH
secretion is unaffected, urinary osmolarity may provide more insight
into the turnover of water. Hypertonic urine points to dehydration
while a diluted urine indicates haemodilution.   The equation
defining free water clearance (positive or negative).

$$C_{osm} = \frac{U_{osm} \times V}{P_{osm}} \qquad C_{H_2O} = V - C_{osm}$$

Where: C_{osm} = osmolar clearance (ml/min)

V_{osm} = urinary osmolarity (mOsm/l)

P_{osm} = plasma osmolarity (mOsm/l)

V = urinary flow (ml/min)

C_{H_2O} = free water clearance (ml/min)

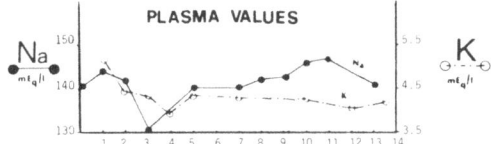

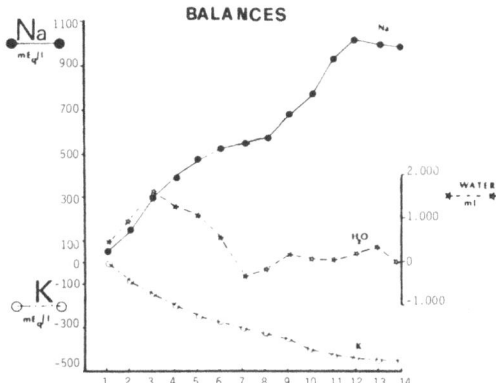

Fig. 1. Total parenteral nutrition (TPN)
        in a patient with acute vascular
        encephalopathy. A positive
        sodium balance and a neutral
        water balance are associated
        with a low or normal serum
        sodium. The serum potassium is
        constant in the presence of a
        negative K balance: this reflects
        a profound intracellular depletion,
        since the plasma equilibrium of K
        is maintained by a transfer of
        sodium in the intracellular com-
        partment and a concomitant
        sodium-potassium exchange.

predicts the amount of water that has been lost or retained by the
renal route. Urea usually makes a significant contribution to
urinary osmolarity. This metabolite diffuses across the biological
membranes and distributes uniformly inside and outside cells. As a
consequence, changes in its concentration do not significantly
influence the division of water between ECF and ICF and leave the
extra-intracellular water gradient unaffected. Electrolytes, on
the contrary, are actively transported through cellular membranes.
Therefore, it is worth distinguishing between the electrolyte and
non-electrolyte fraction of urinary osmolarity.

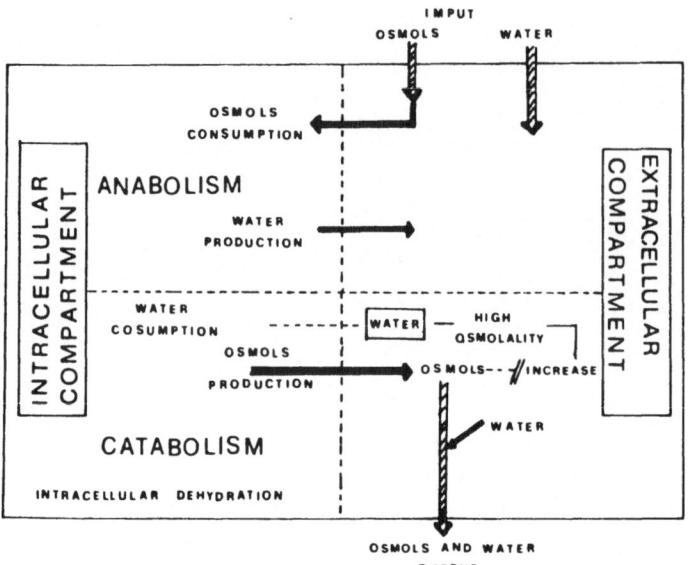

Fig. 2.   Effects of anabolism and catabolism
          on the production or consumption of
          osmols and water.

The nutritional state also interferes with the turnover of
water through the fundamental relationship between anabolism,
catabolism, water balance and osmolar balance (Figure 2).   Anabolism
leads to water production and sequestration of osmotically active
substances for cellular synthesis, whilst catabolism results in
release of osmols as waste products and in consumption of water.

Potassium Physiology

As has been stated above, the plasma potassium concentration
(PPC) is a doubtful index of potassium deficiency or repletion,
since some factors may alter the relationship between PPC and the
total body potassium stores.

To understand the regulation of PPC it is useful to distinguish
between potassium capacity, concentration and content.   The capacity
for potassium depends on the size of cell mass and the amount of
intracellular protein and other anions that can work as potassium
acceptors.   A decreased content: capacity ratio will result in K
depletion and hypokalaemia.   However, normokalaemia may coexist
with reduced potassium content if cell capacity is correspondingly
reduced.   On the other hand, hyperkalaemia can occur in the
presence of normal or reduced potassium content if capacity is
proportionally more impaired.

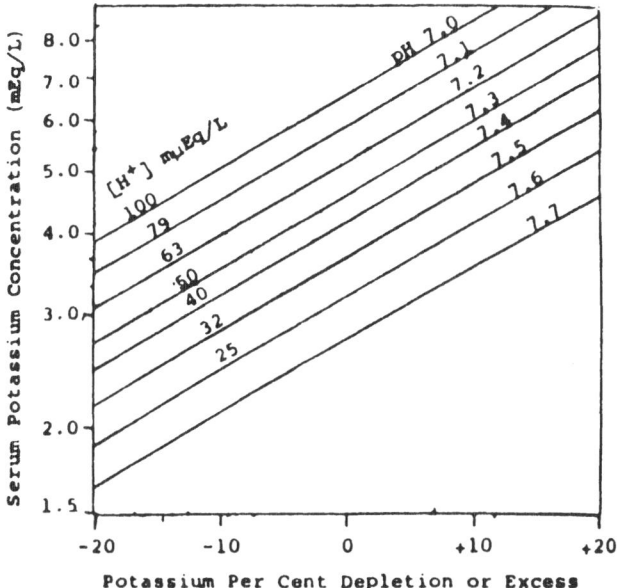

Fig. 3.   From Scribner and Burnell,1956

Plasma pH is known to affect the extra-intracellular distribution of K. A shift of potassium from ECF to ICF occurs when pH increases, so that hyperkalaemia may be present without depletion of total body potassium.   On the other hand acidosis normokalaemia or even hyperkalaemia may be associated with a marked potassium deficiency.   The underlying concept is that when hydrogen ions enter the cells, potassium moves out, possibly because the two cations compete for intracellular acceptor sites.

These principles have been elaborated and summarized by Burnell and Scribner[2] in a nomogram that can be used to predict the degree of potassium deficiency as a function of plasma potassium concentration and pH (Figure 3).

An often overlooked aspect of potassium metabolism is the potassium magnesium interrelationship in the cell.   The intracellular conentration of K generally correlates with that of Mg (Figure 4) while the replacement of K evokes a concomitant increase of Mg stores (Table 2).[3]   Therefore, magnesium must be supplied along with K during treatment of potassium deficits, otherwise Mg will be driven from bone and plasma into the ICF and skeletal reserves will be depleted.

Phosphorus

Underestimation of phosphorus abnormalities is not uncommon.

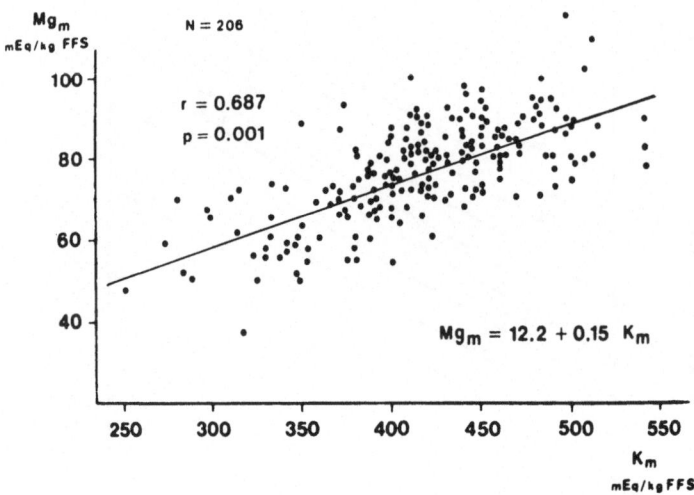

Fig. 4.   Relationship between muscle potassium
          (K_m) and muscle magnesium (Mg_m) in
          206 patients.

in clinical practice and can be explained by the following reasons:
a) phosphorus abnormalities are rarely isolated as electrolyte
   disorders;
b) the physiology of phosphorus has been studied less and later than
   that of other ions;
c) hypophosphataemia is almost always a clinically silent disorder,
   only extreme depletion being associated with well-defined symptoms.
As a consequence, the most frequent cause of hypophosphataemia is
iatrogenic, i.e. the omission of phosphorus from total parenteral
nutrition (TPN) (Figure 5).

Table 2. Plasma (p) concentration and muscle (m)
         content of potassium and magnesium
         before and after potassium repletion.

	before	after	p	21 controls
K_p mEq/l	3,34 ± 1,12	4,59 ± 0,42	.001	/
Mg_p mEq/l	2,04 ± 0,26	1,97 ± 0,26	.025	/
K_m mEq Kg/FFS	★★★ 359,6 ± 49,0	★ 445,2 ± 40,3	.001	451,5 ± 29,3
Mg_m mEq/KgFFS	★★★ 65,0 ± 12,4	★ 82,6 ± 13,6	.001	80,4 ± 6,2
p vs controls: ★ n.s. ★★ .01 ★★★ .001				

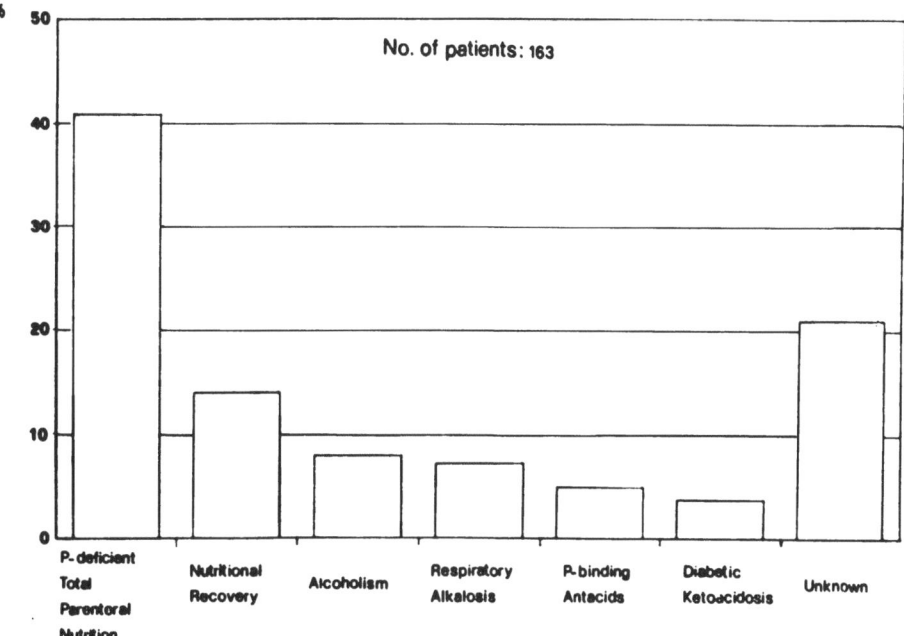

Fig. 5. Main causes of acute hypophosphataemia

However, it has been observed that phosphorus depletion has a
considerable incidence and represents, per se, a trigger for other
electrolyte abnormalities (Table 3).[4]   Muscle composition is
abnormal in the course of hypophosphataemia: the content of water,
sodium and chloride is increased, while that of potassium is
decreased (Figure 6).[5] This can be due, at least in part, to a
reversible inhibition of the Na-K pump, which has been demonstrated
in the red cells of hypophosphataemic subjects (Figures 7 and 8).[6]

Senescence from the viewpoint of Water and Electrolyte Physiology

The body undergoes significant changes in chemical anatomy with
advancing age. The water fraction of body weight progressively
decreases as a result partly of increasing body fat and partly of
decreasing lean body mass (Figure 9).[7]   Inasmuch as the muscle
tissue is reduced, total exchangeable potassium parallels the fall
of water content (Figures 10 and 11). Direct measurements of
muscle composition detect a drop in the potassium concentration.
The potassium changes are associated with increasing concentration
of sodium (Figure 12) and chloride,[8] indicating that the ratio
ECF/ICF is raised.

Since the cellular capacity for ions and radicals is reduced,
this makes the elderly more vulnerable to acute variations of
balance. The diminished cellular reserves should be carefully

Table 3. Muscle water, electrolytes and total phosphorus
          in hypophosphataemic patients with (A) and
          without (B) malnutrition and in control subjects

Muscle values	Controls No.	AHPP			P			
		A	B	A+B	A+B vs C	A vs C	B vs C	A vs B
TW M ±SD	21	3.243 0.170	3.572 0.295	3.773 0.372	3.675 0.344	0.001	0.001	0.001	NS
ECW	11	0.701 0.155	0.872 0.193	1.154 0.316	1.013 0.293	0.005	0.05	0.001	0.05
ICW	11	2.594 0.256	2.694 0.260	2.630 0.471	2.658 C.369	NS	NS	NS	NS
Cl_m	11	90.8 16.9	113.7 26.4	139.0 37.6	126.4 34.2	0.005	0.05	0.005	NS
Na_m	21	127.0 17.6	204.2 48.8	266.5 52.5	235.4 58.8	0.001	0.001	0.001	0.02
K_m	42	456.0 31.3	412.9 32.5	345.0 32.5	378.9 47.0	0.001	0.001	0.001	0.001
Mg_m	32	40.5 3.3	36.8 6.1	31.3 4.3	34.1 5.9	0.001	0.02	0.001	0.05
P_m	11	270.8 29.0	253.8 28.7	176.1 28.1	214.8 48.7	0.005	NS	0.001	0.001
$[Na]_i$	11	10.0 6.6	25.0 18.0	34.0 25.0	30.0 2.2	0.01	0.025	0.01	NS
$[K]_i$	11	175.0 16.4	153.5 17.7	133.0 19.6	143.2 21.0	0.001	0.01	0.001	0.025
$[Mg]_i$	11	31.1 2.8	27.1 5.3	24.1 4.3	25.6 5.0	0.005	0.05	0.001	NS
$[K]_i/[K]_e$	11	39.7 5.3	34.9 5.3	33.1 6.3	34.0 5.7	0.01	0.05	0.001	NS

A = AHPP without malnutrition;   B = AHPP with malnutrition
FFS = muscle fat free solids; TW, ECW, ICW = total extracellular and
intracellular water, Kg/KgFFS.
Cl_m, Na_m, K_m, Mg_m, P_m = mM/KgFFS ; $[Na]_i$, $[K]_i$, $[Mg]_i$ = mM/KgICW

evaluated when planning a replacement therapy and calculating the
deficits. The skeletal mass is also decreased in the elderly and
the buffering capacity of bone for hydrogen ions is impaired. On
the one hand this results in greater vulnerability to acid-base
imbalance and, on the other hand, imposes a greater caution in the
correction of these abnormalities.

In addition to changes in chemical anatomy, some homeostatic
regulatory mechanisms are impaired in old age, the most important
being kidney function. Plasma creatinine rises with age in spite
of diminished production by the muscle tissue, indicating a fall of
filtration rate.[9] The ability to concentrate and dilute the urine
is impaired, even earlier and more significantly than glomerular
filtration, thus affecting the rapidity with which a water load is
excreted. The urinary ammonia production is similarly lowered[10]
so that acid loads are less promptly eliminated.

Nutrition is the next important point.[11] The old are exposed

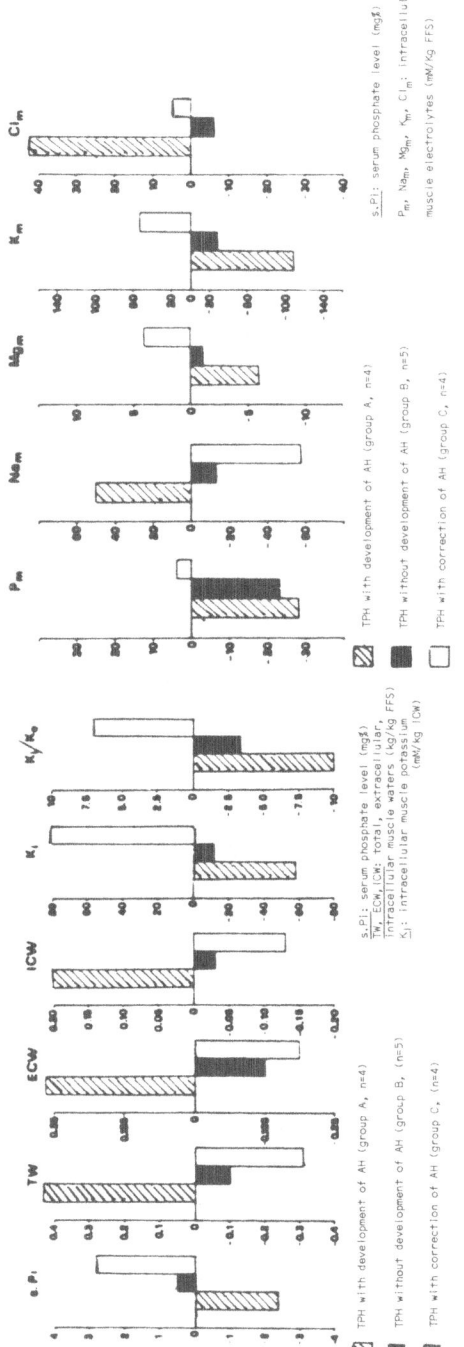

Fig. 6. Development and correction of acute hypophosphataemia (AH) during total parenteral nutrition (TPN). The figure compares the changes of serum phosphorus (sPi) to the changes in the composition of muscle tissue. Total tissue water (TW), extracellular water (EW), intracellular water (IW), muscle sodium (Na_m) and potassium (K_m) are differently affected according to whether treatment results in correction of development of AH.

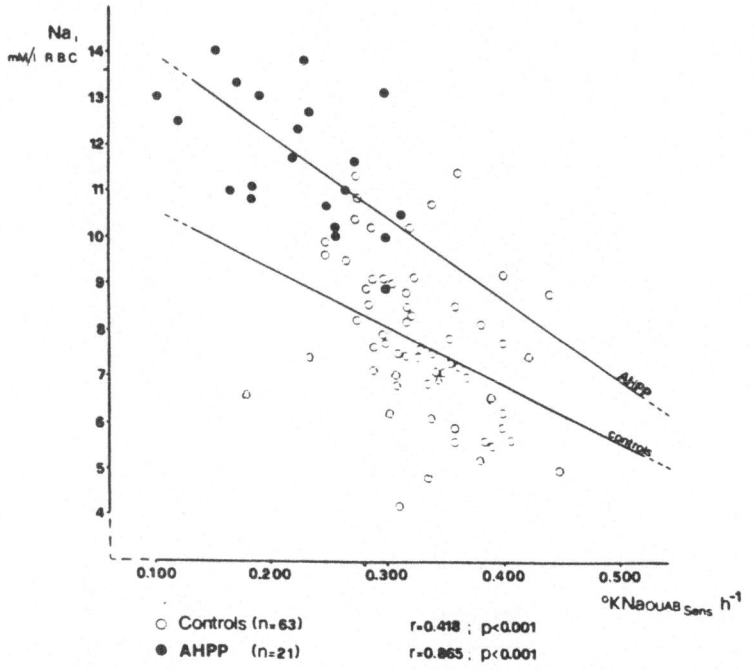

Fig. 7. Relationships between ouabain-sensitive rate
of Na efflux and intracellular sodium
concentration in erythrocytes from patients
with hypophosphataemia and from control subjects.

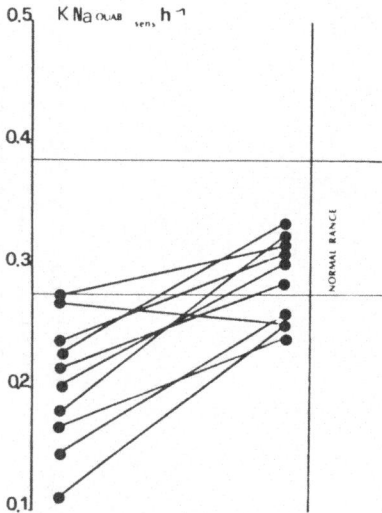

Fig. 8. Changes in ouabain-sensitive rate constant of Na
erythrocyte efflux after normalization of serum
phosphorus.

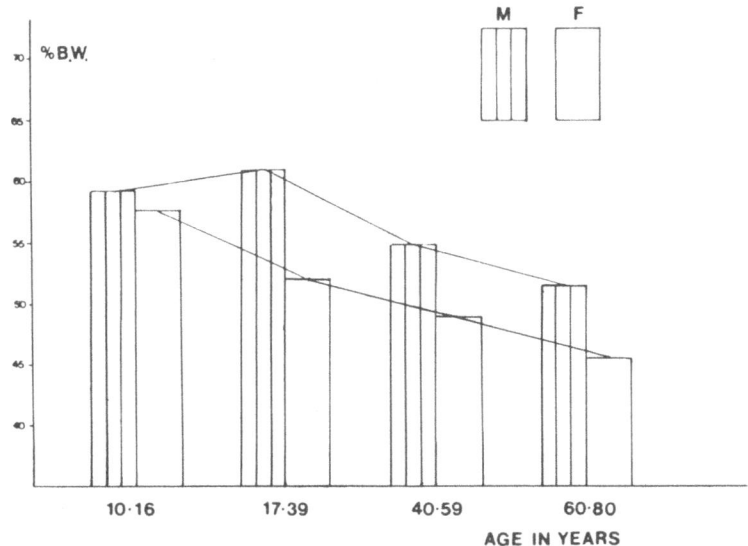

Fig. 9. Total body water at various ages in normal
men and women.

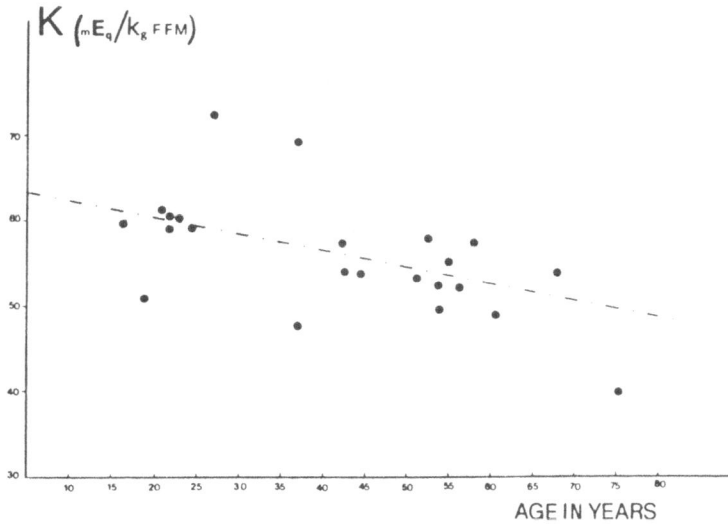

Fig. 10. Exchangeable potassium adjusted for the lean
body mass at various age levels.

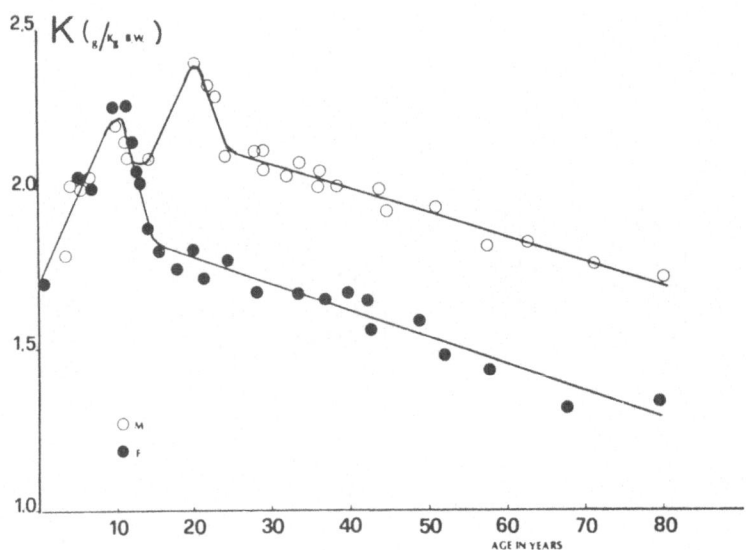

Fig. 11.   Body potassium content as a function of age
           in normal men (above) and women (below).
           (From Anderson and Langham, 1959 modified)

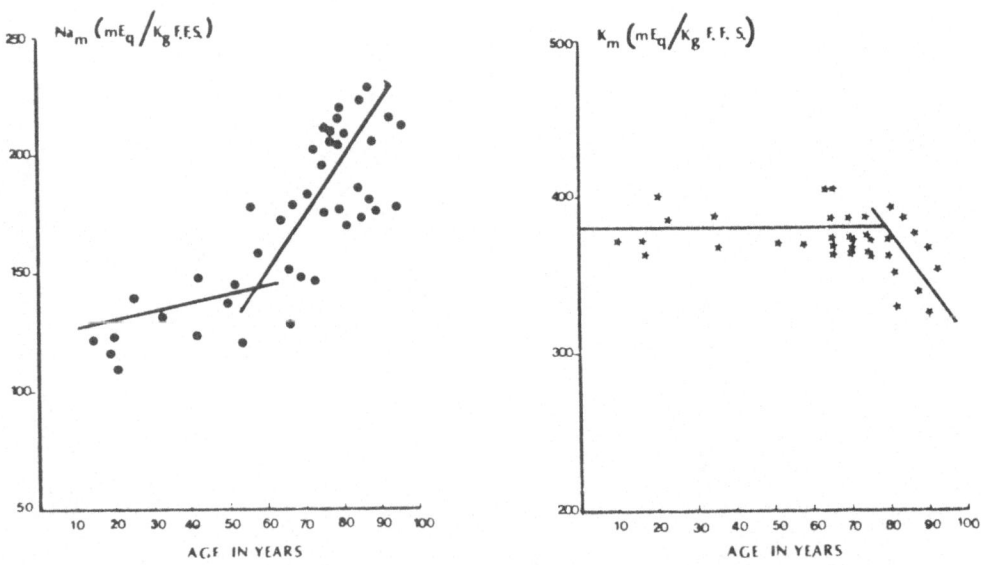

Fig. 12.   Na and K muscle contents at different ages in normal
           subjects (from Dubois, 1972)

to nutritional failure.   Disorders affecting locomotion,
difficulties of language and all conditions that result in
restricted mobility and restrained communication may limit access
to food and lead to malnutrition.   A number of disorders may also
alter the feeling of hunger and thirst.   Because of the altered
body composition, the nutritional requirements of patients should
not be calculated on the basis of the actual or even the ideal body
weight.   A better evaluation can be approached by measuring the
body fat as a function of skinfold thickness or by estimating the
muscle mass by creatinine excretion.   The external balance of
nitrogen can further help to establish whether the patient is in a
catabolic state.

The situation is even more dramatic in stroke, the incidence
of which is higher in old age.   Here failure to communicate and
difficulties of locomotion become absolute.   Such patients lose
the ability to signal their own needs and are in a dependent state.

Finally, it should be remembered that a variety of diseases
affecting mineral metabolism are more frequent in this age, the
most common being cardiac insufficiency, liver cirrhosis, diabetes
and pyelonephritis secondary to prostatic hypertrophy.

All these items make the elderly more prone to derangements of
fluids and electrolytes and create problems for the correct
assessment of needs and for proper therapy.

Common Clinical Patterns: Hyperosmolar Syndrome

A loss of water in excess of sodium and other solutes is the
usual cause of this syndrome.   In old age it is typically observed
in comatose or disorientated patients whose water intake is
inadequate and whose "insensible" water losses are underestimated.
Actually, in conscious subjects, thirst is such a powerful adaptive
mechanism that hypertonicity is not observed unless the patient is
unable to drink or has limited access to water.

Hypernatraemia is the general hallmark of this condition.   The
only exceptions are hyperglycaemia in the course of uncontrolled
diabetes mellitus and iatrogenic administration of extracellular
solutes (i.e. mannitol).   These solutes draw water out of cells,
thereby diluting the plasma sodium.   Hypertonic dehydration may
then occur in the presence of normo-or hyponatraemia.

Cell dehydration involves predominantly the central nervous
system.   Neurological manifestations are nearly always present and
may dominate the pricture.[12]   Depression of the sensorium may vary
from lethargy and confusion to stupor and coma and is roughly
related to the degree of hypertonicity (Figure 13)[13] and to the
rapidity of development of the osmolar gradient.

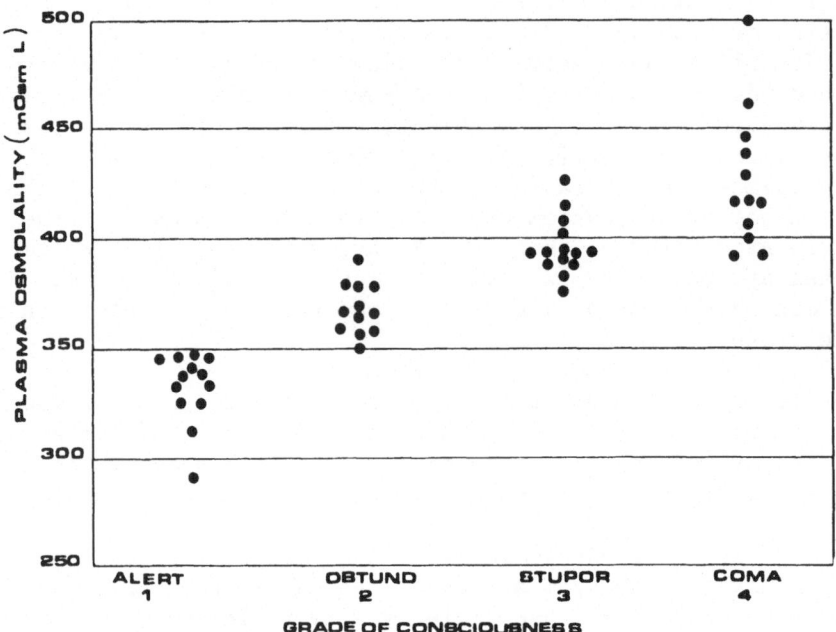

Fig. 13   Relationship between the state of consciousness
          and plasma osmolarity

The vulnerability of the brain to dehydration may be explained
by the property of the blood - brain barrier, which prevents fluxes
of solutes, while permitting inflow and outflow of water. Sodium
and other ions take hours to equilibrate.   Therefore, dehydration
causes the cells to lose volume and, if this occurs sufficiently
rapidly, the sudden shrinking of the brain exerts mechanical
traction on dural blood vessels, producing cerebral bleeding and
subdural haemorrhages.   Thus, a vicious circle may be established
in comatose patients suffering primarily from cerebrovascular
disease; failure to drink leads to desiccation which, in turn, is
responsible for further brain damage.

It has been suggested that the brain may have a protective
mechanism for preservation of cell volume by generating osmotically
active substances ("idiogenic osmols") from material not previously
contributing to the osmotic activity.[14]   Such substances are as
yet unidentified.

In a comatose patient, additional mechanisms may contribute to
the maintenance of hypertonicity.   In the undernourished subject,
catabolism drives osmols from the intracellular to the extracellular
fluid, promoting diuresis and thus further dehydration.   In a
patient receiving enteral or parenteral hypernutrition, the excess
protein or aminoacids will be converted to urea and the excretion of

the latter results in sustained osmotic diuresis. The resulting
urine will be hypotonic for sodium. Hence, hypernatraemia will be
accentuated and the large diuresis will leave the physician with
the false impression that the patient's hydration is preserved.

Water replacement using hypotonic solutions is the basic
therapy of hyperosmolar states associated with water depletion.
To calculate the magnitude of the deficit, the serum sodium should
be considered as if it were distributed in the total body water.
Rehydration must be very gradual, otherwise the increased attraction
for water provided by idiogenic osmols will then cause the brain to
swell. To be safe, restoration should proceed over a period of no
less than 48 hours.

Hypoosmolar Syndromes

Tbe hypoosmolar syndromes may be classified into two broad
categories, according to whether a predominant depletion of salt or
retention of water is the primary event.

In the first case (hypotonic dehydration, hypovolaemic hyponatraemia) salt may be lost through a renal (diuretics, adrenal
insufficiency, salt losing nephropathy) or extrarenal (vomiting,
diarrhoea, laxatives, burns, etc.) route. In this syndrome,
symptoms of ECF contraction are dominant. Hypotension and tachycardia are signs of circulatory impairment and prerenal kidney
insufficiency is signalled by reduced glomerular filtration rate and
oliguria. The skin dryness usually associated with dehydration may
not be apparent, because of the usual age-related changes. Hypotonic dehydration is treated by the administration of isotonic or
hypertonic saline. If the neurohypophysis and kidney have retained
normal function, spontaneous adjustment can be expected; otherwise,
the impact of the basic alterations on the homeostatic response must
be considered.

The second clinical pattern of hyponatraemia consists of hypo-
osmolarity with normal or expanded ECF volume (dilution hyponatraemia, water excess syndrome, hypervolaemic hyponatraemia). This is
caused by a primary excess of total body water, usually due to
inability to excrete a free water load. The impairment of renal
function may be of organic origin (chronic renal failure), but, in
a significant proportion of patients, the function of a structurally
intact kidney is affected by AVP secretion (inappropriate ADH
syndrome).

In this condition hypovolaemia is absent and the clinical
pattern will be characterized by the signs of disturbed brain
function due to overhydration of the CNS. The symptoms will be very
different according to whether hyponatraemia and water excess have
developed rapidly or slowly. Chronic hyponatraemia is generally

quite well tolerated, while acute water intoxication is a life-
threatening situation that must be regarded as a medical emergency.
Here, the symptoms are similar to those observed during hyperosmolar
encephalopathy (confusion, lethargy, headache, stupor, coma). The
treatment will also be different for chronic and acute overhydration.
In a slowly developing water excess syndrome, negative water balance
through restriction of fluid intake is usually effective.   However,
in the acutely intoxicated patient this is an absolutely inadequate
treatement and the osmolarity of the ECF must be promptly increased
by administration of hypertonic solutions (saline and/or mannitol).
The main draw-back of this therapy is the likelihood of precipitating congestive heart failure.   Therefore, it is usually a safer
and more effective approach to associate the administration of
hypertonic saline with "loop" diuretics, like furosemide and
ethachrinic acid, which impair the ability of the kidney to produce
concentrated urine: water will then be excreted in excess of sodium
while the latter is being administered intravenously. Excessive
hypervolaemia will be prevented while plasma sodium concentration is
being raised.

Hypophosphataemia - Phosphate Depletion

Hypophosphataemia is one of the most frequent electrolyte
abnormalities in the elderly and yet one of the least recognized.[15]
Usually chronic phosphate depletion results from starvation and cell
catabolism and is accompanied by a concommitant loss of cell protein
and nitrogen.

If total parenteral nutrition does not include an adequate
amount of phosphorus and urinary excretion is unchanged, a negative
balance of phosphorus is established. When the plasma level falls
below 0.5 mM/m, the EC/IC gradient makes phosphorus leak out of cells
and intracellular depletion ensues.

Acute hypophosphataemia, on the contrary, occurs during recovery
from starvation and is usually the result of rapid influx of
phosphorus into the cells.[16]   When nutrition is corrected without
supplying phosphorus, the latter is driven into the cells by concommitant uptake of other nutrients, restoration of cell water and
resynthesis of protein and glycogen. The intracellular stores then
increase at the expense of the extracellular fraction. The more
severe the initial intracellular deficit, the more marked will be
the fall in plasma phosphorus during recovery.

Clinical manifestations of the disease include neurological
signs and symptoms that may be so polymorphic as to mimic a variety
of neurologic diseases.[17]   The blood cells are among the most
affected since they are directly exposed to the fall of plasma
concentration.   Haemolytic anaemia and impaired leucocyte and
platelet function have been reported.   ATP deficiency may explain

these abnormalities.   Regeneration of ATP from ADP depends on the
availability of phosphate.   This is at least partly demonstrated by
the finding of failure of ATP-ase dependent Na-K pump in the red cell
of hypophosphataemic subjects (Fig. 7 - 8).   Prevention can be
easily accomplished by routinely including some amount of phosphate
in the composition of intravenous alimentation and therapy.   If
phosphate depletion and hypophosphataemia are already present, the
amount of phosphorus to be supplied must obviously be larger
(figure 14) and can be approximately calculated by the following
formula:

$$gP = 0.74 \times (P_n - P_a) + (0.15 \times \Delta)$$

where: gP is the amount of phosphorus to be supplied (in grams) to
restore a normal plasma level (P_n); P_a is the actual plasma concentration before therapy and Δ is the per cent of deviation from the
ideal weight.

The danger of forcing phosphate restoration is generally low,
except in renal failure, where the risk of producing hyperphosphat-
aemia is inversely proportional to the residual glomerular
filtration rate.

Potassium Disorders

Potassium deficiency is very rarely due to reduced intake,
since all the foods contain some potassium and the normal kidney is
able to drastically reduce the output.

Excessive loss of potassium is therefore the usual cause.
Extrarenal dissipation of K mainly occurs through the gastro-
intestinal tract.   Fistulas, intractable vomiting, protracted
diarhoea and laxative abuse are among the most common causes.
Urinary loss of K can be due to primary renal disease, like the
congenital form of renal tubular acidosis or, more frequently, to
an excessive mineralcorticoid effect, as in primary or secondary
hyperaldosteronism, Cushing's syndrome and other forms of adrenal
hyperplasia.   Prolonged administration of diuretics may also
create a negative body balance of the cation.   The manifestations
of hypokalaemia predominantly involve the neuromuscular function.
A raising of the IC:EC ratio of potassium concentration increases
the resting membrane potential of excitable cells and thereby
depresses their excitability.[18]   Changes in performance of striated
and smooth muscle include generalized "muscle weakness" that can
progress to paralysis, decreased motility of the bowel, ileus and
hypotension.   Changes in cerebral function may vary from apathy
and lethargy to drowsiness, disorientation and confusion.   Typical
ECG findings are: a flattening and lowering of the T wave,
appearance of a prominent U wave, prolongation of P-R interval.
Atrial arrhythmias are similar to those observed in the presence of
digital toxicity.

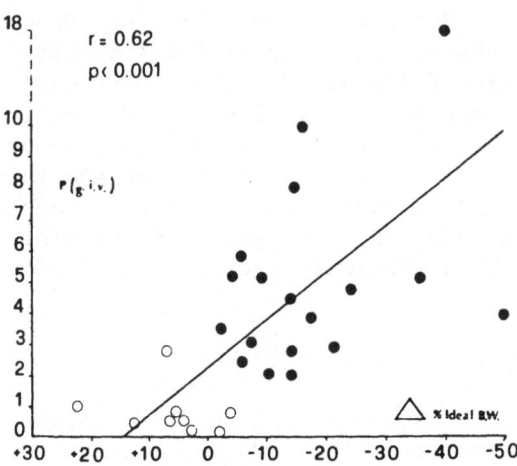

Fig. 14. Relationship between the
         amount of intravenous phosphate
         required to obtain a normal
         plasma Pi and percentage devia-
         tion from the ideal body weight.

     A positive potassium balance is usually due to decreased
excretion by the kidney.   The underlying renal abnormality may be
organic (as in acute and chronic renal failure), pharmacological
(mineralcorticoid antagonists, amiloride, etc.) or functional
(adrenal insufficiency) in origin.   However, hyperkalaemia without
a net increase of body potassium may result from a transcellular
shift of K secondary to acidosis, cellular breakdown and loss of
cell mass (decreased content: capacity ratio).   The increase of
extracellular potassium concentration lowers the resting membrane
potential and enhances the excitability of muscular and nervous
tissue.   The earlier and most threatening manifestations are those
concerning the electrical activity of the heart.   High and peaked
T waves, depression of the S-T segment, disappearance of P waves
and prolongation of QRS complex may herald more dangerous
arrhythmias like ventricular tachycardia and fibrillation.
Immediate therapy is aimed at decreasing the plasma concentration
of potassium, so preventing membrane depolarization and the potentially lethal cardiac arrhythmias.   This can be accomplished in a
threefold way.   The administration of bicarbonate ions raises blood
pH and stimulates the cellular uptake of K; the infusion of insulin
(along with glucose to prevent hypoglycaemia) also induces a
cellular inflow; finally, the depolarizing effect of K can be
counteracted by administering Ca salts (gluconate) and increasing
the extracellular concentration of calcium.

The second step of therapy is aimed at reducing total body potassium. Synthetic resins having the property of sodium exchangers can be administered orally along with sorbitol (to prevent constipation and to induce dissipation by mild diarrhoea). If the gastrointestinal tract is not accessible, or in patients having renal failure, potassium withdrawal may be accomplished by peritoneal dialysis or haemodialysis. Chronic diseases that lead to starvation and decrease of body potassium capacity involve specific risks for the management of potassium homeostasis. In long-lasting diabetes, liver cirrhosis, chronic renal failure, etc., potassium deficiency is associated with depletion of muscle mass. The danger inherent in these conditions is that the changes of K balance are insufficiently buffered by the decreased potassium capacity and result in exaggerated fluctuations of the plasma potassium concentration. Thus, an acute potassium load (as it occurs in association with blood transfusions, administration of aldosterone antagonists, catabolic states, etc.) may rapidly reverse the picture from a state of hypokalaemia to acute hyperkalaemia. In treating the latter, care should be taken lest the rapid removal of potassium from plasma should lead again to profound hypokalaemia.

The items treated above are only examples that have been chosen to emphasize the link between the nutritional state and electrolyte physiopathology in the elderly. They would stress the need for a comprehensive clinical judgement as a basis for diagnostic interpretation and for therapy.

REFERENCES

1. M. H. Maxwell, C. R. Kleeman, "Clinical Disorders of Fluid and Electrolyte Metabolism," McGraw Hill Book Company, New York (1980).
2. B. H. Scribner, J. M. Burnell, Interpretation of the serum potassium concentration, Metabolism 5:478 (1956).
3. A Montanari, L. Borghi, M. Canali, A. Curti, G. Bucciero, P. Perinotto, A. Novarini, A. Borghetti, Effets du potassium et des protéines cellulaires sur le magnésium musculaire chez des patients atteints de troubles hydro-electrolytiques, Rev.Franc.End.Clin.6:531 (1979).
4. A. Montanari, L. Borghi, A. Curti, M. Mergoni, E. Sani, G. F. Elia, M. Canali, A. Novarini, A. Borghetti, Skeletal muscle cell abnormalities in acute hypophosphatemia during total parenchima nutrition, Min.Elect.Med.(in press) (1982).
5. L. Borghi, A. Curti, M. Canali, M. Mergoni, E. Sani, A. Montanari, A. Novarini, A. Borghetti, Relationships between muscle K, Mg and Na content and acute hypophosphatemia (AH) with and without phosphate depletion in man, Magnesium Bulletin 2:154 (1981).

6. L. Borghi, M. Canali, E. Sani, A. Curti, A. Montanari,
    A. Novarini, A. Borghetti, Erythrocyte sodium transport in
    acute hypophosphatemia in man, Min.Elect.Met. (in press)
    (1982).

7. R. M. Pierson, D. H. Lin, R. A. Philips, Total body potassium
    in health: effects of age, sex, height and fat, Am.J.Physiol.
    226:206 (1944).

8. J. Dubois, Water and electrolyte content of human skeletal
    muscle. Variations with age, Rev.Europ.Etudes Clin.et Biol.
    17:505 (1972).

9. J. W. Rowe, R. Andres, J. Tobin, A. H. Norris, N. W. Shock,
    The effect of age on creatinine clearance in man: a cross-
    sectional and longitudinal study, J.Ger. 31:155 (1976).

10. B. N. Agarwal, G. G. Cabebe, Renal acidification in elderly
    subjects, Nephron 26:291 (1980).

11. J. P. Grant, P. B. Custer, J. Thurlow, Current techniques of
    nutritional assessment, Surg.Clin.N.Amer.61:437 (1981).

12. C. M. Cosey, A. I. Arieff, Disorders of sodium and water
    metabolism and their effects on the central nervous system,
    in:"Sodium and water homeostasis,"B. M. Brenner, J. H.
    Stein, eds., Churchill Livingstone, New York (1978).

13. M. Fulop, H. Tannenbaum, N. Dreyer, Chetotic hyperosmolar coma,
    Lancet 2:635 (1973).

14. A. I. Arieff, C. R. Kleeman, Studies on mechanisms of cerebral
    edema in diabetic comas: effects of hyperglycemia and rapid
    lowering of plasma glucose in normal rabbits, J.Clin.Invest.
    52:571 (1973).

15. M. Betro, R. Pain, Hypophosphatemia and hyperphosphatemia in a
    hospital population, Brit.Med.J. 1:273 (1972).

16. R. A. Kreisberg, Phosphorus deficiency and hypophosphatemia,
    Hosp.Pract. 29:121 (1977).

17. J. P. Knochel, The pathophysiology and clinical characteristics
    of severe hypophosphatemia, Arch.Int.Med. 137:203 (1977).

18. J. P. Knochel, Rhabdomyolysis and effects of potassium
    deficiency on muscle structure and function, Cardiov.Med.
    3:247 (1978).

LONG TERM TREATMENT WITH ORAL ANTIDIABETICS

F. S. Feruglio

General Clinical Medicine Department
University of Trieste, Italy

In 1959, at the VIII Congress of the Gerontology and Geriatric Medicine Society which was held during the 60th Congress of the Internal Medicine Italian Society, Greppi[1] and his colleagues presented a report on "Senile Diabetes", in which hereditary, constitutional, endocrine, metabolic, vascular and therapeutic aspects were highlighted. It is significant that in a period when there still were nosographic uncertainties and at the beginning of therapy with oral hypoglycemic agents, these veteran gerontologists had already drawn the attention to a problem which still today presents controversial aspects.

That report, which anticipated problems which are still under discussion, is even more significant not only for its interesting research cues, and its still valid original remarks, but also because in the interesting book by Freeman on "Clinical Features of the Older Patient" in 1964,[2] there is no chapter on diabetes, although there are some remarks on insulin and on metabolism and glucose tolerance. Diabetes is indirectly mentioned in the chapters dealing with oral, vascular, renal, ocular complications, but no specific therapeutic advice is given.

Later on, the topic of diabetes in old patients was included in the literature dealing with old-age diseases. I wish to recall here the chapter by Butterini[3] in the Italian Textbook on Gerontology and Geriatric Medicine edited by Antonini and Fumagalli, and the chapter on Diabetes by Fitzgerald[4] in the "Textbook of Geriatric Medicine and Gerontology" by Brocklehurst. Although the definition of senile diabetes no longer exists, interest in this disease is still high. In fact, all agree that the incidence of diabetes increases after the 5th decade of life till over the 7th.

This leads to considerable problems both of a clinical and nosological nature.

DEFINING DIABETES

The core of the discussion consists in trying to answer the
question on the "glucose tolerance" in old persons. In fact all
definitions of diabetes are based right on this point: from those
which were valid till some years ago, classifying diabetes into
potential, latent, asymptomatic and clinical, to the new ones
proposed by the National Diabetes Data Group (N.D.D.G.) of the
U.S.A.,[5] i.e.: type I diabetes (insulin-dependent), type II
diabetes (non-insulin-dependent), and diabetes associated with
another disease. This group also reports some conditions of
"impaired glucose tolerance", a "gestational diabetes", a condition
of "previous impaired glucose tolerance" and a "potential impaired
glucose tolerance", i.e. those cases which may be recorded in
population groups at statistical risk.

Although it has repeatedly been stated that there is an
increasing incidence of diabetes in patients 60, 70, and 80 years
old, and that glucose tolerance progressively decreases with age,
the classification proposed by the N.D.D.G. does not seem to take
these data into account. It therefore becomes useful to try to
give an explanation of a classification which, if on one hand, has
the merit of simplifying many aspects of the problem, on the other
hand seems to neglect the progressive aging of the population.

Besides a different glucose tolerance in old patients, we may
observe the three following situations:

1) patients affected by type I diabetes since young age, who
    become old;
2) patients affected by type II diabetes occurring before old age;
3) patients whose diabetes is first diagnosed in old age.

At his time, Greppi had noted a fact which may still be found
today, i.e. some diseases (including diabetes), which are present
before old age, clinically disappear with aging. He insisted that
there were improvements and even cases of "recovery" after decades.
mainly in extreme old age. The changing disease behaviour as
time goes by cannot be neglected, but in view of the importance of
correct therapy, appropriate knowledge and controls are needed.
It is universally recognized that most elderly patients (as most
diabetics in general) are affected by type II diabetes, but this is
not an absolute criterion, as it is possible to find type I diabetes
appearing for the first time in old age.

A controversial point which has already been mentioned, is to establish the clinical or laboratory criteria showing the presence of a real diabetes, for which appropriate therapy is required.

It has often been pointed out that, in many cases of diabetes, and especially in old patients, the basic guideline symptoms for a diagnosis - polydipsia, polyuria, polyphagia, fatigue, muscular weakness, loss of weight, frequent infections and inflammatory processes - are either lacking or less pronounced.

Sense of thirst and hunger are often reduced in old patients, polyuria can be masked by frequency of micturition due to prostate hypertrophy or cystitis or prolapse in females. Weight loss is also frequent in old persons who have reduced food intake, and finally asthenia, atrophy and muscular hypotonia, if not well evaluated are attributed to old age decay. All these phenomena which are due to various causes, are relatively frequent among elderly patients. However, in some cases, they occur months, and sometimes years before the onset of clinical diabetes.

Symptoms such as frequent infections of the urinary tract and of the skin, with a tendency to recurrency, resistance to therapy and candidosis superimposed on bacterial infections are more significant. Vulvar itching in females is also symptomatic, although not pathognomonic. The presence of eye troubles, peripheral or cerebral vascular complications already call for an accurate diagnostic control, even in cases of ignored diabetes.

The diagnosis can clearly be defined when fasting blood sugar values are within the pathological range and there is postprandial glycosuria. In these cases, the diagnosis is not in doubt, but there could be some interpretation difficulties in borderline cases for which an oral glucose tolerance load is needed. In this connection, the criteria indicated by the N.D.D.G. must again be taken into consideration: these criteria refer to middle-aged patients, while things are different in elderly patents.

The limits indicated by the N.D.D.G. for the diagnosis of diabetes concern fasting venous blood sugar measured by enzymatic method. <100 mg/dl values are obtained in healthy persons, <120 mg/dl in individuals with reduced glucose tolerance, and >120 mg/dl show a diabetic state. For venous blood plasma, corresponding values are: <115 mg/dl for healthy subjects, <140 mg/dl for reduced glucose tolerance and >140 mg/dl for diabetics. While as to diabetic diagnosis there is no controversy even in old patients, it is worth pointing out both the indicated value for reduced glucose tolerance and the result of an oral (or intravenous) glucose loading.

Table 1.   Upper limits of plasma glucose during fasting and after
           oral glucose loading (ranked for age effects)*

Age	50	51 – 60	61 – 70	71 – 80	80+
Fasting (mg/dl)	125	135	145	155	165
1 hour	195	205	215	225	235
2 hours	140	150	160	170	180
3 hours	125	135	145	155	165

* Based on US Public Health Service criteria. 10 mg/dl added
   for each decade over age 50. If whole blood venous glucose is
   measured, decrease level by 15%.   (From T. G. Skillman)

For type I diabetics, the insulin therapy is obligatory; for
the others, in oral therapy, two types of drugs are used:
sulfonylureas and derivatives, and diguanides.

Sulfonylureas

Sulfonylureas have a common basic formula, i.e.:

$$R_1 - \underset{\underset{O}{\|}}{\overset{\overset{O}{\|}}{S}} - \underset{}{\overset{H}{N}} - \overset{\overset{O}{\|}}{C} - \underset{}{\overset{H}{N}} - R_2$$

and their hypoglycemic properties stem from the radicals R_1 and R_2;
they are rapidly absorbed by the proximal intestinal tract and are
carried into the blood by protein complexes.

When they are released by the carrier proteins, they diffuse
into the tissues where they act.   The acute administration of
sulfonylureas causes a release of insulin from the beta cell with a
stimulus in the first phase of cell secretion, for a direct action
on the plasma membrane.   It is probable that the mechanism that
induces insulin secretion is a redistribution of calcium ion within
the beta cells.

In long term administration, the stimulus to insulin release
does not seem to persist, and the glucogon level does not seem to
decrease either.   The most significant effect in long term sulfonyl-
urea administration is represented by an increase in the number of

This review is carried out with the purpose of identifying cases of unknown or mild diabetes and therefore calls for a careful evaluation.   After adequate nutrition and physical activity in the days before the tolerance test with 75 g glucose, the limit values proposed by the N.D.D.G. are: at the first hour after sugar administration on venous plasma, <200 mg/dl for healthy subjects, >200 mg/dl for reduced glucose tolerance, and >200 mg/dl for diabetics.   The limits proposed at the second hour are <140 mg/dl, from 140 to 200 mg/dl, and above 200 mg/dl respectively.   If whole venous blood sugar is measured, the normal values at the first hour are <180 mg/dl, >180 mg/dl, and >180 mg/dl respectively, and at the second hour <120 mg/dl, from 120 to 180 mg/dl, and above 180 mg/dl respectively.

In order not to make any evaluation mistake, in old patients, due to a reduced glucose tolerance, Davidson[6] suggests an adjustment of 10 mg/dl for each decade of life over age 50; therefore the above-mentioned values will have to be adjusted on the basis of the indicated coefficient (see Table 1).

If both values at the first and second hour are high, glucose intolerance is confirmed and the patient should be warned against the effects of obesity and infections.   Diabetes should not be diagnosed, and consequently no drug therapy should be undertaken as long as there is no repeated confirmation of fasting hyperglycemia. In addition, it should be taken into account that most elderly patients in whom the glucose tolerance test gives a result of impaired glucose tolerance do not necessarily develop diabetes; on the contrary, the levels could return to normal values.

It will be noticed that the presence of glycosuria has not been mentioned for confirming the diagnosis of diabetes.   This diagnosis parameter is very important in elderly patients, whose impaired renal function may cause a reduced glucose loss, even in the presence of hyperglycemia.   However, when glycosuria is found, this must be carefully evaluated, not only because it is a clinical-therapeutic point of reference which needs to be adjusted, but also because it represents a quantifiable loss of food energy components.

TYPES OF ANTIDIABETIC DRUG

Whenever diabetes mellitus is dianosed in old patients, it will be necessary to choose the type of therapy by taking into account clinical, biological, and laboratory data.   As we have already mentioned, most of these old diabetics are affected by type II diabetes, and therefore the use of oral hypoglycemic agents seems to be justified.

insulin receptors on human monocyte membranes.   In addition, long
term administration increases the insulin hypoglycemic effect.
These drugs therefore seem to be more indicated to improve the
insulin receptor capacity in those subjects who still maintain
hormone production.

Sulfonylureas may be classified according to their pharmacological characteristics and different dosages.   These depend on
both the drug itself and on the metabolism and excretion patterns,
and these factors should be taken into account during their use.

The plasma half-life ranges from 1.5 h. for glyquidone, to
2.5 - 4 h. for glypizide, 5.8 h. for most compounds, and 36 h. for
chlorpropamide.   These data give some indications on the frequency
of administration during the day for the short-life compounds.
A single administration will be sufficient for compounds having a
half-life of several hours (chlorpropamide and glycazide).

Also the metabolic and excretory pathways may vary depending
on the type of drug.   In the urine there are inactive metabolites
of the glybenclamide, glyclazide, glypizide and tolbutamide, and
active metabolites of the acetohexamide, chlorpropamide, glycodiazine
and tolazamide.   It is clear that in the various pathological
conditions, different drugs will have to be used.   If renal
excretory function is impaired, liver metabolized drugs (tolbutamide)
should be used in preference; while in liver disorders, drugs which
do not depend on this metabolic pathway (chlorpropamide, tolazamide,
etc...) should be used.

Compounds have also been classified into first and second
generation drugs.   The basic difference between these compounds
consists in the single doses which are about 100 times higher for
first generation drugs if compared with second generation ones.
There does not seem to be any difference in the action and efficacy,
so that the choice of the drug will be determined by various
factors such as cost, tolerance, availability, direct experience of
the prescribing physician, etc....

In a small number of cases, gastrointestinal disturbances,
varying from dyspepsia to diarrhoea, are recorded.   These inconveniences may be overcome by administering the drug after meals, and
by resorting to one single administration per day, or by choosing
another compound (at low dosage).

Skin reactions are common and disappear when drug administration
is suspended.   Sometimes water retention of the hyponatremic
dilutional type may be observed.   Frequent bursts of heat and
flushing can occur when drugs are taken at the same time as alcoholic
beverages; this phenomenon is most frequent with chlorpropamide.

Diguanides

In the basic structure

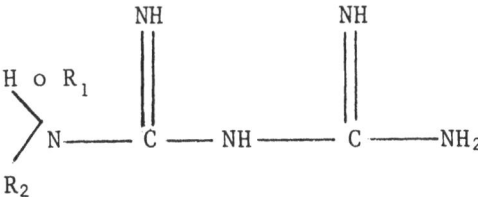

the replacement of one or more hydrogen atoms in one of the amino-groups causes the pharmacological effect observed in the various compounds.

A good concentration of diguanides is recorded, especially of metformin, in the intestine, and of the phenformin in the liver. Protein binding is almost nil for metformin, which is not metabolized and is excreted unchanged in the urine.

The presence of these substances in the liver affects its function. In fact, a delayed glucose absorption and a reduced gluconeogenesis have been shown; these effects, together with an increased peripheral glucose concentration, especially in the muscle, are responsible for their hypoglycemic action. The pharmacological effect is recorded only in cases of hyperglycemia, while the serum glucose of non-diabetics remains unchanged.

Because of these effects, which are associated with anorexia, diguanides are indicated in those cases needing weight loss. Besides this action, a reduction of serum triglycerides is seen when levels are elevated. The use of diguanides, especially of metformin, in this field, was extended also to non-diabetic cases. Their action is dose-dependent and requires high dosages (up to 2-3 g/d. for metformin).

Plasma half-life is 1.5 h., and therefore short intervals between administrations are needed. The average dosage is 50 mg for phenformin (do not exceed 100 mg), and 100 mg for metformin.

Side effects of diguanides are more frequent than those of sulfonylureas. In 10 - 15 per cent of the patients we find dose-dependent anorexia, nausea and diarrhoea, so that the therapeutic dosage cannot always be reached. Reports of lactic acidosis occurring during treatment with diguanides have remarkably reduced their use in general; for this reason they were banned in the U.S.A. It seems that lactacidemia may be caused more by phenformin than by metformin. In any case, in the presence of renal failure or of precarious tissue perfusion conditions (cardiorespiratory failure) this drug must not be administered.

THERAPY

From a theoretical viewpoint, therapy for diabetes in elderly patients is not different from that in other ages, but some characteristics need to be kept in mind, because, if neglected, there could be serious consequences.

First of all, there must be a certain diagnosis of diabetes mellitus according to the above-mentioned criteria and parameters. In the case of insulin-dependent diabetes and ketosis in thin patients, insulin therapy is undertaken.   In the other cases of type II diabetes, a strict preliminary diet prescription must be followed.   The calorie intake must take into account the ideal weight (not the actual one, especially in the case of obese patients), and the physical activity that the patient is able to carry out.   The latter should be encouraged and considered as an integral part of therapy, (see Figure 1).

However, as far as the mechanism of action of the sulfonyl- ureas and their effect on peripheral receptors are concerned, we should recall that high-calorie diets decrease the number of insulin receptors while low-calorie ones allow the restoration of the number of receptors even when they do not induce any weight loss.   This mechanism confirms that diet is of basic importance for correct diabetes therapy, which should be supplemented by administration of oral hypoglycemic agents.   Also obesity reduces the number of insulin receptors, and therefore weight loss achieved through a lower calorie intake, plays an important role favouring the therapy itself.

Oral hypoglycemic therapy will have its effects if the number of insulin receptors is insufficient, and if insulin secretion is relatively active.   It too often happens that oral hypoglycemic therapy is used in non-diabetic, but only hyperglycemic, patients, in a useless competition between drug and food administration. This irrational behaviour is completely useless, and sometimes dangerous.   In fact, if the patient does not accept a correct dietary approach, we shall not have his active collaboration which is indispensible for effective therapy.

Sometimes, drug administration leads the patient to believe that a "pill" is sufficient to treat diabetes, so that he does not realize that this is only an adjuvant in the treatment of hyper- glycemia.   In addition, if the patient does not feel himself bound to a food-drug therapy, he will be led to make those common mistakes such as consuming meals with excessive, alternating with lack of calories.

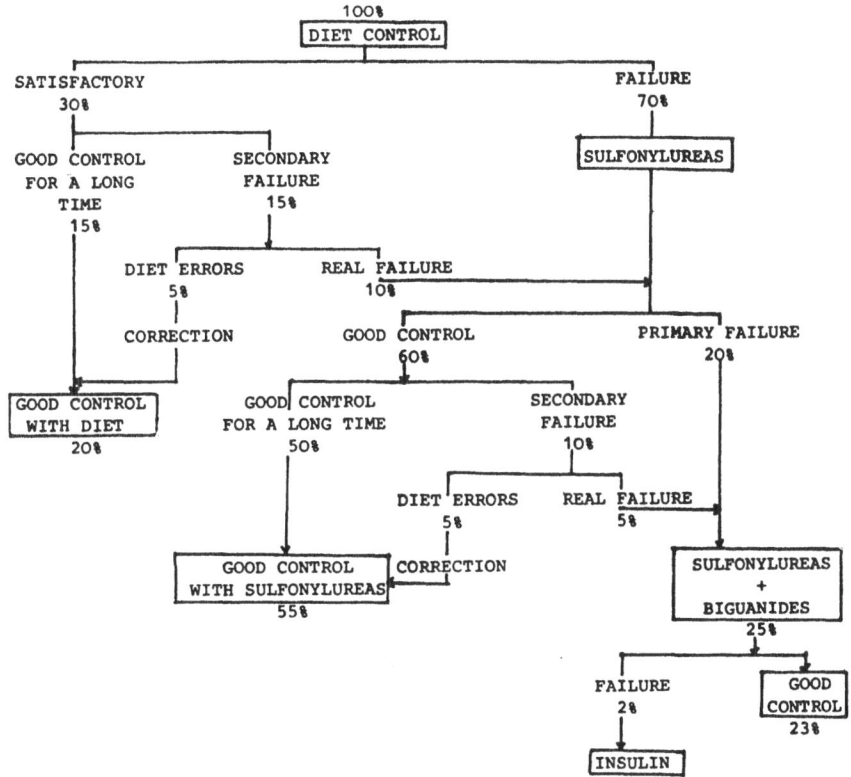

Fig. 1. Results of diabetes type II therapy in patients
         of all ages (From Romani[8] - modified)

On the other hand, by continuing a constant drug therapy, phases
of uncontrolled hyperglycemia will alternate with dangerous hypoglycemia. It is therefore necessary to call the attention of the
physician and health professional to their responsibility of repeatedly giving all due information to the patient and his relatives in
order to encourage them to participate in the therapeutic programme
in the most complete and rational way.

Finally, it is worth mentioning that there is no proof that the
administration of sulfonylureas in cases of glucose intolerance has
any preventive effect against a possible evolution towards diabetes.
Moreover, it should be recalled that, according to some research,
oral hypoglycemic agents seem to be responsible for an increased
death rate from acute cardiovascular diseases. Thanks to more
in-depth analyses, it has been shown that such statements are not
completely true; however, it is true that drug administration is at
least useless if not dangerous.

Table 2.   Sulfonylureas

Compound	Plasma half-life hr	Form excreted by kidneys	Single dose mg	Daily dose g
TOLBUTAMIDE	5 - 8	inactive metabolite	500	2 - 4
CHLORPROPAMIDE	36	parent drug	250	1 - 2 once
TOLAZAMIDE	8	active metabolite	200	1 - 3
ACETOEXAMIDE	2 - 5	active metabolite	500	1 - 2
GLICLAZIDE	12	inactive metabolite	80	1 - 3 once
GLIBENCLAMIDE	6	inactive metabolite	5	1 - 3
GLICICLAMIDE	-	-	500	1 - 3
GLISOLAMIDE	-	-	5	1 - 3
GLISOSSEPIDE	-	-	4	1 - 3
GLICODIAZINE (Glymidine)	5 - 8	active metabolite	500	1 - 3
GLIPIZIDE	2.5 - 4	inactive metabolite	5	1 - 3
GLIPENTIDE*	-	-	5	1 - 3
GLICOPIRAMIDE*	-	-	250	1 - 3
CHLOREXAMIDE*	-	-	250	1 - 3
GLIBORNURIDE*	8	inactive metabolite	25	1 - 2
GLUQUIDONE*	1.5	inactive metabolite	30	3 - 4
GLYBUZOLINE*	-	-	250	1 - 3

*(not available in Italy)

Table 3. Biguanides

Compound	Plasma half-life hr	Form excreted by kidneys	Single dose mg	Daily dose g
FENFORMIN	3	active metabolite	25	1 - 3
METFORMIN	1.5	active metabolite	500	1 - 3
BUFORMIN	-	-	100	1 - 4

Choice of Drug (see Tables 2 and 3)

     All authors agree to give preference to tolbutamide or its
derivatives as the drug of first choice. The initial dosage will
be the minimal, and will be gradually increased in the following
days and weeks to control patients' glycemia and weight. The
faster metabolization of these compounds and their administration
in more frequent doses during the day will prevent the risks of
hypoglycemia. If it is deemed useful to keep one single daily
administration (problems of health care, forgetfulness of old
patients, etc...) chlorpromazide can be administered, preferably in
the morning or at midday, so that it may have its hypoglycemic
effect during the day, when the patient is awake and has his meals
regularly.

     In the U.S.A., diguanides are almost unknown, while they are
rarely used as isolated hypoglycemic drugs in Europe. They were
reintroduced in cases of obesity and hypertriglyceridemia, and are
often used in association with sulfonylureas, for the enhancement
of their actions. Each single dose of these two drugs is usually
half the usual dose. These are drugs of second choice, which are
administered when the sulfonylureas have been shown to be insufficient to control diabetes or have given rise to side effects.(Table 4).

Therapy Assessment

     Therapy with oral hypoglycemic agents needs regular controls,
every 2 - 3 months. If in addition morbid phenomena occur,
especially of an inflammatory, vascular or infective nature, there
must be immediate or constant control. At each visit, the body
weight trend, both a weight increase and decrease, will have to be
taken into account. In the former case, the diet programme will

Table 4. Use of biguanides plus sulfonylureas

	each tablet mg		each tablet mg
FENFORMIN	30	CHLORPROPAMIDE	125
"	25	GLIBENCLAMIDE	2.5
"	12.5 - 25	"	2.5 - 5
"	30	"	2
METFORMIN	350	GLICLAMIDE	150
"	400	TOLBUTAMIDE	125

have to be changed, in the latter case, besides this measure, it will be necessary to find out other possible causes.

The basic tests will consist in the complete urine test and in the fasting blood sugar test.   If non-compliance to the dietary therapeutic plan on the part of the patient is suspected, a good diagnostic aid is glycosilate haemoglobin dosage, which informs us of any possible glycemic rate increase in the 10 - 15 days before the visit.   It should also be kept in mind that in elderly patients, the HbA1c is slightly higher when compared with the normal values of young patients.

During each visit, the patient shall be invited to report on any disorder or trouble and on the results of the glycosuria test, to be carried out daily or weekly by means of the appropriate dextrostick.   Special attention must be given to the early discovery of renal failure, which is often revealed before the chemical and biological urinary tests, by a blood pressure increase. In addition, the eye, as the site of early retinal vascular lesions and of lens opacity, should be examined at least once every year.

Hypoglycemia is a phenomenon which can for long be missed in elderly patients, as it escapes recognition due to its irregular and temporary character.   Hypoglycemia usually occurs during the night, with tremor, sweating, anxiety, hot and cold sensations, weakness and palpitations which could be followed by a confusional state.   The latter may not be noticed at all by the patient, or may be attributed by the doctor to other causes (cerebral vascular sclerosis, etc...).   Hypoglycemia usually occurs when the patient does not have his evening meal or is an alcohol abuser.   In the former case, it will be necessary to inform him of the possibility of replacing the meal with a certain amount of carbohydrate (sugar) or of not taking drugs without an adequate food intake.   Thus here is again the general rule of involving the patient and his relatives in a health education programme.

Administration of all other drugs must be reduced to the minimum dosage in order to avoid problems caused by reduced visual acuity, tremors, amnesia, difficulties in differentiating drugs. All these are quite common, and often lead to mistakes, especially if the patient lives alone.

Besides this general rule, it is known that some drugs induce an enhanced action of sulfonylureas, due to a competitive action on the carrier proteins.   The most commonly used drugs in obese and old patients are the following: clofibrate, dicoumarol, sulfafenazole, beta blockers, high dosage of salicylates and phenylbutazone.

It is therefore clear that if on one hand, the definition of senile diabetes rightly lost its specific nosological features, on the other hand, in view of its high incidence, its peculiar characteristics, and all relevant health care problems for elderly patients, this disease is of paramount interest, and should not be either overlooked or neglected.

REFERENCES

1. E. Greppi, L. Adezati, F. M. Antonini, and S. Signorelli, Diabete senile. Atti 60 Congr.Soc.Med.Int. L. Pozzi, ed., Roma (1959).
2. J. T. Freeman, "Clinical Features of the Older Patient," C. C. Thomas, Springfield, Illinois (1964).
3. U. Butturini, Alterazioni del Metabolismo glicidici nel vecchio, in: "Gerontologia e Geriatria vol. 2," F. M. Antonini and C. Fumagalli, eds., A. Wassermann (1976).
4. M. G. Fitzgerald, Diabetes, in: "Textbook of Geriatric Medicine and Gerontology," J. C. Brocklehurst, ed., Churchill Livingstone, Edinburgh and London (1973).
5. National Diabetes Data Group, Classification and diagnosis of diabetes mellitus and other categories of glucose intolerance, Diabetes 28:1039 (1979).
6. M. B. Davidson, The effect of aging on carbohydrate metabolism: a review of the English literature and a practical approach to the diagnosis of diabetes mellitus in elderly, Metabolism 28:688 (1979).
7. T. G. Skillman, Oral agents and the elderly diabetic, Geriatrics 34:41 (1979).
8. J. D. Romani, Etude comparative des traitments oraux du diabéte, Sem.Hôp.Paris 45:1768 (1969).

INSUFFICIENCY OF HAEMOACTIVE PRINCIPLES IN THE ELDERLY

AND ITS TREATMENT

A. Ventura, U. Senin and E. Mannarino

II Medical Clinic
University of Perugia, Italy

One of the most common problems and one of the most difficult to resolve in the elderly is that posed by haematological disorders.

Although the haemopoietic tissue conserves its mitotic activity throughout life, that is it is continuously renewed, certain haematological disorders of old age may arise from a progressive reduction of mitotic activity in the pluripotent haemopoietic stem cells. However, when Harrison[1] transplanted bone-marrow stem cells from old mice, he did not observe any loss of their reproductive potentiality. It would therefore seem that the origin of haematological problems of old age due to compromised haemopoietic tissue must be sought elsewhere.

An investigation on the uptake of colloidal Tc^{99m} by the bone marrow which demonstrated that the haemopoietic matrix is markedly reduced in old age,[2] excited much interest when first published and it was hypothesized that the degree of this reduction might be responsible for the increase of haematological disorders during old age. Today we know that such a reduction in the matrix does not lead to any significant drop in blood components below the normal range, but only to a drop in their reserve pool in the bone marrow.[3] Other sources or factors must, therefore, be responsible for deviating haemopoietic activity from the norm in old age.

Erythropoiesis starts with the uncommitted stem cell, passes through various metabolic and morphological stages and ends with the mature erythroblast. Erythropoietin is the regulator which initiates the first differentiation step, its function is to stimulate stem cell colonization and so, ultimately, to regulate the various stages of erythroblast maturation and proliferation and participate in haemoglobin synthesis.

189

Reports on the behaviour of the erythropoietin compartment in the aged are limited, but they all agree with Harrison's[1,4] finding in mice that erythropoietin production is normal and the response of the bone marrow to this regulator is within the norms. Other factors that intervene in the regulation of erythropoiesis at a later stage of maturation, beginning with the proerythroblast, such as vitamin B_{12} and folic acid, which play a part in the synthesis of nucleic acids and globulin, are, on the other hand, well-documented.

VITAMIN B_{12}

There is a mass of data on the vitamin B_{12} compartment in old age, but, partly due to lack of uniformity in the series and the diverse methods used, they do not always agree.[5] Taken together, however, they provide convincing evidence that the liver reserves do not undergo any significant changes during aging, but that the plasma content is subject to, sometimes considerable, alterations.[6] This fact stimulated considerable interest in cerebral deterioration in the elderly during the 60s and 70s, but attempts to treat this condition with vitamin B_{12} did not lead to any worthwhile results. However, administration of vitamin B_{12} to non-anaemic aged persons with low plasma levels of this vitamin markedly improved their psychological and physical performance.[7]

In cases of true vitamin B_{12} deficiency, where both liver and plasma levels of this compound are decreased, reduced uptakes due to a defect in gastrointestinal activity is thought to be the main underlying cause.[6] In this type of deficiency, the elderly patient develops classical pernicious anaemia with a characteristic picture of megalocytes and macrocytes, and occasional immature forms in the peripheral blood and of megaloblastosis in the bone marrow. The correction of this condition requires careful treatment with B_{12} at a dose rate of 1,000 gamma on alternate days for four days and then with a monthly maintenance dose of 250-1,000 gamma. This should be given parentally in order to by-pass the frequently present digestive defects.

FOLIC ACID

The analysis of the behaviour of folic acid in the plasma and liver of the elderly has given rise to divergent results. Only the research carried out by Varadi[8] on the erythrocyte content of folic acid has revealed constantly reduced levels, average 16 per cent in old age. However, the same holds true for folic acid as for vitamin B_{12} in that a drop in the levels of this acid does not seem to be implicated in specific haematological alterations. Only rarely has megoblastic anaemia exclusively due to folic acid deficiency been reported; this disorder normally requires an associated B_{12} deficiency. The therapeutic use of folic acid is reserved for rare cases of specific deficiency and it should be borne in mind that its use may aggravate the nerve damage so often present in this period of life.

VITAMIN B_6

Of the remaining B complex vitamins, increasing insight is being gained into the specific role vitamin B_6 plays in erythropoiesis. Its influence on the insertion of the Fe atom in the porphyrin nucleus suggests that a deficiency of this vitamin may induce an anaemic state. Rose et al[9] demonstrated constantly reduced blood levels of this vitamin, as well as of its active metabolite, pyridoxal 5'-phosphate, in elderly subjects. Such reduction does not usually give rise to haematological disorders. However, if because of either reduced intake or defective conversion to the active form, the vitamin B_6 levels fall below the critical threshold, 30 - 40 per cent of the norm, the uptake of Fe for haemoglobin synthesis is blocked and sideroblastic anaemia results. This is, however, susceptible to therapy with B_6.[10] Sideroblastic anaemia is characterized by an anomalous accumulation of Fe around the nucleus (ring cells) and in the mitochondrial crests of the erythroblasts. The therapy is parenteral pyridoxin at a dose rate of 50-250 gamma every two months. Desferrioxamine is also useful in dispersing the anomalous accumulation of iron.

IRON

We now arive at the last haematinic factor to intervene, Fe, since when this is absent there is no core on which to construct the haemoglobin molecule. The behaviour of Fe undergoes a change in old age. The tissue levels of iron increase and should act as a reserve; the elderly must, therefore, be protected against decreased intake.[11] However, despite the raised content, the iron is not easily detached by the administration of desferrioxamine and, consequently, a large part of the Fe is not available for haemoglobin synthesis, as has been shown by several researchers.[12,13,14] In fact, a higher incidence of iron-exchange disturbances exist in old age than in youth, 38.7 per cent against 14.3 per cent and most of these (68.4%) are due to iron deficiency. This defect is mainly the result of chronic blood loss, an uptake anomaly or reduced dietary intake.[11] Once such a deficiency is established, and this commonly happens, it passes through various metabolic and clinical stages. The first change is a drop in tissue reserves without any evident clinical outcome.[15] A further reduction quantitatively compromises the cytochrome-C and aconytase enzymes and results in the clinical picture known as "iron deficiency without anaemia":[16]

the main signs and symptoms of which are headache, fatigue, a loss
of physical and mental capacity, tiredness, depression, anorexia.
A still greater drop leads to a reduction in both enzymatic and
haemoglobin iron and the appearance of the clinical picture described
as "anaemia of the blood"[15] with the signs and symptoms of tiredness,
palpitations, buzzing in the ears, depression, confusion, anaemia,
transitory ischaemic attacks, pallor, tachycardia, dyspneoa and
oedema.   The morphological picture is hypochromatic and microcytic
elements in the peripheral blood and hyperplasia with a blocking
of the maturative processes in the bone marrow.   Should the
deficiency become even more pronounced, there is the clinical picture
of "anaemia of the tissues"[15] characterized by typical signs and
symptoms of glossitis, pharyngeal oesophagitis, gastritis, hypo-
achylia, enterocolitis, cheilitis of the corners of the mouth,
flaking of the nails and atrophy of the taste buds; all of which
indicate tissue asphyxia due to reduced oxygen availability.

Therapy for iron deficiency is, obviously, by intramuscular
injection of iron, at least in the initial stages, and then by oral
administration of delayed-release complex salts.

A further point that should be taken into account is that
reduced levels of haemoglobin and enzymatic iron can provoke
symptoms generally considered to be an expression of coronary,
cerebral and peripheral vessel involvement in the elderly.   Such
symptomatology is often mistakenly attributed to sclerosis of the
arteries rather than to an iron deficiency, especially when there
is a symptomatic picture of depression, mental confusion, angina
pectoris, cardiac insufficiency, myocardial infarction, intermittent
claudication and transitory cerebral ischaemic attacks.   Such an
error means that long-standing anaemia is often not recognized as
such and that dyspneoa, fatigue and oedema of the ankle are diagnosed
as symptoms of old age rather than early signs of a potentially
reversible anaemia.   Therefore, the physician should never lose
sight of the fact that these are often the presenting symptoms of an
iron deficiency.

REFERENCES

1. D. E. Harrison, Normal production of erythrocytes by mouse marrow
        continuous for 13 months, Proc.Natl.Acad.Sci.USA 70:3184
        (1973).
2. S. Ventura, U. Senin, E. Mannarino, Iron pathology in elderly
        people, in: "Nutritional Status of the Elderly," Excerpta
        Medica (1982) (in press).
3. D. E. Hyams, The blood, in:"Geriatric Medicine and Gerontology,"
        J. C. Brocklehurst, ed., Churchill Livingstone, London (1978).
4. D. E. Harrison, Defective erythropoietin responses of aged mice
        not improved by young marrow, J.Gerontol. 30:286 (1975).

5. L. Kass,"Pernicious anemia. Major problems in internal medicine," Vol. III. Saunders, Philadelphia (1976).
6. I. Bird, M. R. P. Hall, R. O. K. Schade, Gastric histology and its relation to anaemia in the elderly, Gerontology 23:309 (1977).
7. J. Pearce, E. Miller, "Clinical Aspects of Dementia," Bailliere-Tindall, London (1973).
8. S. Varadi, A. Elwis, Folic-acid deficiency in the elderly, Brit.med.J. ii:410 (1966).
9. C. S. Rose, P. György, M. Beutler, Age differences in vitamin B status of 617 men, Am.J.Clin.Nutr. 29:847 (1976).
10. P. Larizza, Anemie sideroacrestiche, Min.Med.55:2074 (1964).
11. L. Hallberg, H. G. Harwerth, A. Vannotti, eds., "Iron deficiency: Pathogenesis, Clinical Aspects, Therapy," Academic Press, London (1970).
12. S. Ventura, L. Coli, M. Cappelletti, Ricerche sulla desferiox-amina B.I. Haematologica 40:161 (1964).
13. S. Ventura, F. Orlandi, Eritropoiesis in porphyrias. International Symposium on the Normal and Pathological metabolism of Porphyrias, Saint Vincent, giugno 1965, Rass. Minerva Medica, (1965).
14. S. Ventura, A. Del Favero, L. Coli, U. Senin, G. Abbritti, Aspetti clinici e metabolici della porfiria cutanea tarda. Symposium aulla porfiria cutanea tarda. Roma, 24 ottobre 1970, Editrice Universo - Roma, (1970).
15. E. Beutler, V. F. Fairbanks, The effects of iron deficiency, in: "Iron in Biochemistry and Medicine II." A. Jacobs and M. Worwood, eds., Academic Press, London (1980).
16. A. Vannotti, "Iron metabolism - An International Symposium," Springer Verlag, Berlin (1964).

PLASMA LIPIDS AND FREQUENCY OF HYPERLIPIDAEMIA IN THE ELDERLY:

INDICATIONS AND COST/BENEFIT RATIO OF HYPOLIPIDAEMIC TREATMENT

R. Fellin,* G. Valerio,** T. Beroldin,**
A. L. Angelini,** O. De Candia,** G. Crepaldi*

* Institute of Clinical Medicine
  University of Padova, Italy

** Geriatric Hospital
   Padova, Italy

INTRODUCTION

In recent years substantial progress in the study of atherosclerosis has led to the definition of some now well known risk
factors, and hence to the possibility of carrying out realistic
programs of prevention and therapy. These risk factors include
the hyperlipaemias, common metabolic disorders of the plasma lipoproteins which may be primary with genetic origins, or secondary
to endocrine disease, organic disease, drug therapy, or dietary
habits. The interest in these alterations, however, has mostly
been focused on the young and middle age groups, overlooking the
elder age groups (Kritchevsky).[1] Moreover, the data supplied by
the Lipid Research Clinic Programs of the National Heart, Lung,
and Blood Institute (U.S.A.) refer to plasma lipid values for age
and sex for quinguennial groups, but only up to 70 years of age.[2]
Nonetheless, the increase in the population of elderly persons,
which has been recorded in these years, and the consequent growth
of geriatric pathology, raise, among others, the problem of
defining the metabolic alterations present in this age group, and
among these, the dyslipidaemias.

Variations in Plasma Lipids with Aging

One aspect of the problem regards the variations in plasma
lipids with age, a subject on which not all workers agree. In

an epidemiologic survey on 1118 subjects from 25 to 79 years of age[3]
it was recorded that women over 70 years of age had higher average
cholesterol and triglyceride levels than men.   Other studies that
did not go beyond 70-79 years of age have shown an increase in
cholesterol levels up to 60-69 years, and a slight decrease in both
men and women in the successive decade.[4]   According to Woldow,[5]
plasma cholesterol and triglyceride levels rise up to 70 years of
age, then fall, and at around 80 years of age they return to the
same levels found in the third decade.   Other workers[6],[7]report
instead that the triglyceride level tends to show a stable increase
with increasing age.

     The capacity to remove triglycerides from the plasma, as
evaluated on IVFTT (intravenous fat tolerance test), diminishes with
increasing age (Fellin et al).[8]   This result was upheld by Kekki,[9]
who studied over 100 subjects from 17 to 62 years of age, and found
a decrease in plasma lipoprotein lipase concentrations with advancing
age, as well as an age-correlated incapacity to eliminating TG from
the plasma.   This finding had been reported by Reaven[10] in an
experimental study conducted on rats, where the increase in TG levels
with age was attributed to a decreased capacity of the animals of
both sexes to remove TG from the plasma.

     However, data on the manner with which age influences the
mechanisms of cellular afflux and efflux of cholesterol, its system
of production and removal, are lacking.   Most of the cholesterol
is present in tissues with a poor cholesterol turnover, while the
rapidly exchangeable cholesterol pool represents less than 20 per
cent of the total body cholesterol; the metabolic slowdown, which
occurs in old age, seems to be followed by a delay in exchange and
turnover, with consequent cholesterol accumulation in the blood and
tissues.[1]   In any case, there is evidence of an increase in plasma
cholesterol with age significantly associated with a decrease in
maximal aerobic capacity in elderly subjects,[11]studies on cholesterol
LDL catabolism in the rat have shown that the removal of this
plasma lipid fraction is significantly reduced by age and by the
induction of hypothyroidism.[12]

     In advancing age, physical, hormonal, and enzymatic factors
concur in regulating the behaviour of the plasma lipids, and the
lack of definitive data in this regard is not surprising.

Prevalence of Hyperlipaemia in the Elderly Person

     The problem of lipid metabolism in the aged should also be
viewed in relation to cerebral or peripheral vascular manifestations of arteriosclerosis, and in regard to its possible correlations with particular longevity.   To this end, some workers have
studied the liporotein pattern of octogenarian kindreds, and
initially obtained a confirmation of the existence of familial

lipoproteinaemia (hyper-alpha, and hypo-beta) acting as a predisposing factor for longevity.[13] More recent investigations[14] have not upheld these findings, leaving the question open.

Various epidemiological studies, in part already mentioned,[5,6] have furnished evidence for the presence of dyslipidaemias in the elderly, reporting a frequency of 14 per cent in one group (diagnosis of hyper-lipoproteinaemia) and 30 per cent in the other (lipid pattern abnormalities). Horsey et al[15] noted that 58 per cent of his elderly patients with ischaemic heart disease had an abnormal lipoprotein pattern, compared to 70 per cent of controls.

However, it must be kept in mind that absolute quantitative definitions for hyperlipaemia do not exist; statistical definitions are employed,[16] and the limit for establishing when hyperlipidaemia begins is arbitrary. An evaluation of the lipid alteration must be correlated with the patient's life style, the presence or absence of associated disease, and in some cases, the distribution of the plasma lipid levels in family members.[2] In geriatric patients, it is reasonable to assume that forms of dyslipidaemia related to endocrine alterations (diabetes, hypothyroidism), dietary factors (excess caloric and/or alcohol intake), pharmaceutical factors (administration of oestrogens, glucocorticoids, thiazides, beta-blockers), and specific organic disease (liver, kidney disease) acquire greater significance. Therefore, the importance of the primary forms related to transmitted genetic disorders in lipid metabolism would lessen, but not necessarily disappear; it should instead be recalled that elderly patients have shown a particular capacity to tolerate genetic hyperlipaemia.[2]

Primary and secondary prevention of arteriosclerosis in man with lipid-lowering drugs

In the field of clinical pharmacology, the section on lipid-lowering therapy does not currently contain substances which are universally accepted as absolutely efficacious and devoid of side effects. On the other hand, interest in this type of preparation is kept alive by different, often contrasting results which in turn express favourable opinions or doubts on the drug used.

The need for an evaluation of the risk/benefit ratio in lipid-lowering therapy stems from some basic considerations. The first regards the demonstration that the increase in the plasma cholesterol levels constitutes a major risk factor for atherosclerosis;[17,18] the importance of hypertriglyceridaemia as a risk factor is not as clear. Epidemiologic surveys[19,20] have found an association between TG levels and the frequency of atherosclerotic lesions, but the mechanisms by which hypertriglyceridaemia favours the onset of these lesions have not been definitely explained. However, the finding that a TG increase accompanies a higher risk for athero-sclerosis has also been shown in prospective-type studies.[20]

The second consideration emerges from some recent observations indicating that primary prevention of atherosclerosis is possible through a reduction in plasma lipids, a reduction that may also be achieved by drugs.[21]   The third consideration derives from recent reports of severe, harmful effects probably related to the use of lipid-lowering drugs, and has thus increased the scepticism surrounding these preparations.   Nonetheless, the need for lipid-lowering treatment appears warranted in patients with hyperlipo-proteinaemia; the object is to carry out primary prevention of atherosclerosis in asymptomatic, hyperlipaemic patients, or to obtain a reduction in the number of infarcts (secondary prevention) in hyperlipaemic patients with a previous history of infarction. Some wide-scale studies concluded in the 1970's, despite variations in results, seem to indicate the possibility of reaching this goal.

Secondary prevention studies with clofibrate terminated in 1971 (The Scottish Study;[22] The Newcastle Study[23]) have demonstrated a reduction in deaths and in the number of re-infarcts in patients with a history of angina (with or without previous infarct); the effect was independent of the lipid levels, and the lipid-lowering effect of the drug.   The final results of the Coronary Drug Project[24] (secondary prevention) were published in 1975, and with regard to clofibrate (1.8 g/daily) did not uphold the findings of the Scottish and Newcastle Studies; the only positive result was a significant reduction in non-fatal infarcts in patients treated with nicotinic acid (3 g/daily).   This study also included treatment with two different dosages of oestrogens (5 mg and 2.5 mg/daily), and D-thyroxine (6 mg/daily).   Oestrogen therapy was interrupted because of an increased frequency of phlebothrombosis, pulmonary emboli, lung carcinoma, and non-fatal infarcts, compared to controls. D-thyroxine was halted because of a mortality increase due to myocardial infarct (Coronary Drug Project Research Group).[25,26] These three studies on the whole do not seem to support the 'lipid hypothesis' for atherosclerosis.

Primary prevention studies include the investigation by Krasno and Kidera[27] (clofibrate), the WHO survey[28] (clofibrate), the multicentric study with colestipol,[29] and the Clinics Type II Coronary Primary Prevential Trial (colestiramine) sponsored by the National HLI (USA), which is still in course.

Krasno and Kidera[27] studied the employees of an American airlines company, and showed that the frequency of non-fatal myo-cardial infarct was 1.8 per cent yearly in a group of subjects with an average age of 47.5 years treated with clofibrate for 39 months, while untreated age-matched controls showed a frequency of 6.6 per cent yearly.   In the multicentric study with colestipol[29] a significant difference in mortality from cardiovascular disease was observed after three years between hypercholesterolaemic subjects treated with colestipol (1.5%) and subjects receiving

placebo (4%).   In addition, Kuo et al[30] reported a reduction in
xanthomas and coronary atherosclerotic lesions in hypercholesterolaemic patients treated for 7 years with colestipol.

In the WHO-sponsored study,[28] about 5000 male subjects ranging
in age from 30 to 59 years with no signs of ischaemic heart disease
were treated for 5 years with clofibrate (1.6 g/daily). Besides a
9 per cent reduction in cholesterol levels, there was a 25 per cent
reduction in the number of non-fatal infarcts in the subjects with
higher cholesterol levels.   This study underlines once more the
importance of even a slight reduction in cholesterol levels for
the prevention of ischaemic heart disease in a random population.

The modality of lipid-lowering treatment depends closely upon
the characteristics of hyperlipaemia (Crepaldi et al).[31]
However, dietetic therapy represents the initial, obligatory step
before any type of pharmacological intervention.   Substitution of
the usual diet with an appropriate diet, in most cases, is sufficient for reducing or normalizing TG and cholesterol levels in
patients with hypertriglyceridaemia (types III, IV, and V); a diet
low in cholesterol and saturated fatty acids is not similarly
efficacious in subjects with hypercholesterolaemia (types IIa and
IIb) (Briani and Crepaldi).[32]   In these forms, in fact, a 5 - 15
per cent reduction in cholesterol may be obtained with diet. When
the decrease is not sufficient for bringing plasma cholesterol
levels back to normal levels, it becomes necessary to combine
drugs with the diet.

The ideal lipid-lowering agent should be able to reduce the
LDL cholesterol levels (atherogenic lipoproteins) (Goldstein and
Brown),[33] and increase (or at least not reduce) HDL cholesterol
levels, since an anti-atherogenic role has been assigned to this
lipoprotein class (Gordon et al).[34] A drug endowed with these
characteristics has not yet been found, even though several
substances are available which reduce CH levels significantly.
In addition, it has been shown that the combined use of different
drugs occasionally brings about better results, compared to the
use of higher doses of a single preparation. Furthermore, the
entity of the lipid-lowering response that may be achieved with
treatment (be it only dietetic or also pharmacologic) varies
remarkably in relation to the aetiopathogenesis of hyperlipoproteinaemia.   In fact, genetic forms respond less well to
treatment compared to environmental forms or, in any case, not
strictly genetic forms.

1) Exchange resins

Colestiramine and colestipol constitute the prototype of this
cetegory of drugs.   They are not absorbed in the intestine, and
their mechanism of action consists in binding the biliary acids,

Table 1. Most commonly used drugs in the treatment of hyperlipoproteinaemia

DRUG	DOSAGE	TYPE OF HYPERLIPOPROTEINAEMIA	SIDE EFFECTS	DRUG INTERACTIONS
Cholestiramine/ Colestipol	12-32 g/day	11a	– nausea – dyspepsia – constipation – intestinal occlusion – hyperchloraemic acidosis – hypertriglyceridaemia	reduced absorption of: – thyroxine – thiazide diuretics – digitalis glycosides – oral anticoagulants – phenobarbital
Clofibrate	1.5-2 g/day	11a, 11b III IV, V	– nausea – diarrhoea – weight gain – weakness – decreased libido impotence – skin rash – leukopenia – myopathy – cardiac arrhythmias – liver function abnormalities – cholelithiasis – increased incidence of various tumors	enhancement of the effects of: – oral anticoagulants – oral hypoglycaemic agents

DRUG	DOSAGE	TYPE OF HYPERLIPOPROTEINAEMIA	SIDE EFFECTS	DRUG INTERACTIONS

Nicotinic Acid	3-6 g/day	IIa, IIb
		III
		IV, V
			- cutaneous flush
			- pruritus
			- cutaneous hyper-pigmentation
			- nausea, dyspepsia
			- activation of peptic
			  ulcer
			- liver function
			  abnormalities
			- cardiac arrhythmias
			- toxic amblyopia
			- hyperuricaemia
			- hyperglycaemia
				synergy with:
				- thiazide diuretics
				
				enhancement of the
				effects of:
				- antihypertensive
				  drugs
				- ganglionic blocking
				  agents
				- vasodilators

thus increasing their excretion with the faeces.   In this way,
an interruption in the biliary acid enterohepatic cycle is
obtained, and an increase in cholesterol catabolism through this
metabolic pathway is provoked.   The mechanism by which the exchange
resins increase LDL removal is not clear, and may involve an
increase in the number or the activity of the peripheral lipoprotein
receptors.   Colestiramine is the drug of choice in the treatment
of the pure hypercholesterolaemias (type IIa) where a constant
effect is achieved with a 15 - 20 per cent decrease in plasma
cholesterol (Ryan and Jain;[35] Fellin et al;[36] Fellin et al[37]).
The side effects observed during treatment with anionic exchange
resins are summarized in Table 1.   These resins also bind some
anticoagulants, cardioactive glycosides, and thyroxine, and thus
interfere with their absorption.   Lastly, the possibility that an
increase in plasma TG may occur during treatment in some patients
should be kept in mind (Havel and Kane).[38]

2) Clofibrate

Although clofibrate is the lipid-lowering agent that has
been most widely used, its mechanism of action is not yet completely
clear; it seems to act on VLDL synthesis and catabolism (Grundy et
al).[39]   Clofibrate is efficacious in reducing cholesterol levels
in primary hypercholesterolaemias, and TG levels in endogenous
hypertriglyceridaemia (types III, IV, and V) (Crepaldi et al).[40,41]
However, its use in patients with hypertriglyceridaemia or mixed
hyperlipidaemias (type IIb) may bring about an increase in LDL
cholesterol (atherogenic lipoprotein fraction).   The most common
side effects of clofibrate are listed in Table I.   Although
significant side effects were not reported in the long-term study
mentioned above (Crepaldi et al),[40,41] it should be recalled that
prolonged treatment (and lipid-lowering therapy should theoretically last indefinitely) with this type of drug is not devoid of
risk.   The WHO cooperative study on primary prevention of
ischaemic heart disease (cited above) not only reported a
reduction in the incidence of non-fatal infarcts in treated
subjects, but also a greater number of deaths from liver, biliary,
and intestinal disease including neoplasia in the same subjects,
compared to controls.   This observation may have resulted from a
chance association, since the number of deaths was very small
(19 in about 5000 subjects), but it could be attributed to
clofibrate as such, or its effects, which favour steroid excretion
through the hepatic-biliary route.   Although a conclusive result
regarding the relationship between clofibrate and onset of
neoplasia could not be reached, the study, however, confirms a
significantly higher number of cholecystectomies for calculosis
among the treated subjects.

The latest study by this group of WHO workers (Oliver et al)[42]
reports the most recent findings regarding mortality in the

subjects treated.   The total observation period is now 9.6 years, of which 5.3 years were under trial and 4.3 after its conclusion. Hypercholesterolaemic subjects treated with clofibrate currently account for 25 per cent more deaths that non-treated subjects with the same CH levels.   Particular causes of death that might explain this difference have not been identified, so, it is possible that the higher mortality might be due to a long-term toxic effect of clofibrate, the consequence of a reduction in the body cholesterol pool, or even to a coincidence.

3)Nicotinic acid and derivatives

Nicotinic acid at high dosages (from 2 to 4 g daily) strikingly reduces plasma CH and TG levels (Parsons),[43]   This drug seems to be able to act at various steps of lipid metabolism; even if its antipolytic activity is assigned a basic role in its lipid-lowering effect (Carlson et al).[44]

The reduction of lipid levels is rapid, persistent and proportional to the dose; however, an increase in HDL has been observed during treatment with this preparation.   On the other hand, nicotinic acid exerts some side effects which make it poorly tolerable (cutaneous vasodilatation, pruritus, gastrointestinal disorders) (Table 1) (Parsons).[43] To obviate these inconveniences, some nicotinic acid derivatives have been prepared which undergo a slow conversion into nicotinic acid only after ingestion (beta-piridylcarbinol, exanthinol nicotinate, etc.).

Recently, nicotinic acid derivatives have been associated with clofibrate, and a good lipid-lowering effect has been observed (Briani et al).[45]   Nicotinic acid was the only drug that gave positive results in the Coronary Drug Project.[25]

Other drugs

Many other drugs have been tested for the treatment of hyper-lipidaemia (Table 2). D-thyroxine, the dextro-isomer of thyroxine is a powerful lipid-lowering agent (Searcy et al),[46] but may provoke the onset of signs of hyperthyroidism, and at high doses (over 6 mg/daily) it may bring about episodes of angina or myocardial infarct (Stamler et al).[47]   The use of this preparation is therefore limited to hypercholesterolaemic patients with no signs of vascular disease.

The search for substances that are able to exert better lipid-lowering effects with a continually minor incidence of side effects has kindled interest for new drugs that might be suitable for use in patients with hyperlipoproteinaemia. Among these, tiadenol, bezafibrate, procetofene, metformin, and probucol are worth mentioning.

Table 2. Other drugs used in the treatment of hyperlipoproteinemia

DRUG	DOSAGE	TYPE OF HYPERLIPOPROTEINEMIA	SIDE EFFECTSS	DRUG INTERACTIONS
Clofibrate derivates: - simfibrate - bezafibrate - procetofene or fenofibrate	1-1.5 g/day 0.4-0.8 g/day 0.1-0.4 g/day	IIb, III, IV, V IIa, IIb, III, IV, V IIa, IIb, III, IV, V	- gastrointestinal disturbances - weakness	
Tiadenol	2-3 g/day	IIa, IIb, IV	- gastrointestinal disturbances - increase in plasma transaminase activities	
Probucol	0.5-1 g/day	IIa	- diarrhoea, nausea - flatulence, abdominal pain	
β - sitosterol	9-18 g/day	IIa	- constipation - diarrhoea, nausea, vomiting	
D - thyroxine	4-8 g/day	IIa	- angina pectoris - cardiac arrhythmias - nausea, diarrhoea - hyperthyroidism	- enhancement of the effects of coumarine anticoagulants

DRUG	DOSAGE	TYPE OF HYPERLIPOPROTEINAEMIA	SIDE EFFECTIVE	DRUG INTERACTIONS
Neomycin	1.5-2 g/day	IIa	– nausea, diarrhoea – steatorrhoea – malabsorption – ototoxicity – nephrotoxicity	reduced absorption of: – digitalis glycosides – fat soluble vitamins
Heparin derivatives	18-36 mg/day	IV, V	——	——
Metformin	1.5-3 g/day	IV, V	– anorexia – vomiting – diarrhoea – B_{12} deficiency – hyperlactacidaemia	enhancement of the effects of: – coumarinanticoagul- ants – sulphonamides – phenylbutazone
P.A.S.	6-8 g/day	IIa	– gastrointestinal disturbances	——
Halofenate	0.5-4.5 g/day	IIa	——	——
Tibric acid	0.5-1 g/day	IV, V	——	——

Tiadenol reduces plasma cholesterol levels by acting mainly
on the LDL without inducing significant variations in alpha-
cholesterol (Baggio et al).[48]   It has also shown an important,
but less pronounced triglyceride-lowering effect (Crepaldi et al).[49]

Bezafibrate in type IIa patients has a cholesterol-lowering
effect equal or superior to clofibrate action, and this effect is
obtained with much smaller doses (600 mg daily) than are commonly
employed with clofibrate (2 g daily) (Crepaldi et al).[50] The CH-
lowering effect is due to a reduction in LDL-CH, while HDL-CH is
not involved.   In addition, bezafibrate exerts a pronounced TG-
lowering activity, which is superior on the average to clofibrate-
induced action.   In patients with hypertriglyceridaemia under
treatment with this drug, an increase in HDL-CH was observed.

Procetofene is a lipid-lowering substance which somewhat
resembles clofibrate in structure, but differs in its therapeutic
potency and specificity.   In fact, procetofene seems to be able
to correct both hypercholesterolaemia and hypertriglyceridaemia in
most patients with types IIa, IIb, and IV (Rouffy et al).[51] In this
context, metformin too is very interesting, since along with its
well-known anti-diabetogenic action, it exerts a good TG-lowering
effect (Sirtori et al).[52]

Probucol has a biphenol structure, and has been in commerce for
many years in the USA.   It has shown an important CH-lowering
activity in animals and man, but it does not act on TG levels
(Barnhart et al).[53] Important side effects have not been reported,
but high doses in dogs seem to favour the onset of epinephrin-
induced atrial fibrillation.

Hyperlipidaemia and Arteriosclerotic Complications in the Aged

The studies cited refer in every case to young or middle-aged
patients.   It remains to be defined if dyslipidaemia is also a
risk factor for the development of arteriosclerotic lesions in the
elderly patient as well; if, in other words, correlations exist
between the lipid pattern and arteriosclerotic disease in the
geriatric patient, and what is their significance.

The conclusions reached by several workers are contrasting.
Some stress the fundamental importance of dyslipidaemia control in
the prevention of arteriosclerosis, and specify that prevention
should start at an early age.   According to these workers,
screening for dyslipidaemia in the asymptomatic elderly person does
not seem warranted and the same may be said of therapeutic attempts
to lower plasma lipid levels in these patients, whereas there is no
reason to believe that an individual who is already symptomatic for
arteriosclerosis will benefit from lipid-lowering therapy.

In reference to the CH plasma fractions, it has been demonstrated that in the last decades of life, the LDL fraction falls
gradually from a peak reached at around 65 years of age, while the
HDL fraction is stable from 65 years onwards (Nicholson et al).[54]
Moreover, the importance assigned to the HDL fraction in predicting
the incidence of ischaemic heart disease should be kept in mind.
While the relationship between HDL-CH and coronary disease has been
widely documented, studies regarding HDL-CH and other forms of
arteriosclerosis do not abound (Fellin et al;[55] Eder and Gidez[56]).
From a review of the Framingham data and other studies, weak but
consistent negative correlations between HDL and peripheral
arteriosclerotic disease seem to emerge, as well as between HDL
and stroke (Gordon et al).[57]

According to Fellin et al,[58] the metabolic component
(diabetes and hypercholesterolaemia) in elderly women seems to play
a significant role in acute cerebral vascular disease.   Other
studies have shown a significant association between CH and TG rise,
and the presence of ischaemic heart disease and/or cardiac
arrhythmias in aged patients (Horsey et al).[15] Woldow[5] had previously
defined an increased risk of death from coronary disease in hyperlipidaemic individuals in their 8th and 9th decade of life, since
he considered dyslipidaemia as significant a risk factor in the
geriatric age as in the young age group, with consequent need for
treatment.   Schneider[59] correlated the HDL-CH levels of a group of
elderly patients (80 and 90 years of age) with the presence of
clinically proven peripheral arterial disease, and demonstrated
that this last is associated with lowest levels of HDL-CH.   This
finding, according to this worker, favours a rational basis for
prospective and prevention studies of dyslipidaemias in the aged
individual.

ORIGINAL CONTRIBUTION

In this review of lipids in old age, we wish to add a personal
observation with the intention of bringing new data to this field,
which, as we have seen, is still insufficiently studied.   To this
end, we carried out a retrospective investigation from the
clinical records of elderly, hospitalized persons.   The objective
of this study was to evaluate the cholesterol and TG levels in
relationship to age and sex, as well as the frequency of hyperlipidaemia and diabetes in this age population.

Material and Methods

The records of 1488 patients over 60 years of age (677 males,
811 females) who had been admitted to the 3rd Emergency Division
of the Geriatric Hospital of Padova from January 1, 1981 to July
31, 1982 were studied.   All clinical and laboratory data were

obtained from their clinical records.   The patients were separated
according to sex, and then divided into 4 age-clases; three of which
were closed (60-69, 70-79, 80-89), and one was open (90 and over)
(Table 3).   The following data were drawn from the clinical records:

> Laboratory: cholesterol and TG levels determined at admission,
> and electrophoretic pattern of plasma lipoproteins in hyper-
> lipaemic subjects.
> Clinical findings: history, diabetes mellitus, TIA or stroke
> (current or past), myocardial infarct (current or past),
> clinically evident peripheral arterial disease (claudication
> and/or ischaemic lesions), liver disease (acute hepatitis,
> cirrhosis, jaundice, alcoholic liver disease), neoplasia,
> thyroid disease, and nephrosis.

On the basis of arbitrary cut-off points, patients with a total CH
over 240 mg/dl were defined as hypercholesterolaemic; patients with
TG levels over 190 mg/dl were considered hypertriglyceridaemic.
Hyperlipoproteinaemias were typed by lipoprotein electrophoresis.

In a second phase of the study, we excluded all patients who
presented thyroid, renal, and neoplastic disease (see characteristics listed above), and selected a sub-population of 543 males and
695 females (total 1238), who were divided for sex, and separated
into age groups, omitting the fourth (open) group.   This sub-
population was then further sub-divided into four categories
(A, B, C, and D) as follows:

> A: diabetic patients without vascular complications;
> B: diabetic patients with vascular complications;
> C: non diabetic patients without vascular complications;
> D: non diabetic patients with vascular complications.

By 'vascular complications' we meant the presence of one or more of
the following pathologies in the clinical record: TIA; stroke;
myocardial infarct, and peripheral arterial disease.   Comparisons
between average CH and TG levels in each age group were carried out
by correlating the four categories as follows: A vs B, C vs D,
A vs C, B vs D.

Total cholesterol and plasma TG levels had been determined
enzymatically according to Roschlau et al[60] and Wahlefeld[61]
respectively (reagents were purchased from Boehringer, Mannheim).
Electrophoresis of the lipoprotiens had been performed using a
modification of Noble's method (Pagnan et al).[62]   Statistical
analysis of the data was accomplished using Student 't' test, and
the X^2 test.

Table 3. Cases, sex and age groups

AGE	MALES	FEMALES	TOT.
60-69	150	131	281
70-79	335	358	693
80-89	175	298	473
>90	17	24	41
tot.	677	811	1488

RESULTS

Figure 1 illustrates the average values (M+SE) of total cholesterol (CT) and plasma triglycerides (TG) for all the patients studied, sub-divided according to sex and age-group. Both lipid parameters show a clear tendency to decrease with increasing age, except for the last female class.   CT in the females is invariably significantly higher than in the males; the same tendency was observed for TG, but it was not statistically significant.

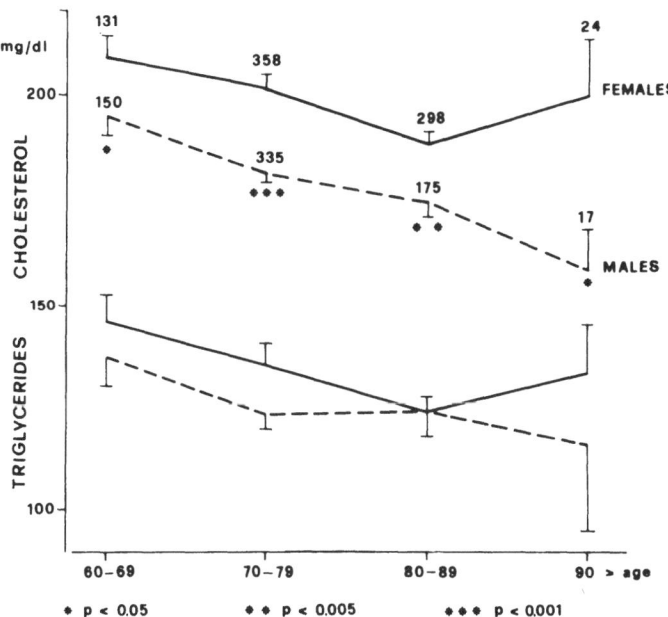

Fig. 1. Plasma cholesterol and triglyceride
         concentrations (mg/dl; m+s.e.) in
         677 males and 811 females divided
         into 4 age-classes.

Figure 2 shows the frequency rates of the various hyperlipidaemia phenotypes and diabetes for age, class, and sex.   A higher
frequency of hyperlipaemia and diabetes is evident in the females,
and the finding is nearly constant in all age classes.   Table 4
and Figure 3 summarize these parameters, which are separated only
according to sex.   The greater, statistically significant frequency
of type IIa hyperlipidaemia (p <0.01), hyperlipidaemias in toto

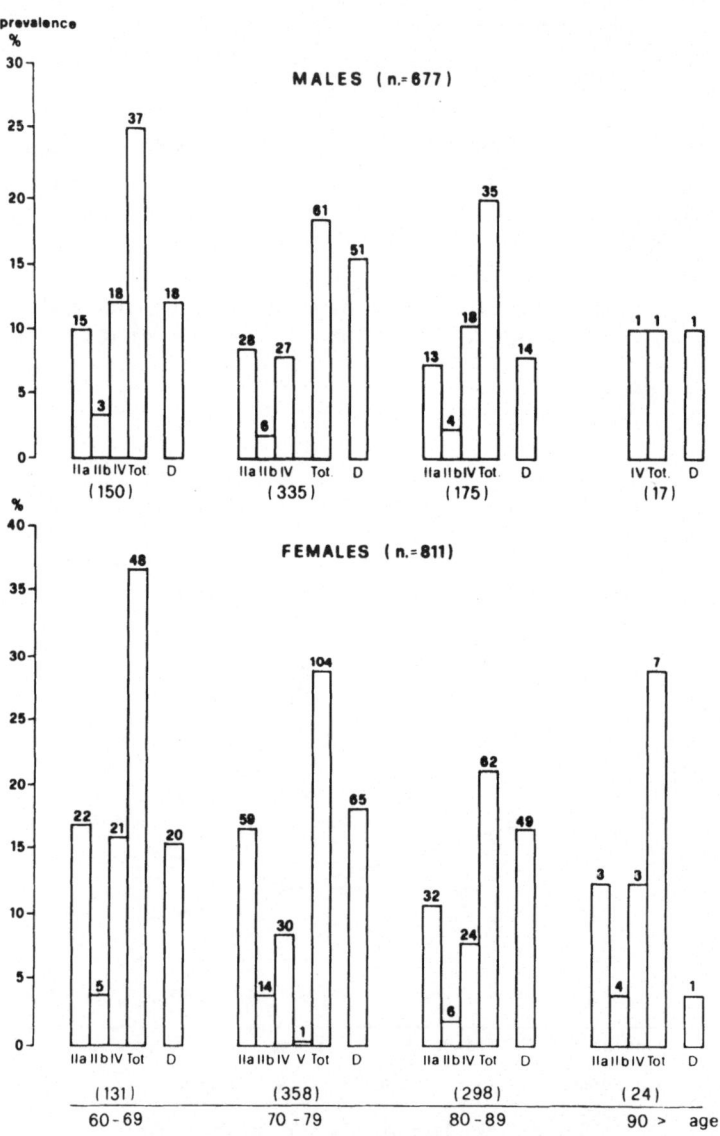

Fig. 2.   Prevalence of diabetes and of different
          phenotypes of hyperlipidemia in the
          patients divided according to sex and age.

Table 4. Prevalence of various phenotypes of hyperlipoproteinaemia in 1488 patients. Mean plasma values of cholesterol and triglycerides (m±S.E.,mg/dl) in different phenotypes distinct by sex; association (in percentage) of hyperlipaemia and diabetes in hyperlipaemic patients.

	Cholesterol		Triglycerides		Prevalence rate		Association with diabetes	
	M	F	M	F	M	F	M	F
Hypercholesterolaemia (IIa)	226±4	273±3	125±5	125±3	8.3%	14.3% *	16%	12%
Combined Hyper-lipidaemia	295±8	282±5	223±7	217±5	2.2%	3.2%	33%	19%
Hypertriglycerid-aemia (IV)	224±6	213±5	292±14	283±13	9.4%	9.6%	27%	32%
Total					20%	27.3% *	23%	20%

* $p < 0.01$

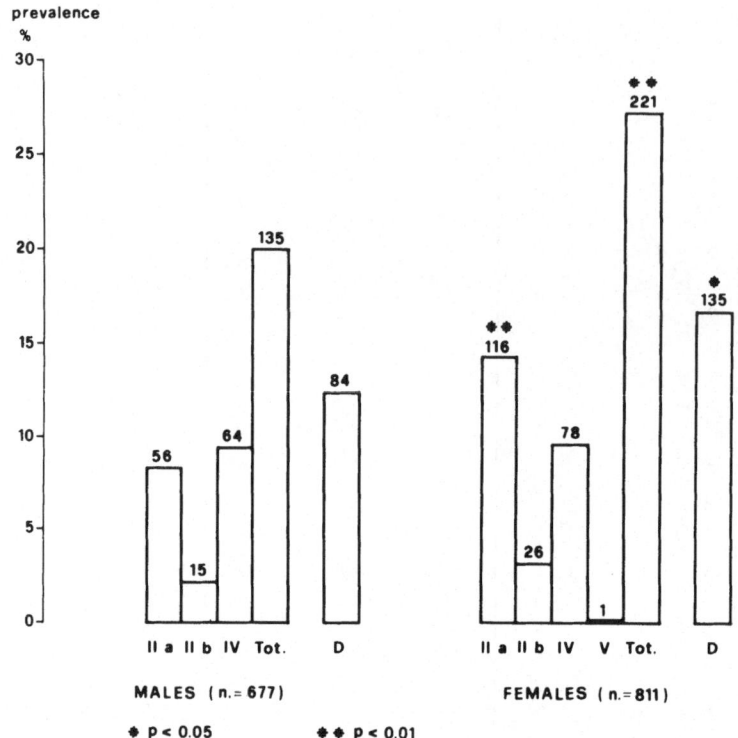

Fig. 3. Prevalence of diabetes and of different
         phenotypes of hyperlipidaemia in 677 males
         and 811 females over 60 years of age.

(p <0.01) and diabetes (p <0.05) in females (14.3%, 27.3%, 16.3%
respectively) compared to males (8.3%, 8.0%, 12.4% respectively) is
noteworthy.

    Phenotype IIa is more frequent in females, and type IV in males.
The average CT and TG levels in the different phenotypes found do
not show sex-related differences.

    Twenty per cent of the female, and 23 per cent of the male
hyperlipaemic patients are diabetic.   The frequency rates are
clearly higher than those found in the remaining population (15.4%
vs 11%), and statistically significant for the males (p <0.01).

    Figures 4 and 5 show the trend of CT and TG levels respectively,
in the sub-population selected with the previously described
criteria (A, B, C and D), and the statistical evaluation.   It
appears worthwhile to stress that CT and TG levels in group C are
almost constantly lower than in the other groups; moreover, lipid
values in groups A and C tend to be lower than in their corresponding comparison groups (B and D).

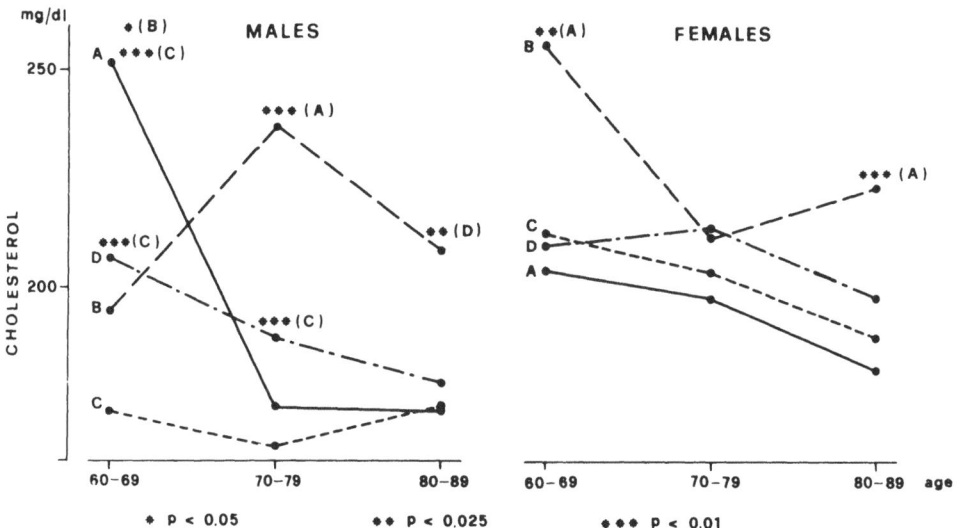

Fig. 4. Plasma cholesterol concentration (mg/dl; m±s.e.) in a
sub-population selected with the criteria indicated in
"Material and Methods".

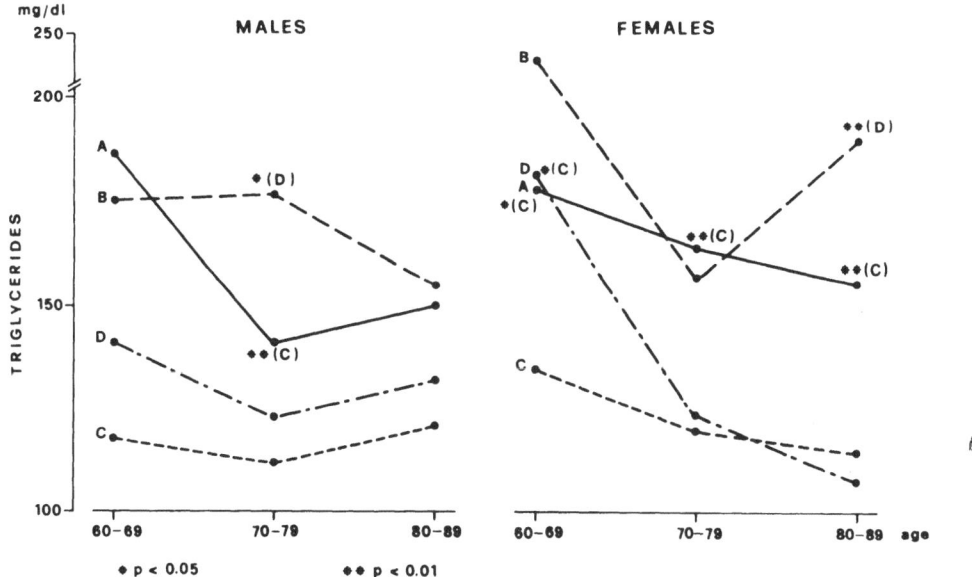

Fig. 5. Plasma triglyceride concentrations (mg/dl; m±s.e.) in
a sub-population selected with the criteria indicated in
"Material and Methods".

DISCUSSION

Only in recent years, and with an unexplainable delay, has
sufficient interest been shown by various workers in the vast chapter
on lipid metabolism in the aged.   In the international literature,
in fact, only lately have studies appeared which face the problem
of the physiopathology, epidemiology, and treatment of hyperlipidaemias in old age in a systematic manner.   The cause of all this
might be due to merely 'methodologic reasons' (Schneider),[59] or
more likely to the fact that many workers consider the hyperlipaemias a negligible risk factor in old age (Descovich et al).[63]
This attitude would have excluded frankly old people from many
epidemiological investigations, and would have contributed to a
certain laxity in checking lipid risk factors in aged persons.
The problem instead is wide-scale, and certainly deserves more
profound study; as Goto[64] notes, an evident increase in 'coronary'
deaths corresponds to aging of the population and to increase in
average lipid levels (diet).   The results of our study, which aimed
at contributing to the definition of the problem of geriatric lipid
metabolism, are in general agreement with recent observations of
several workers regarding the previously known tendency of lipid
values to decrease with aging (Woldow);[5] Harrill et al;[6] Nicholson
et al[54]).   It is also unanimously held that lipid values, and
specifically cholesterol, are constantly higher in aged females,
compared to males; in addition, the females show a greater frequency
of hypercholesterolaemia.   The finding of a high rate of diabetes
mellitus in the population studied, and its frequent association
with different hyperlipidaemia phenotypes is also meaningful.

After an initial denial of the problem of what clinical
significance should be attributed to hyperlipidaemia in the aged in
terms of 'vascular risk factor' (Kannel et al),[17,65]very recently
opposing conclusions have been reached.   Various workers, in fact
(Woldow;[5] Glueck et al;[13] Schneider;[59] Descovich et al[63]), concede
the same significance of the vascular risk factor of hyperlipidaemia
in geriatric patients that is accepted for younger age groups.
In this sense, our results are in perfect agreement with the findings of the workers cited.   In fact, in our series, patients with
vascular complications (classes B and D) almost constantly present
higher C and TG levels, compared to age peers without vascular
complications (classes A and C).   This result corresponds with our
previous finding (Buzzolo et al)[66] that CT and TG levels were
significantly higher in a group of 42 elderly patients with myocardial infarct than in a control group, and the hyperlipidaemic
frequency was 23.7 per cent.   While our study could not quantify
the duration of the hyperlipaemias detected, and thus evaluate their
aetiopathogenic role in the genesis of the vascular complications,
it does fully confirm their clinical significance also in elderly
patients.   It is hoped that further investigations will better
define the problem and contribute to the development of valid treatment schemes for the elderly patient.

CONCLUSION

Despite the various reports, and the recent interest in the regulation of lipid metabolism in the aged individual, and its clinical implications, studies regarding contingent treatment of dyslipidaemias in old age are very rare. Some workers (Cucinotta et al) [67]have judged the possibility of dietary treatment as very difficult and poorly operative in the elderly individual since social, economic, and psychological factors come into play. Although more recent investigations (Lees and Lees;[68] Levy;[69] Kane and Malloy[70]) into the clinical pharmacology of the dyslipid- aemias do not furnish precise data regarding the aged patient in order to evaluate the risk/benefit ratio, it seems to be useful to refer to the adverse effects and the pharmacologic interactions of the lipid-lowering drugs (Tables 1 and 2). If these effects are important in young and adult subjects, they assume greater relevance in old age, when the patient is subjected to more than one therapy, and is more vulnerable to adverse reactions and side effects. It is immediately clear how the side effects present in many of the above mentioned drugs tend to intensify a multiplicity of disorders often already present in the elderly patient, and there exist many possibilities of pharmacological interactions with drugs widely used in the treatment of the patient (diuretics, digitalis derivatives, antihypertensives). Thus, the potential risks related to drug treatment of dyslipidaemia in old age emerge, and explain at least in part, the reason for the lack of clinical studies on this topic.

If the drugs cited appear more contraindicated than perhaps necessary, dietetic therapy lends itself to different considerations, once the frequency of dyslipidaemia in the elderly and the necessity of treating it are defined on the basis of clinical and epidemio- logical findings. Diet, in fact, remains a corner stone in the treatment of dyslipidaemia (Conner and Connor;[71]Lees and Lees[68]).

It has been shown that diet (Harrill et al)[6] which reflects economic, educative and living (home or institution) factors, influences the plasma lipid levels in the sense that a lower intake of fats and calories accompanies a decrease in cholesterol and triglyceride levels in elderly patients. Nonetheless, systematic studies regarding the control of dyslipidaemia in the elderly population through the use of drugs or with dietary treatment are currently lacking.

From these considerations, it emerges that a thorough study of lipid metabolism in the aged individual is needed, particularly in relation to the cholesterol afflux-efflux balance in the cells, the enzyme situation, the hormone picture, and the variations with age in the HDL-LDL fractions. Moreover, an epidemiologic and clinical definition of the dyslipidaemias in old age should be

formulated with specific reference to the secondary forms and to
the possible study of 'survivors' of the primary forms.   In
addition, the weight embodied by the dyslipidaemias as atherosclerotic risk factors in old age should be completely and
critically evaluated.   Finally, the effects of diet should be
carefully recorded with a view to the more or less realistic
possibilities of its application for dyslipidaemia in this group
of subjects, compared to the use of lipid-lowering drugs.

REFERENCES

1. D. Kritchevsky, How aging affects cholesterol metabolism,
    Postgrad Med.63:133 (1978)
2. R. J. Havel, Symposium on Lipid Disorders, Med.Clinics N.Amer.
    66:319 (1982).
3. P. D. S. Wood, M. P. Stern, A. Silvers, G. M. Reaven and J.
    von der Groeben, Prevalence of Plasma Lipoprotein Abnormal-
    ities in a free-living population of the Central Valley,
    California, Circulation 45:114 (1972).
4. M. Werner, R. E. Tolls, and J. V. Hultin, Influence of sex and
    age on the normal range of eleven serum constituents,
    J.Clin.Chem.Clin.Biochem. 8:105 (1970).
5. A. Woldow, Hyperlipidemia and its significance in the aged
    population, J.Amer.Geriat.Soc. 23:407 (1975).
6. I. Harrill, C. Jansen, and J. Barthrop, Serum cholesterol and
    triglycerides and hyperlipoproteinemia in elderly women,
    J. Gerontol. 33:347 (1978).
7. M. S. Greenfield, F. Kraemer, T. Tobey, and G. Reaven, Effect
    of age on plasma triglyceride concentrations, in: "Man.
    Metabolism," 29:1095 (1980).
8. R. Fellin, F. Bellavere, E. Manzato, G. Briani, G. Baldo, and
    M. R. Baiocchi, Aspetti del metabolismo nel vecchio, Estratto
    dagli Atti del XXIII Congresso Nazionale Trieste, 27-29 Set-
    embre 1976, Giorn Gerontol. Suppl LIX:74 (1976).
9. M. Kekki, Plasma triglyceride turnover in 92 adult normolipaemic
    and 30 hypertriglyceridaemic subjects – the effect of age,
    synthesis rate and removal capacity on plasma triglyceride
    concentration, Ann.Clin.Research 12:64 (1980).
10. G. M. Reaven, Effect of age and sex on triglyceride metabolism
    in the rat, J.Gerontol. 33:368 (1978).
11. S. Matter, B. A. Stamford, and A. Weltman, Age, diet, maximal
    aerobic capacity and serum lipids, J.Gerontol. 35:532 (1980).
12. M. Sykes, W. M. Cnoop-Koopmans, P. Julien, and A. Angel, The
    effects of hypothyroidism, age, and nutrition on LDL
    catabolism in the rat, Metabolism 30:733 (1981).
13. C. J. Glueck, P. S. Gartside, P. M. Steiner, M. Miller, T.
    Todhunter, J. Haaf, M. Pucke, M. Terrana, R. W. Fallat, and
    M. L. Kashyap, Hyperalpha- and hypobeta- lipoproteinemai in
    octogenarian kindreds, Atherosclerosis 27:387 (1977).

14. H. Heckers, W. Burkard, F. W. Schmahl, W. Fuhrmann, and D. Platt, Hyper-alpha-lipoproteinemia and hypo-beta-lipoprotein- emia are not markers for a high life expectancy - Serum lipid and lipoprotein findings in 103 randemly selected nonagenarians, Gerontology 28:176 (1982).

15. J. Horsey, B. Livesley, and J. W. T. Dickerson, Aged men and ischaemic heart disease: Serum cholesterol and triglyceride levels, Age and Ageing 9:154 (1980).

16. J. L. Goldstein, and M. S. Brown, in: "Harrison's Principles of Internal Medicine," 407 IX edition, McGraw Hill (1980).

17. W. B. Kannell, W. P. Castelli, T. Gordon, and P. M. McNamara, Serum cholesterol, lipoproteins, and the risk of coronary heart disease, Ann.Intern.Med.74:1 (1971).

18. F. H. Epstein, Epidemiologic aspects of atherosclerosis, Atherosclerosis 14:1 (1971)

19. H. J. Albrink, J. W. Meigs, and E. B. Man, Serum Lipids, hypertension, and coronary artery disease, Am.J.Med.31:4 (1961).

20. L. A. Carlson, and L. E. Bottiger, Ischaemic heart disease in relation to fasting values of plasma triglycerides and cholesterol, Lancet 1:865 (1972).

21. R. Jr. Barndt, D. H. Blankenhorn, D. W. Crawford, and S. H. Brooks, Regression and progression of early femoral atherosclerosis in treated hyperlipoproteinemic patients, Ann.Intern.Med.86:139 (1977).

22. The Scottish Study (1971), A secondary prevention trial using clofibrate, Research Committee in: "The Scottish Society of Physicians," Brit.med.J.4:775 (1971).

23. The Newcastle Study, Trial of clofibrate in the treatment of ischaemic heart disease, (Physicians of the Newcastle upon Tyne Region), Brit.med.J. 4:767 (1971).

24. Coronary Drug Project Research Group, Clofibrate and niacin in coronary heart disease, J.Amer.Med.Assoc. 231:360 (1975).

25. Coronary Drug Project Research Group, Findings leading to discontinuation of the 2.5 mg/day estrogen group, J.Amer.Med.Assoc. 226:652 (1973).

26. Coronary Drug Project Research Group, The coronary drug project: findings leading to further modifications of its protocol with respect to dextrotyroxine, J.Amer.Med.Assoc. 220:996 (1972).

27. L. R. Krasno, and G. J. Kidera, Clofibrate in coronary heart disease, J.Amer.Med.Assoc. 219:845 (1972).

28. M. F. Oliver, J. A. Heady, J. N. Morris,and J. Cooper, A cooperative trial in the primary prevention of ischaemic heart disease using clofibrate, Brit.Heart J. 40:1069 (1978).

29. A. E. Dorr, K. Gundersen, J. C. Schneider, T. W. Sprencer, and W. B. Martin, Cholestipol hydrochloride in hyper- cholesterolemic patients-effect on serum cholesterol and mortality, J.Chron.Dis. 31:5 (1978).

30. P. T. Kuo, K. Hayase, J. B. Kostis, and A. E. Moreyra, Use of combined diet and colestipol in long-term (7-7.5 years) treatment of patients with type II hyperlipoproteinemia, Circulating 59:199 (1979).

31. G. Crepaldi, R. Fellin, G. Baggio, and E. Manzato, Drug treatment of primary hyperlipemias. Atherosclerosis Review 7:203 (1980).

32. G. Briani, G. Crepaldi, Le diete ipolipidemizzanti. Giorn. Arteriosclerosi 1: 147 (1976).

33. J. L. Goldstein, M. S. Brown, The low density lipoprotein pathway and its relation to atherosclerosis, Ann.Rev.Biochem. 46:897 (1977).

34. T. Gordon, W. P. Castelli, M. C. Hjortland, W. Kannell, and T.R. Dawber, High density lipoprotein as a protective factor against coronary heart disease, Am.J.Med. 62:707 (1977).

35. J. R. Ryan, and A. Jain, The effect of Colestipol or Cholestiramine on serum cholesterol and triglycerides in a long term controlled study, J.Clin.Pharmacol.-New Drugs, 12:268 (1972).

36. R. Fellin, G. Briani, P. Balestrieri, G. Baggio, M. R. Baiocchi, and G. Crepaldi, Long term effects of colestipol (U-26, 597A) on plasma lipids in familial type II hyperbetalipoproteinemia, Atherosclerosis 22:431 (1975).

37. R. Fellin, G. Baggio, G. Briani, M. R. Baiocchi, E. Manzato, G. Baldo, and G. Crepaldi, Long term trial with colestipol plus clofibrate in familial hypercholesterolemia, Atherosclerosis 29:241 (1978).

38. R. J. Havel, and J. P. Kane, Drug and lipid metabolism, Ann.Rew. Pharmacol. 13:287 (1973).

39. S. M. Grundy, E. H. Ahrens, and G. Salen, Mechanism of action of clofibrate on cholesterol metabolism in patients with hyperlipidemia, J.Lipid Res.13:531 (1972).

40. G. Crepaldi, R. Fellin, and G. Briani, Human hyperliprotienemia. Principles and methods, in: "Advances in experimental medicine and Biology,"R. Fumagalli, G. Ricci, S. Gorini, eds., Plenum Press, New York (1973).

41. G. Crepaldi, D. Fedele, R. Fellin, and G. Briani, Long term clofibrate treatment in familial hyperlipoproteinemias, in: "Atherosclerosis IV. Proceedings of the Fourth International Symposium on Atherosclerosis, West Berlin, Germany, October 24-28 1973," G. Schettler, and A. Weizel, eds., Springer Verlag, New York (1974).

42. M. F. Oliver, J. A. Heady, N. Morris, and J. Cooper, Cooperative trial on primary prevention of ischaemic heart disease using clofibrate to lower serum cholesterol: mortality follow up, Lancet 2:379 (1980).

43. W. B. Parsons Jr., Use of nicontinic acid to reduce serum lipids levels, J.Am.Geriat.Soc.10:850 (1962).

44. L. A. Carlson, L. Oro, and J. Ostman, Effect of a single dose of nicotinic acid on plasma lipids in patients with hyperlipoproteinemia, Acta Med.Scand. 183:457 (1968).

45. G. Briani, G. Valerio, R. Fellin, M. R. Baiocchi, G. Baggio, G. Baldo, E. Manzato, F. Beccaro, and G. Crepaldi, Trattamento delle iperlipoproteinemie primitive tipo IIa e IIb con l'associazione Clofirato + Xantiol/icoltinato, Giorn.Arterioscl.5:61 (1980).

46. R. L. Searcy, D. A. Hungerford, and M. Y. Lowe, Effects of dextrothyroxine on serum lipoproteins and cholesterol levels, Curr.Ther.Res. 10:177 (1968).

47. J. Stamler, M. Best, and J. D. Turner, The status of hormonal therapy for the primary and secondary prevention of athero-sclerotic coronary heart disease, Prog.Cardiovasc.Dis. 6:220 (1963).

48. G. Baggio, G. Briani, R. Fellin, S. Martini, M. R. Baiocchi, and G. Crepaldi, Effect of Tiadenol on the concentration and composition of serum lipoproteins in familial hypercholest-erolemias, Artery 5:486 (1979).

49. G. Crepaldi, G. Briani, U. Senin, U. Montaguti, A. Capurso, and A. Bondioli, Multicenter trial with tiadenol in primary hyperlipidemias, in: "International Conference on Athero-sclerosis, Milan Italy, November 1977," L. A. Carlson, R. Paoletti, and G. Weber, eds., Raven Press, New York (1978).

50. G. Crepaldi, R. Fellin, G. Baggio, and E. Manzato, Effect on lipid, lipoprotein and apoprotein levels of different hypo-lipidemic agents in type II hyperlipoproteinemia, in: "Drugs Affecting Lipid Metabolism," R. Fumagalli, D. Kritchevsky, and R. Paoletti, eds., Elsevier, (1980).

51. J. Rouffy, C. Dreux, Y. Goussault, R. Dakkak, and F. J. Renson, Antilipidemic Drugs. Part 5: Evaluation of the hypolipidemic effect of LF-178 in 191 patients affected by the atherogenic form of endogenous hyperlipoproteinemia (types IIa, IIb, III, and IV), Arzneim Forsch 26:901 (1976).

52. C. R. Sirtori, E. Tremoli, M. Sirtori, F. Conti, and R. Paoletti, Treatment of hypertriglyceridemia with Metformin: effectiveness and analysis of results, Atherosclerosis 26: 583 (1977).

53. J. W. Barnhart, J. A. Sefranka, and D. D. McIntosh, Hypocholest-erolemic effect of 4.4 (isopropydenedithio) bis (2.6-di-t-buthylphenol) (probucol), Am.J.Clin.Nutr. 23:1229 (1970).

54. J. Nicholson, P. S. Gartside, M. Siegel, W. Spencer, P. M. Steiner, and C. J. Glueck, Lipid and lipoprotein distribut-ions in octo- and nonagenarians,Metabolism 28:51 (1979).

55. R. Fellin, E. Manzato, F. Bernard, G. Valerio, M. R. Baiocchi, S. Martini, D. Tempesta, and G. Crepaldi, Fattori di rischio dell'arteriosclerosi periferica. Ruolo delle lipoproteine ad alta densita (HDL), Giorn.Gerontol.8:465 (1981).

56. H. A. Eder, and L. J. Gidez, Symposium on lipid disorder, Med.Clinics N.Amer.66:431 (1982).

57. T. Gordon, W. B. Kannell, and W. P. Castelli, Lipoprotein, cardiovascular disease and death, Arch.Intern.Med. 141:1128 (1981).

58. R. Fellin, R. Brentari, G. Valerio, S. Martini, E. Manzato, G. Baggio, A. Meggio, R. Mezzena, and G. Crepaldi, Studio dei fattori di rischio di TIA i ICTUS in uno popolazione geriatrica, Giorn.Gerontol. 11:805 (1982).

59. J. Schneider, High-density lipoproteins and peripheral vascular disease in octo- and nonagenarians, J.Am.Geriat.Soc.28:215 (1980).

60. P. Roschlau, E. Bernt, and W. Gruber, IX Congresso Intern. Chimica Clinica, Toronto 1975. Estratto n. 1 P. Trinder, Ann.Clin.Biochem. 6:24 (1969).

61. A. W. Wahlefeld in H. V. Bergmeyer: Methoden der enzymatischen Analyse. 3° Ed. Vol. II: 1878, Verlag Chemie Weinheim (1974).

62. A. Pagnan, R. Fellin, and G. Crepaldi, Studio elettroforetico delle lipoproteine plasmatiche su agarosio nei soggetti normali ed in diverse condizioni morbose, Acta.Med.Patav. 31:71 (1971).

63. G. C. Descovich, S. Rimondi, Z. Sangiorgi, A. Gaddi, M. S. Benassi, C. I. Cordaro, G. Copparoni, G. Cavallo, N. Trivelli, C. Ceredi, G. Dalmonte, A. Dormi, P. Perini, G. L. Magri, L. Aluigi, G. Mannino, and S. Lenzi, Lo studio di Brisighella: i fattori di rischio cardiovascolare nella popolazione sopra i 60 anni. Corso di aggiornamento sulla cardiopatia ischemica, Chieti 1981, S. Lenzi, F. Cuccurullo, P. Puddu, F. Possati, eds., Compositori, Bologna (1981).

64. Y. Goto, Hyperlipemia and atherosclerosis in Japan, Athero-sclerosis 36:341 (1980).

65. W. B. Kannel, T. Gordon, and T. R. Dawber, Role of lipids in the development of brain infarction: the Framingham Study, Stroke 5:679 (1974).

66. S. Buzzolo, P. Naliato, E. Tomat, S. Martini, R. Fellin, and G. Crepaldi, Fattori di rischio di infarto del miocardio nell'anziano, Giorn.Gerontol. 30:287 (1982).

67. D. Cucinotta, M. Mancini, G. Soncini, and M. Passeri, II Tiade-nolo nella terapia degli stati dislipidemici in senescenza, Giorn.Gerontol. 25:389 (1977).

68. A. M. Lees, and R. S. Lees, "Handbook of Drug Therapy," Miller and Greenblatt, eds., Elsevier, New York (1979).

69. R. I. Levy, "The Pharmacological Basis of Therapeutics," Goodman and Gilman, eds., MacMillan, New York (1980).

70. J. P. Kane, and M. J. Malloy, Symposium on lipid disorders, Med.Clinics N.Amer. 66:537 (1982).

71. W. E. Connor, and S. L. Connor, Symposium on lipid disorders, Med.Clinics N.Amer. 66:485 (1982).

METABOLIC BONE DISEASE AND FRACTURE OF THE FEMORAL NECK

A. N. Exton-Smith

Department of Geriatric Medicine
School of Medicine
University College, London

The two common forms of metabolic bone disease in old age are
osteoporosis and osteomalacia. As in the case of many other
diseases of old age they may occur together. Their clinical
importance lies in the fact that each condition leads to a
diminution in strength of the bone and consequent increased liability to fracture. The aim of this presentation is to assess the
role of metabolic bone disorder in the aetiology of femoral neck
fracture.

OSTEOPOROSIS

The amount of cortical bone in individual patients and the
changes which occur with advancing age can be most easily monitored
by means of cortical width measurements on standard radiographs.
Such measurements have been performed on the femur, tibia, clavicle,
radius and other bones[1] but the site most commonly chosen is the
mid-point of the second metacarpal.[2] The length (L) is measured
with a millimetre rule and at the mid-point the diameters of the
periosteal envelope (D) and of the medullary canal (d) are measured
with a Vernier caliper. Although cortical thickness (D-d) has been
used as an index, the transverse cross-sectional cortical area is
a better measure of tubular bone mass than cortical width. It is,
therefore, customary to calculate medullary and total area on the
assumption that the bone is cylindrical and derive a value for the
cross-sectional cortical area (CA). In the metacarpal this measurement has been shown to correlate well with the ash content as
determined by incineration.[3] A derived index relating cortical
area to surface area (CA/SA) has been used in our studies since this
ratio compensates for the differing skeletal size between individuals
and between the sexes.[4,5]

Aging in the Skeleton

Using the ratio cortical area/surface area percentile ranking curves for the normal population of men and women between the ages of 2 and 85 have been constructed.   These curves not only indicate the variance but also the distribution about the median values (see Figure 1).

The features of bone development and loss as revealed by these measurements are:

1.   The curves for the percentile ranks (10, 25, 50, 75 and 90) remain roughly parallel with age; that is, the variance remains unchanged with age and the distribution at each age group is normal.
2.   There is a rapid increase in the amount of bone during the period of growth up to the age of 17, but the increase continues at a slower rate for another 12 years or so after increase in height has ceased.
3.   Loss of bone is steady after about the age of 45, occurring more rapidly in women than in men.
4.   There are some individuals in old age who have more bone than others at the time of skeletal maturity.
5.   If there is a critical level of bone mass below which the clinical manifestations of osteoporosis develop then individuals with an amount of bone in the lower percentile ranges will reach this level at an earlier age than those in the higher percentiles. Indeed, individuals with bone mass in higher percentile ranges, even though they lose bone in middle and old age, may never reach the critical level.

The main practical use of the percentile ranking curves is that by measurement of metacarpal cortical index (CA/SA) the skeletal status of the individual patient can be determined and related to that of the general population of corresponding age and sex.

Amount of Bone in Old Age

Osteoporosis is not a single disease entity but it is the end result of a number of processes which lead to a diminution of the amount of bone in the skeleton.   Some of the factors concerned are:

Amount of bone at maturity.   Although some reservation must be placed on the interpretation of cross-sectional data, the fact that the percentile ranking curves remain parallel with age suggests that the amount of bone in adult life and old age is determined by the skeletal status at maturity.   Those individuals who have an amount of bone in the higher percentile ranges can afford to lose bone with aging and never develop the clinical manifestations of osteoporosis during their lifetime.   Skeletal development in earlier life may be

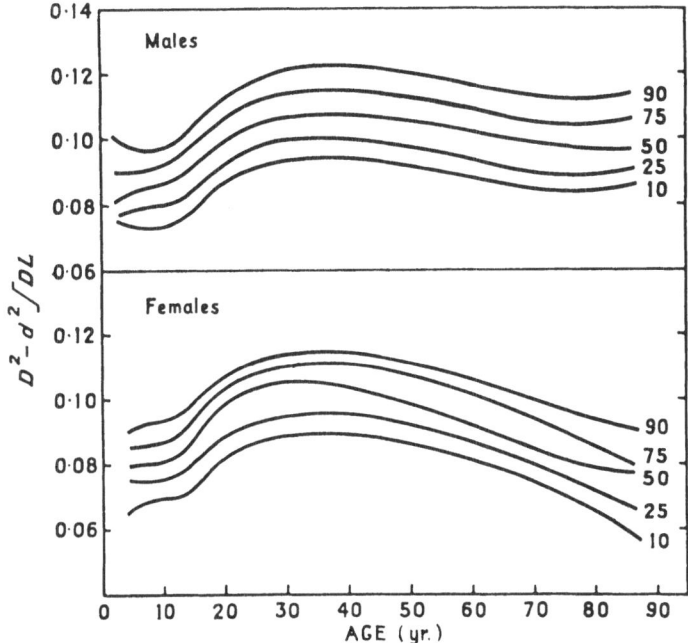

Fig. 1.   Percentile ranking curves for the amount
          of bone in males and females aged 3 - 82
          years

influenced by physical activity, disease, nutrition and by endocrine,
racial and other factors.

Poorer skeletal development of girls. By the age of 17 the
cortical area/bodyweight ratio is 20 per cent lower in girls than
in boys.[5]   This relatively lower bone mass in girls may be one of
the factors accounting for the higher incidence of osteoporosis in
older women.

Excessive bone loss. The most important physiological factor
responsible for increased bone loss in women during middle life is
the menopause.   This rapid loss occurs in the first five to ten
years after the menopause and the rate of loss can be reduced by the
administration of oestrogen.   Excessive bone loss also occurs in
a number of pathological conditions, for example, following partial
gastrectomy, immobilisation and the administration of corticosteroids.

Whatever be the relative importance of these factors responsible
for excessive bone loss it seems likely that people with densely
calcified bones in early life will have some protection against
developing the clinical syndrome of osteoporosis when, in later life,
part of the bone mineral is lost.

OSTEOMALACIA

Osteomalacia is a generalised disease of bone characterised by deficient calcification of a normal bone matrix. Histological examination reveals an increase in the amount of osteoid, that is, non-calcified matrix around the bony trabeculae. It is a disease produced by lack of vitamin D.

Osteomalacia is not confined to any age group, but in old age it is not uncommon since there are many conditions which can give rise to it. Anderson and his colleagues[6] in Glasgow found an incidence of 16 per cent in a group of elderly women who had been newly admitted to a geriatric department and who had possible clinical indications of osteomalacia (see Table 1).

In old age there may be multiple factors which lead to vitamin D deficiency and consequent production of osteomalacia. In any one individual several factors may co-exist and the more important are:

Lack of Exposure to Sunlight

Stamp and Round[7] have shown seasonal variations in both young and old subjects in plasma levels of 25-hydroxyvitamin D when measured by radiostereo assay techniques. Old people have much

Table 1 Possible indications for diagnosis of
        osteomalacia (Anderson et al, 1966)

Vague and generalised pain

Low backache

Muscle weakness and stiffness

Waddling gait

Skeletal deformity

Bone tenderness

Malabsorption states

Long confinement indoors

Malnutrition

---

lower levels than those found in younger subjects.   The conclude
that summer sunlight is an important and possibly the chief
determinant of vitamin D nutrition in Britain.   Stephen[8] has
analysed the effects of age on the winter and summer levels of 25-
hydroxyvitamin D based on estimations made on 960 old people living
at home.   The regressions on age (between 60 to 95 years) showed a
linear decline.   The summer levels were higher than the winter
levels in the younger age groups but the rate of decline was faster
in the summer levels and by the age of 95 the summer levels were
approximately equal to the winter levels in each sex.   These findings emphasize the importance of sunlight exposure which diminishes
with advancing age.

Dietary Vitamin D Deficiency

    The housebound elderly represent the largest single group
especially at risk of developing vitamin D deficiency.   In addition
to lack of sunlight exposure they often have very low nutrient
intakes; Exton-Smith, Stanton and Windsor[9] in a nutritional study
of the housebound showed that the vitamin D intakes were significantly lower than those of active old people.

Malabsorption

    Conditions causing malabsorption are not uncommon in old age.
In some cases the features of osteomalacia can overshadow all other
manifestations of malabsorption.   Enteropathy due to rye and wheat
gluten sensitivity similar to that causing childhood coeliac disease
is still seen in patients in their seventies.   The stagnant loop
syndrome due to jejunal diverticulosis is a well recognised cause of
osteomalacia.   Duodenal diverticula are usually regarded as
innocuous but in a series of 15 elderly patients with large primary
duodenal diverticula reported by Clark[10] two had osteomalacia.

Gastrectomy

    Osteomalacia may develop five to ten years after gastrectomy
and it is a complication of the polya-type gastrectomy more
frequently than of the Billroth type.   It is believed to be due to
a diminished intake and/or defective absorption of vitamin D.   The
development of osteomalacia, however, is not related to the severity
of the steatorrhoea which in some cases is minimal or absent.

Hepatic Disorders

    Liver and biliary tract disease can lead to osteomalacia due to
impaired absorption of vitamin D and to a reduction in its conversion
to 25-hydroxyvitamin D.

Renal Impairment

Fraser and Kodicek[11] drew attention to the importance of
adequate renal function for the conversion of 25-hydroxyvitamin D
to the active 1,25-dihydroxyvitamin D.   Thus osteomalacia can occur
in renal failure but it is uncertain to what extent impaired metabolism of vitamin D can be attributed to an age-related physiological
decline in renal function, which is believed to affect all individuals.

It is important to make a diagnosis of osteomalacia and to
distinguish it from osteoporosis with which it is so often confused.
Unlike many other conditions in old age osteomalacia readily responds
to treatment by correcting vitamin D deficiency.

FEMORAL NECK FRACTURE

Alffram[12] made an epidemiological analysis of fracture of the
hip involving 1664 cases observed over a 13 year period in the population of Malmo.   In both males and females the incidence was
negligible below the age of 50 and it apparently doubled for each 5
year increment after the age of 60.   The incidence in females was
2.4 times that in males.   It is generally recognised that osteoporosis is the major metabolic bone disorder accounting for the
rising incidence of femoral neck fracture with age.   More recently
it has been shown that a proportion of patients with fracture of the
femoral neck have increased amounts of osteoid as revealed by
histological examination of bone biopsy specimens.

Aaron and her colleagues[13] in Leeds have shown that 20 to 30
per cent of women and about 40 per cent of men with fracture of the
femoral neck have histological evidence of osteomalacia on bone
biopsy.   Later they showed[14] that the proportion with osteomalacia
varied with the season of the year.   The highest proportion of
histological specimens with abnormal calcification fronts (43 per
cent) was observed in February to April and the lowest (15 per cent)
in October to December.   The highest frequency of abnormal osteoid
covered surfaces (47 per cent) was observed in April to June and
the lowest (13 per cent) in October to December.   These authors
concluded that variation in hours of sunshine is responsible for the
seasonal change in the incidence of femoral neck fractures and
possibly for osteomalacia in the elderly population as a whole.
The significance of vitamin D deficiency as an important factor in
fracture of the proximal femur has been confirmed in our own studies[15]
even though the classical features of osteomalacia are absent.

We have further investigated the relative contributions of
osteoporosis and osteomalacia to the etiology of fracture of the
femoral neck.[16,17] An examination was carried out on 384 patients

(70 males and 314 females) with fracture of the proximal femur admitted to two hospitals in London and three in Manchester. Their ages ranged from 50 to 100 years with a mean of 76 years for the males and 79 for the females.

Metacarpal Cortical Bone

The degree of osteoporosis was assessed by calculation of the individual values for the metacarpal cortical index for the fractured femur patients using morphometry of the second metacarpal. The regression of the mean MCI values against age have been plotted on the percentile ranking curves for the general population. For female fracture patients the regression line lies between the 25th and 50th percentiles for the general population and for male patients between the 10th and 25th percentiles. In each sex the mean of bone in fractured femur cases was found to be significantly less than that for the general population of similar age (P <0.001). The absolute values of MCI in individual fractured femur cases can be expressed as percentiles of the values for the general population of corresponding ages, thereby eliminating the effect of age. The resultant frequency distributions are shown in Figure 2. The distributions are skewed towards the lower percentile values, particularly in male patients.

Trabecular Osteoid Area

The mean proportion of iliac trabecular bone matrix area occupied by osteoid in a series of microscopic fields was calculated. The values have been expressed as osteoid area percentages for fractured femur and control cases shown in Table 2.

Table 2 shows that there is a significantly greater amount of osteoid in fractured femur biopsy specimens than in those of controls, particularly in men. The mean osteoid area for the male fractured femur patients was significantly higher P < 0.05) than that for the female patients. Inspection of the distribution of the trabecular osteoid areas in the control series showed an asymptotic curve tending towards zero cases at the 2.0 per cent trabecular osteoid level. This level of osteoid area was therefore taken as the upper limit of normal. Significantly more (P <0.01) of fractured femur patients, both male and female, had excessive osteoid according to this definition compared with the numbers of controls.

Metacarpal Cortical Bone and Trabecular Osteoid Area

The fractured femur patients with osteomalacia (trabecular osteoid area > 2 per cent) had mean MCI percentile values of 45.8 for the males and 39.4 for the females; corresponding values for patients without osteomalacia were 1.5 and 34.0. Thus, patients with osteomalacia tended to have higher MCI values (that is, they

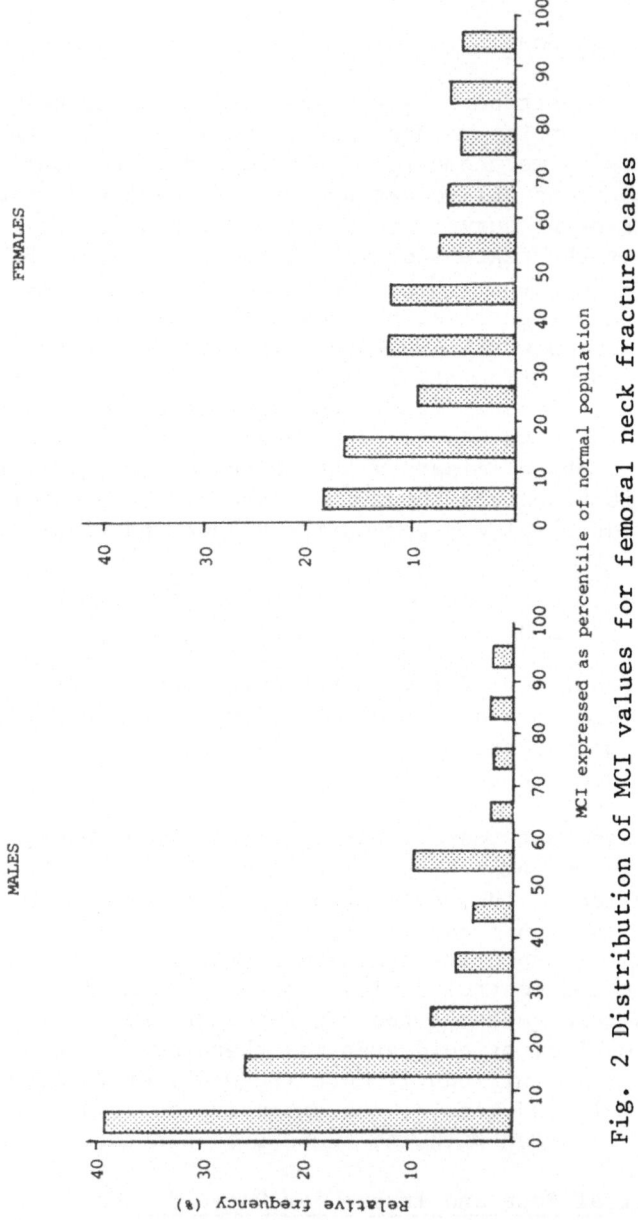

Fig. 2 Distribution of MCI values for femoral neck fracture cases

Table 2. Osteoid area percentage for fractured femur and control case

	Fractured Femur	Controls	$P <$
Males	5.60	0.71	0.002
Females	2.54	0.44	0.05
All subjects	2.96	0.59	0.004

had less osteoporosis) although the differences were not quite significant.

Conclusion

Although osteoporosis is of major importance in the etiology of this type of fracture there are some individuals who sustain fractures with MCI values in the upper percentile ranges (see Figure 2). These findings are difficult to interpret on the basis of cross-sectional studies. It is possible that there has been a marked decline in the quantity of bone if their skeletal status was very good at maturity. We shall not know whether this is the explanation until we have information on the rate of bone loss derived from longitudinal studies. Such studies are also required to determine the cause of the greatly diminished bone quantity commonly found in fractured femur cases.

Osteomalacia is an additional factor in the pathogenesis of femoral neck fracture and there is evidence that it may play a more important role in those patients with relatively little osteoporosis. For the fracture series the mean trabecular osteoid area was significantly greater than that in controls and a significantly higher proportion of them had osteomalacia with an iliac trabecular osteoid area greater than 2.0 per cent. The incidence of osteomalacia in our studies of patients with fracture of the proximal femur is similar to that reported in Leeds by Aaron and her colleagues.[13] It seems likely that the impaired vitamin D status of patients with femoral neck fracture is a reflexion on the decreased out-of-doors activity of the patients prior to their fracture.

REFERENCES

1. A. B. Horsman, Bone Mass, in: "Calcium, Phosphate and Magnesium Metabolism," B. E. C. Nordin, ed., Churchill Livingstone, London (1976).
2. E. Barnett and B. E. C. Nordin, The radiological diagnosis of osteoporosis, Clin.Radiol.2:166 (1960).
3. C. I. Gryfe, A. N. Exton-Smith, and R. J. C. Stewart, Determination of the amount of bone in the metacarpal, Age & Ageing 1:213 (1972).
4. A. N. Exton-Smith, P. H. Millard, P. R. Payne, and E. F. Wheeler, Pattern of development and loss of bone with age, Lancet, ii:1154 (1969).
5. C. I. Gryfe, A. N. Exton-Smith, P. R. Payne, E. F. Wheeler, Pattern of development of bone in childhood and adolescence, Lancet i:523 (1971).
6. I. Anderson, A. E. R. Campbell, A. Dunn, and J. B. M. Runciman, Osteomalacia in elderly women, Scot.Med.J. 2:429 (1966).
7. T. C. B. Stamp, and J. M. Round, Seasonal changes in human plasma levels of 25-hydroxyvitamin D, Nature 247:546 (1974).
8. J. M. L. Stephen, Unpublished data, 1980.
9. A. N. Exton-Smith, B. R. Stanton, A. C. M. Windsor, Nutrition of Housebound Old People, King Edward's Hospital Fund, London (1972).
10. A. N. G. Clark, Deficiency states in duodenal diverticular disease, Age & Ageing 1:14 (1972).
11. D. R. Fraser and E. Kodicek, Unique biosynthesis by kidney of a biologically active vitamin D metabolite, Nature 288:764 (1970).
12. P. A. Alffram, An epidemiological study of cervical and trochanteric fractures of the femur in an urban population, Acta.orthop.scand.Supp. 65 (1964).
13. J. E. Aaron, J. C. Gallagher, J. Anderson, L. Stasiak, E. B. Longton, B. E. C. Nordin and M. Nicholson, Frequency of osteomalacia and osteoporosis in fractures of the proximal femur, Lancet i:229 (1974).
14. J. E. Aaron, J. C. Gallagher, and B. E. C. Nordin, Seasonal variation of histological osteomalacia in femoral neck fractures, Lancet i:84 (1974).
15. J. M. Faccini, A. N. Exton-Smith, and A. Boyde, Disorders of bone and fracture of the femoral neck, Lancet i:1089 (1976).
16. J. C. Brocklehurst, A. N. Exton-Smith, S. M. Lempert-Barber, L. P. Hunt, and M. K. Palmer, Fracture of the Femur in old age: A two-centre study of associated clinical factors and the cause of the fall, Age & Ageing 7:7 (1978).
17. P. J. Cook, A. N. Exton-Smith, J. C. Brocklehurst, and S. M. Lempert-Barber, Fractured Femurs, Falls and Bone Disorders, J.Roy.Coll.Physns. 16:45 (1982)

THE PROPHYLAXIS OF SENILE OSTEOPOROSIS

M. Passeri and E. Palummeri

Department of Geriatrics and Gerontology
School of Medicine, University of Parma
"G. Stuard" Hospital, Italy

It is well known that in elderly people there occurs a gradual loss both in trabecular and in compact bone substance; this situation is clearly observed in bone biopsy, in radiological pictures, and by densitometry. The continuous bone turnover (both as regards bone proteins and mineral content) takes place throughout life, even in later years, although at a slower rate. The clear relationship between phospho-calcium balance and bone is based on this continuous turnover and renewal process, so that during the whole life-span[34] the skeleton represents a great store, where the organism can deposit and can take out minerals, according to the needs of the individual.[4,10,12,21,24,29,44,46,62,65,67]

PATHOGENESIS OF BONE LOSS

It is by now well known that in the elderly organism the bone resorption processes are not completely compensated by their accretion counterpart; the consequence is manifested as the above mentioned bone loss (osteopenia), which does not uniformly affect the whole skeleton. It occurs first in trabecular bone, and afterwards in the compact bone. The spine is affected more severely, especially in women, than the peripheral skeleton.[17,27,40,47,79]

Bone loss is caused by genetic and racial factors, by local aging processes of the tissue (in particular the alterations in protein synthesis), by senile extra-bone organic function, which we will return to later, and also by environmental situations and style of living.[10,11,67,68] It must be made clear that this is not a pathological situation, but a normal state: that is physiological aging. The skeleton meets metabolic needs of the elderly subjects and fits its support function.

231

Aging may cause functional disturbances only when bone structure, owing to extreme old age, becomes so deficient that it does not allow a normal functioning, with regards to the capacity of holding up the body. Apart from the occasional case which reaches this extreme stage, elderly people do not feel the shrinking of their bone mass, when it remains within the limits of a "senile physiological osteopenia".

The organism copes with a very different situation when the bone loss is excessive or too rapid, according to age. In this case the bone structure is no longer able to be completely functionally efficient.[10,68] Metabolic function, which balances the phospho-calcium turnover and maintains acid-base equilibrium by the release of bases (Na^+, K^+, Ca^{++}, Mg^{++}), is efficient; on the other hand support function grows more and more precarious and finally reaches a real and dramatic insufficiency. This happens both in the case of osteoporosis itself, or when, due to a lack of vitamin D, the mineralization is even more deficient with the production of osteomalacia. Malacic features are detectable in at least one-third of senile pathologic osteopenia.

Later the signs and symptoms of pathologic osteopenia occur: aching bones, bone deformity, fractures. The latter are most frequent in post-menopausal and old age in white women, and particularly in the axial skeleton, at the hip, and also at peripheral bones.[40 59] This situation is of the utmost seriousness owing to the secondary invalidity, which handicaps the elderly. In civilised countries inhabited by white or yellow people as in USA or in Japan, more than 25 per cent of postmenopausl women are affected by osteoporosis; after 65 years the incidence of femoral fractures increases 3 times with a mortality higher than 15 per cent.[17,52,57]

One may affirm that senile pathologic osteopenia (and in particular post-menopausal and senile osteoporosis) represents one of the most serious and widespread disorders capable of causing irremediable damage and grave suffering to the patient, and at the same time a nursing and financial burden to the state.[61] Marcus[57] affirms that "it is no exaggeration to regard osteoporosis as a public health issue of enormous magnitude".

The consideration of the pathogenesis of the illness is of the greatest importance, followed by the diagnostic means which permit early identification of an excessive or too rapid bone loss.

In actual fact many pathogenic factors of senile osteoporosis are an integral part of the normal functional behaviour of the senile organism, but it is obvious that the exaggeration of one or more of these may determine or initiate the illness, which usually has a multifactorial genesis.

To the familial and racial factors which surely are of great significance in determining the intrinsic aging process of bone, can be added other general organic functional factors, which affect the skeleton.   It is sufficient here to mention some for their importance: calcium absorption defects, functional modification within endocrine system, tendency for metabolic and respiratory acidosis, blood flow defects, lowering of muscle tone etc.[10,44,47,49]

Impaired Calcium Absorption

Calcium absorption impairment in the elderly is a well known fact; it is a consequence of changes in the mucosa of the small intestine, as well as of a lack of adaptive response owing to the insufficient production of the active hydroxylated metabolite of vitamin D (1-25(OH)$_2$D$_3$).   The intestine of old people is no longer capable of varying its absorption ability according to the quantity of calcium ingested or to the needs of the organism, as happens in the young.[12,28,38,60,76,78] If to this is added the fact (especially noted in old people's homes) that the diet of the elderly is often lacking in the calcium necessary to maintain balance (1 g a day and more in post-menopausal women),[46] one realizes the pathogenetic importance of these absorption and adaptation intestinal defects.

Dietetic Calcium Deficiency

The situation of calcium dietetic deficiency has in fact been underlined on various occasions (e.g. Nordin's[62] research on cortical bone, that of Caniggia and Gennari[12] and that of Marcowic[57]), according to whom the initial appearance of osteoporosis is the point at which the calcium balance becomes negative.   This occurs in white and yellow races (Finnish women, who have a calcium intake on an average of 1300 mg daily, do not manifest vertebral fractures; this is not true of Japanese women whose calcium intake averages 400 mg daily).[57] The coloured races or central American populations, who take very little calcium show no signs of osteoporosis.[29] It is certain that at least for our race adequate calcium intake is extremely important.

Obviously calcium intake cannot be considered on its own but together with other dietetic factors and with the evaluation of calcium urinary loss.   It is noteworthy to remember for instance, the importance of dietetic phosphate.   When the optimal ratio Ca:P=1:1 is not maintained in food and the phosphate content of the diet is excessive a hyperphosphatemia can occur, and afterwards a kind of "nutritional hyperparathyroidism", accompanied firstly by hypocalciuria and later by hypercalciuria.[9]   According to Lutwak,[52] this is quite common in the USA, where the ratio Ca:P in food is on an average 0.36, partly due to the excessive use of phosphate

additives.   On the other hand, phosphate deficiency leads to
urinary calcium loss.

Protein Intake

Dietary proteins are important too.[48] It is well known that a
high protein diet causes hypercalciuria and negative calcium
balance.[1,3,53,58]   This clearly occurs when pure proteins or
aminoacids are added to the diet: a 100 per cent increase of protein
intake causes a 50 percent increase in calcium urinary excretion.[57,81]
In actual fact this is less evident when the high protein diet
involves meat intake; the latter contains phosphates which reduce
calcium excretion.[37,86]

It is not clear why proteins cause hypercalciuria: it could
depend either on the increase of glomerular filtration rate as a
consequence of the need to excrete more nitrogen, or on the metabolic
acidosis from a high protein diet, or also on an excessive production
of ammonium ions which decrease the tubular reabsorption of calcium,
or a stimulation of glucagon production.[23,57]   In any case, it is
certain that dietetic factors may have influence on the phospho-
calcium balance and bone loss.

Endocrine System

The endocrine system is surely of great importance. It is by
now well known that sex hormones,[14] especially in women, play a
significant part in calcium absorption and increase bone mass. It
is certain that bone loss in women occurs in the vertebrae from the
age of 20, but this is more marked after the menopause.[57,79]
In fact, this phenomenon is met with after this age not only in the
vertebrae but also in the compact bone of the limbs. Measurements
of the mineral content carried out on the radius and the ulna show
that bone density in women up to the menopause is maintained, while
there is a sudden deterioration when ovarian functioning ceases.
In the female the graph representing mineralization behaviour with
aging looks in fact like a line broken by an angle at the age of
fifty; on the other hand in the male this shows a more gradual and
continuous decrease after 25 - 30 years (Fig.1). Albright's
original hypothesis, according to which a lack of oestrogens has a
specific influence on the bone protein matrix, is considered to be
dated: it is now acccepted that, besides acting favourably on
protein anabolism in general, the sex hormones not only influence
the phospho-calcium balance, but in particular indirectly calcium
intestinal absorption. Such hormones (e.g. testosterone), possibly
through parathyroid hormone,stimulate kidney 1α-hydroxylase activity
so that hydroxylation of $25-OH-D_3$ takes place with transformation
to $1\alpha-25(OH)_2D_3$, the hormonal product which stimulates calcium
intestinal absorption.[26,27,28]

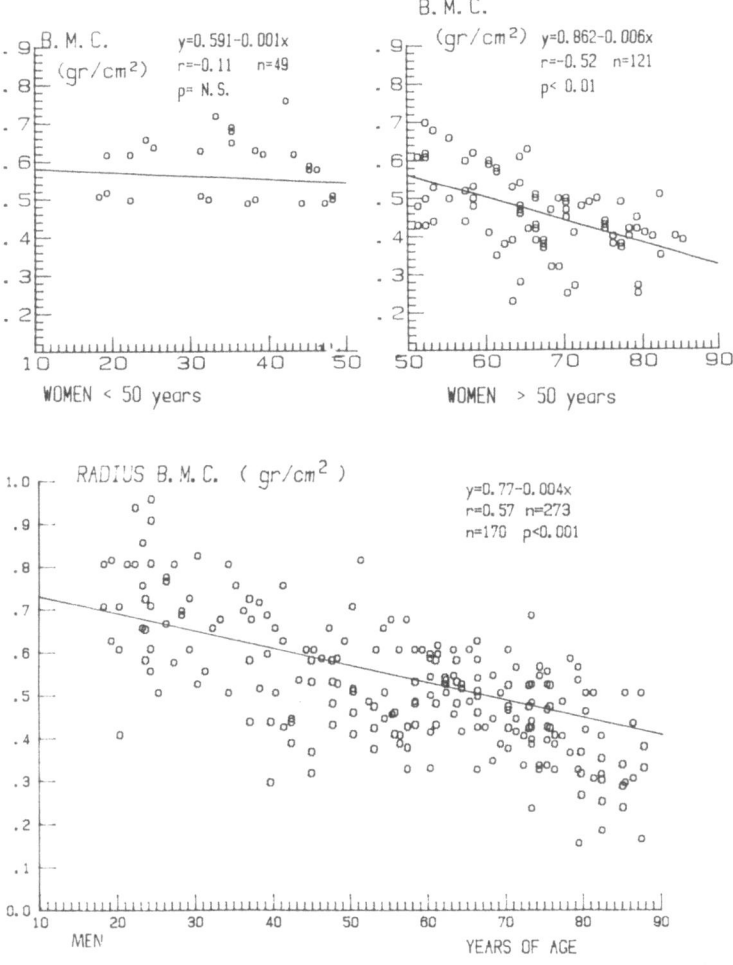

Fig. 1.   Correlations between age and bone mineral
          content.

     In spite of their importance in this field sex hormones are
not considered calcium active hormones in the strictest sense as
are parathyroid hormone (PTH), calcitonin (CT) and $1\alpha-25(OH)_2D_3$.
Many workers have shown that with aging blood PTH tends to increase.
[19,25,70]   Bone resorption is consequently stimulated owing to
osteoclast activation.   In actual fact it is possible to see a
tendency for an increase in basal serum PTH values in the elderly
(even if not highly significant), as our research has shown (Fig.2).
We cannot give much meaning to this fact because there is a wide
individual variability and probably hormonal blood basal level has
not such a high value as daily PTH production variations and
peripheral organ responsiveness.

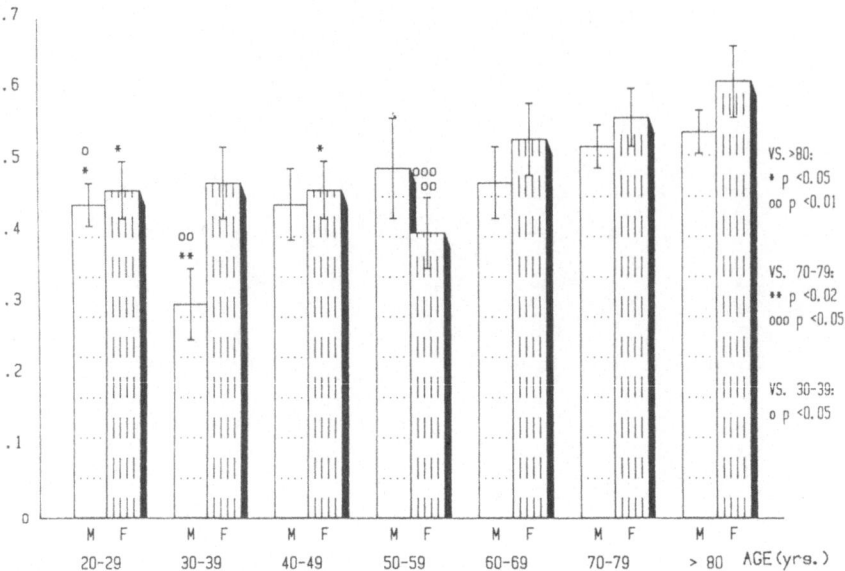

Fig. 2.   Parathyroid hormone - Basal Levels (MV +/- SEM)
           Males = M   Females = F   n. = 252

Together with this tendency to an increase in blood basal PTH, a lowering of CT values are frequently detectable in the elderly and in particular in women, a fact which has been underlined by many authors[6,16,20,36,39,70,84,87,88 91] and also reported by ourselves (Fig. 3).   The lowering of blood CT may induce a reduction in the protective action antagonizing bone resorption which is character- istic of the hormone.   Even now the reported data are however very uncertain, the differences in values hardly significant, owing to a high individual variance.

So it is not appropriate to attach great importance to CT basal decrement and to PTH basal increase in the elderly.   Daily secret- ory rhythm variations[63] (described also by us),[70] (Fig. 4) some possible peripheral hormone receptor variations and especially the alterations in the hormone production in response to appropriate stimulation are definitely of a greater pathogenetic importance.

Especially for CT it has been clearly shown that the response to hypercalcemic stimulus is lowered in the elderly.[6,20,36,39,59,90] What is more in aged people the entero-hormonal-CT does not function so well as in the young following the consumption of a meal.[22,64,68, 69,71]   This mechanism is of value in order to maintain proper post- meal calcium homeostasis and to avoid and prevent post-meal blood calcium variations.   While in the young a CT post-meal rise takes place, in the elderly CT response is minimal or completely absent (Fig. 5).

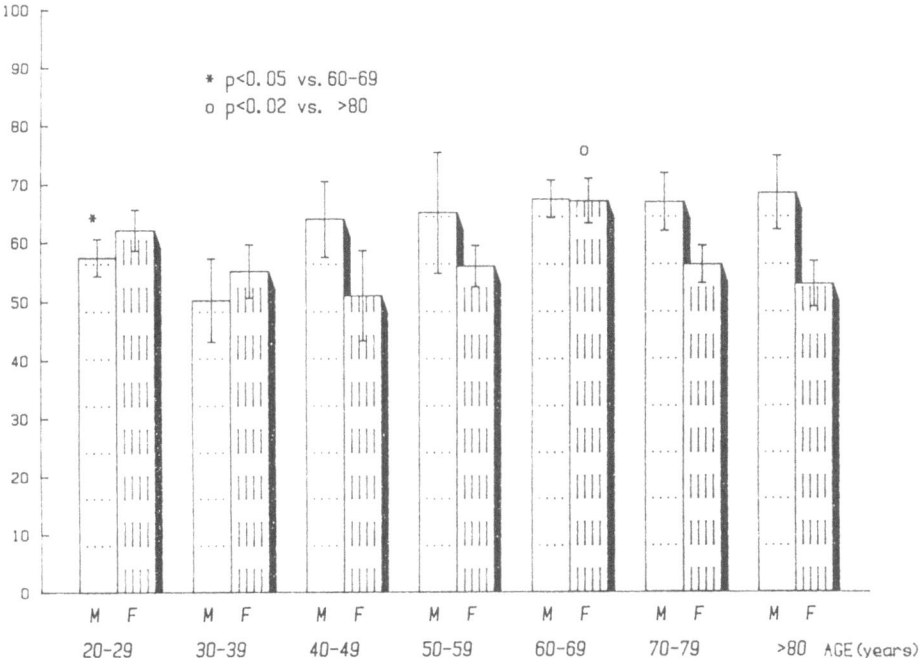

Fig. 3. Calcitonin – Basal Levels (MV +/– SEM)(pg/ml) n. = 252

Vitamin D

A low blood level of dihydroxylated vitamin D is associated
with the global CT deficit, as a consequence both of reduced
stimuli on kidney 1α-hydroxylase, (exerted by sex hormones), and of
poor sensitivity of the same kidney enzyme to PTH action.[28,66,68,76,85]
To this has to be added both the frequent insufficiency of
skin exposure to the action of the sun's U.V. rays, and the lower
oral vitamin D consumption and intestinal absorption capacity in
the elderly.

As regards 25(OH)D_3 reports show it to be low in the blood of
aged people. Caniggia[12] found a high 25(OH)D_3 blood level in
cases of high-turnover osteoporosis, and attributed to this finding
a pathogenetic significance.[85]

The tendency to metabolic general and local acidosis[42,45]
and the arterial and venous impaired perfusion in the elderly should
not be forgotten, since the reduced oxygen and nutrient supplementation leads to bone resorption.[66]

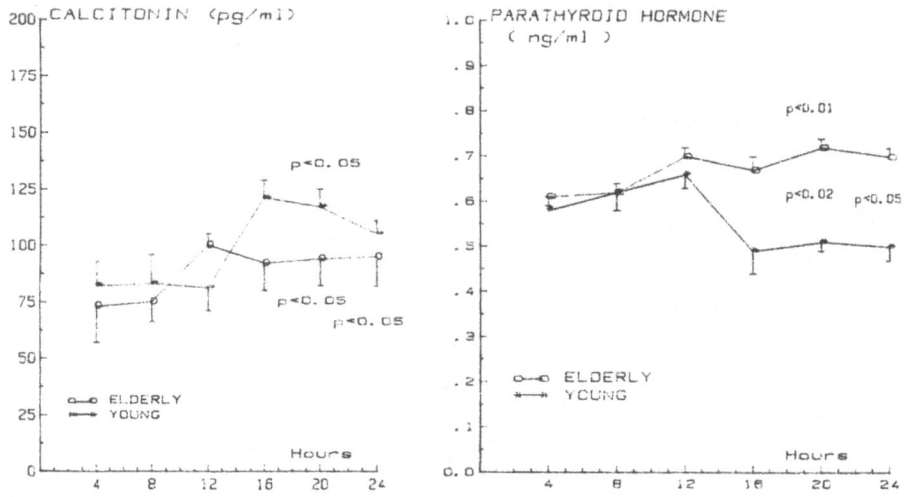

Fig. 4. Circadian rhythm

Mechanical Stress Factors

Probably, even more important is the defect in mechanical stimuli (pressure, traction on the skeleton) which produce piezo-electrical forces at bone crystal surfaces;[89] these forces exert a stimulant action on osteoblast activity. It is also known that the accretion of hydroxy-apatite crystals tends to take place along the direction of the mechanical forces, and that the latter provoke the synthesis of cAMP and ornitin-decarboxylase in the bone cell, and the local production of prostaglandins (PGE$_2$) and of some proteins connected with bone and cartilage accretion, such as BMP (bone morphogenetic protein, containing hydroxyproline and α- carboxyglutamic acid) and CGP(cartilage growth factor).[92]

The severe osteoporosis accompanied by hypercalciuria, which in astronauts follows a period of reduced gravity, and the osteo-porosis which very rapidly affects paralysed limbs, confirms this view. The importance of mechanical forces is also confirmed by our group's investigations carried out on a number of athletes (weight lifters) and on hemiplegic patients.[15] In the former, radius bone mineral content directly correlates with muscle power (lifting on bench) and with muscle mass (arm circumference) (Fig. 6). During rehabilitation of hemiplegic patients there occurs rapid restor-ation of bone mineralization in the paralysed limb (Fig. 7).

All the above mentioned factors have a role in the age-related physiological loss of bone tissue; one or more of them may become important in the pathogenesis of senile pathological osteopenia when they become exaggerated and/or predominant.

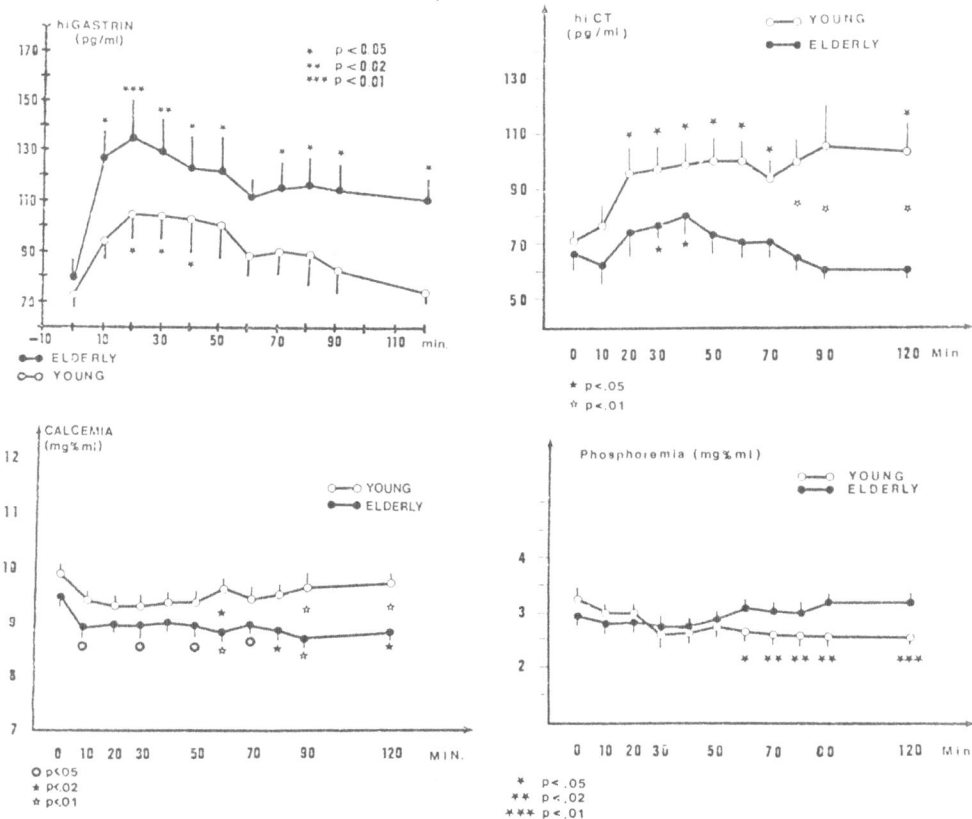

Fig. 5. Serum immunoreactive gastrin, serum immunoreactive
calcitonin, serum calcium and phosphorus after protein
meal. (Young = 9; Elderly = 21 - MV + SEM)

PREVENTION OF BONE LOSS

Since the damage caused by pathological senile osteopenia
e.g. crush fractures)[18,43,79] is serious and often irreversible,
and if one considers the difficulty of therapeutic success in
improving skeletal damage and in obtaining a noteworthy restoration
of bone mass, it is clear that prevention of these morbid forms is
essential. It has to be carried out with determination and
accuracy; it has to be started in time, and to be continued for a
long period and it must be tenaciously followed by the physician.

Two possible periods may be recognized in the prophylaxis of
osteoporosis: a properly named primary prophylaxis and a secondary
one, devoted to preventing severe secondary damage.

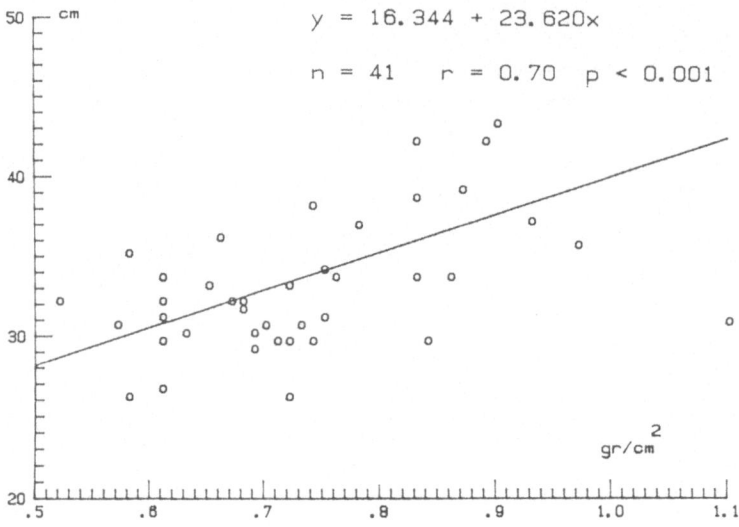

Fig. 6.   Correlation between B.M.C. of the radius and
          upper arm circumference in weight lifters

Primary Prevention

     Primary prophylaxis consists in the necessity of inducing first
of all in the physician, and in people at large, an awareness of the
hazard represented by osteoporosis.   The physician first has to
convince himself that it is necessary to think about the conditions
that may have a secondary damaging effect on the bone in its
anatomical and functional status.   He has to be aware that he
provokes a heavy iatrogenic injury when he does not take care to
convince his patients about the need to prevent with all means
possible loss of bone mass, from which may arise a pathological
osteopenia (osteoporosis, osteoporomalacia, osteomalacia) in the
elderly.

     The physician has always to consider and evaluate the possibil-
ity that from errors in daily living style, or from secondary effects
of organic illnesses, or from side effects of therapeutic intervention, bone might be damaged.

     He must request all the laboratory and instrumental investigat-
ions available for the early recognition of reduction in bone mass:
this is of increasing importance nowadays, when the diagnostic
methods[5,35,55] are becoming more reliable.   Mineralometric methods
(particularly those suitable for detecting not only radius and ulna,

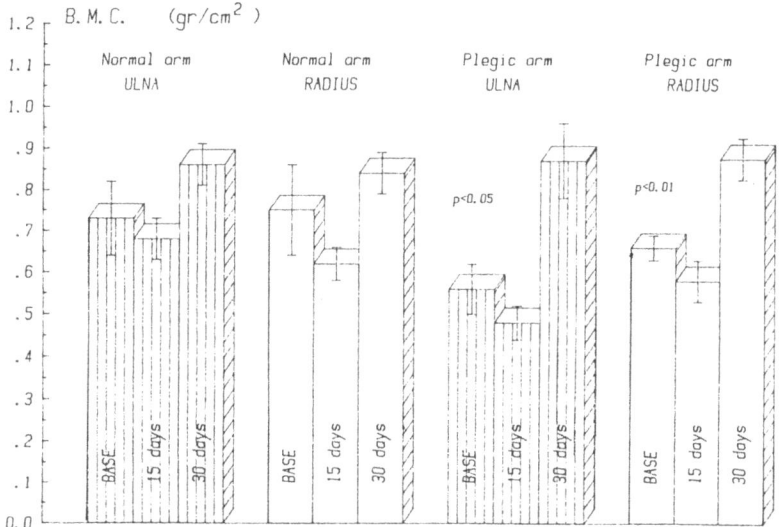

Fig. 7. B.M.C. of normal and plegic arm before and
during rehabilitation treatment

but also spine, and femur mineralization), isotopic kinetic methods,
and biopsy evaluations, may offer the possibility of recognizing
modest loss of bone mass and therefore of intervention at a much
earlier stage than could be done some years ago, when it was
necessary to depend upon radiological evidence.

Health education is also fundamental in order to promote better
knowledge of osteoporosis and of demineralizing illnesses in general.
It should make people aware of the frequence of osteoporosis in post-
menopausal women and in particular in those who have thinner bones.
The practice of evaluating bone status by periodical mineralometric
estimations should be made more widespread. Patients should understand osteoporosis and discuss its possible occurrence with the
physician, from whom they can obtain the necessary advice both as
regards life-style habits and possible preventive medical interventions.

As regards living activities a proper diet, movement, physical
exercise and exposure to sunlight are essential.[33,38,57]

The diet should contain on average 1 g. calcium/day. This
amount has to be higher in women undergoing the menopause. The diet
must therefore be rich in milk, cheese, vegetables, fruit following
the calcium content of the foods shown in tables 1 and 2. It is

Table 1. Calcium and phosphorus content of some milks

	Calcium (mg%gr)	Phosphorus (mg%gr)
Whole Cow Milk	119	93
Skim Cow Milk	122	97
Sheep Milk	180	
Goat Milk	141	106
Buffalo Milk	190	
She-Ass Milk	80	
Condensed Milk	293	235

necessary also to take particular care of this when the calcium content in drinking water is low as often happens with mineral water; the use of oligo- or medio-mineral waters is becoming more and more widespread and such water has a very low calcium content (Table 3). It has been demonstrated that the incidence of fractures is higher in populations drinking calcium poor waters (Fig. 8).

It is also of some importance to consider the observations mentioned above, according to which a dietary excess of proteins provokes hypercalciuria and skeletal demineralisation, due to a mechanism not yet well clarified.[1,3,23,37,48,81] This hazard is real when pure proteins or aminoacids are added to food; it is less serious when the diet is rich in meat, owing to the fact that meat is also rich in phosphate, which reduces the hypercalciuric effect of proteins. On the other hand the consideration of the quantity of phosphate ingested is important too,[9] since it is necessary to maintain as near as possible a P:Ca ratio equal to one in the diet.

Together with the diet some habits such as smoking or drinking have a great importance: it has been demonstrated that alcoholism and smoking damage the bone. Respiratory acidosis, hypoxia and vessel damage of smokers may disturb the skeleton. The mechanism through which an excess of alcohol acts on bone metabolism is probably due to a disturbance in liver 25-hydroxylation of vitamin D: it has been observed that alcohol has an inducing effect on the cytochrome system (cytocrom P-450) responsible for this reaction. Consequently $25-OH-D_3$ blood levels increase, while bone and muscle vitamin D stores decrease (in the rat).[30] On the other hand the entero-hepatic cycle of vitamin D is altered, with a reduced resorption of $25-OHD_3$ derivatives excreted through the bile. The chronic changes in alcoholics, with their gastric atrophic hypochloridria and chronic duodenitis also induce vitamin D and calcium malabsorption.[83] Osteopenia and in particular osteomalacia is a frequent condition in alcoholic cirrhosis.

Table 2.   Calcium, phosphorus, Ca/P ratio of
          some foods

	CALCIUM (mg%gr)	PHOSPHORUS (mg%gr)	Ca/P
BEL PAESE CHEESE	604	480	1.26
EMMENTHAL CHEESE	700	860	0.81
MOZZARELLA CHEESE	280		
SHEEP'S MILK CHEESE	470		
BUTTER-MILK CURD	189	196	0.96
PARMESAN CHEESE	1277	734	1.74
HEN EGG	324	1072	0.30
LEAN BEEF	11	171	0.06
LEAN VEAL	13	221	0.06
BOLOGNA SAUSAGE	21	180	0.12
HAM	22	180	0.12
BOILED HAM	18	164	0.11
WURSTEL	19	130	0.15
LOBSTER	60	242	0.25
CODFISH	28	191	0.15
MACKEREL	24	240	0.10
TUNNY-FISH	31	218	0.14
RED MULLET	25	132	0.19
CRACKERS			
CORN FLAKES	9	59	0.15
COMMON BREAD	13		
MACARONI	17	142	0.12
RICE	12	118	0.10
FRESH BEANS	64	77	0.83
HARICOT BEANS	157	332	0.47
PEAS	30	96	0.31
ARTICHOKES	40	90	0.44
CARROTS	41	30	1.36
ONIONS	80	40	2.00
POTATOES	9	42	0.21
TOMATOES	7	17	0.41
ORANGES	8	26	0.31
APPLES	7	7	1.00
GRAPES	16	18	0.89

Table 3.   Calcium content (mg/l) of commonly
          consumed "mineral waters" in Italy

BOARIO.....................	172
CRODO......................	520
FERRARELLE.................	448
FIUGGI.....................	14
LEVISSIMA..................	17
LORA RECOARO..............	38
SANGEMINI..................	345
SAN PELLEGRINO.............	212

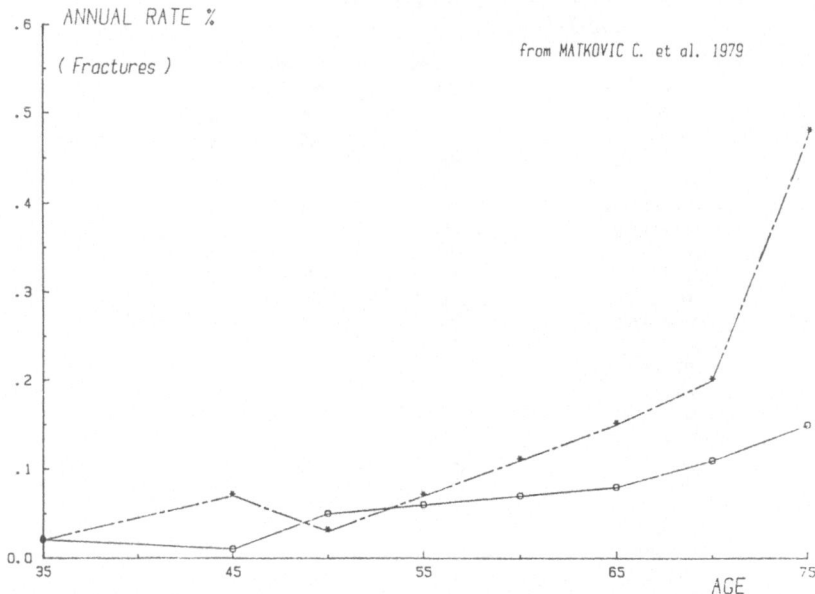

Fig. 8. Incidence of fractures in populations drinking
calcium poor (*) and rich (O) waters.

The reduced exposure of the skin to ultra voilet (U.V.) is of
great importance in the elderly.[26]   Sun exposure has to be
recommended in the adult and in the elderly; alternatively vitamin
D supplementation is necessary especially at a time of year in
which the vitamin D blood content is particularly low (in spring)
so as to avoid periods of hypovitaminosis D.

As regards the importance of physical activity it is certain
that good muscle tone is beneficial, whatever the age, to the
skeleton.   In the young this may increase bone mass putting the
individual in a more advantageous position, when physiological
bone loss later takes place.

In the elderly physical activity is useful in maintaining
bone trophism; avoidance of the hypokinetic syndrome is of the
greatest importance, but physical activity has to suit the
individual and not exceed his capabilities.   Some reports, according to which running is followed by bone loss in women, do not
negate this view because the phenomenon was observed in post-
menopausal or/ in amenorrhoeic females, who had obviously an
oestrogenic defect and a hyperprolactinemic status; probably they
were also hypocalcitoninemic, as it is well know that calcitonin
inhibits prolactin secretion.[82]

Some aspects connected with pathology outside the skeleton deserve a particular mention.   They have to be emphasized for their influence on the skeleton.   It is widely known that in many diseases (acromegaly, Cushing's syndrome, kidney or liver mal-functioning, intestinal malabsorption, acidosis, pneumopathy, arthrosis, arthritis, etc.)[71] considerable bone impairment takes place.   Treatment of the fundamental illness has a favourable effect on the bone, reducing or preventing bone impairment.

In particular it is necessary to treat chronic illnesses which after a long time may damage the bone: it is appropriate to carry out anti-osteoporotic therapy together with treatment of the fundamental illness.   We refer firstly to the metabolic disorders (diabetes, gout) particularly when these are followed by metabolic acidosis which suppresses kidney 1α-hydroxylase and therefore vitamin D activation and calcium absorption.[42]   Osteoporosis is very common and often severe in diabetics.[10]   Nutrition illnesses are of great importance, both undernutrition and obesity.   Lastly, it has been reported that patients with hypertension have low serum concentrations of ionized calcium.

Besides treating the above mentioned morbid states the physician must avoid causing iatrogenic damage to the bone.   The damage caused by cortisone therapy[56] is well known; mainly it is a consequence of the impairment of calcium absorption resulting from reduced activation of vitamin D due to depression of 1α-hydroxylase activity provoked by cortisol.   The activity of heparin and heparin-oids on bone phosphatase[7,10] (with secondary osteoporosis); the damage caused by barbiturates and other liver enzyme inducers are also well known.   Less clear are the effects on bone of anti-rheumatic drugs, thiazide diuretics, theophyllin, cytotoxic agents, calcium antagonists etc., when these substances are used for years as often happens in the therapy of the elderly.   It is possible that many of them cause bone damage.

Furthermore the physician commonly does not think about the possible bone loss which may be due to a rich diet in fibre or to laxatives given for long periods to treat chronic constipation, or to treatment of cirrhosis with lactulose, or to the administration of guar-gum in diabetics.   All these treatments may cause problems with absorption of calcium, vitamin D and other nutrients.

In the field of primary prophylaxis there are other possibil-ities such as water fluoridation or chronic fluoride supplementation. We have known for many years that in those countries where there are waters rich in fluoride, osteoporosis is very rare, the fluoride fixes itself to hydroxyapatite and inhibits bone resorption; fluoride and calcium are useful in osteoporosis.[41,77,80,94]

Finally if one considers the importance of sex hormone deficiency in the pathogenesis of osteoporosis the question of oestrogen supplementation clearly arises[2,14,27,50,51,74,75,93], and one has to wonder whether oestrogen preventive therapy must be given to all menopausal patients or reserved only for a small number of particular cases.   It is known that ovarian deficiency in women, whatever their age, causes calcium loss.   Thus one must take the opportunity of intervening with a preventive oestrogen therapy at the very beginning of the menopausal period.[28]   The tendency to try to reduce secondary disturbances of the menopause with oestrogen (0.75 - 1.25 mg daily) or mixed (oestrogen-progesterone) therapy meets with the above mentioned opportunity.   For this reason it is useful to be generous in such therapy, using preferentially long acting preparations or better, mixed hormone administration, for example, oestrogen-progesterone (ethinyl-estradiol 0.05 mg + norgestill 0.5 - 21 days a month), oestrogen-testosterone (4 mg + 90 mg/every month).   This therapy has to be continued for years (almost 6 - 8) after the menopause.   The use of progesterone alone has also been proposed (20 - 100 mg/day). In the male, and also in some females, prolonged therapy with amabolic steroids may be useful.

Secondary Prevention

This consists of the prevention of bone fractures when the oestoporotic or osteomalacic process has taken place.   It is in practice the main therapy of osteoporosis and includes all the above mentioned methods (from health education to sex hormone therapy) together with calcium, fluoride administration, vit. D supplementation ($1-25(OH)_2D_3$ or fluoride substituted derivatives of vitamin D); therapy with calcitonin is also useful as many clinical trials have shown.   Diphosphonate therapy may have some advantages, but it has to be carried out discontinuously, owing to the possibility of secondary osteomalacic damage.

In any case, for the primary or secondary prevention of osteopenic damage it is essential to create in the minds of the physician and of the patient an awareness which permits the right evaluation of the dangers presented by pathological osteopenia (and in particular senile and post-menopausal osteoporosis), so that all efforts be made in order to prevent them.

REFERENCES

1. L. H. Allen, E. A. Oddoye, and S. Margen, Protein-induced-hyper calciuria: a longer term study, Am.J.Clin.Nutr. 32:741 (1979).
2. J. F. Aloia, Estrogens and exercise in prevention and treatment of osteoporosis, Geriatrics 37:81 (1982).

3. C. R. Anand and H. M. Linkswiler, Effect of protein intake on calcium balance of young men given 500 mg calcium daily, J.Nutr. 104:695 (1974).

4. C. D. Arnaud, Calcium homeostasis: regulatory elements and their integration, Fed.Proc. 37:2558 (1978).

5. J.-P Aubert, and G. Milhaud, Méthode de mesure des principales voies du méthabolisme calcique chez l'homme, Biochim. Biophys.Acta 39:122 (1960).

6. L. A. Austin, and H. Heath, III, Calcitonin. Physiology and pathophysiology, New.Eng.J.Med. 304:269 (1981).

7. L. V. Avioli, Heparin-induced osteopathy: an appraisal, Adv.Exp.Med.Biol. 52:375 (1975).

8. L. V. Avioli, Postmenopausal osteoporosis: prevention versus cure, Fed.Proc. 40:2418 (1981).

9. R. R. Bell, H. H. Draper, D. Y. M. Tzeng, Physiological responses of human adults to foods containing phosphate additives, J.Nutr. 107:42 (1977).

10. U. Butturini, and M. Passeri, Osteopatie senili, Relaz.XVI Congresso Soc.It.Gerontol.e Geriatria, Castellammare di Stabia 1969, Suppl. XL, part I, Giorn.Gerontol. (1969).

11. U. Butturini and M. Passeri, "Fondamenti di Gerontologia e Geriatria," Oppici Ediz., Parma (1972).

12. A. Caniggia, and C. Gennari, "Fisiopatologia clinica del metabolismo fosfo-calcico," Ediz. Pozzi, Roma (1976).

13. A. Caniggia, and A. Vattimo, Kinetics of 99^m Technetium-tin-methylen-diphosphonate in normal subjects and pathological conditions: a simple index of bone metabolism, Calcif. Tissue Int. 30:5 (1980).

14. C. E. Cann, H. K. Genant, B. Ettinger and G. S. Gordon, Spinal mineral loss in oophorectomized women, J.Amer.Med.Assoc. 244:2056 (1980).

15. G. Cervellin, R. Di Ruvo, A. M. Malagnino, E. Palummeri, and D. Provvedini, Importance of physical activity in the main-tenance of bone trophism: densitometric study in a group of young athletes (weight lifters) (in press).

16. C. H. Chusnut, D. J. Baylink, K. Sisom, W. B. Nelp, B. and B. A. Roos, Basal plasma immunoreactive calcitonin in post-menopausal osteoporosis, Metabolism 29:559 (1980).

17. S. W. Cohn, A. Vaswani, J. Zanzi, and K. J. Ellis, Effect of aging of bone mass in adult women, Am.J.Physiol.230:143 (1976).

18. P. S. Cook, A. N. Exton-Smith, J. C. Brocklehurst, and S. M. Lempert-Barber, Fractured femurs, falls and bone disorders, J.Roy.Coll.Phys.Lond. 16:45 (1982).

19. D. H. Copp, and R. V. Talmage, Endocrinology of calcium metab-olism, Proceedings of the VIth Parathyroid Conference, Vancouver, Canada, June 12-17, 1977, Excerpta Medica, Amsterdam/Oxford (1979).

20. L. J. Deftos, M. H. Weisman, G. W. Williams, Influence of age and sex on plasma-calcitonin in human beings, New Eng.J.Med. 302:1351 (1980).

21. W. De Guembecker, C. Guembecker and J. Duriez, Variation due contenu mineral osseux en fonction de l'âge. Etude par la méthode de Cameron, Nouv.Presse Med. 10:3831 (1981).

22. R. Di Ruvo, E. Palummeri, G. Cervellin, M. Marchesi, D. Provvedini, C. Monica, and M. Passeri, Rapporti tra increzio ne di calcitonina e di gastrina dopo carico orale di calcio in giovani ed in anziani, Giorn.Clin.Med. 62:381 (1981).

23. F. H. Epstein, Dietary protein intake and the progressive nature of kidney disease, New.Eng.J.Med. 307:652 (1982).

24. A. N. Exton-Smith, P. H. Millard, P. R. Payne, and E. F. Wheeler, Pattern of development and loss of bone with age, Lancet 1:1154 (1969).

25. M. A. Fiatarone, K. Steel, and R. H. Egdahl, Hyperparathyroidism in the elderly, J.Am.Ger.Soc. 39:343 (1981).

26. J. C. Gallagher, B. L. Riggs, J. Eisman, A. Hamstra, S. B. Arnaud, and H. F. De Luca, Intestinal calcium adsorption and serum vitamin D metabolites in normal subjects and osteoporotic patients: effect of age and dietary calcium, J.Clin.Invest. 64:729 (1979).

27. R. P. Gallagher, Age-related changes in calcium metabolism in perimenopausal women and their relationship to the development of osteoporosis, NIH-Endocrine Society, Conference on Endocrine Aspects of Aging, Bethesda, Md. October (1979).

28. J. C. Gallagher, B. L. Riggs and H. F. De Luca, Effect of estrogen on calcium absorption and serum vitamin D metabolites in postmenopausal osteoporosis, J.Clin.Endocrinol. 51:1359 (1980).

29. S. M. Garn, "The Earlier Gain and Later Loss of Cortical Bone," C. C. Thomas, Springfield (III) (1970).

30. M. Gascoin-Barré, Interrelationship between vitamin D_3 and 25-OH-D_3 during chronic ethanol administration in the rat, Metanolism 31:67 (1982).

31. C. Gennari, F. M. Chierichetti, C. Vibelli, G. Francini, E. Maioli, and F. Gonnelli, Acute effects of salmon, human and porcin calcitonin on plasma calcium and cyclic AMP-levels in man, Curr.Ther.Res. 30, 1024 (1981).

32. C. Gennari, Intestinal calcium absorption in post-menopausal osteopaties, Bibl.Nutr.Diet (1982) (in press)

33. W. Gevers, Calcium: the managing director, S.A.Med.Tyd. 18:406 (1981).

34. C.J. Gryfe, A. N. Exton-Smith, P. R. Payne, and E. F. Wheeler, Pattern of development of bone in childhood and adolescence, Lancet 1:523 (1971).

35. H. E. Gruber, and D. J. Baylink, The diagnosis of osteoporosis, J.Am.Ger.Soc. 29:490 (1981).

36. H. Heath, G. W. Sizemore, Plasma calcitonin in normal man. Differences between men and women, J.Clin.Invest. 60:1135 (1977).

37. M. Hegsted, S. A. Schuette, M. B. Zemel, Urinary calcium and
     calcium balance in young men as affected by level of
     protein and phosphorus intake, J.Nutr. 111:553 (1981).
38. R. P. Heaney, P. D. Saville, and R. R. Recker, Calcium
     absorption as a function of calcium intake. J.Lab.Clin.Med.
     85:881 (1975).
39. C. J. Hillyard, J. C. Stevenson, and I. MacIntyre, Relative
     deficiency of plasma-calcitonin in normal women, Lancet 1:
     961 (1978).
40. J. Jasani, B.E.C. Nordin, D. A. Smith, Spinal osteoporosis and
     the menopause, Proc.Roy.Soc.Med. 58:441 (1975).
41. J. Jowsey, B. L. Riggs, P. J. Kelly, and D. L. Hoffman,
     Effect of combined therapy with sodium fluoride, vitamin D
     and calcium in osteoporosis, Am.J.Med. 53:43 (1977).
42. H. Kawashima, J. A. Kraut, K. Kurokawa, Metabolic acidosis
     suppresses 25-hydroxyvitamin D_3- 1 α-hydroxylase in rat
     kidney, J.Clin.Inv. 70:135 (1982).
43. J. S. Keene, C. A. Anderson, Hip fractures in the elderly.
     Discharge prediction with a functional rating scale,
     J.Amer.Med.Assoc. 248:564 (1982).
44. H. P. Kruse, F. Kuhlencordt, Pathogenesis and natural course
     of primary osteoporosis, Lancet 1:281 (1980).
45. J. Leman, J. R. Litzow, and E. J. Lennon, Studies of the mech-
     anism by which chronic metabolic acidosis augments urinary
     calcium edcretion in man, J.Clin.Inv. 46:1318 (1967).
46. F. Lenzi, A. Caniggia, Fisiopatologia e clinica delle osteo-
     porosi diffuse, Relaz. LXIII Cong.Soc.It.Med.Int. Pozzi Edit.
     Roma (1962).
47. F. Lenzi, A. Caniggia, R. Guideri, C. Gennari, G. Bertelli,
     L. Cesari, G. Ravenni, M. Rubegni, L'osteoporosi senile,
     Minerva Medica Edit., Torino (1966).
48. A. A. Licata, E.Bov, F. C. Bartter, F. West, Acute effect of
     dietary protein on calcium metabolism in patients with
     osteoporisis, J.Gerontol. 36:14 (1981).
49. T. S. Lindholm, O. S. Nilsson, S. A. Eriksson, T. C. Lindholme,
     B. R. Kyhle, Changes in bone histomorphometry and bone
     mineral during treatment of osteoporosis with 1α-hydroxy-
     vitamin D_3 and calcium, Acta Vitamin.Enzymol. 3:170 (1981).

50. R. Lindsay, A. MacLean, A. Kraszewski, D. M. Hart, A. C. Clark
     and J. Garwood, Bone resorption to termination of oestrogen
     treatment, Lancet 1:1326 (1978).
51. R. Lindsay, J. M. Atkin, J. B, Anderson, D. M. Hart, E. B.
     McDonald, and A. C. Clark, Long term prevention of post-
     menopausal osteoporosis by oestrogens, Lancet 1:1038 (1979).
52. L. Lutwak, Metabolic and biochemical considerations of bone,
     Ann.Clin.Lab.Sci.5:185 (1975).
53. R. A. McCance, E. M. Widdowson, and H. Lehman, The effect of
     protein intake on the absorption of calcium and magnesium,
     Biochem.J. 36:686 (1942)

68. M. Passeri, Osteopenie senili ed autosufficienza: un grave problema di diagnostica, terapia e prevenzione, Giorn.Clin. Med. 62:365 (1981).

69. M. Passeri, E. Palummeri, G. Soncini, C. Monica, P. G. Ferretti, R. Galvagni, Inibizione della increzione di gastrina esercitata negli anziani dalla calcitonina, Giorn. Clin.Med. 58:321 (1977).

70. M. Passeri, E. Palummeri, D. Provvedini, M. Cecchettin, Gli ormoni calcio-attivi, Giorn.Gerontol.11:763 (1979).

71. M. Passeri, E. Palummeri, D. Provvedini, Calcitonin-gastrin interactions in bone diseases in old age, II Int.Symp. on Calciotropic Hormones (1980).

72. A. Pecile, V. R. Olgiati, G. Luisetto, F. Guidobono, C. Metti, D. Ziliotti, Calcitonin and control of prolactin secretion, "Calcitonin 1980".

73. E. D. Pellegrino, R. M. Blitz, The composition of human bone in anemia. Observation on the reserve function of bone and demonstration of a labile fraction of bone carbonate, Medicine 44:397 (1965).

74. R. Punnonen, A. Lukola, Oestrogen-like effects of ginseng, Brit.med.J. 281:1110 (1981).

75. R. R. Recker, P. D. Saville, and R. P. Heaney, Effect of oestrogens and calcium carbonate on bone loss in postmenopausal women, Am.Intern.Med. 87:649 (1977).

76. M. L. Ribovich, and H. F. De Luca, Adaptation of intestinal calcium absorption: parathyroid hormone and vitamin D metabolism, Arch.Biochem.Biophys.188:157 (1978).

77. B. L. Riggs, G. F. Hodgson, D. L. Hoffmann, P. J. Kelly, K. A. Johnson, and D. Taves, Treatment of primary osteoporosis with fluoride and calcium, J.Amer.Med.Assoc. 243:446 (1980).

78. B. L. Riggs, P. J. Kelly, V. R. Kinney, Calcium deficiency and osteoporosis: observations in one hundred and sixty-five patients and critical review of the literature, J.Bone Joint Surg. 49A:915 (1967).

79. B. L. Riggs, H. W. Wahner, and R. L. Dunn, Differential changes in bone mineral density of appendicular and axial skeleton with age. Relationship to spinal osteoporosis, J.Clin.Inv. 67:328 (1981).

80. B. L. Riggs, E. Seeman, S. F. Hodgson, D. R. Taves, W. M. Fallon, Effect of the fluoride/calcium regimen on vertebrate fracture occurrence in postmenopausal osteoporosis, New Eng.J.Med. 306:446 (1982).

81. W. G. Robertson, P. J. Heyburn, M. Peacock, The effect of high animal protein intake on the risk of calcium stone-formation in the urinary tract, Clin.Sci. 57:285 (1979)

82. E. Rolandi, A. Sannia, G. M. Milesi, and T. Barreca, Effects of salmon calcitonin administration on serum prolactin levels in man, IRCS Med.Sci.7:64 (1979).

54. D. A. McCarron, Low serum concentration of ionized calcium in patients with hypertension, New.Eng.J.Med. 307:226 (1982).

55. D. H. Manicourt, S. Orloff, J. Brauman, and A. Schontens, Bone mineral content of the radius: good correlations with physicochemical determinations in iliac crest trabecular bone of normal and osteoporotic subjects, Metabolism 30:57 (1981).

56. S. C. Manolagas, and D. C. Anderson, Adrenal steroids and the development of osteoporosis in oophorectomized women, Lancet 2:8143 (1974).

57. R. Marcus, The relationship of dietary calcium to the mainten- ance of skeletal integrity in man. An interface of endocrinology and nutrition, Metabolism 31:93 (1982)

58. S. Margen, J.-Y. Chu, N. A. Kaufmann, and D. H. Calloway, Studies in calcium metabolism; I. The calciuretic effect of dietary protein, Am.J.Clin.Nutr. 27:584 (1974).

58b. V. Matkovic, K. Kostial, I. Simonovic, Bone status and fracture rates in two regions of Yugoslavia, Am.J.Clin.Nutr. 32:3 (1979).

59. G. Milhaud, M. Benezech-Le Fevre, M. S. Moukhtar, Deficiency of calcitonin in age related osteoporoses, Biomedicine 29:272 (1978).

60. D. A. Norman, J. S. Fordtran, L. J. Brinkley, J. H. Zerwekh, M. J. Nicar, S. M. Storwing, and C. Y. C. Pak, Jeunal and ileal adaptation to alterations of dietary calcium, J.Clin.Invest. 67:1599 (1981).

61. B. E. C. Nordin, International patterns of osteoporosis, Clin.Orthop. 45:17 (1966).

62. B. E. C. Nordin, G. Aaron, R. Speed, and R. G. Crilly, Bone formation and resorption and the determinant of trabecular bone volume in post-menopausal osteoporosis, Lancet 2:267 (1981).

63. C. Y. A. Pak, Parathyroid hormone and thyrocalcitonin: their mode of action and regulation, Ann.N.Y.Acad.Sci. 179:450 (1971).

64. E. Palummeri, D. Provvedini, D. Cucinotta, L. Cella, G. Cervellin, R. Di Ruvo, M. Marchesi, M. Passeri, Rapporti tra increzione di gastrina e calcitonina dopo pasto pro- teico in giovani ed in anziani, Giorn.Clin.Med. 62:394 (1981).

65. A. M. Parfitt, C. Matthews, D. Rao, Impaired osteoblast function in metabolic bone disease, in: "Osteoporosis: recent advances in pathogenesis and treatment," H. F. De Luca, H. M. Frost, W. S. S. Jee, C. C. Johnson jr. A. M. Parfitt, eds., Univ. Park Press, Baltimore (1981).

66. M. Passeri, Aspetti di fisiopatologia e terapia della osteo- porosi denile, Relaz. Congr.S.Marino, Ottobre 1970, Incontri Medico-Chirurgici v. VIII (1970).

67. M. Passeri, La fisiopatologia del ricambio osseo nei rapporti con la patogenesi dell'osteoporosi senile, La Settimana Medica 63:2 (1975).

83. P. D. Saville, Alcohol-related skeletal disorders, Am.N.Y.Acad.
     Sci. 252:287 (1975).
84. I. M. Shamonki, A. M. Frumar, I. V. Tataryn, Age-related changes
     of calcitonin secretion in females, J.Clin.Endocrinol.Metab.
     50:437 (1980).
85. D. M. Slovik, J. S. Adams, R. M. Neer, M. F. Holik, J. T. Potts,
     Deficient production of 1α-25-dihydroxyvitamin D in elderly
     osteoporotic patients, New Eng.J.Med. 305:372 (1981).
86. H. Spencer, L. Kramer, D. Osis, Effect of a high protein (meat)
     intake on calcium metabolism in man, Am.J.Clin.Nutr. 31:2167
     (1978).
87. J. C. Stevenson, G. Abeyasekera, C. J. Hillyard, K. G. Phang,
     I. McIntyre, Calcitonin and the calcium regulating hormones
     in post menopausal women; Effect of oestrogens, Lancet 1:693
     (1981).
88. J. C. Stevenson, and M. I. Whitehead, Post menopausal osteopor-
     osis, Brit.med.J. 285:585 (1982).
89. A. F. Stewart, M. Adlar, C. M. Byers, G. V. Segre, A. F.
     Broadus, Calcium homeostasis in immobilization: an example
     of resorptive hypercalciuria, New Eng.J.Med. 306:1140 (1982)
90. H. Mc A. Taggart, C. H. Chestnut, J. L. Ivey, D. J. Baylink,
     K. Sisom, M. E. Hube and B. A. Roos, Deficient calcitonin
     response to calcium stimulation in postmenopausal osteopor-
     osis, Lancet 1:475 (1982).
91. R. V. Talmage, S. A. Grubb, N. Hirotashi, C. J. Van der Wiel,
     Evidence for an important physiological role for calcitonin,
     Proc.Soc.Acad.Sci. USA 1:609 (1980).
92. M. R. Urist, M. A. Conover, A. Lietze, J. T. Trifitt, R.De
     Lance, Partial purification and characterization of bone
     morphogenetic protein, in: "Atti della VIII "Parathyroid
     Conference," Excerpta Medica, Amsterdam (1981).
93. N. S. Weiss, C. L. Ure, H. J. Ballard, A. R. Williams, J. R.
     Daling, Decreased risk of fractures of the hip and lower
     forearm with postmenopausal use of oestrogen, New Eng.J.Med.
     303:1195 (1980).
94. I. Wolinsky, A. Simkin, K. Guggenheim, Effects of fluoride on
     metabolism and mechanical properties of rat bone, Am.J.Phy-
     siol. 233:46 (1972).

TREATMENT OF POST-MENOPAUSAL OSTEOPOROSIS

A. Caniggia

Director of General Clinical Medicine
University of Siena, Italy

I should like to thank Professor Barbagallo-Sangiorgi for
inviting me to speak again on a topic I discussed here some years
ago.   During this time our knowledge of the pathophysiology of post-
menopausal osteoporosis has advanced consistently and several
therapeutic trials have been undertaken.   The main scheme I
proposed then has not changed substantially: nevertheless at present
it must be integrated with more recent results, particularly those
in the field of vitamin D metabolites and I shall communicate now
the status of the art.

Post-menopausal osteoporosis is a very common pathological
condition characterized by a reduction in bone mass that is mainly
the result of bone resorption.   A spongy appearance of cortical
bone is observed as a consequence of the presence of large resorption lacunae, and a thinning of the trabeculae of cancellous bone
may also be apparent.   No osteoid seams are present, which are
characteristic of osteomalacia.   Thus bone mass is reduced but the
mineralisation of the remaining bone is normal.

The skeletal changes, that are mostly evident in spine, lead to
a progressive brittleness of bones with increased liability to
fractures; the painful crush fractures of vertebrae lead to kyphotic
deformity of the spine and reduction in stature.

It has been largely demonstrated that oophorectomy and menopause accelerate bone loss in women.[1-3]   Estrogen lack leads to a
negative calcium balance with a secondary increase in bone resorption for homeostatic purposes.   Estrogen therapy decreases
significantly bone resorption with a positive transformation of
calcium balance.[4,5]

In 1963 we demonstrated an impairment of intestinal radio-calcium transport in post-menopausal osteoporosis;[6] this observation has been largely confirmed thereafter.[7,14]   Subsequently we demonstrated that the impaired intestinal calcium transport returned to normal after a 6 month treatment with an oestrogen-gestogen combination;[15] these results have been recently confirmed.[16] The possibility that estrogen lack might be responsible for the impaired absorption of calcium with secondary loss of bone was then considered and the positive therapeutic effect of estrogens on post-menopausal osteoporosis did find its rationale.

VITAMIN D METABOLITES

It is widely known that vitamin D has not a direct action on its targets.   Cholecalciferol, that is the vitamin D_3 synthesized by the skin and largely dependent on exposure to ultraviolet light, is converted in the liver to 25 hydroxyvitamin D_3 (25OH vitamin D) which in turn must be converted in the kidney to 1.25 dihydroxy-vitamin D_3 (25$(OH)_2$vitamin D) the most effective and rapidly acting of the vitamin D metabolites.[17]   1,25$(OH)_2$vitamin D promotes the active transport of calcium in the gut: its plasma level is dependent upon the calcium needs of the organism; particularly 1,25$(OH)_2$ vitamin D is actively synthesized in hypocalcemic conditions due to nutritional calcium defects, with a consequent adaptation of the intestinal calcium transport.[17]

In post-menopausal osteoporosis the serum level of 1,25$(OH)_2$ vitamin D is lower than in normal age-matched controls,[18,20] whereas the vitamin D status measured by the serum 25OH vitamin D level appears to be excellent: the levels of 25OH vitamin D are not only higher than in age-matched non-osteoporotic controls but higher than in young normal women as well.[19]   This could be accounted for by an increased activity of liver 25-hydroxylase through an inadequate product-inhibition of this enzyme.

Similarly the increased serum levels of 24,25$(OH)_2$ vitamin D in our post-menopausal osteoporotic women support this hypothesis:[20] it is largely known that there is a reciprocal relationship of 1,25$(OH)_2$ vitamin D and 24,25$(OH)_2$ vitamin D in the sense that under circumstances that stimulate the production of 1,25$(OH)_2$ vitamin D there is a repression of the ability to produce 24,25$(OH)_2$ vitamin D and vice versa.[21]

Every physician knows that vitamin D given as such is ineffective in the treatment of post-menopausal osteoporosis. On the other hand post-menopausal osteoporotic women given 1 microgram of 1,25$(OH)_2$ vitamin D^3 daily, that is the physiological dose, showed within 10 days a dramatic improvement of the intestinal transport of radiocalcium.[22]   A similar treatment with 24,25$(OH)_2$

vitamin D was ineffective so that the presence of an hydroxyl in
1 position seems to be necessary.   According to these findings we
proposed the hypothesis that estrogen lack could lead to a decrease
of renal alpha-hydroxylase activity with secondary impairment of
the intestinal calcium transport.[24]

We have recently performed a clinical trial on a long-term
treatment with physiological doses of $1,25(OH)_2$ vitamin D_3
administered orally in women with symptomatic post-menopausal
osteoporosis.[20] The improvement in motility and pain was appreciable
in all cases and frequently very impressive.   While receiving
$1,25(OH)_2$ vitamin D intestinal calcium absorption rose dramatically
within 2 months to the upper values of the normal range.

The fasting levels of plasma calcium did not increase significantly except at the beginning of treatment; no changes occurred
in phosphate and serum alkaline phosphatase.   On the contrary a
significant and persistent increase in urinary calcium excretion
was apparent and this reflected ongoing intestinal calcium
absorption many hours following the calcium containing meals.   The
rise in urinary calcium could not be accounted for by a stimulation
of bone resorption: an increased bone resorption should have
increased both urinary calcium and hydroxyproline and the 24 hour
urinary excretion of hydroxyproline did not change whereas the
urinary calcium/hydroxyproline ratio did rise significantly.
Last but not least photon absorptiometry, which has become widely
accepted for evaluation of bone mineral content, showed in these
patients an increase in bone mineral content at the end of the
treatment.

These results support the hypothesis that decreased calcium
absorption and the inability of osteoporotic patients to adapt to
the low calcium diet, which is very common in these women,[25]
can be accounted for by inadequate transformation of 25(OH) vitamin
D to $1,25(OH)_2$ vitamin D so that the administration of physiological
doses of $1,25(OH)_2$ vitamin D_3 could be considered as a substitution
therapy.

CALCITONIN

Calcitonin was first used in man in the treatment of ideopathic
hypercalcemia and Paget's disease because of its inhibitory action
on bone resorption.   The increased bone resorption responsible for
the decrease in the total mass of bone, known to exist in post-
menopausal osteoporotic women, suggested the possibility that
calcitonin might be of therapeutic value in this disease.   In a
first trial we had been able to demonstrate a substantial reduction
in bone turnover rate as assesssed by[47] calcium kinetics,[29] and
this has been more recently confirmed by the kinetics of [99m]Tech-

netium-tin-methylene diphosphonate.[30,31]   In both instances values
fell after the treatment and these results were particularly evident
in patients with high-turnover osteoporosis who improved dramatically
after calcitonin administration.   In these patients the rapid therapeutic effect was probably due to the inhibitory action of calcitonin
on bone resorption.

Nevertheless a deeper insight showed that long-term calcitonin
treatment led to a better utilisation of dietary calcium.[29]
Recently we have confirmed those early results and have demonstrated
that at the end of long-term treatments with calcitonin the post-
menopausal osteoporotic women (as well as Paget's disease patients)
showed a significant increase in the urinary cyclic AMP/creatinine
ratio, a parameter which reflects closely parathyroid activity.[37]
The increase in the urinary cyclic AMP/creatinine ratio has been
accounted for by a secondary parathyroid stimulation as a consequence
of decrease in plasma levels induced by calcitonin: indeed our
patients showed a mild but significant decrease in fasting plasma
calcium levels.   As an alternative hypothesis a direct action of
calcitonin on renal cells might be suggested: but calcitonin does
not increase directly the urinary excretion of cAMP[32] and an
increase in the urinary cyclic AMP/creatinine ratio has been
observed as well in patients treated with etidronate, another hypocalcemic agent used in Paget's disease and osteoporosis.[23]

Thus it is tempting to speculate that the effect of calcitonin
on calcium balance and on intestinal calcium absorption may be
induced through secondary stimulation of parathyroid secretion
which activates the renal 1-alpha-hydroxylase.   A favourable effect
of treatment with 1-30 fragment of human parathyroid hormone in
post-menopausal osteoporosis has been reported,[33] and this was
probably obtained through stimulation of the activity of renal
1-alpha-hydroxylase.

Recently low levels of serum calcitonin have been found in
women with post-menopausal osteoporosis,[34,35] and an interrelationship between estrogens and calcitonin production has been postulated.[36]   Should this view be confirmed, estrogen lack could be
considered as responsible for decreased calcitonin secretion and
calcitonin treatment as a substitution therapy.   In any case a
twofold effect of calcitonin in post-menopausal osteoporosis has
been demonstrated:
(i)   a direct inhibitory action on bone resorption;
(ii)   an indirect stimulation of calcium absorption through a
        secondary effect on parathyroid secretion and its stimulat-
        ory action on renal 1-alpha-hydroxylase.
In the light of present knowledge the bone loss of post-menopausal
osteoporotic women can be accounted for by a homeostatic response
to deficiency of the intestinal calcium transport; this is due to
a negative effect of estrogen lack on renal 1-alpha-hydroxylase.

METHODS OF TREATMENT

Many forms of treatment have been advocated:
(i)   Estrogen administration is effective in stimulating the
       activity of 1-alpha-hydroxylase, thus promoting the synthesis
       of $1,25(OH)_2$ vitamin D_3.

The untoward effects of a long-term estrogen treatment are well
known though probably over-emphasized.

(ii) Calcitonin administration presents a twofold effect that is
     shared by diphosphonate treatment: an indirect stimulation of
     the activity of 1-alpha-hydroxylase through a secondary
     increase in parathyroid hormone and a symptomatic effect on
     bone resorption.

Untoward side effects sometimes noted are flushings and nausea
and the development of antibodies when porcine or salmon calcitonins
are used.

(iii) $1,25(OH)_2$ vitamin D_3 administration is a very effective sub-
      stitution therapy.

Despite transient phases of hypercalcemia and persistent
increases in urinary calcium excretion no adverse effects on the
kidney have been found when physiological doses of this metabolite
were administered; nevertheless monitoring of plasma calcium levels
and renal function are necessary.

SUMMARY

The pathophysiology of post-menopausal osteoporosis has been
considered in the light of the more recent advances in vitamin D
metabolites.   Bone loss can be accounted for by a deficiency in
intestinal calcium transport; this is due to an impaired synthesis
of $1,25(OH)_2$ vitamin D_3 resulting from the negative effect of
oestrogen lack on the activity of the renal alpha-hydroxylase.

Many forms of treatment have been advocated: mainly estrogens,
calcitonin and $1,25(OH)_2$ vitamin D_3.   The rationale of these
therapeutic trials has been discussed.

REFERENCES

1. H. W. Wahner, B. L. Riggs, and J. W. Beabout, Diagnosis of
    osteoporosis: usefulness of photon absorptiometry at the
    radius, J.Nucl.Med. 18:432 (1977).

2. R. R. Recker, Photon absorptiometry in the investigation and
    treatment of metabolic bone disease, Am.J.Roentgenol.
    131:543 (1978).

3. M. R. A. Khairi, and C. C. Johnston, Exponential bone loss in
    post-menopausal women, Am.J.Roentgenol. 131:543 (1978).

4. F. Albright, and E. C. Reifenstein, "The Parathyroid Glands and
    Metabolic Bone Disease," Williams and Wilkins Co., Baltimore
    (1948).

5. B. E. C. Nordin, R. Speed, J. Aaron and R. G. Crilly, Bone
    formation and resorption as the determinant of trabecular
    bone volume in post-menopausal osteoporosis, Lancet 2:277
    (1981).

6. A. Caniggia, C. Gennari, V. Bianchi, and R. Guideri, Intestinal
    absorption of 45-Ca in senile osteoporosis, Acta Med.Scand.
    173:613 (1963).

7. A. Caniggia, C. Gennari, L. Cesari and S. Romano, Intestinal
    absorption of 45-Ca in adult and old human subjects,
    Gerontologia 10:193 (1964-65).

8. V. Parson, N. Veall, and W. J. H. Butterfield, The clinical use
    of orally administered 47-Ca for the investigation of
    intestinal calcium absorption, Calcif. Tissue Res. 2:83
    (1968).

9. J. R. Bullamore, R. Wilkinson, J. C. Gallagher, B.E.C. Nordin,
    and D. H. Marshall, Effects of age on calcium absorption,
    Lancet 2:535 (1970).

10. J. Szymendera, R. P Heaney, and P. D. Saville, Intestinal
    calcium absorption: concurrent use of oral and intravenous
    tracers and calculation by the inverse convolution method,
    J.Lab.Clin.Med.79:570 (1972).

11. C. C. Aleviazaki, D. C. Ikkos, and P. Singhelakis, Progressive
    decrease of true intestinal calcium absorption with age in
    normal man, J.Nucl.Med.14:760 (1973).

12. P. Ireland, and J. S. Fordran, Effect of dietary calcium and
    age on jejunal calcium absorption in humans studied by
    intestinal perfusion, J.Clin.Invest. 52:2672 (1973).

13. B. E. C. Nordin, R. Williams, D. H. Marshall, J. C. Gallagher,
    A. Williams, and M. Peacock, Calcium absorption in the
    elderly, Calcif.Tissue Res. 21:422 (1976).

14. J. C. Gallagher, B. L. Riggs, J. Eisman, A. Hamstra, S. B.
    Arnaud, and H. F. De Luca, Intestinal calcium absorption and
    serum vitamin D metabolites in normal subjects and osteo-
    porotic patients: effect of age and dietary calcium. J.Clin.
    Invest. 64:729 (1979).

15. A. Caniggia, C. Gennari, G. Borrello, M. Bencini, L. Cesari,
    C. Poggi, and S. Escobar, Intestinal absorption of 47-
    Calcium after treatment with oral oestrogen-gestogens in
    senile osteoporosis, Brit.med.J. 4:30 (1970).

16. J. C. Gallagher, B. L. Riggs, and H. F. De Luca, Effect of
    estrogen on calcium absorption and serum vitamin D metabol-
    ites in post-menopausal osteoporosis, J.Clin.Endocrinol.
    Metab. 51:1359 (1980).

17. A. W. Norman, "Vitamin D," Academic Press Inc., New York, San Francisco, London (1979).

18. B. L. Riggs, J. C. Gallagher, H. F. De Luca, A. J. Edis, P. W. Lambert, and C. D. Arnaud, A syndrome of osteoporosis, increased serum immunoreactive parathyroid hormone, and inappropriately low serum 1, 25-dihydroxy-vitamin D, Mayo Clin.Proc.53:701 (1978).

19. F. Lore, G. Di Cairano, A. M. Signorini, and A. Caniggia, Serum levels of 25-hydroxyvitamin D in post-menopausal osteoporosis, Calcif.Tissue Int. 33:467 (1981).

20. A. Caniggia, R. Nuti, F. Lorè, and A. Vattimo, Vitamin D and its metabolites in post-menopausal osteoporosis: effect of a long-term treatment with 1,25-dihydroxyvitamin D_3. Intern. Congress on the Steroid Hormones, Jerusalem (Sept. 6-11th 1982) (In press).

21. I. T. Boyle, R. W. Gray, J. L. Omdahl, and H. F. De Luca, Calcium control of the in vivo biosynthesis of 1,25-dihydroxy-vitamin D3. Nicolaysen's endogenous factor, in: "Endocrinology," S. Taylor, ed., Heineman, London (1971).

22. A. Caniggia, and A. Vattimo, Effects of 1,25-dihydroxycholecalciferol on 47-Calcium absorption in post-menopausal osteoporosis, Clin.Endocrinol. 11:99 (1979).

23. A. Caniggia, F. Lorè, R. Nuti and A. Vattimo, The rationale of calcitonin treatment in postmenopausal osteoporosis, in: "Calcitonin 1980," A. Pecile, ed., Excerpta Medica, Amsterdam-Oxford-Princeton, Milano (1980).

24. A. Caniggia, C. Gennani, and A. Vattimo, The intestinal transport of 47-Ca in post-menopausal osteoporosis, in: "Osteoporosis II," ed., U. S. Barzel, Grune and Stratton, New York-San Francisco-London (1978).

25. A. Caniggia, Medical problems in senile osteoporosis, Geriatrics 20:300 (1965).

26. G. Milhaud, and J. C. Job, Thyrocalcitonin effect on idiopathic hypercalcemia, Science 154:794 (1966).

27. J. C. Haddad, S. J. Birge, and L. V. Avioli, Effects of prolonged thyrocalcitonin administration on Paget's disease of bone, New Eng.J.Med. 283:549 (1970).

28. N. J. Y. Woodhouse, P. Bordier, M. Fisher, G. F. Joplin, M. Reiner, D. N. Kalu, G. V. Foster, and I. MacIntyre, Human calcitonin in the treatment of Paget's bone disease, Lancet 2:1139 (1971).

29. A. Caniggia, C. Gennari, M. Bencini, L. Cesari, and G. Borrello, Calcium metabolism and 47-Calcium kinetics before and after long-term thyrocalcitonin treatment in senile osteoporosis, Clin.Sci. 38:397 (1970).

30. A. Cannigia, and A. Vattimo, Kinetics of [99]mTechnetium-tin-methylene diphosphonate in normal subjects and pathological conditions: a simple index of bone metabolism, Calcif.Tissue 30:5 (1980).

31. A. Caniggia, and A. Vattimo, Kinetics of ^{99}mTechnetium-methyl-ene-diphosphonate (^{99m}Tc-Sn-MDP) in women with post-meno-pausal osteoporosis, in: "Molecular Endocrinology," I. MacIntyre and Szelke, eds., Elsevier/North Holland Biomedical Press (1979).

32. A. Caniggia, C. Gennari, A. Vattimo, P. Nardi, R. Nuti, and M. Galli, Early effects of synthetic bovine parathyroid hormone and synthetic salmon calcitonin on urinary excretion of cyclic AMP, phosphate and calcium in man, Calcif.Tiss.Res. 20:209 (1976).

33. J. Reeve, D. Williams, R. Hesp, P. Hulme, L. Klenerman, J. M. Zanelli, A. J. Darby, G. W. Tregear, and J. A. Parsons, Anabolic effect of low doses of a fragment of human para-thyroid hormone on the skeleton in postmenopausal osteo-porosis, Lancet 1:1035 (1976).

34. G. Milhaud, A. Julienne, C. Desplan, C. Calmettes, M. Benezech, M. S. Moukhtar, Dosaggio radioimmunologico della calcito-nina e dell'ormone paratiroideo nell'osteoporosi, in: "Atti del Simposio Internazionale sulle applicazioni te-rapeutiche della calcitonina," E.Polli, ed., Capri (1977).

35. J. C. Stevenson, G. Abeyasekera, I. MacIntyre, G. Lane, and M. I. Whitehead, The interrelation of 1,25 dihydroxyvitamin D and calcitonin during normal pregnancy, in: "Fifth Workshop in Vitamin D,"Historic Williamsburg, Virginia, Abstracts (1982).

36. J. C. Stevenson, G. Abeyasekera, C. J. Hillyard, K. G. Phang, I. MacIntyre, S. Campbell, P. T. Townsend, O. Young, and M. I. Whitehead, Calcitonin and the calcium-regulating hormones in postmenopausal women; effect of oestrogens, Lancet 1:693 (1981).

37. A. Caniggia, R. Nuti, and M. Galli, The 24th urinary cyclic adenosine 3',5' monophosphate/creatinine ratio: a useful approach to the diagnosis of parathyroid disorders and function, J.Endocr.Invest. 4:281 (1981).

THE ROLE OF THE PHYSICIAN IN FEMORAL NECK FRACTURE

R. E. Irvine

Consultant Physician
Department of Medicine for the Elderly
Hastings Health Authority

A major concern of the British National Health Service in the past decade has been the rapid growth in orthopedic waiting lists. In 1972 these lists contained 67,231 names. By 1979 the number had grown to 130,959. The lists are growing at 10 per cent a year and the average waiting time for an elective orthopedic operation is two years. Such figures represent a disturbing picture of human suffering and major social cost. Two years is too long to wait if you are old. It is too long also if you are younger and potentially able to work.[1] The reasons for this situation were considered by an influential committee chaired by Professor R. Duthie of the Orthopedic Department at Oxford. Its report was published last year.[2]

DUTHIE REPORT

The lengthening waiting lists were attributed to two main factors, the aging of the population and advances in surgical technology. There are now fewer younger and more older patients in every orthopedic department. Because of this there are more patients also with fractured neck of femur. Advances in technology have created a huge demand for joint replacement. The report comments on the dilemma which affects every surgical department. This is to balance emergency work against elective surgery. They point out that there is little prospect of a massive injection of resources into orthopedic departments. The problem, as with the rest of the National Health Service, is to make better use of the resources already available. This involves collaboration between the orthopedic surgeon and the physician in geriatric medicine, who have much to gain if they can work together. A hospital where

they do so will provide a better service, not only for the patients
with fractures but also for those who need elective surgery. The
field of geriatric orthopedics covers all forms of fracture and the
physician's help may by sought also in the assessment of candidates
for joint replacement and in their post operative management. But
the main area of collaboration is the field of femoral neck fracture.

EPIDEMIOLOGY

The most reliable studies of the epidemiology of fractured
neck of femur have been made in Newcastle-upon-Tyne.[3,4] Table 1
compares the Newcastle figures for 1976/77 with those for Hastings
1981. Figures for England and Wales in 1977 are shown for
comparison.[5] The reliability of the national figures, which depend
on hospital activity analysis (HAA), has been questioned recently.[6]

The most striking difference between the Hastings figures and
the others is the greater proportion of patients operated on, 97
per cent in Hastings against 70 per cent nationally and 82 per cent
in Newcastle. Death in hospital was more common in Newcastle.
The Hastings and national figures are much the same. The length
of stay in the orthopedic wards is much shorter in Hastings.
Patients with fractured neck of femur often move from one ward to
another in the course of their stay and the total stay for patients
in Hastings is 33 days against 59 days in Newcastle. There is no
comparable figure from the national statistics. The important
difference between Hastings and Newcastle is the smaller proportion
of orthopedic beds occupied by patients with femoral neck fracture.
This allows the orthopedic department some leeway for elective
surgery. The differences between the figures for Hastings and the
figures for England and Wales in this respect are not striking.
Indeed the proportion of orthopedic beds occupied in Hastings is
greater than the national average. This can possibly be attributed
to the fact that Hastings is in the center of a retirement area
where the proportion of people of 65 in the population is almost
twice the national average.

POLICY IN HASTINGS

In Hastings 27 per cent of the population are over 65 and 11
per cent are over 75. In 1981 there were 270 patients with 275
femoral neck fractures. The average age of the patients was 82.
The relatively favorable Hastings figures spring from a surgical
policy developed by Devas[6] twenty years ago and by fine operative
surgery by all the orthopedic surgeons in Hastings.

These principles are shown in Table 2. The most important
point is that no patient is too old to be denied the relief of pain

Table 1. Fractured neck of femur - some comparisons

	Eng/Wales	Newcastle	Hastings
Operated	70%	82%	97%
Died in hospital	16%	33%	18%
Orthopedic days of stay	37	57	11
Overall hospital stay	?	59	33
Orthopedic beds occupied	18%	43%	23%

Sources 3, 4, 5

and the improvement in mobility that may be expected to follow surgery. Only those who are likely to die within a day or two are regarded as too ill for operation.

The operation itself must be one which will allow the patient to bear weight immediately. Methods of treatment which do not allow immediate post operative weight bearing, like the traditional Smith Peterson pin, are not used.

The third principle is that although the hip itself must be efficiently treated it is the patient as a whole that matters. This is where the physician can play a vital role. To understand the patient as a whole implies the need for a full medical, functional and social assessment by someone familiar with the techniques of geriatric medicine. In Hastings one of the medical staff from the department of medicine for the elderly pays a weekly visit to the orthopedic wards to look at any patients who are causing problems.[7] The most common problems are prolonged

Table 2. Fractured neck of femur

Principles of Management

Surgery for all unless moribund

Operation on next list

Immediate weight-bearing post-op

The patient not the part

Medical Assessment

confusion and failure to regain mobility. If this crucially
important stage in the patient's illness is mishandled another long
stay patient will have been created. If it is handled well the
patient may yet return to the community. The most important reason
for confusion after operation is pre-existing mild dementia made
worse by the injury itself, the operation, the anesthetic, post
operative sedation and overdosage with analgesics. In addition the
confusion may reflect a urinary or respiratory infection, a
pulomary embolism, a myocardial infarct or a stroke. A very
important and often overlooked cause is fecal retention. An
important function of the physician's visit is to identify, in
consultation with the orthopedic ward staff, the patients who should
be transferred to the geriatric orthopedic unit, (GUO).[8]

GERIATRIC ORTHOPEDIC UNIT

In Hastings the accident department, the surgical admissions
unit and the orthopedic wards and operating theaters are in one
hospital, the Royal East Sussex, while the medical and geriatric
wards are in another, St. Helen's, two and a half miles away. In
1962 a surgical ward in St. Helen's Hospital, which had been closed
for lack of staff, reopened and ten beds were allocated to geriatric
orthopedics. Since then the ward has been called the geriatric
orthopedic unit. The principles on which the ward operates have not
changed since they were first described in 1964.[9] They are
summarised in Table 3.

The fundamental principle is that the department of medicine for
the elderly, formerly the Hastings geriatric unit, undertakes to
provide the day to day care of selected patients transferred to the
ward at the decision of the surgeon. The surgeon does not have to
plead with the physician to take the patient. If there is a bed
in the unit the surgeon can choose who to send.

The other essential feature is the combined ward round carried
out each week by one of the orthopedic surgeons and one of the
consultant physicians from the department of medicine for the elderly.
Both the physician and the surgeon are empowered to act for their
colleagues and the round is attended also by the junior medical staff
from each department since it is regarded as an important educational
experience. The remedial therapist, the nurses, and the social
worker, a most vital contributor, also attend.

The joint round is expensive in professional time but it
ensures that the patient receives full consideration from both a
surgical, medical, and a social point of view and that nothing is
left undone which could help the patient to return to the community.
It is important to emphasize that the unit is for treatment and
not convalescence. The iller the patient the sooner she should be

Table 3. St. Helen's Geriatric Orthopedic Unit

Priciples

Surgeon selects patients

Physician supplies care

Combined ward round

Treatment not convalescence

Joint responsibility till fracture healed

Separate unit for lighter patients

transferred to the unit, provided she can stand a journey of $2\frac{1}{2}$ miles by ambulance.

The patient stays in the ward for as long as the surgical problem is contributing to her need for care. When this is no longer the case, the department of medicine for the elderly undertakes to remove the patient from the special unit into its ordinary geriatric ward. I have stressed the role of the unit in the management of patients with femoral neck fractures but one third of the patients who pass through the unit have other orthopedic and medical problems suitable for combined care.

BEXHILL UNIT

Since 1980 we have had a second unit for geriatric orthopedics in the geriatric rehabilitation unit at Bexhill. This takes orthopedic patients who formerly went to a convalescent home, now closed as a result of the financial cuts. Most of the patients who come to this unit are of low dependency. Only about a quarter have a femoral neck fracture. Most are recovering from elective surgery, mainly joint replacements. These patients too receive combined medical and surgical assessment. The clinical assistants who run the Bexhill unit under the direction of the consultant physicians do a combined round with the orthopedic surgeons every week. The consultant physicians in geriatric medicine here see only selected problems. The results are very good. In 1981 all 48 patients who went to the Bexhill unit with a fractured neck of femur were discharged in an average of 13 days from transfer.

MEDICAL ASSESSMENT

In these two units the staff of the department of medicine for
the elderly compile a complete medical history, consulting the
relations and the general practitioner if necessary as well as the
patient. A full medical examination is undertaken with special
reference to pressure areas, the rectum and the bladder. It is
particularly important to examine the nervous system and the plantar
responses need careful attention. The patient's hearing and
eyesight must be considered. To give the patient the best opportunity to communicate it is important that he has his spectacles,
false teeth and hearing aid. A vital part of the medical
examination is the assessment of the mental state.

Medical Problems

In a consecutive series of 50 patients with femoral neck
fractures studied in our unit by Campbell[11] brain failure was the
commonest medical problem and was present in more than a third of
patients. Next came other neurological diseases such as stroke and
Parkinsonism. Heart disease, respiratory disease, arthritis,
alimentary tract disorders and diabetes were all important also.
It is worth noting that 12 per cent of Campbell's series had
malignant disease. Every series of patients with femoral neck
fracture contains a fair proportion of patients in whom the fracture
is pathological due to bony infiltration by malignant deposits.

Walking Difficulties after Surgery

The main reason for the transfer of patients from the acute
orthopedic ward to the geriatric orthopedic unit is that the patient
is slow to get moving after the operation. The reasons why a
patient fails to walk after operation embrace the whole of medicine
but Wright and Fenwick[12] provide a scheme which I have found helpful.
The main causes of difficulty in walking are, they say, weakness,
stiffness, loss of balance and pain.

Pain

All patients go through some pain in the first few days after
operation but pain after the first week raises questions about the
surgery. It is most important that the surgeons should review any
patient with persistent pain, since this may indicate need for the
surgery to be revised.

Weakness

Any patient unable to walk because of general weakness is
seriously ill either with an infection of heart failure or pulmonary

embolism.   Weakness may also be due to postural hypotension and
sometimes to metabolic or electrolyte disturbance.   Patients too
weak to walk after an operation are a challenge to the physician.

Stiffness

Siffness after operation usually implies the presence of a
stroke, Parkinsonism or arthritis.   All old people stiffen up
quickly when they are immobilised for any length of time.   The best
way to prevent stiffness is the earliest possible mobilisation.

Loss of Balance

A major reason for poor mobilisation after operation is loss of
balance.   This may be related to the fall which led to the fracture.
Many people with vertebrobasilar arterial disease lose their sense
of verticality and believe themselves to be upright when in fact
they are leaning backwards.   Some authorities advocate raising the
heels for people in this state.   We find that a short period of
training with a low walking frame to encourage the patient to bend
forward, is usually sufficient.   Dementia is also an important
cause of falling and immobility as well as of failing memory.
Although balance may remain precarious, most elderly patients are
able to regain mobility.

REHABILITATION

Rehabilitation depends on accurate functional assessment.
Functional assessment by the doctor is a rather crude affair and
must often depend upon the reports of other people more than his own
observations.   This is one of the reasons why a team is essential
for rehabilitation.   All members of the team need to be considering from the beginning how the patient will cope when she gets home.
The social worker is an essential member of the rehabilitation team
and her contribution is indispensable.   The physiotherapist is the
specialist in walking but the occupational therapist has an equally
important role to play, particularly in assessing the patient's
ability to dress and to perform the activities of daily living.
The therapist will note what aids the patient may need to increase
her independence.   A growing tendency among occupational therapists
is to accompany the patient on a home visit, to assess the situation
prior to discharge.

The nurses are the only people who are with the patient
throughout the twenty-four hours and their observations are
indispensable also.   They are the first to notice the patient's
response to drugs and any change in their condition.   The ward
sister herself is the hub of the rehabilitation team.

TERMINAL CARE

A particular skill of the nurse, which usually exceeds that of the doctor, is to recognise when the patient is dying. Care can be greatly improved if the team realises that the patient is dying. When this moment is reached further investigation and efforts at rehabilitation become inappropriate. A new set of priorities come into existence, the priorities of terminal care. It is important then that the doctors prescribing should ensure that the patient is relieved of distress both physical and mental.

CONCLUSION

I hope I have convinced you that the physician in geriatric medicine and his team can give assistance to the orthopedic surgeon and that their close co-operation is likely to lead to a shorter hospital stay for the patient and better use of health service resources. A particular benefit to the orthopedic surgeon is likely to be an increase in the number of beds available for elective surgery such as joint replacement. The department of geriatric medicine gains also. Regular visits from the orthopedic surgeons are of great value and there are always patients in the ordinary geriatric wards who can benefit by the advice of an orthopedic specialist. Finally the opportunity for regular mutual consultation strengthens friendship and collaboration throughout the hospital, maintains clinical enthusiasm, ensures the best deal for the patient and makes the best use of limited resources.

SUMMARY

Elderly patients with femoral neck fractures have so many concomitant medical problems that orthopedic surgery alone, however skilful, is not enough. The patients need also the assistance of a physician trained in geriatric medicine. A system of collaborative care which has been running for 20 years in Hastings is described.

REFERENCES

1. Editorial, Heddle's Hip List, World Medicine 17:5 (1982).
2. DHSS, "Orthopaedic Services: Report of a Working Party to the Secretary of State for Social Services", H.M. Stationery Office, London (1981).
3. J. Grimley Evans, D. Prudham, and I. Wandless, A prospective study of fractured proximal femur: incidence and outcome, Public Health, London 93:235 (1979).

4. J. Grimley Evans, Fractured proximal femur in Newcastle-upon-
    Tyne, Age and Ageing 8:16 (1979).
5. A. Fenton Lewis, Fractured neck of femur: changing incidence,
    Brit.med.J. 283:1217 (1981).
6. J. J. Rees, Accuracy of hospital activity analysis data in
    estimating the incidence of proximal femoral fracture,
    Brit.med.J.284:1856 (1982).
7. M. B. Devas, "Geriatric Orthopaedics," Academic Press, London
    (1977).
8. R. Sainsbury and K. G. F. Benton, When a geriatrician can
    contribute to orthopaedics, Geriat.Med. 11:54 (1981).
9. R. E. Irvine and M. B. Devas, Fractured neck of femur in
    elderly women, in: "Age with a Future," P. From Hansen, ed.,
    Munksgaard, Copenhagen (1964).
10. R. E. Irvine and T. M. Strouthidis, Medical care in geriatric
    orthopaedics, in: "Geriatric Orthopaedics," M. B. Devas, ed.,
    Academic Press, London (1977).
11. A. J. Campbell, Femoral neck fracture in elderly women,
    Age and Ageing,5:102 (1976).
12. W. B. Wright and G. M. Fenwick, The fractured femur: why call in
    the geriatrician? Injury 9:282 (1978).

DIGITALIS METABOLISM IN AGED PATIENTS

F. I. Caird

University Department of Geriatric Medicine
Southern General Hospital
Glasgow

Digitalis metabolism, in particular that of digoxin, has been studied in the elderly more than perhaps that of any other drug and has been almost completely elucidated.

The absorption of digoxin appears to be unchanged in the elderly[1,2] as is its transport in the serum. Some 30 per cent is protein-bound, and it does not appear that this is substantially altered by age.[3] The volume of distribution of digoxin is substantially reduced in the elderly, from approximately 500 to 300.[4,5] This results from a lower average body weight and a reduction from approximately 7-8 to 5-6 1/kg body weight.[5,6] This change probably results from a reduction in the volume of muscle in the elderly, since this forms quantitatively the main receptor for digoxin.[7] Three studies have shown that this volume is much increased in thyrotoxicosis, including in the elderly.[5,8,9] The reasons for this are not known, though it is tempting to speculate that there is an increase in the amount of the muscle sodium-Na/K ATPase, possibly the principal site of digoxin uptake.

The principal alteration in digoxin metabolism is a reduction in its renal excretion, mainly as a result of the low creatinine clearance of elderly patients taking digoxin who have in addition to any age-related reduction in creatinine clearance, a disease-related reduction also.[10] Renal excretion by tubular secretion also exists, but does not appear to be of any great importance.[11] Also reduced is the hepatic excretion of digoxin, from 50-60 ml/min in young people to approximately 20 ml/min in the elderly.[10,12] The effects of these falls in both renal and hepatic excretion are greatly to reduce the steady state dosage of the drug required to give rise to adequate serum levels. Maximum maintenance dosage in the elderly should not exceed 0.25 mg/day.[4]   The latter authors showed that knowledge of the pharmacokinetics of digoxin was adequate to explain at least the effects of three loading doses.

The pharmacodynamics of digoxin have been barely worked out in the elderly. Overall pharmacodynamic changes may well be less important than the pharmacokinetic. Reid et al[5] showed that as far as the extra-cardiac toxic effects are concerned, and they are of great importance, there is little or no overlap between toxic and non-toxic serum concentrations. What is not known is whether "normal" serum concentrations of 1-2.5 mg/ml produces the actions of digoxin in the elderly. Probably, and as far as is known at present, the effects of digoxin in the elderly are the same as in the young, concentration for concentration.   It is not known whether the observation of impaired inotropic but unimpaired cardiotoxic response in elderly animals[13,14]   also applies to man.

REFERENCES

1. B. B. Taylor, R. D. Kennedy, and F. I. Caird, Digoxin studies in the elderly, Age & Ageing 3:79 (1974).
2. B. Cusack, J. Kelly, K. O'Malley, J. Noel, J. Lavan, and J. Morgan, Digoxin in the elderly: Pharmacokinetic consequences of old age, Clin.Pharmacol.Ther. 25:772 (1979).
3. S. Wallace and B. Whiting, Some clinical implications of the protein binding of digoxin, Br.J.Pharmacol. 1:325 (1974).
4. F. I. Caird and R. D. Kennedy, Digitalisation and digitalis detoxication in the elderly, Age & Ageing 6:21 (1977).
5. J. Reid, R. D. Kennedy, and F. I. Caird, Digoxin kinetics in the elderly, Age & Ageing, In the press.
6. J. K. Aronson and D. G. Grahame-Smith, Monitoring digoxin therapy. II. Determinants of the apparent volume of distribution, Br.J. Clin.Pharmac. 4:223 (1977).
7. J. E. Doherty, W. H. Perkins, and W. J. Flanigan, The distribution and concentration of tritiated digoxin in human tissues, Ann.Intern.Med.66:116 (1967).
8. G. M. Shenfield, J. Thompson, and D. B. Horn, Plasma and urinary digoxin in thyroid dysfunction, Eur.J.Clin.Pharmacol. 12:437 (1977).
9. J. R. Lawrence, D. J. Sumner, W. J. Kalk, W. A. Ratcliffe, B. Whiting, K. Gray, and M. Lindsay, Digoxin kinetics in patients with thyroid dysfunction. Clin.Pharmacol.Ther. 22:7 (1977).
10. M. A. Roberts and F. I. Caird, Steady-state kinetics of digoxin in the elderly, Age & Ageing 5:214 (1976).
11. E. Steiness, Renal tubular secretion of digoxin, Circulation 50:103 (1974)
12. B. Whiting, J. R. Lawrence and D.J. Sumner, Digoxin pharmacokinetics in the elderly, in: "Drugs and the Elderly", J. Crooks and I. H. Stevenson, eds., MacMillan, London and Basingstoke (1979).

13. G. Gerstenblith, H. A. Spurgeon, J. P. Frohlich, M. Weisfeldt, and E. G. Lakatta, Diminished inotropic responsiveness to ouabain in aged rat myocardium, Circ.Res. 44:517 (1979).
14. T. Guarnieri, H. Spurgeon, J. P. Froehlich, M. L. Weisfeldt, and E. G. Lakatta, Diminished inotropic response but unaltered toxicity to acetylstrophanthidin in the senescent beagle, Circulation 60:1548 (1979).

DIGITOXIN IN ELDERLY PATIENTS WITH RAPID ATRIAL FIBRILLATION

N.G. Dey, C.M. Castleden and J.E.F. Pohl*

Department of Geriatric Medicine
The General Hospital, Leicester
*Department of Medicine
University of Leicester

A preliminary study showed that digitoxin pharmacokinetics were unaffected by aging (Table 1), and that the half-life was sufficiently long to maintain plasma concentrations within the therapeutic range for a week after a single intravenous injection of 0.6 mg digitoxin (Fig.1).[1] These results raise the possibility that elderly patients needing a digitalis preparation could be treated with weekly doses of digitoxin. The aim of this study was to assess the safety and efficacy of such therapy in elderly patients with atrial fibrillation and the following three case studies are presented as a preliminary report.

SELECTION OF PATIENTS

Elderly patients with rapid atrial fibrillation had their ventricular rate controlled with digitoxin. When they were in a clinically stable condition and on a maintenance dose of digitoxin for at least one month, those able to understand the nature and possible complications of the study were asked if they would like to take part. At this time all were living independently in the community and had no evidence of mental impairment. Blood urea, creatinine, potassium, liver function tests and thyroid function were within normal limits for the local laboratory. In every case the patient's General Practitioner have his permission and the study as a whole had Local Ethical Committee approval.

METHOD

The patient's suitability and clinical state was assessed as an

Table 1. Pharmacokinetic parameters for digitoxin estimated in six
          young and six old subjects from a single intravenous dose
          of 0.6 mg of digitoxin

	Half-life (days)		Apparent volume of distribution (1/kg)		Clearance ($1 \text{ kg}^{-1} \text{ day}^{-1}$)	
	Mean	s.e. mean	Mean	s.e.mean	Mean	s.e.mean
Young	10.0	0.5	0.64	0.03	0.045	0.002
Old	8.3	0.8	0.62	0.02	0.054	0.006

$$p > 0.05 \text{ throughout}$$

outpatient, and control of ventricular rate was confirmed by an E.C.G.
A plasma sample was taken before and after that day's normal digitoxin
dose to assess drug compliance and measure the normal range of
digitoxin concentrations in the patient.   The patient was then asked
to omit digitoxin for six days after which she was admitted to the
Coronary Care Unit.   No food or drink, apart from water, was allowed
from midnight until three hours after the patient had received an
oral dose of digitoxin equal to seven days of the maintenance dose
taken previously.   Plasma was taken for estimation of digitoxin
concentration by the method described previously[1] at 0, 1, 1½, 2, 4
and 6 hours through an intravenous cannula kept patent with hepar-
inised saline.   The patient was monitored by continuous electro-
cardiography for six hours and assessed clinically at each sampling
time for evidence of digitalis intoxication.   The patient was allowed
home in the afternoon and was visited next morning at home when she
underwent a further clinical examination.   The patient took no
further digitoxin for six days and was then re-admitted to the
Coronory Care Unit and procedure was repeated.   Six days later the
patient resumed the daily maintenance dose of digitoxin.

RESULTS

Case 1

    Mrs F. K. aged 81 presented with atrial fibrillation and a
ventricular rate of about 150/minute, due to ischaemic heart disease.
She took salbutamol for chronic obstructive airways disease.   She
was digitalised with digitoxin (3 doses of 400 μg, 200 μg and 200 μg
over 36 hours), and then given 100 ug daily.   Her ventricular rate
settled around 72/minute at rest.

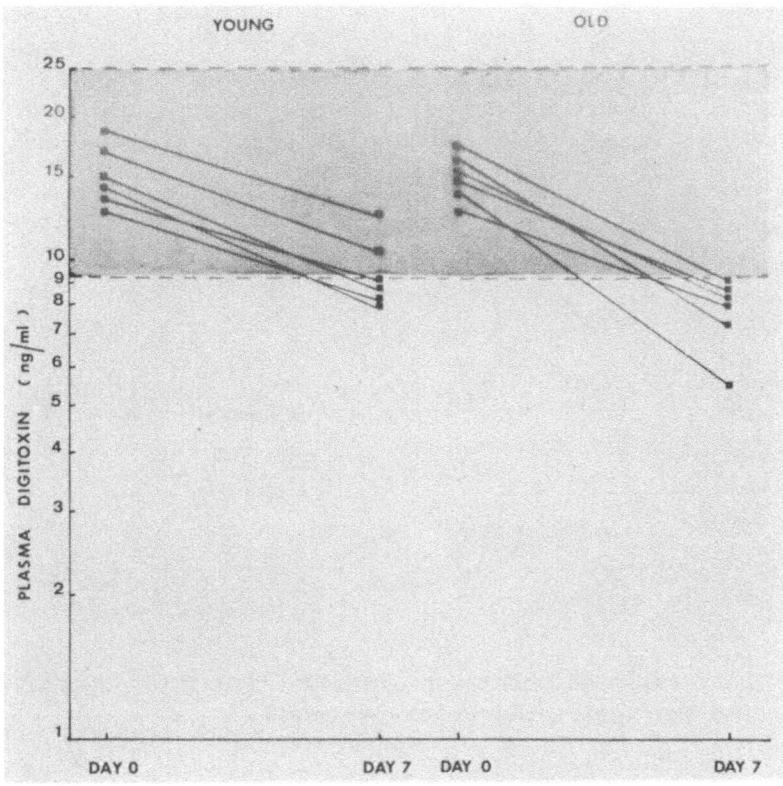

Fig. 1.   The fall in digitoxin levels over 7 days
          in six young and six old subjects
          calculated from the β elimination phase
          following a single 0.6 mg i.v. dose of
          digitoxin. The shaded area indicates the
          therapeutic range (reproduced with
          permission[1]).

Figure 2 shows her plasma digitoxin concentration before and
two hours after her normal daily 100 μg dose, and following 700 μg
once a week for two weeks. It also indicates the ventricular rate
during this period. The patient did not show any adverse effects
clinically or electrocardiographically during the study, and was
able to be discharged home in the afternoon following her weekly
dose.

After completing the study the patient resumed her usual
maintenance dose of 100 μg of digitoxin and has remained in well
controlled atrial fibrillation.

Case 2

Mrs. W.F. aged 87 presented with atrial fibrillation with a

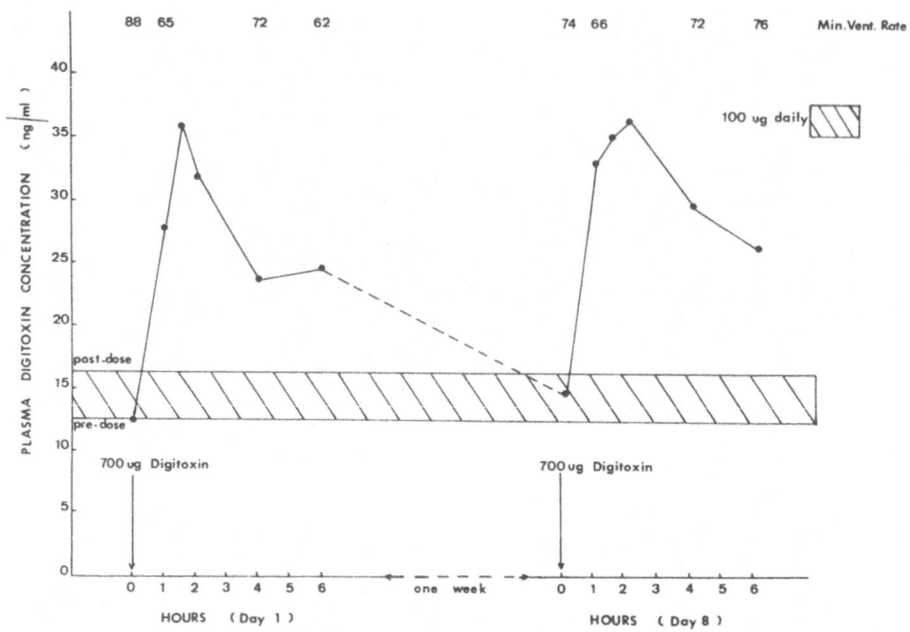

Fig. 2. The relationship between plasma digitoxin concentration,
         time and ventricular rate in Case 1.

ventricular rate of about 150/minute.   She was in marked congestive
cardiac failure despite taking one tablet of moduretic.   She had
had an attack of acute bronchitis 6 weeks previously.   She was
digitilised with digitoxin as in Case 1 and her ventricular rate was
maintained around 84/minute, on a maintenance dose of 100 µg/day.
Whilst in the ward she developed a left femoral embolus which
required embolectomy and anticoagulation with heparin and warfarin.
She also developed an auto-immune haemolytic anaemia which was
treated with prednisolone.

        All these problems had resolved by the time of the study.   After
the first dose of 700 µg she felt nauseated, vomited and needed 10 mg
of metoclopramide intravenously.   On a second occasion she was given
350 µg and no adverse effects occurred.   One week after the study
she continued on a maintenance dose of 50 µg.

        Figure 3 shows her plasma digitoxin concentration on her
maintenance daily dose of 100 µg and before and after 700 and 350 µg,
and the ventricular response during this period.

Case 3

        Mrs. L.P. aged 80 was a known hypertensive on sotalol, methyl-

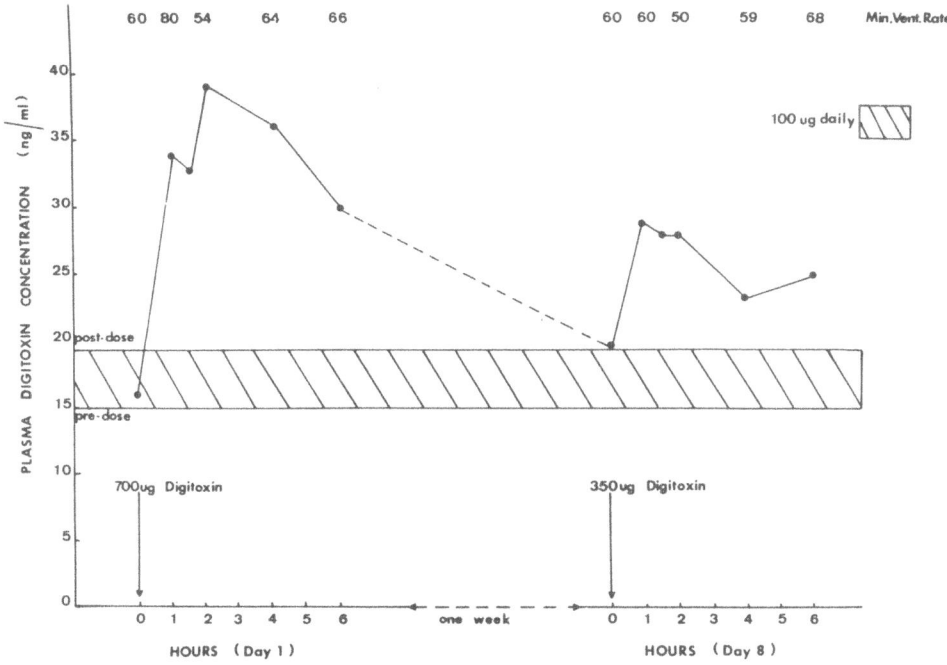

Fig. 3.   The relationship between plasma digitoxin concentration,
time and ventricular rate in Case 2.

dopa and hydrochlorothiazide. She presented with exertional
dyspnoea and atrial fibrillation with a ventricular rate about
120/minute.   She was not in overt cardiac failure.   She was
digitalised with digitoxin as in the previous two cases and her
ventricular rate settled at about 80/minute.   In view of her
relatively high plasma concentration on maintenance dosage, only
350 μg were given at weekly intervals.   She completed the study
without any side effects and continued her maintenance dose one
week later.

Figure 4 shows her plasma digitoxin concentrations on her
maintenance dose and following the two doses of 350 μg, and the
ventricular response during this period.

DISCUSSION

These case reports provide preliminary evidence that atrial
fibrillation can be safely and effectively controlled on weekly oral
digitoxin therapy.   If larger studies confirm these results, then
physicians have an alternative to daily digoxin therapy.   Toxicity

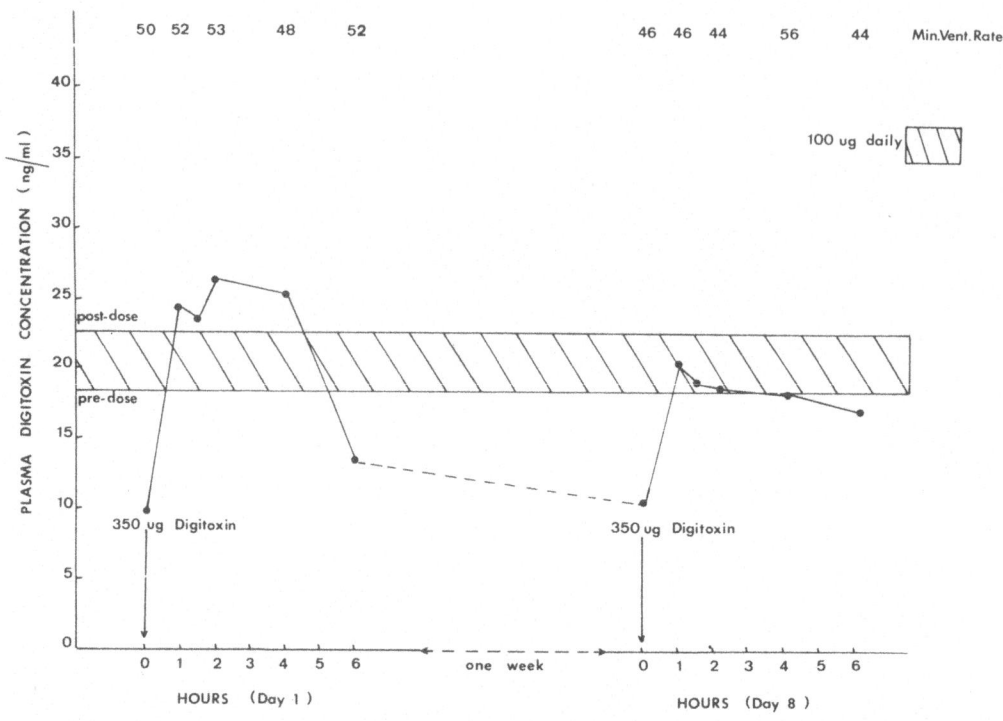

Fig. 4. The relationship between plasma digitoxin concentration,
         time and ventricular rate in Case 3..

to this drug remains high in elderly patients[2,3] because correct
dosage is difficult to achieve especially if the patient is in sinus
rhythm.   This is because muscle mass fails with aging so that blood
urea and serum creatinine values are often within normal limits even
though glomerular filtration rates are less than 50 per cent of that
in young adults.[4]   Chamberlain and his colleagues[5] also pointed out
that many patients with atrial fibrillation have latent digoxin
toxicity which readily becomes manifest if the clinical situation
reduces the glomerular filtration rate such as in myocardial
infarction or dehydration.   Finally, plasma digoxin concentrations
may be altered by co-administrations of certain anti-arrhythmic drugs
Such interactions do not occur with digitoxin,[7] nor is its elimination which is primarily by hepatic metabolism, so clearly related
to glomerular filtration rate, hepatic function[8] or aging.[1]

    For these reasons digitoxin in general may be valuable in many
other patients with atrial fibrillation, and may be particularly
relevant as a weekly dose in those living alone and known to be
unreliable tablet takers.   Most physicians recognize such patients
who are admitted to hospital a number of times in cardiac failure
with untouched bottles of tablets.   A therapeutic trial of weekly

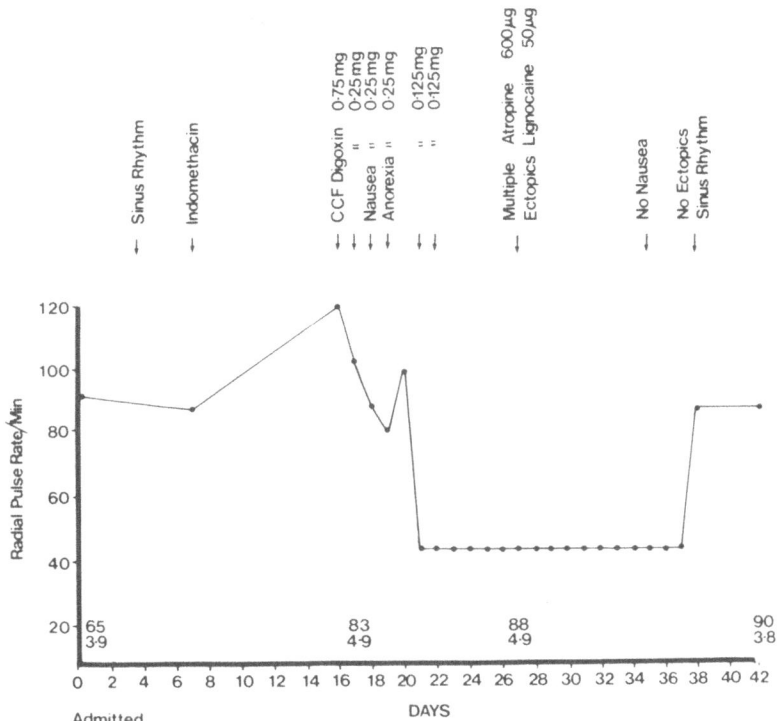

Fig. 5.   The clinical course of a subject with
          digoxin toxicity.

digitoxin in such patients could be undertaken easily in hospital,
and then subsequent doses given by the District Nurse. Possibly
a long-acting diuretic such as chlorthalidone could also be used in
these patients if necessary.

      One drawback to digitoxin is that its toxicity would be more
prolonged than that of digoxin even though digoxin toxicity can last
many days in some patients (Figure 5).   However, digitoxin toxicity
should not occur so frequently once doctors become familiar with its
dosage provided that they use it only in patients with atrial
fibrillation so that the ventricular rate could indicate the correct
dosage.   Since many centres now measure plasma digitoxin concentrations, these provide a further indication of correct dosage.   The
normal therapeutic rage is 9-25 ng/ml.   Two patients exceeded this
reaching concentrations above 35 ng/ml.   Although such concentrations
were only temporary, lasting a matter of a few hours, one patient
felt nauseated and was sick.   No cardiac toxicity was noted but
clearly weekly dosing will not be possible in all patients.   A
solution might be to give two doses a week or possibly two on the
same day separated by at least four hours.   Since such toxicity can

be predicted following a maintenance dose by the plasma digitoxin
concentration, and there seems a good correlation between this
concentration and toxic symptoms, centres without facilities for
cardiac monitoring, and immediate estimation of plasma digitoxin
concentrations should not undertake this therapy until further
results and guidelines are available.

REFERENCES

1. M. A. Donovan, C. M. Castleden, J. E. F. Pohl, and C. A. Kraft,
    The effect of age on digitoxin pharmacokinetics, Br.J.Clin.
    Pharmacol.11:401 (1981).
2. N. Hurwitz, Predisposing factors in adverse reactions to drugs,
    Brit.med.J. 1:536 (1969).
3. T. W. Smith and E. Haber, Digoxin intoxication : the relationship
    of clinical presentation to serum digoxin concentration,
    J.Clin.Invest.49:2377 (1970).
4. J. M. Hansen, J. Kampmann, and H. Laursen, Renal excretion of
    drugs in the elderly, Lancet 1:1170 (1970).
5. D. A. Chamberlain, R. J. White, M. R. Howard, and T. W. Smith,
    Plasma digoxin concentrations in patients with atrial
    fibrillation, Brit.med.J. 3:429 (1970).
6. M. Wandell, J. R. Powell, D. Hager, P. E. Fenster, P. E. Graves,
    K. A. Conrad, and S. Goldman, Effects of quinine on digoxin
    kinetics, Clin.Pharm.Ther.28:425 (1980).
7. H. R. Ochs, J. Pabst, D. J. Greenblatt, and H. J. Dengler,
    Noninteraction of digitoxin and quinidine. New Engl.J.Med.
    303:672 (1980).
8. D. Perrier, M. Mayersohn and F. I. Marcus, Clinical Pharmaco-
    kinetics of digitoxin, Clin.Pharmaco 2:292 (1977).

THE USE OF THE ANTIARRHYTHMIC AGENTS IN THE ELDERLY

P.U. Carbonin, M. Di Gennaro, R. Bernabei, A. Cocchi,
L. Carosella, M. Pahor and I. Taddei

Chair of Gerontology
University of Cattolica, Rome

Similar disturbances of cardiac rhythm occur in all ages, even if more frequently in the elderly.[1,2,3] As a consequence, the diagnostic problems and the therapeutic principles concerning the etiology, the recognition, the quantification and the choice of treatment of cardiac arrhythmias are generally identical for young and old patients.

Since these arguments have been recently reviewed,[4,5,6] we believe it more important to focus our attention on some topics specifically related to the use of antiarrhythmic agents in the elderly. In this connection, the main questions which need to be answered seem to be the following: (1) the possible existence of age-related alterations in cardiac electrophysiology that can explain the higher incidence of cardiac arrhythmias in the elderly subject; (2) the possible differences in the response to anti-arrhythmic agents between the young and aging heart; (3) the age-induced modifications in pharmacokinetics of antiarrhythmic drugs; (4) the specific therapeutic interventions in the elderly patient in order to avoid the frequent side effects of antiarrhythmic agents.

AGE-RELATED CHANGES IN CARDIAC ELECTROPHYSIOLOGY

Recent studies[1,7] have reported that a healthy elderly population is much more prone than a young one to alterations in some cardiac electrophysiological parameters and to cardiac rhythm disturbances. In these works,[1,7] the absence of cardiac abnormalities was checked by clinical examination and non invasive tests (treadmill exercise and thallium scintigraphy). Furthermore, microprocessor-assisted high resolution ECG and 24 hour-Holter ECG have revealed that: (1) a positive significant correlation exists between age and P-R and P-H intervals; (2) No significant correlation is present between the age and basal heat rate; (3) old subjects are characterized by a high incidence of premature beats and complex arrhythmias of both supraventricular and ventricular origin; (4) high degree A-V block, profound sinus bradycardia, abnormal sinus pauses and sinus arrest are rare in the normal elderly subject.

Therefore, the high incidence of cardiac arrhythmias, observed in healthy elderly populations cannot be explained only by the age-related increase in cardiac and extracardiac diseases, but it is necessary to suppose that age, per se, importantly influences the electrophysiology of the heart. This suggestion has noteworthy clinical implications. In fact, on the one hand it becomes mandatory to exercise great caution in treating cardiac arrhythmias in the elderly subject, since they may not represent a pathological finding; on the other, it is likely that the age-induced alterations in cardiac electrophysiology can modify the heart responsiveness to antiarrhythmic agents.

Some experimental results confirm that cardiac electrophysiology is altered by age. However, these studies are sparse and often inconsistent and, thus, it is difficult to draw definitive conclusions in this field. Lakatta[8] did not find significant differences in the characteristics of the ventricular muscle fiber action potential between young and senescent rats. On the contrary, Goldberg and Roberts,[9] Rosel et al,[10] and Rumberger and Timmermann[11] observed that age significantly modified the action potential of rat atrial cells, of canine Purkinje fibers and guinea pig papillary muscle, respectively. All these authors[9,10,11] observed a decrease of the action potential amplitude and duration, probably due to a variation in the slow inward current, that is in the calcium influx during the plateau phase of action potential.[12]

Age changes in cardiac cell calcium kinetics have also been demonstrated by means of other techniques. Guarnieri et al[13] showed that the inotropic response to catecholamines is depressed in the normal aging myocardium, and believed that this phenomenon was due to an alteration in transsarcolemmal calcium entry. Froelich et al,[14] and Guarnieri et al[13] observed a decreased rate of calcium uptake by sarcoplasmic reticulum of ventricular muscle fibers of senescent rats.

Since the cytosolic calcium overload is known to play a main role in the genesis of abnormal automaticity,[15,16] it cannot be excluded that alterations in calcium kinetics, similar to (or different from) those observed by Guarnieri et al[13] and Froelich et al,[14] may also be responsible for the tendency of the aging heart to develop arrhythmias.

Our recent results[17] seem to confirm this suggestion. In fact,
the ventricular arrhythmias provoked by reperfusion or by reoxygen-
ation (for the original methods see ref[18],[19]) are significantly more
frequent in the isolated heart of senescent rats in comparison with
that of young animals (Figure 1). This finding, associated with a
significant increase of the diastolic pressure (figure 2),
emphasizes the possible correlation between the incidence of
ventricular arrhythmias and the intracellular calcium load in the
aging heart.   In fact, the increase in diastolic pressure (contract-
ure) is an expression of the cytosolic calcium load.[20]

Actually, however, this suggestion only represents a useful
working hypothesis, the validity of which requires many more data
for its confirmation.   In fact, further and more detailed studies
made with more sophisticated techniques (for example, voltage clamp
technique), are necessary to characterize the possible age-induced
changes in the ionic currents and the electrical and metabolic
correlations, as well as to verify the age-interferences in other
possible mechanisms or arrhythmias and the importance of the
alterations, found in the isolated preparations, in the in vivo heart.

AGE-RELATED CHANGES IN CARDIAC RESPONSE TO ANTIARRHYTHMIC DRUGS

Since antiarrhythmic agents interfere with the transsarcolemmal
ionic currents, they have been classified according to their effect
on the action potential.[4]   This classification, however, does not
seem to be completely satisfactory for several reasons.   In fact,
even if some similar electrophysiological properties permit the
subdivision of the antiarrhythmic drugs into various classes, each
drug has its own unique characteristics.[6] Besides, the direct
electrophysiological effects can be complicated by influences on the
autonomic nervous system, as is the case of the paradoxical
arrhythmogenic effect of quinidine and dysopyramide due to their
anticholinergic action.[6] Furthermore, the fundamental mechanisms
underlying cardiac arrhythmias are not yet well known, and, thus,
it is likely that the antiarrhythmic action of some agents is not
only mediated by their known effect on the ionic currents, but also
by interferences with other mechanisms.[21]

All these considerations make it difficult to recognize the
exact meaning (and the practical importance) of the age-induced
changes (see Table 1) in the electrophysiological response to the
antiarrhythmic agents observed in isolated cardiac preparations.
According to Goldberg and Roberts[23]it is impossible to find precise
clinical implications in their experimental results "since studies
on the effectiveness of these agents in older patients are lacking.
Aside from general caution of carefully monitoring older individuals
on antiarrhythmic therapy, no clear systematic studies have come
to our attention. This is an area most fertile for investigation."

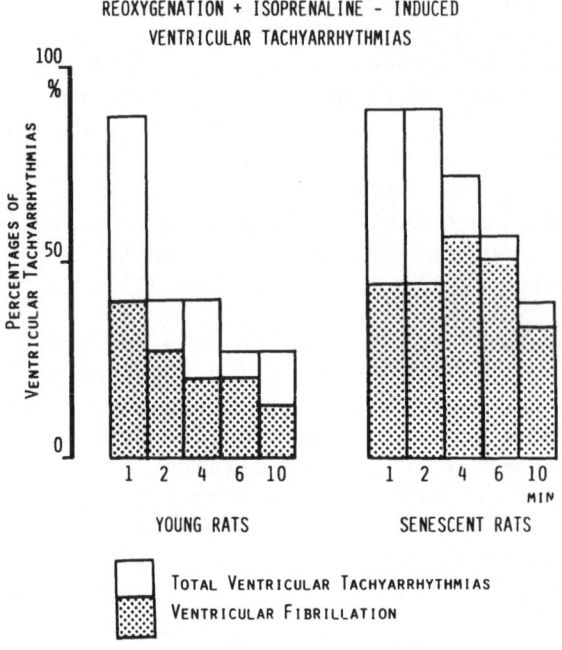

Fig. 1.   Percentages of ventricular
          arrhythmias during the reoxygen-
          ation of the Langendorff-perfused
          rat heart from young (2 months)
          and senescent (25 months) rats.
          The experiment consists in perfus-
          ing (after 15 min of stabilization)
          the isolated heart with an anoxic
          medium for 15 min and, successively,
          with a medium gased with 95% O_2 and
          5% CO_2 (reoxygenation). During
          anoxia, the perfusion was made with
          a glucose-free-medium. During
          anoxia and reoxygenation, 10^{-6}M
          isoprenaline was added to the medium.
          (From Pahor et al,[17] reproduced with
          permission).

There are several studies on the increased sensitivity of older
patients to digitalis, which is a useful antiarrhythmic agent in the
treatment of some supraventricular arrhythmias.   However, it is not
clear why lethality and toxicity of digitalis are age-related.
This fact might be due to one or more of the following causes:
age-induced variations in the pharmacokinetics of polar glycosides
(the non polar glycoside digitoxin seems to have normal kinetics
even in advanced age[24]); age-related changes in the number and

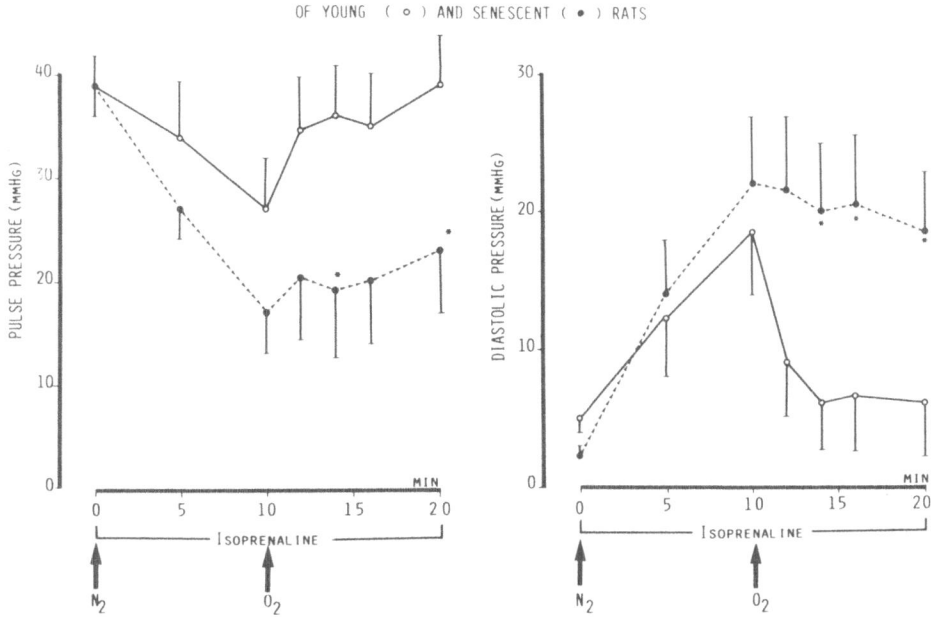

Fig. 2. Pulse pressure and diastolic pressure in mmHg of the
         isolated rat heart from young (n=15) and senescent (n=18)
         rats. The conditions of the experiment are the same as
         for Fig. 1.

function of the specific digitalis receptors; and finally, age-
related modifications of electrical and mechanical response to
cardiotonic steroids.   Actually, the last possibility has not been
extensively investigated.   Anyway, it is important to stress that
no parameter permits us to recognize clinically hypersensitivity
to digitalis and, in other words, to distinguish a priori those
patients who may develop toxicity in the presence of a therapeutic
digitalis plasma level.

PHARMACOKINETICS OF ANTIARRHYTHMIC DRUGS IN THE ELDERLY

     With the exception of digitalis glycosides (see ref[27,28]), the
kinetic studies of antiarrhythmic drugs are not numerous, and only
limited to a small number of substances and mostly in normal aged
subjects.

     As to the first A class[4] of the antiarrhythmic agents (useful
against the arrhythmias of supraventricular and ventricular origin,
see Tables 2, 3), studies of aged individuals have been performed

Table 1. Age-related changes in electrophysiological responses to the antiarrythmic agents (from Roberts and Goldberg,[9] Rosen et al,[10] and Goldberg and Roberts[22])

| | TRANSMEMBRANE POTENTIAL CHARACTERISTICS | | | | | |
	MPD	AP AMPLITUDE	V$_{MAX}$	AP DURATION	PHASE 1	PHASE 2
QUINIDINE	±	±	±	±	±	±
LIDOCAINE	±	±	±	+	+	+
TETROTODOXIN	±	±	±	+	±	±
PROPANOLOL	±	±	±	+	±	±
PHENYTOIN	±	±	+	−	+	−
CALCIUM ENTRY BLOCKERS (AHR-266 and VERAPAMIL)	±	+	+	+	+	+

± No age-related changes; + Age-related increase in response; − age-related decrease in response.

Abbreviations: MPD = MAXIMUM DIASTOLIC POTENTIAL; AP = ACTION POTENTIAL; V$_{MAX}$ = MAXIMUM RATE OF RISE OF PHASE 0 ACTION POTENTIAL.

Table 2. Characteristics of the antiarrhythmic drugs commonly used for the treatment of supraventricular arrhythmias*

DRUG	MAIN ELECTR.PHYSIOL. PROPERTIES	HEPATIC METABOLISM	BIOAVAIL-ABILITY	T 1/2	MAIN DRUG INTERACTIONS	MAIN ADVERSE EFFECTS
DIGOXIN$	Indirect vagal effect	Non Predominant	70%	36 H.	Quinidine Verapamil Diuretics	Arrhythmias GI symptoms Neurol.symptoms
VERAPAMIL	Block of I_{SI}	Predominant	25%	4-8H	Digoxin	Hypotension Constipation AV block
PROPRANOLOL$	Inhibition of sympathetic discharge	Predominant	25-50%	4-8H	Verapamil	Depression of Myocardial Contractility
DISOPYRAMIDE	Quinidine-like action	Non Predominant	80%	4-8H	?	Hypotension Anticholinergic Arrhythmias
QUINIDINE	Block of INA ↑E.R.P. in A.V.P.F.	Predominant	80%	6-7H	Enzyme Inductors Digoxin Alpha Blocking D	Hypotension Cinchonism Anticholinergic Arrhythmias
AMIODARONE	↑A.P.D. IN A.V.P.F.	?	Erratic	Very Long	Digoxin Warfarin	Corneal deposits Dysthyroidism Hypotension Arrhythmias

*   The table is a partial synthesis of data reported by Opie,[4] Bigger,[6] Einchelbaum and Somogyi,[30] Riva et al,[31] Padrini et al,[32] Klein et al,[33]
$ For other cardiotonic steroids see ref. 28
§   See Table 4 for the characteristics of other β-blocking agents.
Abbreviations: INA=Fast Inward Current, ISI=Slow Inward Current, APD=Action Potential Duration
ERP=Effective Refractory Period, A.V.P.F.= Atrial, Ventricular, Purkinje Fibers,
A.=Action, D.=Drugs

Table 3. Characteristics of the antiarrhythmic drugs commonly used for the treatment of ventricular arrhythmias*

DRUG	MAIN ELECTR.PHYSIOL. PROPERTIES	HEPATIC METABOLISM	BIOAVAIL-ABILITY	T 1/2	MAIN DRUG INTERACTIONS	MAIN ADVERSE EFFECTS
LIDOCAINE	Block of I_{NA} A.P.D. in P.F.	Predominant	0	100/200 Min	Drug influencing on hepatic flow enzyme inductors	CNS
MEXILETINE	Lidocaine-like action	Predominant	80%	10-12H	?	on CNS
PROCAINAMIDE	Quinidine-like action	50%	80%	4H	?	ECG changes Hypotension Lupus-like synd.
PHENYTOIN	Lidocaine-like action	Predominant	Erratic	24H	enzyme inductors	on CNS Megaloblastic Anaemi Osteomalacia
BRETYLIUM	↑ A.P.D. in P.F.	No	?	6-10H	?	Hypotension
AMIODARONE	SEE TABLE 2					
PROPRANOLOL	SEE TABLE 2					
DISOPYRAMIDE	SEE TABLE 2					
VERAPAMIL	SEE TABLE 2					

*This table is a partial synthesis of data by Opie,[4] Bigger,[6] Garfein[29] and Shanks.[33]
Abbreviations: as in Table 1.

with quinidine by Stevenson et al[35] and Ochs et al.[36]   These
authors[35] [36] observed a prolongation of plasma half-life and a
reduction of total body clearance of the drug, and suggested that a
reduction of quinidine clearance in old individuals is likely to be
due to an impairment of hepatic metabolism.   Stevenson et al,[35]
in fact, demonstrated a reduced induction response in old subjects
treated with both quinidine and the enzyme inductor dichlorophena-
zone.   Comacho et al[37] observed that the renal excretion of
procainamide and its main active metabolite is strictly age-related.

As to the first B class[4] of the antiarrhythmic agents (active
against ventricular arrhythmias, see Table 3), Triggs[38] showed a
significant increase in the distribution volume and a prolongation
of the half-life of lidocaine in elderly persons.   Similar results
have been obtained by Nation et al.[39]   The metabolism of phenytoin
in the elderly has been studied by Hayes et al[40] who noted an age-
related decrease of the drug clearance.   According to Hayes et al.[40]
however, the study of phenytoin is complicated by the fact that the
drug kinetics depend upon the level and binding-capacity of serum
albumin.

As to the second class,[4] including β-blocking agents active
against supraventricular and ventricular arrhythmias (see Tables
2,3,4), Castleden et al[43] showed an increase of serum level and of
bio-availability of propranolol in old subjects, and suggested that
these variations were due to a reduced hepatic biotransformation of
the drug.   Similar results have been obtained by Bianchetti et al,[44]
whilst Barber et al[45] found minor and not significant differences
in propranolol kinetics between young and old individuals.
Generally, it has been seen that plasma drug level, renal and body
clearances and plasma half-life of those β-blockers, which are
predominantly excreted unchanged in the urine (atenolol, practolol,
sotalol and nadolol), increase in renal insufficiency.[46,47,48]
Therefore, it is likely that the kinetics of the same β-blocking
agents is also impaired in the elderly because of the age-related
decrease of renal clearance.   Actually, Barber et al[45] found an
increase in AUC_{0-12} and a reduction in total body clearance of
atenolol in 5 elderly persons compared with 6 young subjects, and
observed that the atenolol body clearance was significantly correlated with creatinine clearance.   More recently, however, Rubin et
al[49] did not observe any difference in atenolol kinetics between
young and aged individuals.   It is likely that the discrepancy
between the results of Barber et al[45] and Rubin et al[49] is due to
the small sample size and to differences in atenolol bio-availability of the subjects studied.

As to the third class,[4] it is mainly consituted by amiodarone,
a useful agent in the treatment of supraventricular and ventricular
arrhythmias (Tables 2, 3).   In a recent study of young and old
patients with recurrent paroxysmal supraventricular tachycardia

Table 4.　Characteristics of the commonly used β-blocking agents*

DRUG	TYPE OF ELIMINATION	T 1/2$ H.	BODY CLEARANCE (L/min)	% OF DOSE ELIMINATED UNCHANGED IN URINE	TOTAL URINARY EXCRETION (% OF DOSE)	β-BLOCKING& ACTIVITY	INTRINSIC SYMPATHETIC ACTIVITY	MEMBRANE STABILIZING ACTIVITY
ALPRENOLOL	Hepatic	2-4	1.2	1	90	0.3	+	+
ATENOLOL	Renal	6-9	0.10	40	40	1	0	0
METOPROLOL	Hepatic	3-4	1.1	3	95	1	0	0
OXPRENOLOL	Hepatic-Renal	2	0.6	50	70-95	0.5-1	+	+
PINDOLOL	Hepatic-Renal	2	0.6	40	90	6	+	+
PRACTOLOL	Renal	5-13	0.14	90	90	0.3	+	0
PROPRANOLOL	Hepatic	3-6	1	1	90	1	0	+
SOTALOL	Renal	5-13	0.11	90	90	0.3	0	0
TIMOLOL	Hepatic-Renal	4-5	0.4	20	65	6	0	0
NADOLOL	Renal	14-17	0.15	90	15-21	0.3	0	0

* This table is a partial synthesis of data reported by Heikkila et al,[41] Dreyfuss et al,[42]
$ Generally, the biological half-life is longer than the plasma half-life
& Activity of propranolol : 1. The reported values are referred to the β-blocking activity observed in vitro, marked differences. From these values may be noted in the in vivo studies.

(PSVT) and chronically treated with amiodarone (Table 5), Padrini et al[32] noted an increase in plasma level of the drug and a decrease in the metabolite/amiodarone ratio in a group of elderly patients. In the same group, the main side effect of the drug, i.e. corneal deposits, occurred more frequently (Table 5).   Interestingly, Padrini et al[32] observed a positive correlation between amiodarone and digoxin in both young and aged patients treated with both drugs. The digoxin serum concentration, however, never exceeded the therapeutic range.   Moysey et al,[50] on the contrary, found toxic serum levels of digoxin after acute administration of amiodarone.

As to the fourth class,[4] it includes verapamil and other calcium antagonists (but less active than verapamil), useful against supraventricular and, probably, ventricular arrhythmias.   We are not aware of any studies of verapamil kinetics in the elderly.

In conclusion, even if the limited studies do not allow us to have a complete picture of the clinical pharmacokinetics of antiarrhythmic agents in the elderly, the complex of data show the tendency of the aged subject to a reduced elimination of these drugs. This fact is evident for those agents which are characterized by a predominant hepatic metabolism and excretion,[35,36,38,39,43,44] and also for those which are mainly eliminated unchanged by the kidney.[37,45]   Therefore, it is likely that the adverse side effects typical of all these substances occur more frequently in the elderly patient.

However, it is not easy to establish if the increased frequency of side reactions to antiarrhythmic drugs is only due to kinetic alterations or, additionally, to an augmented sensitivity of the target organs.   For example, the toxic effects of lidocaine on the CNS are actually more frequent in the elderly.[6,29]   Anyway, in the absence of measurements of the drug plasma levels in aged patients, it is impossible to ascribe the responsibility for this fact exclusively to an impairment of the drug kinetics and metabolism. On the other hand, it must be emphasized that the aged brain is particularly prone to the toxic effect of many drugs,[52] and that many antiarrhythmic agents can provoke neurological disorders (Table 6).   Furthermore, the anticholinergic effect of some antiarrhythmic drugs, such as dysopyramide and quinidine,[6]can worsen glaucoma and symptomatic prostatic hypertrophy, two diseases which frequently occur in the elderly.

THERAPEUTIC DECISIONS IN THE ELDERLY

Several unresolved problems (important for both young and aged patients) complicate the decision of how and when to treat the cardiac arrhythmias.

Table 5. Results observed after Amiodarone treatment of young and aged cardiac patients with recurrent episodes of supraventricular tachycardia*

PATIENTS	MEAN AGE	MEAN WEIGHT (kg)	% OF CASES WITH SUPPRESSION OF SVT	MEAN WEEKLY DOSE (mg)	MEAN DURATION OF THERAPY (MONTHS)	$\dfrac{\text{PDC}}{\text{W.D./B.W.}}$ (/mg/kg)	$\dfrac{\text{M.P.H.}}{\text{A.P.H.}}$	$T_3{}_R$ (ng/ml)	% OF CASES WITH CORNEAL DEPOSITS
< 60 yrs (N=17)	45.8 (23-59)	67.7 (50-78)	78%	1121 (1000-1400)	44	0.029 ± 0.014	1.23 ± 0.014	0.35 ± 0.13	N=9 (53%)
> 60 yrs (N=15)	69.3 (61-82)	72.3 (52-86)	93%	1028 (600-1400)	53	$0.048\pm0.019^{\&}$	$0.86\pm0.27^{\$}$	0.40 ± 0.15	N=13$ (93%)

* From Padrini et al,[32] Values are expressed as Mean $\pm$ SD

$ P <0.05; $ P <0.025(2x2 contingency tables with Yates correction); & P <0.01

Abbreviations: PDC=Plasma Amiodarone Concentration; W.D.=Weekly Dose; B.W.=Body Weight;
               M.P.H.=Metabolic Peak Height; A.P.H.=Amiodarone Peak Height.

As mentioned above, knowledge about the origin of the
arrhythmias and the mechanism of action of antiarrhythmic agents is
still insufficient, and, thus, the choice of the appropriate
treatment must be almost exclusively based on empirical criteria.
As a consequence, the "ideal" antiarrhythmic drug for efficacy,
safety and handling has not yet been found, and the majority of the
old and new agents are characterized by a great variability in the
individual response and by a low therapeutic ratio.[21]

The data from oldest literature were defective in evaluating
the efficacy of the antiarrhythmic treatment. In fact, methods
which could precisely check the natural variability of cardiac
arrhythmias were not available.[29] Furthermore, there was not the
possibility of measuring the antiarrhythmic drug plasma level by
means of exact assay procedures.[29]

More recently, promising results have been obtained with new
techniques, such as long term ECG monitoring, provocative electrophysiological tests and much more reliable drug assay methods.[29]
However, the information obtained with these new methods is still
incomplete so that, for example, we cannot yet exclude the possibility that different therapeutic plasma levels of each antiarrhythmic
drug are necessary for treating different types of arrhythmia (or
the same arrhythmia present in different cardiac diseases).[29]

The physician faces even more complicated problems when using
antiarrhythmic agents in the elderly patient. In fact, older
individuals develop more easily adverse pharmacological reactions
because of the increased sensitivity of the target organs, the
frequent coexistence of other cardiac and extracardiac diseases and
a decreased capacity for elimination of drugs. In spite of this,
clinical trials specifically programmed for evaluating the efficacy
and tolerance of antiarrhythmic agents in the elderly, have never
been made. In particular, there is a complete absence of systematic
studies made with the simultaneous evaluation of the drug serum
concentration and the therapeutic-toxic effects in order to assess
if different dosage regimens (or different therapeutic interventions)
must be chosen for optimizing antiarrhythmic therapy in young and
old patients.

Therefore, great caution is mandatory in the use of antiarrhythmic agents in the elderly in order to avoid the probability of toxic
reactions exceeding the beneficial effects.

First of all, it is important to be accurate in recognizing the
etiology and in establishing the severity of cardiac arrhythmias.
In fact, therapeutic success strictly depends on the elimination of
the fundamental cause of arrhythmias such as in the case of
disturbances of electrolyte balance frequently occurring in the aged
patient. Furthermore, no therapeutic intervention is needed when

Table 6. Central nervous system side effects of the antiarrhythmic
          drugs*

CLASS I A

    QUINIDINE - Headache, Confusion, Tinnitus, Dizziness,
                Disorders of Colour Vision

    PROCAINAMIDE - Depression, Hallucinations, Psychosis

    DISOPYRAMIDE - Psychosis, Hallucinations, Insomnia

CLASS I B

    LIDOCAINE - Paresthesias, Ataxia, Nystagmus, Confusion,
                Convulsions

    MEXILETINE - Tremor, Ataxia, Nystagmus, Confusion, Convulsions

    PHENYTOIN - Nystagmus, Ataxia

CLASS II

    PROPRANOLOL - Fatique, Depression, Nightmares

CLASS III

    AMIODARONE - No Important Effects

CLASS IV

    VERAPAMIL - Headache, Lethargy (rare).

---

* This table is a partial synthesis of data reported in Ref. 4,6,
  29, 34, 51.

any arrhythmia exhibits characteristics of "benignity" for either
its nature or its frequency as tested by long term EC monitoring.

    Secondly, it is crucial to choose the safest treatment in
relation to the patient's pharmacokinetic characteristics and
responsiveness. Everything considered, the suppression of severe
(or life-threatening) arrhythmias by means of different methods of
cardiac pacing, provokes less serious side effects than those
observed during treatment with antiarrhythmic drugs.[3]

    The problem of chronic treatment (or prevention) is much more
complex. It has been demonstrated that the provocative electrophysiological tests with artificial pacing are useful for
distinguishing the "responders" from the "non responders" to any
antiarrhythmic drug.[3,21,54] The chronic dosage regimen of the
same drug can be established on the basis of the patient

characteristics and of that plasma drug concentration which, without causing toxic reactions, is able to abolish the pacing-induced arrhythmias. This rational approach, however, is not possible in all instances for various reasons. First of all, many drug assay methods are complex and/or require long execution times and sophisticated and expensive equipment. Secondly, there is not always a precise correlation between the drug plasma concentration and the antiarrhythmic effect.[21] Finally, with some drugs, such as amiodarone, chronic treatment may be successful even in patients classifiable as non-responders on the basis of a previous provocative test.[21,54]

Sinus node dysfunction is a typical disease of advanced age, and often it is associated with other conduction-excitation-automaticity disturbances of other tracts of the specialized system.[6] In several instances, these disturbances are subclinical and may not be revealed even by a basal electrophysiological test. In these cases, some antiarrhythmic agents, having a marked depressant effect on the specialized system, may reveal the concealed involvement of sinus node function but, simultaneously, cause severe or catastrophic side effects, such as profound sinus pauses or electrical standstill and/or life-threatening arrhythmias. Figures 3 and 4, concerning two old patients treated with amiodarone for recurrent ventricular tachycardia, represent two paradigmatic examples of this important clinical eventuality. Besides, these examples emphasize the necessity of accurate controls with electro-physiological tests and/or with long-term ECG monitoring in order to establish if any antiarrhythmic chronic treatment of elderly patients should be associated with a pacemaker implantation.

Anyway, it seems to be opportune to discuss in some detail the treatment of several arrhythmias frequently occurring in the elderly.

Sinus Bradycardia

Sinus bradycardia during exercise is a constant finding in the elderly, while conflicting results have been obtained about the basal heart rate. However, recent data obtained with long term ECG monitoring, have shown that the resting heart rate of healthy old people is not significantly different from that of the young.[1,2]

The most frequent causes of symptomatic bradycardia in the elderly are represented by cardiac disorders (mainly myocardial infarction) or extracardiac diseases (for example, obstructive jaundice).

Atropine is often effective in controlling sinus bradycardia due to acute myocardial infarction, since it causes frequent side effects in the older patient, it should be employed only when the decrease in heart rate severely impairs ventricular function.

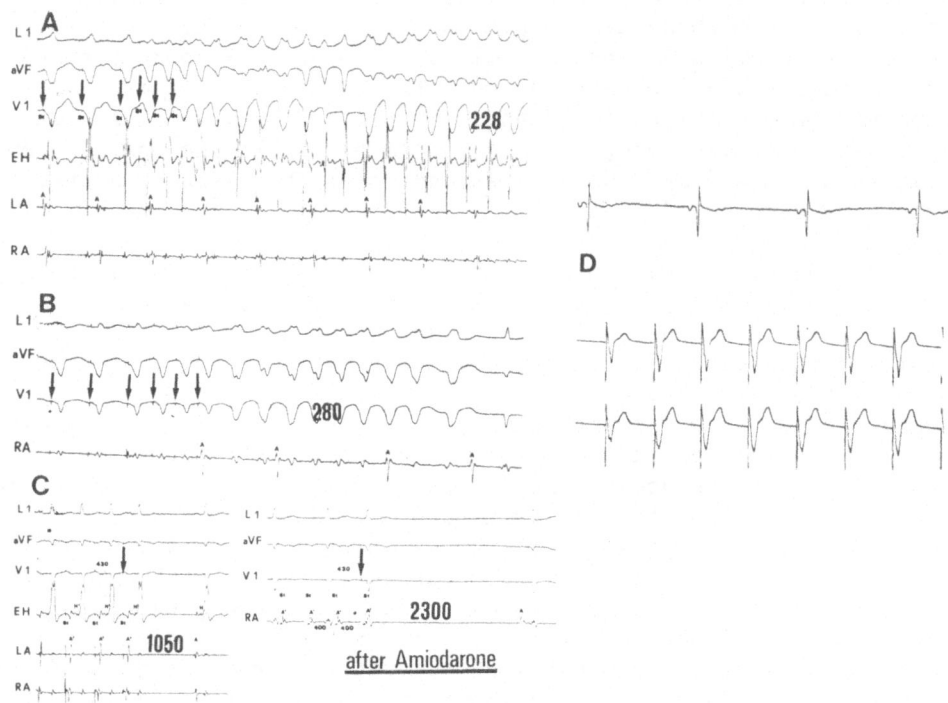

Fig. 3.   70 years old male patient with coronary heart disease and
recurrent episodes of ventricular tachycardia.
In A: increased paced rate (arrows) with basal cycle reduced
from 500 to 220 ms resulted in sustained ventricular tachycardia (basal cycle 228 ms).   In B: after a loading dose of
amiodarone (5 g in 4 days) the same provocative test
resulted in a brief period of ventricular tachycardia
(basal cycle 280 ms).   In C (left side of the panel):
atrial pacing revealed a normal sinus node recovery time
(1050 ms).   In C (right side of the panel): a loading dose
of amiodarone (5 g in 4 days) caused a marked prolongation
of sinus node recovery time (2300 ms).   In D (top recorder):
long term ECG monitoring evidenced profound sinus bradycardia during the treatment with the loading dose of
amiodarone.   After implantation of a ventricular pacemaker,
chronic treatment of amiodarone abolished ventricular
arrhythmias (bottom records in D panel, obtained after a
new 24 hours-Holter ECG).   From Furlanello et al,[54]
reproduced with permission).

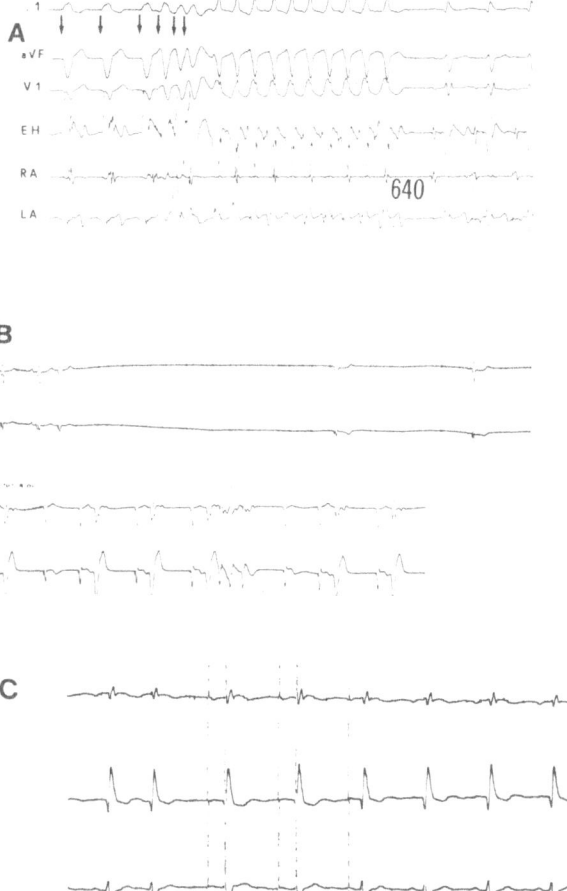

Fig. 4.   72 years old male patient with coronary heart disease and
           episodes of ventricular tachycardia.
           In A: increased paced rate (arrows) with basal cycle from
           500 to 220 ms resulted in a brief period of ventricular
           tachycardia at the end of which a normal sinus node
           recovery time could be seen (640 ms).   In B: 24 hours-
           Holter ECG evidenced marked sinus pauses (top records) and
           complex ventricular arrhythmias (bottom records) after
           1800 mg of amiodarone.   In: records from a 24 hours-Holter
           ECG after implantation of a D.D.D. pacemaker and during
           chronic treatment of amiodarone.   2nd and 3rd beats show
           a double stimulation (atrial and ventricular) by the
           pacemaker; in the 4th beat the atrial stimulation was
           conducted to ventricles.   1st, 6th, 7th and 8th beats are
           normal sinus beats, the 2nd a supraventricular premature
           beat.   (From Furlanello et al,[3]   reproduced with
           permission).

Sinus node dysfunction is very frequent in the elderly.
Systematic Holter ECG monitoring revealed an incidence of 9 per cent
of the sick sinus syndrome in an elderly population.[55]   Permanent
pacing with different types of pacemakers is indicated in bradycardia
-tachycardia syndrome due to sinus node dysfunction and often
results in a marked improvement in the symptoms, but does not
modify the likelihood of these patients developing a stroke.[56]
At postmortem examination, three out of four patients with sick sinus
syndrome have been found to have atrial thrombi.[55]   In patients
with sinus node dysfunction there is also a high percentage (at
least 7 per cent of cases per year) of cardiogenic cerebral
embolization.[55]   These facts explain the frequent association of
sinus node dysfunction with stroke and/or TIA, and justify the
term "cardiogenic neurology" proposed by Abdon[55] for describing
this syndrome.   Therefore anticoagulant therapy to prevent cerebral
embolization should be considered in patients with sinus node
dysfunction.

Atrial Fibrillation

The most important causes of atrial fibrillation in the elderly
are heart failure (more frequently due to coronary heart disease),
thyrotoxicosis, sick sinus syndrome and "lone" atrial fibrillation.

In the case of atrial fibrillation associated with heart
failure, digitalis is the choice of treatment.[58]   Atrial fibrillation due to thyrotoxicosis must be treated with the specific anti-
thyroid therapy.   Atrial fibrillation due to the sick sinus
syndrome is often paroxysmal, but may also be chronic.   In this last
instance, it is characterized by a slow heart rate because of the
frequent coexistence of conduction disturbances at lower levels of
the specialized tissue.   In these patients, as mentioned above,
any antiarrhythmic treatment, if not associated with a pacemaker
implantation, must be avoided owing to the possibility of cardiac
arrest.[6]

In an old study,[57] a particular form of atrial fibrillation
named "lone" has been described.   It was characterized by male
predominance, minimal symptoms, ventricular rate less than 90, by
the low incidence of thromboembolic complications, and the rare
need for any treatment.   Perhaps, the most important reason for
not treating these patients is that lone atrial fibrillation
actually represents the end point of several types of sinus node
dysfunction.

Atrial Flutter

Atrial flutter is frequently associated with cardiac diseases
and is rarely paroxysmal.   In this last instance, it can also occur
in normal subjects.   Usually, atrial flutter severely impairs

cardiac function mainly in the elderly patient and is resistant to
drug treatment.   By contrast, atrial pacing is nearly always
successful in the elimination of this arrhythmia.

In 39 patients over 65 years old, 46 out of 48 episodes of
atrial flutter, consecutively observed by Furlanello et al,[3]
were converted into sinus rhythm (40 cases) or into atrial fibrillation (6 cases) without important complications.   Direct-current
cardioversion, instead, must be made with great caution in the old
patient because of the frequent occurrence of left ventricular
dysfunctions and sinus node insufficiency.[6]   Digitalis alone or
in combination with quinidine and/or amiodarone are useful in the
prevention of atrial flutter.

Paroxysmal Supraventricular Tachycardia (PSVT)

PSVT is nearly always due to re-entry occurring at different
levels (sinus node, atria, AV node, anomalous pathways).[3]   In rare
instances, PSVT has a focal origin due to enhanced normal or
abnormal automaticity.   PSVT often occurs in normal subjects
without cardiac abnormalities and rarely impairs left ventricular
function.   In elderly people, PSVT is rare: only 9 out of 100
consecutive cases of severe recurrent PSVT were over 65 years old.[3]
In all these 100 patients, Furlanello et al[3] studied the prevalence
of the type of the re-entry circuit by means of an electrophysiological examination, and observed a different distribution in young
and old subjects.   In the former, the percentages of the re-entry
circuit at the level of AV node and anomalous pathways were,
respectively, 36 per cent and 57 per cent, and in the latter 67 per
cent and 22 per cent.   PSVT due to Wolff-Parkinson-White syndrome
is also rare in old patients.[59]

When the re-entry circuit occurs in the AV node, digitalis or
verapamil or propranolol or amiodarone can be used.   In cases of
anomalous or accessory pathways amiodarone is the choice of
treatment.[54]   Chronic preventive treatment of PSVT is indicated
only when the recurrent episodes are severe and impair left
ventricular function.[6]

Verapamil and, above all, propranolol should be given with
caution to old patients.   In fact, the normal aging heart is per se
characterized by a reduction in the contractility reserves.[60]
Verapamil, having less depressant effect on myocardial contractility
than propranolol,[61] should be given in preference to the β-blocking
agent in the elderly patient.

As already mentioned, PSVT occurring in the elderly can be
associated with sinus node dysfunction and, then, the suppression
of tachycardia by means of drug treatment or electrical pacing may
be followed by electrical standstill.   In these instances, the

suppression of the arrhythmia should be preceded by the application
of a permanent or temporary ventricular pacemaker.

Ventricular Arrhythmias and Sudden Cardiac Death

Complex ventricular arrhythmias, such as frequent or multiple
premature ventricular beats, bigemini or trigemini rhythms,
ventricular tachycardia, are frequently observed in the elderly.[1,2]
According to a study made with long term ECG monitoring in an
apparently healthy elderly population, complex ventricular arrhythmias are not rare even in the absence of cardiac diseases. This
fact might induce us not to treat these arrhythmias since they do
not seem to be provoked by cardiac or extracardiac diseases. In
this instance, however, the therapeutic decisions are not easy
since, in the absence of adequate data, it seems to be equally
correct to consider as "benign" only the non-complex ventricular
arrhythmias (rare ventricular premature beats). In fact, the
prospective study of Camm et al[2] indicated the possibility of sudden
death even in those asymptomatic subjects in whom a 24 hours ECG
monitoring showed episodes of complex ventricular arrhythmias.
In old patients, Furlanello et al[3] observed that ventricular
tachycardia is nearly always associated with coronary heart disease.
Furthermore, among the asymptomatic subjects with complex
ventricular arrhythmias, it is not easy to distinguish those without
from those with "silent" coronary heart disease. On the other hand,
asymptomatic coronary heart disease is typical of old age.

Ventricular tachycardia should be treated with different methods
of electrical pacing or with various antiarrhythmic drugs (lidocaine,
mexiletine, amiodarone etc).

An unresolved problem is the prevention of coronary sudden death
(CSD). In patients with coronary heart disease, complex ventricular
arrhythmias, mainly if associated with signs of ventricular dysfunction are an important risk factor of CSD.[63,64] Chronic treatment
with amiodarone or other antiarrhythmic drugs or with substances
preventing coronary spasm (calcium antagonists) reduces the incidence
of ventricular arrhythmias,[21,63] but does not with certainty
influence the likelihood of ventricular fibrillation and CSD.[21,63,65]

Conflicting results have been obtained on the incidence of CSD
in old people. The Framingham Study[62] showed an age-related
decrease of CSD, whilst Ruberman and Weinblatt[66] obtained an
opposite result. Data from patients admitted to coronary care
units, [3,67] showed a higher mortality in the group of older
patients; in this group, however, the most frequent causes of death
were not arrhythmias but cardiogenic shock, cardiac rupture, heart
failure and electrical standstill.

Long term treatment with several β-blocking agents seems to be successful in the prevention of CSD.[68] However, conflicting data have been obtained in patients over 65 years of age. Anderson et al[69] observed that alprenolol significantly reduced the mortality in younger coronary patients but increased it in older ones. Instead, the Norwegian Multicenter Study[70] showed a significant reduction of mortality with timolol in both young and old patients. It is not clear if the difference observed between alprenolol[69] and timolol[70] in older patients is due to a different sensitivity of the aging myocardium to the two β-blocking agents or to other causes, such as differences in study design.

Diuretics, especially those with a long-term action, can cause hypokalemia and, in turn, induce severe ventricular arrhythmias and ventricular fibrillation.[71] Elderly individuals are particularly prone to develop hypokalemia secondary to diuretics prescribed for either heart failure or hypertension mainly when they do not have an adequate dietary potassium intake.[72] To avoid hypokalemia and other complications due to diuretics, it is better to reduce (or abolish) the diuretic prescription and to increase the potassium intake. Instead, the combination of thiazide diuretics with potassium sparing diuretics should be used with great caution because it may result in a life-threatening increase of serum potassium (and in ventricular arrhythmias) in old patients and, mainly in those with renal disease.[73]

Ventricular arrhythmias and ventricular fibrillation may be caused by digitalis intoxication which is more frequent in the elderly.[74] There is no doubt that the most effective therapy of digitoxicity is its prevention and/or the timely suspension of digitalis. In fact, antiarrhythmic treatment (or cardioversion) of severe digitalis-induced arrhythmias is nearly always unsuccessful.[75] The use of FAB fragments of digoxin-specific antibodies[76] has ameliorated the prognosis of severe digitalis intoxication but, unfortunately, it is not very easy to get them. An appropriate medical prescription is particularly important for the prevention of digitalis intoxication in the elderly. In this connection, one should not only consider the altered digoxin pharmacokinetic profile of the old cardiac patient[24,27,28] but also the possibility that the association of digoxin with other drugs strongly complicates both its pharmacokinetics and pharmacodynamics and, besides worsens patient compliance (Figure 5).

A significant correlation between the use of tricyclic antidepressants and mortality has been reported.[77,78] Later, the Boston Drug Surveillance Program[79] failed to confirm these results. Anyway, imipramine may develop re-entry arrhythmias because of its depressant effect on action potential phase 0 and on the conduction velocity of the stimulus.[71] Imipramine-induced ventricular arrhythmias might occur more frequently in old individuals since advanced age, altering the drug hepatic metabolism, facilitates

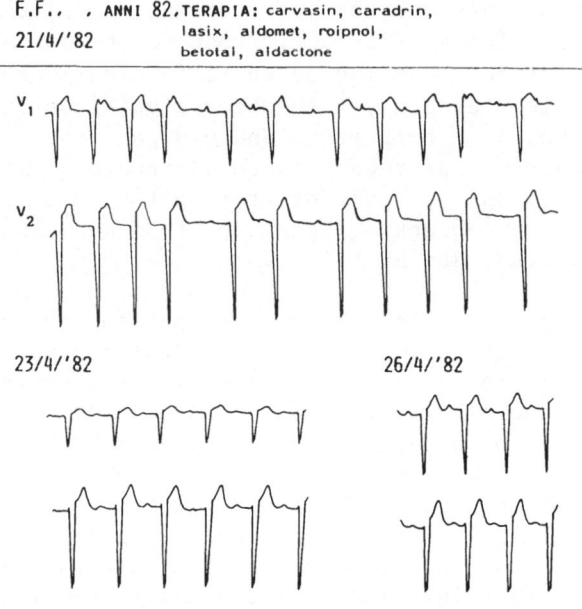

Fig. 5.   82 years old female patient with heart failure and
           coronary heard disease. At the moment of the hospital
           discharge (4/18/1982) the following prescription was made
           for the patient: methylproscillaridin (Caradrin), nitrate
           (Carvasin), furosemide (Lasix), alpha-methyldopa (Aldomet),
           benzodiazepine (Roipnol), vitamin B complex (Betotal) and
           spironolactone (Aldactone). After 3 days from the
           hospital discharge, the patient was again admitted to a new
           hospital (Gerontology Division of Policlinico Gemelli,
           Rome) for pulmonary edema. At the moment of admission
           (4/21/1982), ECG evidenced atrial tachycardia with
           Wenckebach phenomena, after 2 days (4/23/1982) normal sinus
           rhythm with 1st degree AV block and after 5 days (4/26/1982)
           normal sinus rhythm with normal P-R intervals. The patient
           confessed to have mistaken Caradrin for Carvasin and, thus,
           she took 4 pills of Caradrin instead of one each day
           (altogether 3 mg of methylproscillaridin in 3 days). This
           fact easily explains why the patient rapidly developed a
           deterioration of cardiac function and a clear electrocardio-
           graphic picture of digitoxicity.

all the side effects of this group of substances.   Therefore,
it seems to be preferable to treat depressed old patients with
drugs having no cardiac effects.[81]

    Finally, it must emphasized that all the antiarrhythmic agents
can induce ventricular arrhythmias and ventricular fibrillation by
means of different mechanisms.[21]

CONCLUSIONS

The following principles should represent the rational basis of any drug therapy: (1) The pathophysiology of the disease to be treated should be precisely known, (2) The drug treatment should be chosen in order to reverse that pathophysiology, (3) The drug regimen should produce the desired effects without toxicity.

However, when one faces this subject in the field of the antiarrhythmic therapy of the elderly patient, it must be realized that not one of these principles has been actually fulfilled. On the one hand, in fact, the general problems, related to the mechanisms involved in the origin of cardiac arrhythmias, have not yet been resolved; on the other, those aspects, which specifically concern the age influences on the pharmacokinetics and pharmacodynamics of antiarrhythmic agents, need much more clarification. Unfortunately, therefore, only empirical criteria can be suggested and, as a consequence, "primum non nocere" comes to be the essential principle of this therapy.

In this connection, the first suggestion is represented by the discrimination between "benign" and life-threatening arrhythmias and by the consequent decision of not treating the former.

Secondly, it is critical to recognize and eliminate the possible causes of arrhythmias, such,as for example,electrolyte disturbances which are very frequent in the elderly.

Finally, since the individual response is variable and the therapeutic margin of all the antiarrhythmic agents exiguous, it is essential to choose each time the safest and most effective treatment on the basis of the characteristics of patients to be treated. Accurate serial clinical examinations, plasma drug concentration measurements, provocative electrophysiological tests and long term ECG monitoring seem to be the most useful procedures for checking the patient responsiveness and for controlling drug toxicity.

REFERENCES

1. J. L. Fleg, H. L. Kennedy, Cardiac arrhythmias in a healthy elderly population, Chest 81:302 (1982).
2. A. J. Camm, K. E. Evans, D. E. Ward, and A. Martin, The rhythm of the heart in active elderly subjects, Am.Heart J. 99:598 (1980).
3. F. Furlanello, Le aritmie cardiache nell'anziano, G. Gerontol. (in press).
4. H. L. Opie, Antiarrhythmic agents, Lancet 1:861 (1980).
5. J. T. Bigger, Mechanism and diagnosis of arrhythmias, in: "Heart Disease," E. Braunwald, ed., Saunders, Philadelphia (1980).

6. J. T. Bigger, Management of arrhythmias, in: "Heart Disease,"
    E. Braunwald, ed., Saunders, Philadelphia (1980).
7. D. N. Das, J. L. Fleg, and E. G. Lakatta, Effect of age on the
    components of atrioventricular conduction in normal man,
    Am.J.Cardiol. 49:1031 (1982).
8. E. Lakatta, Excitation-Contraction, in:"The Aging Heart,"
    M. L. Weisfeldt, ed., Raven Press, New York (1980).
9. J. Roberts and P. B. Goldberg, Changes in responsiveness of
    the heart to drugs during aging, Fed.Proc.38:1927 (1979).
10. M. R. Rosen, R. F. Reder, A. J. Hordof, M. Davies and P. Danilo,
    Age-related changes in Purkinje fiber action potentions
    of adult dogs, Circ.Res. 43:931 (1978).
11. E. Rumberger and J. Timmermann, Age-changes of the force-
    relationship and the duration of action potential of
    isolated papillary muscles of guinea pig, Europ.J.appl.
    Physiol. 35:277 (1976).
12. H. Reuter, Divalent cations as change carriers in excitable
    membranes, Progr.Biophys. 26:1 (1973).
13. T. Guarnieri, C. R. Filburn, G. Zitnik, G. S. Roth and E.G.
    Lakatta, Contractile and biochemical correlates of
    β-adrenergic stimulation of the aged heart, Am.J.Physiol.
    239:H501 (1980).
14. J. P. Froelich, E. G. Lakatta, E. Beard, H. A. Spurgeon,
    H. A. Weinsfeldt and G. Gerstenblith, Studies of sarco-
    plasmic reticulum function and contraction duration in young
    adult and aged rat myocardium, J.Mol.Cell.Cardiol.10:427
    (1978).
15. R. S. Kass, W. J. Lederer, R. W. Tsien and R. Weingart, Role
    of calcium ions in transient inward current and aftercontr-
    actions induced by strophanthidin in cardiac Purkinje fibers,
    J.Physiol.(London) 281:187 (1878).
16. M. Di Gennaro, R. Valle, M. Pahor, and P. U. Carbonin, Abolition
    of digitalis tachyarrthythmias by caffeine, Am.J.Physiol.
    (in press).
17. M. Pahor, A, Cocchi, M. Di Gennaro, R. Bernabei and P. U.
    Carbonin,Studio delle aritmie da riperfusione nel ratto
    invecchiato, G. Gerontol. (in press).
18. P. U. Carbonin, M. Di Gennaro, R. Valle and R. Bernabei,
    Intracellular calcium and electrogram in ischemic isolated
    heart, Am.J.Physiol.239:H380 (1980).
19. P. U. Carbonin, M. Di Gennaro, R. Valle and A. M. Weisz,
    Inhibitory effect on anoxia on reperfusion - and digitalis-
    induced ventricular tachyarrhythmias, Am.J.Physiol. 240:H730
    (1981).
20. W. G. Nayler, Maintenance of Ca^{++} homeostasis in the myocardium,
    9th World Congress of Cardiology, Moscow, Vol.1 Abst.0072
    (1982).
21. E. Sandøe, E. Damgaard, and B. Sigurd, Clinical efficacy and
    side effects of old and new anti-arrhythmic drugs, 9th World
    Congress of Cardiology, Moscow, Vol.1 abst.0020 (1982).

22. P. B. Goldberg and J. R. Roberts, Age-related changes in rat
     atrial sensitivity to lidocaine, J.Gerontol.36:520 (1981).
23. P. B. Goldberg and J. R. Roberts, Age and responsiveness to
     cardiovascular drugs, in: "Clinical Pharmacology in The
     Aged Patient," L. F. Jarvic, D. J. Greenblatt, D. Harman,
     eds., Raven Press, New York (1981).
24. L. Antico, A. Cocchi, M. Di Gennaro, A. V. Greco and P. U.
     Carbonin, Digitalis metabolism in the elderly, Acta.Med.
     Romana 13:214 (1975).
25. F. Zannad, R. J. Royer, G. Issartel, J. Robert, Digoxin in the
     elderly and in renal failure. Contribution of erythrocyte
     86-rubidium uptake tests, Europ. J. Cardiol. 12:285 (1981).
26. R. Bernabei, G. P. Perna, L. Carosella, P. Di Nardo, A. Cocchi
     and P. U. Carbonin, Digoxin serum concentration measurement
     in patients with suspected digitalis-induced arrhythmias,
     J. Cardiovasc.Pharmacol. 2:319 (1980).
27. F. I. Caird, Digitalis metabolism in aged patients, in: "Ageing
     and Drug Therapy," G. Barbagallo-Sangiorgi, A. N. Exton-Smith,
     eds., Plenum Press (in press).
28. C. M. Castleden, Digoxin therapy in older patients, in: "Aging
     and Drug Therapy." G. Barbagallo-Sangiorgi, A. N. Exton-Smith,
     eds. Plenum Press (in press).
29. O. B. Garfein, Pharmacology of commonly used antiarrhythmic
     drugs and comments on the use of therapeutic drug monitoring.
     Ther.Drug.Mon. 4:1 (1982).
30. M. Eichelbaum, and A. Somogyi, Verapamil distribution in health
     and the diseased state, in: "Calcium Antagonism in Cardio-
     vascular Therapy," A. Zanchetti, D. M. Krikler, eds.,
     Excerpta Med., Amsterdam (1981).
31. E. Riva, M. Gerna, R. Latini, P. Giani, A. Volpi and A. Maggioni,
     Pharmacokinetics of amiodarone in man, J.Cardiovasc.Pharmacol.
     4:264 (1982).
32. R. Padrini, S. Gobbato, G. Inama, M. Ferrari and F. Furlanello,
     Pharmacokinetics and antiarrhythmic activity of amiodarone
     in the elderly, Int.J.Clin.Pharmacol. (in press).
33. H. O. Klein, R. Lang, E. Weiss, E. Di Segni, C. Libhaber,
     J. Guerrero and E. Kaplinsky, The influence of verapamil on
     serum digoxin concentration, Circulation 65:998 (1982).
34. R. G. Shanks, The pharmacology and pharmacokinetics of mexile-
     tine: a review, in: "Management of Ventricular Tachycardia -
     Role of Mexiletine," E. Sandøe, D. G. Julian, J. W. Bell,
     eds., Excerpta Med. Amsterdam (1978).
35. J. H. Stevenson, S. A. M. Salem and A. M. M. Shepherd, Studies
     on drug absorption and metabolism in the elderly, in: "Drugs
     and the Elderly," J. Crooks, J. H. Stevenson, eds.,
     MacMillan, London (1979).
36. H. R. Ochs, D. J. Greenblatt, E. Woo, and T. W. Smith, Reduced
     clearance of quinidine in elderly persons, Am.J.Cardiol
     42:481 (1978).

37. M. Comacho, D. E. Drayer, J. Kluger and M. M. Rerdenberg, Renal
     excretion of procainamide and N-acetyl procainamide in man
     as a function of old age, Clin.Res. 27:599-A (1979).

38. E. J. Triggs, Pharmacokinetics of lignocaine and chlormethiazole
     in the elderly; with some preliminary observations on other
     drugs, in: "Drugs and the Elderly," J. Crooks, J. H.
     Stevenson, eds., MacMillan, London (1979).

39. R. L. Nation, E. J. Triggs, M. Selig, Lignocaine kinetics in
     cardiac patients and aged subjects, Br.J.Clin.Pharmacol.
     4:439 (1977).

40. J. Hayes, Changes in drug metabolism with increasing age.
     Phenytoin clearance and protein binding, Br.J.Clin.Pharmac.
     2:73 (1975).

41. J. Heikkila, A. Jounela, M. Katila, K. Luomanmäki, and M. H.
     Frick, Beta-blockade-selection and use, Ann.Clin.Res.11:267
     (1979).

42. J. Dreyfuss, D. L. Griffith, S. M. Singhvi, J. M. Shaw,
     J. J. Ross, R. A. Vukovich and D. A. Willard, Pharmaco-
     kinetics of nadolol a beta-receptor antagonist: administrat-
     ion of therapeutic single- and multi-dosage regimens to
     hypertensive patients, J.clin.Pharmacol.19:712 (1979).

43. C. M. Castleden, C. F. George, The effect of ageing on the
     hepatic clearance of propranolol, Br.J.Clin.Pharmacol.
     7:49 (1979).

44. G. Bianchetti, D. Brancaccio, F. Forette, J. F. Henry, M. P.
     Hervy, J. Ganansia, R. Gomeni, J. R. Kilborn and P. R.
     Morselli, Pharmacokinetics of SL 75 218 propranolol in old
     age, uraemia and in healthy young adults, Br.J.Clin.Pharmacol.
     9:299 (1980).

45. H. E. Barber, G. M. Hawksworth, J. C. Petrie, J. W. Rigby,
     O. J. Robb, and A. K. Scott, Pharmacokinetics of atenolol
     and propranolol in young and elderly subjects, Br.J.Clin.
     Pharmacol. 11:118 (1981).

46. T. B. Tjandramaga, Altered pharmacokinetics of β-adrenoceptor
     blocking drugs in patients with renal insufficiency,
     Arch.Int.Pharmacodyn.Ther. Suppl. 38:53 (1980).

47. H. Knauf, M. Schafer-Kortin, and E. Mutschler,Pharmakokinetik
     und biologische wirkdoner von β-Rezeptorenblocken bei
     Niereninsuffizienz.Internist 22:616 (1981).

48. J. Herrera, R. A. Vukovich and D. L. Griffith, Elimination of
     nadolol by patients with renal impairment, Br.J.Clin.
     Pharmacol. 8:227 (1979).

49. P. C. Rubin, P. J. W. Scott, K. McLean, A. Pearson, D. Roos,
     and J. L. Reid, Atenolol disposition in young and elderly
     subjects, Br.J.Clin.Pharmacol. 13:235 (1982).

50. J. O. Moysey, N. S. V. Jaggarao, F. N. Grundy and D. A.
     Chamberlein, Amiodarone increases plasma digoxin concentrat-
     ions, Brit.med.J. 282:272 (1981).

51. J. W. Bell, The central nervous system side effects of anti-arrhythmic agents, in: "Management of Ventricular Tachycardia - Role of Mexiletine," E. Sandøe, D. G. Julan, J. W. Bell, eds., Excerpta Med., Amsterdam (1978).

52. R. D. T. Cape, Drugs and confusion states, in: "Drugs and the Elderly," J. Crooks, J. H. Stevenson, eds., MacMillan, London (1979).

53. G. P. Sanna, P. Meoli, D. Pioselli and B. Brusoni, Come condurre una terapia antiaritmica, in: "Aggiornamenti in Cardiologia," G. Cataldo, F. Fovelli, eds., Pozzi, Roma (1981).

54. F. Furlanello, G. Inama, P. Dal Forno, M. Ferrari, M. Padrini, F. Fiorentini, O. Merz, L. Nicosi, G. Vergara and M. Disertori, Amiodarone, in: Attualita in Campo Cardiologico," F. Rovelli, ed., Pozzi, Roma (1982).

55. N. J. Abdon, Clinical considerations in the diagnosis of cerebral ischemia caused by episodic cardiac dysrhythmia, in: "Cerebral Manifestations of Episodic Cardiac Dysrhythmia," H. Bune, ed., Excerpta Med., Princeton (1979).

56. A. J. Fairfax, C. D. Lambert and A. Leatham, Systemic embolism in chronic sinoatrial disorder, New Engl.J.Med. 295:190 (1976).

57. W. Evans, P. Swann, Lone auricular fibrillation, Brit.Heart J. 16:189 (1954).

58. L. H. Opie, Digitalis and sympathomimetic stimulants, Lancet 26:912 (1980).

59. D. Mehiman, M. F. Arnsdorf, Antiarrhythmic therapy for the elderly: a two-edged sword, Geriatrics 34:29 (1979).

60. V. B. Subramanian, M. Bowles, H. Penalver, A. B. Davies, and E. B. Raftery, Verapamil in ischemic heart disease, in: 'Calcium Antagonism in Cardiovascular Therapy," A. Zanchetti, D. M. Krikler, eds., Excerpta Med., Amsterdam (1981).

61. M. L. Weisfeldt, Left ventricular function, in: "The Aging Heart," M. L. Weisfeldt, eds., Raven Press, New York (1980).

62. W. B. Kannel, and H. E. Thomas, Sudden coronary death. The Framingham Study. Ann.New York.Acad.Sci.383:3 (1982).

63. J. T. Bigger, F. M. Weld, and L. M. Rolnitzky, Which post-infarction ventricular arrhythmias should be treated? Am.Heart J. 103:660 (1982).

64. B. Pitt, Sudden cardiac death: role of left ventricular dysfunction. Ann.New York Acad.Sci. 382:218 (1982).

65. R. A. Winkle, Clinical efficacy of antiarrhythmic drugs in prevention of sudden coronary death, Ann.New York Acad.Sci. 382:247 (1982).

66. W. Ruberman and E. Weinblatt, Epidemiology of sudden cardiac death, 9th World Congress of Cardiology, Moscow Vol.1 abst. 0055 (1982).

67. C. A. Latting and M. E. Silverman, Acute myocardial infarction in hospitalized patients over age 70, Am.Heart J.100:311 (1980).

68. A. Hjalmarson, Beta blocking agents: current status in the prevention of sudden coronary death, Ann.New York Acad.Sci. 382:305 (1982).

69. M. P. Anderson, P. Bechsgaard, J. Frederiksen, D. A. Hansen, J. H. Jürgensen, B. Nielsen, F. Pedersen, O. Pdersen-Bjergaard and S. L. Rasmussen, Effect of alprenolol on mortality among patients with definite or suspected acute myocardial infarct-on, Lancet 2:865 (1979).

70. Norwegian Multicenter Study Group, Timolol-induced reduction in mortality and reinfarction in patients surviving acute myocardial infarction, New Engl.J.Med. 304:801 (1981).

71. J. T. Bigger, F. M. Weld, Drugs and sudden death, Ann.New York Acad.Sci. 382:229 (1982).

72. J. L. C. Dall, S. Paulose, and J. A. Fergusson, Potassium intake of elderly patients in hospita-, Gerontol.Clin. 13:114(1971).

73. A. D. Bender, C. L. Carter, and K. B. Hansen, Use of a diuretic combination of triampterene and hydrochlorothiazide in elderly patients, J.Am.Griatr.Soc. 15:166 (1967).

74. A. D. Bender, Drug sensitivity in the elderly, in: "Drugs and the Elderly," J. Crooks, J.H. Stevenson, eds., MacMillan, London (1979).

75. P. U. Carbonin, Diagnosi e trattamento dell'intossicazione digitalica, Atti 39° Cong. Soc. It. Cardiol., Milano, (1978).

76, T. W. Smith, E. Haber, L. Yeatman, V. P. Butler, Reversal of advanced digitalis intoxication with Fab fragments of digoxin-specific antibodies,New Eng.J.Med. 294:797 (1976).

77. D C. Coull, J. Crooks, I. Dengwall-Fordyce, A. M. Scott and R. D. Weir, Amitryptiline and cardiac disease, Lancet 2:590 (1970).

78. D. C. Moir, J. Crooks, W. B. Cornwell, K. O'Malley, I. Dengwall-Fordyce, M. J. Turnbull and R. D. Weir, Cardiotoxicity of amitryptiline, Lancet 2:561 (1972).

79. Boston Collaborative Surveillance Program, Adverse reactions to the tricyclic antidepressants, Lancet 1:529 (1972).

80. L. F. Jarvik, and P. R. Kakar, Aging and response to anti-depressants, in: "Clinical Pharmacology in the Aged Patient," L. F. Jarvik, D. J. Greenblatt, D. Harman, eds., Raven Press, New York (1981).

81. A. W. Gomoll, and J. E. Byrne, Trazodone and imipramine: comparative effects on canine cardiac conduction, Europ. J.Pharmacol. 57:335 (1979).

DIURETIC THERAPY IN OLD PATIENTS

G. Barbagallo Sangicrgi, A. Di Sciacca, G. Frada Jr.,
F. Durante, G. Costanza and G. Cupidi

Department of Medical Pathology
University of Palermo, Italy

The administration of diuretics is indicated in several clinical conditions characterized by more or less severe water and salt retention (congestive heart failure, non-compensated liver cirrhosis, renal syndromes etc.), or in cases in which water and sodium depletion is supposed to improve some pathophysiological aspects of clinical symptoms (e.g. as background treatment in arterial hypertension, or as emergency treatment in left ventricular failure.

The indications and mechanisms of action of diuretics (see Appendix) in aged patients are quite similar and comparable to those known for patients in any other age. The response of an aged organism, however, may be different. and the risk of side effects higher. The biological cost/benefit ratio may therefore not be favourable in all cases.

Several statistical investigations[1,2,3] have shown that diuretics are among the most widely used drugs, largely because they are prescribed in the basic treatment of arterial hypertension even in aged patients. It has also been shown, however, that these drugs most frequently cause unwanted side effects in the elderly.[1,2,3] Such effects are chiefly due to changes concerning the metabolic fate of drugs (involving changes of drug bioavailability, prolonged drug half-life, changes in biological response and water

"This work was supported on a grant from CNR, Italy, Contract No. 82.02087.56"

and salt homeostasis.   A higher frequency of unwanted side effects
from diuretics in the elderly is determined also by the presence of
latent disorders of glucose and uric acid metabolism, by the
peculiar sensitivity of the senile brain, of the myocardium and
of the pressure receptors to alterations in cellular water and salt
concentration.[4,5]

As to water and salt metabolism in the elderly, further aging
brings about not only an impairment of renal function involving a
decrease in glomerular filtration rate and in maximum reabsorption
capacity at the tubular level, but also a variety of anatomical
and functional alterations affecting the different mechanisms
controlling water and salt regulation,[6,7] such as reduced thirst,
reduced response to hypothalamic osmoceptors (associated with a
decreased sensitivity of renal peripheral receptors to the anti-
diuretic hormone[8]) as well as reduced plasma renin activity,
reduced plasma and urine aldosterone concentration, causing the
organism to respond less promptly to acute stimuli.

The response of renal tubules to circulating aldosterone[12]
and the ability of renal sodium excretion to adjust to drastic
dietary changes are therefore less prompt.   Tolerance to large
variations of water intake is impaired as far as both excess and
reduction of water intake are concerned.[11]   Aging involves
moreover a progressive reduction of total body water[13] affecting
chiefly intracellular water content and changes of the electrolyte
pool, without variations of osmolarity.   Sodium concentration in
tissues - particularly as far as its exchangeable quota is
concerned - has been shown to increase, while tissue potassium
depletion has been shown to occur.[14,15,16]

Water and salt homeostasis becomes even more precarious due to
reduced potassium cell pools and to the impairment of the reserve
function of bone tissue.[6]   Reduction of serum albumin and hence
of colloid osmotic pressure, senile changes in capillary and
cellular permeability, changes in the arteriolar and capillary
circulation, in the cellular energetic systems in the enzymes
releasing energy for the potassium and sodium "pump" may also
determine further changes in water and electrolyte metabolism.
In aged patients water and salt metabolism may easily be affected
by disorders developing after any kind of disease, after surgery
or after mistaken or incongruous dietary or drug treatment,
particularly as far as water and salt intake is concerned.
Diuretic treatment, which modifies more less drastically an already
precarious water and salt homeostasis, is among the most important
of these factors.[17]   On the other hand, some of the metabolic
disorders which can be ascribed to diuretic treatment are often
already present, in a more or less latent form, in aged patients
(diabetes mellitus, gout etc).

THE RISKS OF DIURETIC TREATMENT IN THE ELDERLY

Diuretics may be administered in larger or smaller doses, as single dose, or intermittently, or as prolonged treatment. The possible occurrence of unwanted side effects* varies of course according to the mode of administration: we can therefore distinguish acute and chronic side effects.

Sodium Depletion

Intensive diuretic treatment with intravenous or oral massive doses of furosemide or ethacrynic acid, as is given in cases with acute pulmonary oedema or in hypertensive crises, may cause acute sodium depletion, with extracellular dehydration and asthenia, anorexia, disappearance of oedema, loss of body weight, arterial hypotension of the postural type, collapse, muscle cramps, convulsions and drowsiness. Biochemical data reflect blood concentration: owing to the reduction in plasma volume hypernatremia and hyperchloremia are normal during the initial phases, but subsequently both parameters decrease due to water absorption from tissues.

The most interesting aspects in geriatric practice concern the chronic, prolonged, daily administration of diuretics as is carried out in antihypertensive treatment. Inappropriate, prolonged administration of diuretics, even in moderate doses, but associated with a rigorous salt restricted diet may bring about chronic sodium depletion characterized by cellular overhydration and various neuropsychical disorders such as anxiety, vertigo, apathy, mental confusion, drowsiness and even coma. Chronic sodium depletion often causes renal failure with nitrogen retention due to reduction in blood flow and in the glomerular filtrate.

Blood concentration with hyponatremia also occurs, favouring water transfer from the extracellular to the intracellular compartment and hence cellular hyperhydration. Hyperkalemia develops due to the failure of renal exchanges caused by the marked reduction in sodium concentration in the tubules.

---

*We mention unwanted side effects due to allergic and idiosyncratic mechanisms which include cutaneous symptoms of various types, blood changes (agranulocytosis, thrombocytopenia) and changes in various organs (renal tubular necrosis, cholestatic hepatitis, acute pancreatitis and acute pancreatic necrosis, pulmonary allergic reactions etc), disorders of digestion (nausea, vomiting, diarrhoea, gastrointestinal bleeding etc)   Other important side effects include gynecomastia, impotency, uterine bleeding due to spironolactone.   Such side effects are rarely encountered in geriatric practice.

Potassium Depletion

     Prolonged administration of natriuretic diuretics can
frequently determine a variety of other side effects related to
other aspects of water and salt balance (potassium, magnesium,
calcium) as well as to glucose and purine metabolism.   The most
frequent unwanted side effect due to prolonged administration of
diuretics is potassium depletion,* which is particularly important
for its clinical consequences and commands careful evaluation when
administering diuretics to aged patients.   Potassium depletion may
occur also in younger subjects, particularly after prolonged
administration of thiazide diuretics, of furosemide and of ethacrynic
acid. [18,19,20]   Other authors have failed, however, to observe any
significant changes in young subjects.   The mechanism of potassium
depletion due to diuretics is quite complex and depends on various
factors, including:
- excess sodium reaching the distal tubule and determining an
  increase of exchanges of this electrolyte with potassium and
  hydrogen;
- sodium depletion and the reduction of extracellular fluids,
  which determine in turn the activation of the renin-angiotensin
  system and increased production of aldosterone favouring cation
  exchanges (Na^+, K^+, H^+);
- high Cl^- concentration in the distal tubule, which determines an
  increase of transtubular potential and a further transfer of K^+
  from the intraluminal fluid.

     Aged patients are prone to potassium depletion ("subjects with
high potassium depletion risk")[17] because, as has already been pointed
out, ageing is often associated with a progressive impoverishment
of potassium reserve both as a consequence of age itself and of other
factors, such as dietary insufficiency,[23,24] decreased potassium
absorption,[25] increased potassium loss in the urine and in the
faeces (due to laxative abuse).   Moreover, aging causes a decrease
of Na-K-ATPase,[26] an enzyme that releases the energy necessary for
transcellular electrolyte exchanges through the mechanism of the
Na-K pump.

     An episode of diarrhoea and vomiting is often enough to deter-
mine hypokalemia in aged subjects but this condition is also often
caused by drugs and particularly by diuretics.   The clinical
importance of this pathophysiological circumstance is determined by

---

*We consider that the term "hypokalemia" is misleading and semant-
 ically mistaken since it refers to a parameter of the blood
 chemistry which does not really reflect the actual total body
 potassium content.   We consider the term "potassium depletion"
 as more correct and reserve the term "hypokalemia" to mean a
 decreased serum potassium concentration.

its complex and partly non-specific symptoms.   Such symptoms may
be misinterpreted or ascribed to aging or to the diseases requiring
diuretic administration (generally arterial hypertension) and include
deep asthenia, intestinal atonia, neuropsychic disorders (apathy,
psychic depression, disorders of memory and various degrees of
mental confusion), and cardiovascular changes (orthostatic hypotension, various kinds of heart rhythm disorders, reduced tolerance
to cardioactive drugs even with low blood digoxin levels, specific
changes of the repolarization phase.   See Table 1)

   Potassium depletion may also bring about renal symptoms, i.e.
"hypokalemic nephropathy", which is characterized by decreased
tubular concentration and polyuria associated with polydipsia and
insensitive to limitation of water intake and ADH administration,
as well as by insufficient renal regulation of the acid-base balance
with decreased urinary excretion of acid radicals.   In these
cases sodium retention may occur, since renal elimination of this
cation implies an exchange with potassium.   Potassium depletion is
often associated with metabolic alkalosis, depending on hydrogen
and potassium loss.   Metabolic alkalosis is also caused by other
mechanisms such as the decrease in extracellular volume due to
massive diuresis (contraction alkalosis) and elimination of chlorine
instead of bicarbonate (hypochloremic alkalosis).   Acidosis
develops at the cellular level because an aliquot of potassium ions
is replaced by hydrogen ions.

   Risks due to the use of diuretics in patients with liver
cirrhosis may also be ascribed to potassium depletion: diuretics
are in fact the drugs of choice for the treatment of oedema and
ascites but they can frequently trigger portal-systemic encephalopathy.   Saluretics are most frequently responsible for the onset
of the latter syndrome, whose mechanism is quite complex and, as
yet, not completely understood.   Some data suggest that potassium
depletion together with metabolic alkalosis plays an important role
in determining increased permeability of the CNS to ammonium
radicals.   On the other hand the kidneys produce more ammonia.
while hypokalemia causes an increase in glomerular filtration and
hence hyperazotemia as well as a reduction in liver and brain blood
flow, which in turn aggravates liver and brain functional injury.
Since - as we shall see - the feasibility of documenting potassium
depletion through laboratory tests is quite aleatory, attention
must be paid to all clinical symptoms (asthenia, slight neuropsychic
disorders, non specific ECG alterations) in aged subjects, when
they occur after the beginning of diuretic treatment.   Such
symptoms may be ascribed to potassium depletion either if the
patient has received diuretic treatment recently or a long time
before.

   In our opinion potassium depletion - even in a subclinical
stage - is especially important in aged subjects since they are

Table 1. Non specific ECG changes that may
          be related to potassium depletion

- rhythm disorders

- disorders of A-V conduction

- prolongation of the S-T segment

- downsloping of the S-T segment with upward
  concavity

- decreased voltage of the T wave

- inverted T-wave with asymmetric branches

- pathologic V wave

particularly liable to pathological emergencies (acute diseases,
surgery, diarrhoea).   Potassium depletion may also predispose -
a fact usually ignored - to "cascade" heart failure and heart
complications, which are typical of aged patients and aggravate,
sometimes with fatal consequences, pathological conditions which are
"per se" unimportant or may be treated easily.

     In older patients the risk of potassium depletion due to
diuretics is unpredictably potentiated by the frequent use (or abuse)
of laxative favouring a vicious circle: diuretics → potassium
depletion → constipation → laxatives → more severe potassium
depletion.   As is quite evident, aged hypertensive patients undergoing prolonged diuretic treatment are particularly exposed to the
risk of potassium depletion, that may be recognized only through
careful and rational clinical study.

     For these reasons evidence of potassium depletion or at least
early treatment of the tendency to potassium depletion should be
considered of primary importance.   Administration of potassium-
sparing diuretics may be considered useful although it does not
always avert potassium depletion, particularly in low doses. On the
contrary, high doses of these diuretics may even cause hyperkalemia.
A potassium-rich diet is also quite useful, although aged subjects
do not always accept the type and amount of food necessary to attain
the required electrolyte concentration.   It should be borne in mind
that daily urinary potassium excretion, even in the absence of
diuretics, is about 20-30 mEq and that 1g KCl corresponds to 14 mEq
of potassium.

     There are still some controversial problems concerning potassium
depletion from diuretics, i.e.
a) the feasibility of documenting potassium depletion,
b) the timing of its correction with potassium salts.
As to the first point, the risk of potassium depletion should always
be considered as imminent in patients receiving natriuretic agents.

According to some statistics it occurs in 70 per cent of cases.[29]
Other authors[27] observed it in one third of subjects treated with
drug combinations including potassium sparing diuretics.   Such
risks should therefore be borne in mind when starting long term
diuretic treatment in aged patients.

   The problem of documenting potassium depletion offers some
difficulties and is still controversial.   The first problem is that
of the methods useful for evaluating potassium depletion and concerns
the role and the significance of plasma potassium determination and
its correlation to total body potassium content.   A decrease in
plasma potassium of 1 mEq/l is considered to reflect a decrease in
total body potassium content by 10 to 20 per cent.[18,20]   The decrease
in total body potassium, however, is not always associated with
decreased plasma potassium.   In fact plasma potassium concentration
is considered as reflecting only partially and inconstantly total
body potassium levels.[30,31,32]   The long term consequences of
prolonged treatment with depleting substances, such as diuretics,
may therefore not be predicted on the basis of plasma levels.
The same is true for the interpretation of ECG changes and of other
symptoms as due to electrolyte disorders and particularly to hyperkalemia and it is just for this reason that potassium depletion is
ignored in most cases.[33,34,31]

   Other methods yielding more reliable data on intracellular
potassium concentration were therefore devised, such as the
determination of muscle potassium, the determination of leucocyte
potassium, determination of total body potassium by radioisotope
dilution with K^{42} or by neutron activation.   All these methods
yield different results which are not always comparable with each
other and are therefore not suitable for routine use.   For these
reasons in our group research was focussed on red blood cell
potassium determination, which had already been recognized as a
simple and useful tool for evaluating the total body potassium.
[35,36,37,38,39,40,41,42]   We have been determining red blood cell
potassium along with the plasma concentrations in all our patients
for the last few years and we could demonstrate that data obtained
from this determination are much more useful than plasma levels for
detecting potassium deficiency as a long or medium term consequence
of diuretic treatment.

   Red blood cell potassium concentration does not absolutely
correspond to muscle potassium concentration, which is basically
and consistently higher.   On the other hand, since red blood cells
have no nuclei, they cannot be considered as typical cells;[42,43]
alterations in their electrolyte concentration ought therefore not
to be considered as always reflecting similar alterations in other
body cells.   Research work carried out by other authors shows that
there is a functional correspondence[35,36,44]   and that the enzym-

atic regulation system of the sodium-potassium pump is quite
similar and consists of Na-K-ATPase.

We have often reported the results of our experience[47,37,22]
and our large number of cases confirms that red blood cell potassium
determination is more sensitive and useful than serum potassium
determination for detecting potassium depletion. Our most recent
study includes 300 patients in whom a decrease in plasma potassium
was found in 18 per cent of cases, while red blood cell potassium
was shown to be decreased in 55 per cent of cases.[22] Most of these
patients had been treated with saluretic diuretics, generally for
antihypertensive purposes.

A further problem in cases with potassium depletion is when it
should be corrected with potassium salts. Some authors claim that
no treatment is necessary as long as the plasma concentration is
above 3.0 mEq/l, while others maintain that cases with values below
3.5 mEq/l should receive treatment.[21,48,29,20] In our opinion
serum potassium levels cannot be used for establishing which cases
should be treated with potassium salts, except those in whom the
concentration is below 3.5 mEq/l. We usually decide on this matter
on the basis of red blood cell potassium levels, whose acceptable
lower limit has been found to be 95 mEq/l. This limit was shown
to be valid in an investigation on ECG alterations of the repolarization phase and their improvement after potassium salt administration.[28]

Decreased Carbohydrate Tolerance

Potassium depletion is associated with another important
metabolic disorder, i.e. decreased carbohydrate tolerance.[49,50,51]
The onset of diabetes is in fact favoured by saluretic diuretics
via potassium depletion.[52,53,54,55,56,57,58] The relationship
between insulin secretion and potassium depletion is quite evident
also in some of our cases with decreased red blood cell potassium
concentration, in whom the glucose tolerance test, the utilization
rate of intravenous glucose and the blood insulin curve after glucose
administration were improved (Barbagallo-Sangiorgi et al., unpublished data) after treatment with potassium salts.[37] The mechanism
seems to be quite complex and may follow various pathways. The
most important are probably:
a) potassium depletion, which alters peripheral glucose utilization
   and probably also insulin secretion; the latter may also be
   directly influenced by diuretic treatment;
b) excess catecholamine production induced by the decrease in extracellular volume which inhibits liver and muscle glycogenesis;
   the latter is due to phosphodiesterase inhibition involving
   cyclic AMP accumulation and phosphorylase stimulation, this
   enzyme being responsible for the first phase of glycogenolysis.
   Phosphorylase is further stimulated by potassium depletion.

Hyperkalemia

It should also be recalled that treatment with potassium sparing diuretics may bring about an opposite condition i.e. hyperkalemia, particularly in cases in whom the prolonged use of these drugs is associated with excess dietary potassium intake. Hyperkalemia can also be observed in cases with excess diuresis associated with decreased glomerular filtrate, or with sodium depletion caused by decreased tubular sodium concentration impairing exchanges between this electrolyte and potassium and hydrogen. Hyperkalemia is as dangerous as hypokalemia; it determines muscle disorders (flaccid paralysis, abolition of reflexes and increased idiomuscular contraction) and ECG changes, i.e. alterations in the T wave which first becomes high and narrow and is then associated with downsloping of the ST segment. The various degrees of hyperkalemia may cause a widening of the QRS wave, the disappearance of the P wave and heart arrest due to ventricular fibrillation.

Hyperuricemia

Further unwanted side effects of diuretic treatment involve uric acid metabolism; in fact all diuretics may produce a more or less marked increase in uricemia. The onset of acute attacks of gout depends also on congenital predisposition, and aged patients are more frequently liable to these attacks. The mechanism by which uricemia develops depends on the competitive inhibition of uric acid secretion from the proximal and probably also from the distal tubule, and, although indirectly, on the decrease in extracellular volume, which causes an increased reabsorption of uric acid due to the decrease in glomerular filtrate.

Other Side Effects

Most diuretics interfere with calcium elimination, but their mechanism of action is quite different. Benzothiazide produces a decrease in calcium elimination by increasing it reabsorption from the proximal tubule, as induced by chronic contraction of the extracellular volume. Decreased urinary calcium elimination does not usually lead to hypercalcemia, except in patients with renal failure or with accelerated bone calcium reabsorption, in whom calcemia may attain dangerously high levels.

Furosemide, ethacrynic acid, triamterene as well as mercurial diuretics cause instead enhanced calcium elimination by inhibiting its reabsorption from the ascending branch of Henle's loop. Patients show therefore a tendency to hypocalcemia and tetanus crises are possible. Decreased serum magnesium levels have also been observed after prolonged treatment with furosemide and thiazide diuretics. Decreased serum magnesium levels may be responsible for ventricular extrasystoles and increased digitalis toxicity.

Side effects of thiazide diuretics include also the increase in lipoproteins (cholesterol and triglycerides) and this finding raised even more doubts concerning the indiscriminate use of these drugs, which may aggravate vascular risk.[57,58] These changes, which are dose-dependent, were confirmed through investigations carried out in our Institute by Cupidi et al (unpublished data).

CONCLUSIONS

Diuretics may bring about highly dangerous side effects which are inherent in their biological action and depend also on the pathophysiological and metabolic condition of the patients treated. Pathophysiological changes typical of old age easily predispose patients to such unwanted side effects. Thus age, as well as the patient's general condition, should be carefully considered when deciding to start diuretic treatment and the possibility of risks should be evaluated against the advantages that may be expected. Awareness of risks, however, should not preclude diuretic treatment, but rather indicate the need to exert caution and to carry out repeated laboratory and clinical tests.

In patients for whom prolonged diuretic treatment is required (e.g. in antihypertensive treatment) we should check skin conditions, the presence of oedema, monitoring their disappearance and the progressive decrease of body weight as well as arterial blood pressure and it postural variations. Hypovolemia should always be suspected when orthostatic hypotension occurs in subjects in whom arterial blood pressure decreases by 20 - 30 mmHg compared to usual values.

Values of azotemia, creatininemia, uricemia, glycemia, kalemia and particularly red blood cell potassium concentration should be monitored. If natremia attains values lower than 130 mEq/l, diuretic treatment must be discontinued and water intake limited to not more than 0.5 l/day.

It is quite true that modern diuretics involve less risks and that some of these risks such as potassium depletion are attenuated by the administration of low doses of diuretics, by the association with potassium sparing drugs or by the administration of potassium salts, but still such potential risks are quite important in aged patients. Diuretics should therefore be prescribed after careful evaluation of individual cases taking into account all the indications and contraindications.

SUMMARY

Diuretics are among the most frequently prescribed drugs and are extensively used in antihypertensive treatment. They may

involve quite important side effects concerning water and salt
balance, glucose metabolism and uric acid excretion and affecting
particularly aged patients. Attention has been focussed on
potassium depletion, that can be detected by determining red cell
potassium concentration, rather than by plasma levels, which are
often normal even in the presence of ascertained potassium
depletion.

REFERENCES

1. L. J. Christopher, B. R. Ballinger, A. M. M. Shepherd, A.
   Ramsey, G. Crooks, A survey of hospital prescribing for the
   elderly, in: "Drugs and the Elderly," J. Crooks, I. H.
   Stevenson, eds., MacMillan Press, London (1979).
2. N. Hurwitz, Predisposing factors in adverse reactions to drugs,
   Brit.med.J. 1:536 (1969).
3. J. Williamson, Adverse reactions to prescribed drugs in the
   elderly, in: "Drugs and the Elderly," J. Crooks, I. H.
   Stevenson, eds., MacMillan Press, London (1979).
4. G. Barbagallo Sangiorgi, A. Di Sciacca, G. Frada Jr., G. Cupidi,
   G. Costanza, La encefalopatie metaboliche dell'anziano,
   Giorn.Geront. 29:635 (1981).
5. G. Barbagallo Sangiorgi, A. Di Sciacca, G. Frada Jr., G. Cupidi,
   M. Barbagallo, Ill danno jatrogenico da farmaci in geriatria,
   Nuovo Boll.Farmacol.Clin. 3:157 (1981).
6. A. Borghetti, A. Novarini, Fisiopatologia dell'equilibrio
   idricoelettrolitico nella ed acido-base nell'eta senile,
   Giorn.Gerontol. 29:1001 (1981).
7. D. Ceruso, Equilibrio idricoelettrolitico nella senescenza:
   aspetti particolari, G.Clin.Med. 61:177 (1980).
8. J. H. Miller, N. W. Shock, Age differences in renal tubular
   response to antidiuretic hormone, J.Gerontol. 8:446 (1953).
9. C. Flood, C. Gherondache, G. Pincus, J. F. Taix, S. Willonghby,
   The metabolism and secretion of aldosterone in elderly
   subjects, J.Clin.Invest. 46:960 (1967).
10. P. Weidmann, S. Bursztein, J. De Lima, Effect of ageing on renin
    and aldosterone in normal adults, in: "Age as a determinant of
    renal sodium conservation in normal man," Epstein, M.,
    N. K. Hollenberg, eds., J.Lab.Clin.Med. 87:411 (1976).
11. A. Ruol, L. Menozzi, F. Furlanello, Il ricambio idrosalino nell'
    eta senile, Rel.12° Congr.Naz.Soc.It.Gerontol.Ger., Bologna,
    5-7 Aprile 1963, Suppl. 31, Giorn.Geront.
12. M. Epstein, N. K. Hollenberg, Age as a determinant of renal
    sodium conservation in normal man, J.Lab.Clin.Med. 87:411
    (1976).
13. N. W. Shock, Metabolism and age, J.Chron.Dis. 2:687 (1955).
14. F. S. Feruglio, F. Pupita, E. Bartoli, M. Molaschi, La distrib-
    uzione del potassio corporeo nelle varie eta dell'adulto,
    Atti Symp.Intern.Potassio Biol.Med., Siena, 22-23 Maggio
    (1965).

15. T. H. Allen, E. C. Anderson, W. H. Langham, Total body potass-
    ium and gross body composition in relation to age.
    J. Gerontol. 15:348 (1960).
16. M. Lye, Distribution of body potassium in healthy elderly
    subjects, Gerontology 27:286 (1981).
17 A. Novarini, Considerazioni sulla terapia diuretica nello
    anziano, Giorn.Geront. 28:762 (1980).
18. C. Edmonds, B. Jasani, Total body K in hypertensive patients
    during prolonged diuretic therapy, Lancet 1:8 (1972).
19. J. J. Healy, T. J. McKenna, B.S.J. Canning, T. G. Brien,
    G. J. Duffy, F. P. Muldowney, Body composition changes in
    hypertensive subjects in long-term oral diuretic therapy,
    Brit.med.J. 1:716 (1970).
20. P. Wilkinson, H. Issler, R. Hesp, E. Raftery, Total body
    and serum K during prolonged thiazide therapy for essential
    hypertension, Lancet 1:759 (1975).
21. H. J. Dargie, K. Boddy, A. C. Kennedy, P. C. King, P. R. Read,
    D. M. Ward, Total body potassium in long-term frusemide
    therapy: is potassium supplementation necessary? Brit.med.J.
    4:316 (1974).
22. G. Barbagallo Sangiorgi, G. Frada Jr., A. Di Sciacca, G.
    Cupidi, G. Costanza, M. Affronti, E. Putignano, M.
    Barbagallo, Turbe del ricambio idroelettrolitico come causa
    di encefalopatia metabolica nell'anziano, Giorn.Geront.
    30:199 (1982).
23. T. G. Judge, Potassium metabolism in the elderly, Symposia
    Swed.Nutr.Found. 10:86 (1972).
24. C. C. MacLeod, T. G. Judge, F. I. Caird, Nutrition of the
    elderly at home. III. Intakes of minerals, Age and Ageing
    4:49 (1975).
25. P. M. Warren, M. A. Pepperman, R. D. Montgomery, Age changes
    in small-intestinal mucosa, Lancet 2:849 (1978).
26. A. J. Marsh, B. L. Lloyd, R. R. Taylor, Age dependence of myo-
    cardial Na+–K+–ATPase activity and digitalis intoxication
    in the dog and guinea pig, Circ.Res. 48:329 (1981).
27. R. C. Hamdy, J. Tovey, N. Perera, Hypokalaemia and diuretics,
    Brit.med.J. 2:1187 (1980).
28. G. Barbagallo Sangiorgi, G. Costanza, A. Di Sciacca, G. Frada
    Jr., G. Cupidi, R. Malta, M. Affronti, Potassiemia,
    potassio intraeritrocitario e alterazioni della fase di
    ripolarizzazione dell'elettrocardiogramma. Atti del
    "Simposio in tema di cardiopatia ischemica," Catania 14-15
    Novembre (1981).
29. F. G. McMahon,"Management of Essential Hypertension," Futura
    Pub.Co., Mount Kissco, N.Y. (1978).
30. J. A. Graham, J. F. Lamb, A. L. Linton, Measurement of body
    water and intracellular electrolytes by means of muscle
    biopsy, Lancet 2:1172 (1967).

31. F. D. Moore, I. S. Edelman, J. M. Olney, A. H. James, L. Brooks, G. M. Wilson, Body sodium and potassium; inter-related trends in alimentary, renal and cardiovascular disease; lack of correlation between body stores and plasma concentration, Metabolism 3:334 (1954).

32. N. Valentin, K. H. Olesen, Measurement of muscle tissue water and electrolytes, Scand.J.Clin.Lab.Invest. 32:155 (1973).

33. C. T. G. Flear, W. T. Cooke, A. Quinton, Serum-potassium levels as an index of body content, Lancet 1:458 (1957).

34. M. P. Hutt, Effect of disease on erythrocyte and plasma potassium concentrations, Am.J.Clin.Sci. 223:176 (1952).

35. M. Bahemuka, H. M. Hodkinson, Red-blood-cell potassium as a practical index of potassium status in elderly patients, Age and Ageing, 5:24 (1976).

36. M. Bahemuka, H. M. Hodkinson, Red-blood-cell potassium and hand-grip strength in healthy elderly people, Age and Ageing 5:116 (1976).

37. G. Barbagallo Sangiorgi, A. Di Sciacca, A. Pardo, Potassium depletion in aged patients: an evaluation through red-blood-cell potassium determination, Age and Ageing 8:190 (1979).

38. D. W. Boyd, Red-blood-cell potassium and aldosteronism, Lancet 1:594 (1970).

39. G. M. Guest, S. Rapaport, Electrolytes of blood plasma and cell in diabetic acidosis and during recovery, Proc.Am.Diabetic Assoc. 7:97 (1948).

40. W S. Hoffman, H. R. Jacobs, The partition of potassium between the serum and corpuscles in health and disease, J.Lab.Clin. Med. 19:633 (1934).

41. H. Ibsen, Red blood cell potassium as a measure of body potassium in thiazide-treated patients with essential hypertension, Scand.J.Clin.Lab.Invest. 34:161 (1974).

42. D. N. Baron, Down with plastma! Intracellular chemical pathology studies by autoanalysis of cells of solid tissues, erythrocytes and leucocytes, Proc.Roy.Soc.Med. 62:945 (1969).

43. J. Patrick, B. Bradford, A comparison of leucocyte potassium content with other measurements in potassium-depleted rabbits, Clin.Sci. 42:415 (1972).

44. K. V. Johny, J. R. Lawrence, M. W. O'Halloran, Studies on total body serum and erythrocvte potassium in patients on maintenance haemodialysis. The value of erythrocyte potassium status in elderly patients, Nephron 7:230 (1970).

45. I. M. Glynn, Sodium and potassium movements in human red cells, J.Physiol.(London) 134:238 (1956).

46. A. K. Solomon, The permeability of human erythrocytes to sodium and potassium, J.Gen.Physiol.(London) 36:57 (1952).

47. G. Barbagallo Sangiorgi, A. Di Sciacca, G. Frada Jr., G. Costanza, G. Cupidi, E. Putignano, Aspetti di farmacoterapia cardiologica nell'anziano. Atti Congr.Naz. "Il Cuore e l' anziano," Ancona, 19-21 Ottobre (1979).

48. I. K. Ibrahim, A. E. S. Ritch, W. J. MacLenan, T. May, Are
     potassium supplements for the elderly necessary? Age and
     Ageing 7:165 (1978).
49. F. A. Fuhrman, Glycogen, glucose tolerance and tissue metab-
     olism in potassium-deficient rats, Am.J.Physiol. 167:314
     (1951).
50. L. I. Gardner,, N. B. Talbot, C. D. Cook, H. Berman, U. R.
     Concepcion, Effect of potassium deficiency on carbohydrate
     metabolism, J.Lab.Clin.Med. 35:592 (1950).
51. U. Sagild, V. Andersen, P. B. Andreasen, Glucose tolerance
     and insulin responsiveness in experimental potassium
     depletion, Acta Med.Scand.169:243 (1961).
52. A. Amergy, P. Berthaux, W. Birkenhäger, A. Boel, P. Brixko,
     C. Bulpitt, D. Clement, M. Deruyttere, A. De Schaepdryver,
     C. Dollery, R. Fagard, F. Forette, J. F. Henry, J. Hellemans,
     U. Laaser, P. Lund-Johansen, J. MacFarlane, T. Maling,
     A. Mutsers, A. Nissinen, O. J. Ohm, J. Pelemans, A. L.
     Suchehkaye, J. Tuomilehto, J. Willems, Antihypertensive
     therapy in patients above age 60, Acta Cardiol. 33:113 (1978).
53. A. Breckenridge, T. A. Welborn, C. T. Dollery, R. Fraser,
     Glucose tolerance in hypertensive patients on long-term
     diuretic therapy, Lancet 1:61 (1967).
54. J. W. Conn, Hypertension, the potassium ion and impaired carbo-
     hydrate tolerance, New Engl.J.Med. 273:1135 (1965).
55. E. M. Kohner, C. T. Dollery, C. Lowy, B. Schumer, Effect of
     diuretic therapy on glucose tolerance in hypertensive
     patients, Lancet 1:986 (1971).
56. P. J. Lewis, E. M. Kohner, A. Petrie, C. T. Doller, Deterio-
     ration of glucose tolerance in hypertensive patients on
     prolonged diuretic treatment, Lancet 1:564 (1976).
57. M. I. Rapoport, H. F. Hurd, Thiazide-induced glucose intoler-
     ance treated with potassium, Arch.Int.Med.113:405 (1964).
58. R. H. Schwab, J. K. Perloff, R. L. Porus, Chlorthiazide-
     induced gout and diabetes, Arch.Int.Med. 111:123 (1963).
59. R. P. Ames, R. Hill, Elevation of serum lipid levels during
     diuretic therapy of hypertension, Am.J.Med. 61:748 (1976).
60. C. Bengtsson, G. Johnsson, R. Sannerstedt, L. Werrö , Effect
     of different doses of chlorthalidone on blood pressure,
     serum potassium and serum urate, Brit.med.J. 1:197 (1975).

APPENDIX

The main sites of action of diuretic drugs are the following:
- proximal convoluted tubule, where urinary reabsorption of Na, Cl and H_2O takes place (carbonic anhydrase inhibitors, osmolar diuretics, mannitol, urea, secondarily: furosemide and thiazide diuretics).
- ascending branch of Henle's loop, where active Na reabsorption takes place (loop diuretics, such as furosemide, ethacrynic acid and bumetanide).
- proximal tract of the distal tubule, where Na reabsorption takes place, without H_2O reabsorption (thiazide diuretics furosemide)
- distal tract of the distal tubule, the site of the final modulation of Na reabsorption. Na^+ is exchanged against K^+ and H^+, depending on the action exerted by aldosterone (spironolactone, triamterene amyloride).
- collecting tubule, the site of H_2O reabsorption depending on ADH action (ethacrynic acid, furosemide). Saluretic diuretics, such as thiazides, chlorthalidone, furosemide, ethacrynic acid and bumetanide differ from potassium sparing diuretics, such as spironolactone, amyloride and triamterene both as to their biological action and as to unwanted side effects.

SHOULD HYPERTENSION IN THE ELDERLY BE TREATED?

Fabrizio Fabris

Department of Geriatric Medicine
University of Torino, Italy

Any discussion of arterial hypertension in the elderly must
essentially turn on the vexed question of the indications and
limitations imposed on its treatment. We can perhaps oversimplify
the matter by saying that there are two approaches towards antihypertensive management in the aged: intervention, and abstention
wherever possible. Let me say straightaway that I am to be found
in the second camp for both clinical and physiopathological reasons.
From the clinical standpoint, long experience with patients of this
type has convinced me that treatment - by which I mean management
imbued with a certain degree of "aggressiveness" - is often
accompanied by side-effects that greatly cloud the quality of life,
while its influence on the quantity of life is, to say the least,
debatable.   In physiopathological terms, too, there do not seem
to be sure premisses for a rationally effective and useful intervention. The question may thus seem somewhat categorical, and
certainly in opposition to the normal run of things (at any rate to
non-geriatricians).   I shall therefore try to explain the steps
that led me to this conclusion. I must say that I came to the
approach of abstention wherever possible through a long evolution
(or involution, according to one's point of view) starting from the
taken-for-granted equation: "hypertension increases the vascular
risk, therefore correction of hypertension decreases such risk."
This might hold sway as an axiomatic truth were it not for two
considerations that undermine (or ought to undermine) its certainty,
namely: 1) are there indeed cases in which elderly hypertensives
are treated in an energetic manner? 2) has due reflection been
given to the possible association linked with antihypertensive
drug management in the aged? The observation on the part of Jackson
et al[1] of elderly hypertensives brought unconscious to hospital as
emergency patients, yet devoid of symptoms, shortly after

327

commencing antihypertensive therapy, should give us pause.

What has been said above relates to the general position, to a certain extent deliberately placed in contrast with the opposite approach, namely that of treating hypertension in the elderly always and under all circumstances.   It is evident that the question is much more complicated and presents a variety of facets, and above all that it can and must be viewed from different angles.

DEFINING HYPERTENSION

The first point requiring examination is the pressure limits that set the cut-off between normal and high blood pressure in the elderly.   We are fully aware of the truth of Pickering's assertion[2] that hypertension is a quantitative, not a qualitative disease, in which departure from the normal is a question of degree and not of kind, so that any attempt to set its limits is of necessity artificial.   A cut-off is logically artificial. Even so, it is useful for the purposes of classification. An elderly subject displays variations in a number of constants and their frequency is such that the very idea of "normality" takes on a different significance.   I should like to stress this point, because it is poorly understood, or at any rate underestimated in many sectors of the physiopathology of senescence. In other words - and the discussion has, in my opinion, repercussions meaningful and logical on the subject of treatment - a question that is often not put is whether it is or is not correct to refer to the normal state of the young adult.   If it were not - and in my opinion it is not - one could ask whether the theoretical objective of reconverting at least some of the parameters applicable to the elderly towards the ideal values for the young is rationally, i.e. scientifically, correct or otherwise.   Once we regard as normal a situation that is widely represented, we can at least suspect that a situation which is abnormal in accordance with a traditional criterion of reference to the young, though normal when assessed in terms of frequency, represents, following its establishment, a "compensation" or "adaptation" that, were it to be corrected, could result in harm, and not a benefit - for the body itself.   If a biological parameter varies in function of age, a reference to constant values is itself a source of errors.   This is not to say that maintenance in the elderly of biological constants that are close as possible to those noted in the young is not the optimum eventuality. In all probability, this is so, and proves the existence of a disassociation between a person's biological age and what appears on his birth certificate, with the result that the former may be even very much lower than the latter.   Yet should this not be the case, it is still rational, as a rule, to attempt to turn back to ideal values that the organism has altered - and for the worse - over the course of the years?   I believe that certain preliminary conditions must

exist for such an attempt to be beneficial : the feasibility of
the operation itself and, where this is present, the possibility
of its being accomplished in terms of physiological reinstatement,
and not through measures that achieve their objective by altering
- and to a marked degree other constants that are already altered
or in a state of unstable equilibrium.

As a general notion, a question that geriatric practice has
suggested to my mind, one that I feel is a topic of considerable
and hitherto unexplored interest, is whether certain variations
(not all) represent - as would seem to be the case at first sight -
negative phenomena, or might not rather be adaptations that the
senile organism puts into effect to in some way compensate for an
anatomofunctional modification - this indeed occurs! - of a
negative and involutive nature.   As far as the specific subject of
arterial pressure is concerned, a sufficient blood supply may -
at all events conceptually - be influenced or promoted by a higher
tension regimen in the elderly than in subjects in other age groups.

I have heard it authoritatively stated recently by Avogaro
that when dealing with the over-65s it is better to abstain as a
rule with respect to the administration of hypolipaemising drug
therapy.   In agreeing with this approach in principle, I would add
that one is taking for granted that after a certain age the
pathogenetic potential is of less account, and that its correction
is less significant or even of debatable effectiveness as far as
the progress of the disease is concerned, almost as if another
criterion must be applied when assessing the "malignancy" of
hyperlipaemia in the elderly.   A re-examination of certain axioms
is thus in progress even for other conditions.   A further point
that can also be conveniently borne in mind.   The senile
population is a large bracket that covers the sixty-year-old as well
as the ninety-year-old.   The closer one moves towards the extreme
limits of life, in my opinion, the sounder is the notion that the
ongoing clinical situation should be respected, and that caution
should be increasingly observed in the choice of any measure
involving "modification."   And this at the very time when the
boundaries of surgical aggression, for example, are being rightly
extended.

SYSTOLIC AND DIASTOLIC HYPERTENSION

The idea that arterial hypertension is very common in the
elderly is well known and generally accepted.   Only one important
aspect of the question need be mentioned, namely that the systolic
and diabstolic pressure curves, which are parallel until the end
of the presenile period, tend to move apart.   An increase in
systolic pressure, in fact, is accompanied by a slight tendency on
the part of the diastolic curve to fall.   The practical

demonstration of this can be seen in the high frequency of what is
perhaps improperly described as systolic hypertension. This
expression may be accepted, partly because it has entered into
everyday use, though it contains the germ of a contradiction in terms
if we follow the WHO in defining hypertension as one in which
systolic values are $\geqslant$ 160 mmHg and diastolic values $\geqslant$ 95 mmHg.
Other terms have been put forward, such as pseudohypertension and
hypersystolia, but one feels that, provided what has just been said
is kept in mind, the ordinary expression "systolic hypertension" can
conveniently be employed.

As far as the magnitude of the phenomenon is concerned, we can
refer to our own epidemiological data: in a sample population of
1000 patients aged 60 or over consecutively admitted to hospital
for reasons of all kinds, we observed systodiastolic hypertension in
30.8 per cent and systolic hypertension in 24.5 per cent. These
percentages were higher in women. This sample cannot be taken as
representative of the elderly population as a whole, but it certainly
offers a general idea of the size of the phenomenon. More than one
elderly person out of two suffers from high pressure. In 50 per
cent of cases, only systolic hypertension is involved. This second
clinical picture thus acquires, in the aged, the status of a datum
of wide epidemiological significance, to the point that it represents
a characteristic feature of geriatric pathology. It must be made
immediately clear that its physiopathological bases are very
different from those underlying systolic hypertension observed in
other conditions, and in particular during the onset of essential
hypertension in the young adult. Whereas in this case, the increase
in systolic pressure is generally the outcome of a state of hyper-
sympathicotonia with an increase in systolic stroke volume, in the
elderly its cause is to be found in rigidity of the wall of the
aorta and the large vessels emerging therefrom, leading to a
reduction of the Windkessel or bellows effect. As a result, the
amount of pressure that the elastic wall can absorb and store up by
stretching, and then surrender during the diastole, is reduced to
a more or less marked degree, so that it is transferred to the
periphery during the systole. This is an illustration of the well-
known phenomenon whereby a continuous flow of liquid will issue
from a distensible hose connected to an intermittent pump,
whereas a rigid pipe will provide a pulsed flow governed by the
revolution of the pump. Even though the increase in systolic
pressure is the common factor, the physiopathological mechanisms
are radically different in the two situations, and should lead to
the adoption of a different approach to correction, where possible.
In the case of an elderly subject, correction of vessel sclerosis
would be correct, whereas treatment designed to correct hyper-
sympathicotonia and increase the stroke volume would not.

The fundamental distinction must therefore be made between
systolic hypertension alone and systodiastolic hypertension.

The frequency of these two entities in the aged is much the same,
whereas the approach to their management is distinctly different.
In the first case, we substantially refrain from using antihypertensive drugs, apart from mild doses of diuretics; in the
second, however, treatment is carried out, but the drugs are
carefully chosen and their doses tend to be restrained. Having thus
stated our position briefly, we shall return to the subject later.

As far as possible, a distinction should also be drawn between
hypertension with onset in the young adult and continued into
senility, and hypertension observed ab initio in the elderly.
Allowing for the fact that this distinction is not always easy to
make in practice, since the entities are not clearly distinguishable,
and that it thus possesses a significance that is principally
conceptual, it usefully expresses a different clinical approach.
In the first instance, the treatment carried out previously must be
continued, coupled with the use of lower doses: in the second,
the true problem of the proper indication arises, and the selection
of drugs (if any) and their doses.

REGIONAL CIRCULATION

Another aspect of the matter that requires particular attention
and evaluation in the aged is that relating to the condition of the
local circulation, especially the cerebral, coronary, and peripheral
limb vessels. It is important to realise that comparable pressure
values can be encountered in persons with no clear evidence of
cardiovascular alterations and those with cardiac enlargement or
other signs of hypertensive heart disease. The functional status
of the target organs, particularly the kidneys, is obviously of
great importance as well. The question of the local circulation is
one of the outstanding aspects of geriatrics, because it is
extremely frequent - almost the rule - to find that one or more
districts are impared. Impairment of the coronary circulation,
for example, may encourage the choice of a beta blocker, whereas its
use would be ruled out in the event of impairment of the peripheral
circulation. There is evidence indicative of the greater
frequency of cerebral haemorrhage in hypertensives - including those
with high systolic pressures - in relation to the greater frequency
of cerebral aneurysms. Yet it still remains to be shown how far
this phenomenon is due to hypertension and how much to vascular
sclerosis. In addition, the reduction of mean arterial pressure
and heart output capacity beyond certain limits results in a greater
risk of ischaemic episodes caused by thrombi.

All forms of secondary hypertension can be seen in the elderly,
though the most frequent are attributable to chronic urinary
infection and nephrovascular forms induced by arteriosclerotic
stenosis of the renal arteries and their branches. In the first

case, one must remember the predominant importance of antibacterial
treatment, in the second, the preferential position enjoyed by
medical management, especially with beta blockers, as opposed to
surgical correction.   An "acute" form of secondary arterial
hypertension is typically encountered in the elderly during the
course of acute urinary retention.   It is clear that treatment is
directed to the cause.   Recognition firstly arises from directing
one's mind to the possibility of such an eventuality, and this
is more likely to be the case when one has sufficient familiarity
with geriatric pathology.   It goes without saying that the use of
diuretics would have a deleterious effect.   Here, too, there
emerges the desirability - something that we wish to treat as
absolutely fundamental and of top priority - of viewing an aged
subject with an all-encompassing eye, and not sector by sector.

LIMITATIONS OF TREATMENT

     Those who maintain that antihypertensive management should
always be undertaken under all circumstances, even in the elderly,
quote in support of their contention the fundamental Framingham
study.   Hypertension - of this there can be no doubt - is the
primary cardiovascular risk factor, and the logical consequence of
this is that the treatment of hypertension is appropriate.   Yet
one may rightly ask whether, in the case of systolic hypertension
in particular, the vascular risk is more properly associated with
hypertension, or with the arteriosclerosis that is in its turn
responsible for the increase in pressure values.   The objective
sought in the correction of hypertensive states will prove correct
or otherwise depending on the answer given to this question.   If
arteriosclerosis, particularly that of the aorta, is taken as the
prime mover, treatment should be directed to its correction.   As
matters now stand, one cannot speak of the reversibility of the
arteriosclerotic process, at any rate in terms of clinical and
practical implications.   To be more precise, the idea of
reversibility may be envisaged for the florid, "metabolic" stage,
rather than the regressive, stabilised stage of the disease.
It follows from this that antihypertensive management cannot be
seen - in the specific case - as directed in terms of causality of
the increase in risk.   Doubts concerning the utility of reducing
isolated systolic hypertension have been expressed by no less a
person than Kannel,[3] who was the leader of the Framingham study:
"It is difficult to say from this if the fault in the aged lies
with the systolic pressure per se, or if it is only a sign of
atherosclerosis-prone inelastic aged vessels.   Only an examination
of the efficacy of reducing systolic pressure in isolated systolic
hypertension can answer this question."

     A similar opinion can be found in Dyer[4] : "No data have been
reported from a randomized controlled trial on drug treatment of

pure systolic hypertension.   Therefore no scientific statement can
be made at present on the benefit: risk ratio with pharmacologic
treatment of pure systolic hypertension, whatever the age of the
patient."   The co-operative study on the treatment of hypertension
in the elderly (EWPHE), too,has not yet - to the best of my knowledge - led to any final conclusions regarding the prognostic
influence of such treatment.

Moreover, it is well to ask what are the theoretical points of
attack open to the treatment of systolic hypertension in the
elderly.   Stress, in fact, must be laid on certain physiological
features in this form:

- elastic resistance values (aortic compliance) cannot be altered
  by treatment:
- peripheral resistance values are not increased;
- the heart output capacity (systolic stroke volume) is the only
  point that can be altered pharmacologically for the purpose of
  bringing about a decrease in systolic pressure.

However, it is known that cardiac output decreases physiologically in the aged. A further, pharmacologically induced reduction
would thus appear a therapeutic measure ill-suited to the overall
economy of senile haemodynamics.   Indeed, it may even be damaging,
owing to its repercussions on certain local circulations, such
that of the brain.

Ordinary antihypertensive treatments mainly achieve their
objective by reducing peripheral resistance values.   In the absence
of the premiss that these are increased, conventional management is
on the one hand destined to have little effect, and on the other to
cause at the level of the nutritive vessels a negative variation,
which, if insufficiently parried by the distendibility of the major
vessels, may cause, instead of benefits, greater functional
alterations than there were before.

In this connection, one should not lose sight of the significance of orthostatic hypotension, namely a fall in systolic pressure
of the order of < 20 mmHg when standing as opposed to lying down.
This phenomenon increases in frequency with age, particularly where
aging is complicated by diabetes or Parkinsonism, and is observed
in 16 per cent of persons aged 65 to 74 years, and in 30 per cent
of over-75s (Overstall[5]). The dramatic consequences of orthostatic hypotension need hardly be mentioned.   In the case of falls
only, and leaving aside such major consequences as cerebral
ischaemia, it appears that 5 per cent are attributable to a pathogenesis of this type.   Since many hypertensive drugs are particularly effective when the subject is in an upright position, caution in
adopting the pharmacological approach would appear to be amply
justified.   Indeed, for some of these drugs, especially at "high"

doses, an indication for their therapeutic employment in the elderly should be regarded as highly exceptional. This is particularly true of the sympathicoplegics.

PREFERRED DRUGS

Which group of drugs should be preferred? Generally speaking, one may answer: the diuretics. This assertion is only apparently in conflict with the observation of a reduced potassium pool in the elderly. The question is one of the dose employed. The recent literature correctly proposes that diuretics should be used in more restricted doses when treating hypertensives. We agree with this attitude, and have long given it our support in the case of the elderly. The fear of significant side-effects is amply justified not when diuretics are used, but when they are employed at high doses. Between these two therapeutic situations there is a substantial difference. They form, in effect, two separate chapters, united in appearance only. What reasons can be cited in favour of diuretics? On the one hand, the observation – whatever the significance one may wish to attribute to it – that the elderly usually present hypertension with low blood renin values, and hence more susceptible to diuretic therapy; on the other, the concomitance, frequent though not inevitable, of situations of slight or latent cardiocirculatory insufficiency suggesting the use of diuretics.

Which diuretics are preferable? To a certain extent, the answer is a fairly obvious one: those with an action that is not drastic, but slow and protracted. In our experience, the choice of drugs with these characteristics means that the loss of potassium caused by kaliuretics is of no consequence, and can be offset by slight dietary or pharmacological supplements. For reasons of cost, too, though not only on this account, we regard simple potassium savers as drugs of second choice, though their combination with other preparations may be useful.

It is evident from what we have said that we do not consider the beta blockers as the first drugs of choice in the elderly patient. In this age, in fact, one may frequently come across, perhaps in an indistinct form, all the situations that represent contraindications or at all events limitations on their use: the likelihood of bradycardia or conduction disturbances, various severe obstructive respiratory insufficiency, diminished cardiac efficiency and general physical performance. While this is the general rule, there may be individual cases where their use may be even be electively indicated, e.g. when clinically important signs of ischaemic heart disease are also present. In particular, good tolerance and effectiveness have been demonstrated, even by us on the part of the alpha-beta blockers (labetalol). It is not our intention to review the various drugs used to treat hypertension.

We shall confine ourselves to mentioning that alpha methyldopa is worthy of attention for its practicability, even in the elderly. The importance of a low sodium and, as far as may be indicated, a hypocaloric, diet may be taken for granted.

PRINCIPLES OF MANAGEMENT

In an attempt to bring the various situations within a schematic arrangement, we can state our views briefly as follows:

- asymptomatic systolic hypertension: no treatment and periodic checking;
- poorly tolerated systolic hypertension or slight systodiastolic hypertension (diast. 95-100 mmHg): intermittent administration of small doses of diuretics;
- moderate or serious systodiastolic hypertension: diuretics combined with other antihypertensive drugs.

We have for many years been in favour of employing ergot alkaloids in the elderly. We now reiterate this approach, not, of course, in the sense that they can be used in the treatment of hypertension, but as support management, particularly in the forms of systolic hypertension typically found in the elderly. The action of these drugs at the level of cerebral mediation as well has perhaps revived and made clearer their wide clinical favour in recent years. In the aged, though not in this period of life only, of course, one should tend to avoid as far as possible the extra increase in pressure with its possible haemodynamic repercussions, particularly on the brain. We feel that the therapeutic scheme proposed is well suited to this attempt. When a hypertensive crisis is in progress, the measures to be adopted should not differ from those known for the treatment of other age groups. One can at most emphasise that in this case, too, the indication is for diuretics and antivascospastic drugs as the first, dependable course of action.

Compliance on the part of the patient with regard to his treatment will depend, inter alia, on the simplicity of the therapeutic schedules. This is particularly true of the elderly on account of the obvious difficulties they face in following protocols that are complicated or poorly specified. Another extremely important rule when dealing with the aged is that caution should be exercised in the therapeutic attack, which may be followed by a gradual upward adjustment of the doses prescribed. This is advisable for two reasons. In the first place, a pharmacological effect superior to that expected in the young adult may be observed; in the second place, it has been demonstrated experimentally that a sudden fall in arterial pressure is poorly tolerated by the brain, whereas an equivalent fall brought about gradually is

much better tolerated and not complicated by signs of cerebral
ischaemia.

The conclusions to be drawn from these "thoughts" on arterial
hypertension in the elderly can be summerised as follows:

- a preliminary distinction must be made between two groups that
  are certainly separate, though intercommunicating: the systodiastolic and the purely systolic forms.   In the latter situation,
  there are no clinical physiopathological premises in support of
  an "aggressive" approach;
- in each clinical case an assessment must be made of the appro-
  priateness or otherwise of antihypertensive management, account
  being taken of the overall condition of the patient (including
  the extravascular situation);
- some drugs must be excluded as a general rule; doses must usually
  be restricted.

REFERENCES

1. G. Jackson, T. A. Pierscianowski, W. Mahon, and J. London,
      Inappropriate antihypertensive therapy in the elderly,
      Lancet 2:1317 (1976).
2. G. W. Pickering, "Hypertension. Causes, consequences and
      treatment," Churchill Livingstone, Edinburgh, London and
      New York (1974).
3. W. B. Kannel, Importance of hypertension as a major risk factor
      in cardiovascular disease, in: "Hypertension, Physiopathology
      and Treatment," J. Genest, E. Koiw, O. Kuchel, eds.,
      McGraw-Hill, New York (1977).
4. A. L. Dyer, J. Stamler J., R. B. Shekelle, J. A. Schenberger.
      and E. Farinaro, Hypertension in the elderly, Med.Clin.Amer.
      61:513 (1977).
5. P. W. Overstall, Falls in the elderly - epidemiology, aetiology
      and management, in: "Recent Advances in Geriatric Medicine,"
      B. Isaacs, ed., Churchill Livingstone, Edinburgh, London and
      New York (1978).

ANTI-HYPERTENSIVE TREATMENT IN THE OLDER PATIENT

John L.C. Dall

Visiting Professor of Geriatric Medicine
University of Ottawa, Ontario Canada
and
Department of Geriatric Medicine
Victoria Infirmary, Glasgow, Scotland

There are few areas in modern clinical practice that have changed more radically in the last ten years than the management of hypertension. In the 1950s the drugs with which hypertension was being treated were both potent and had severe and very toxic side effects and in these circumstances it was convenient to label the hypertension associated with older patients as 'essential hypertension' or 'benign hypertension'. These two terms allowed the physician to counsel the family that no treatment was indicated. Despite the 'essential' nature of this hypertension or the 'benign' attributes of this hypertension many elderly people died from stroke disease and from cardiac failure and it was inevitable that in the ensuing decades as the management of hypertension became progressively improved with the quality of drug therapy available that physicians in Geriatric Medicine who were addressing the problem of the management of the elderly hypertensive patients should think again as to whether the terms 'benign' and 'essential' were entirely justified.

NEED FOR TREATMENT

In the 1960s when the management of hypertension had become more efficient by the advent of post ganglionic blocking drugs such as guanethidine and methyldopa and the management of cardiac failure had also become greatly improved by the advent of the oral diuretic substances starting with hydrochlorothiazide, physicians were more eager and willing to tackle the problem of managing hypertension. However for the most part they were unwilling to commit a patient to therapy without evidence of what was known as "target-organ damage". This was intended to select hypertension that was of sufficient long standing to have caused either changes in the optic fundi, changes in the heart such as left ventricular hypertrophy or changes in renal function seen either as albuminuria or red cells in the urine micros-copy with or without a rising blood urea or creatinine.   The presence of one or all of these things was taken as an index that the hypertension was causing harm and therefore should be treated.   In the absence of these factors hypertension was still not treated it was still labelled "benign" or even "labile" hypertension because the systolic pressure tended to fluctuate depending on whether it was recorded in the recumbant position or when erect.   In the latter part of the 1960s information became available from the 'Framingham Cohort Study'[1] in the United States to show that systolic hypert-tension was an important entity and that there was a relationship between systolic hypertension and the likelihood of further cardio-vascular events either stroke or heart failure.   The paper from Colandrea and his colleagues[2] at a retirement community in Miami, Florida, showed that stroke occurred suddenly and with devastating consequences and often was the first event in a patient who had previously had asymptomatic hypertension.   This finding was compar-able to the findings of the Framingham study and it became increas-ingly obvious that target-organ damage as a method of selecting people for treatment of hypertension must include the brain as a target-organ and further that if the target-organs were to be protected from damage then it might be necessary to start treatment before extensive damage had occurred.

If therefore there are good clinical grounds for the treatment of hypertension to prevent stroke disease and to prevent cardiac failure there are also considerations in the prevention of the atherosclerotic process.   Almost all the many works relating to atherosclerotic changes in vessel walls acknowledge hypertension as a factor.   Many papers have been produced as to the effect of alter-ing the plasma lipids, the cholesterol and throughout all of these sophisticated investigations, none of which has yet successfully shown any ability to reduce the risk of atherosclerosis, the role of blood pressure has been neglected; and yet it can be shown in the piece of aorta resected from a young subject with coarctation of the aorta there is gross atherosclerotic change on the proximal side of the lesion and this is absent on the distal side of the lesion.   The blood, and thus, the lipid content and the cholesterol content is the same on both sides of the lesion and the only significantly different factor is the level of hypertension.   Systolic hypertension has probably been underestimated in its role in the causation of atherosclerosis but it should certainly no longer be underestimated in its role not only in this respect but also in the etiology of stroke disease.

It is important to determine what kind of hypertension is to be

treated in the elderly. It would be misleading to pretend that
target-organ damage is no longer relevant in the management of
elderly patients who may have stroke disease before there is significant evidence of left ventricular hypertrophy, renal damage or
fundal change but it is necessary to look at hypertension in its
diastolic and systolic components in the elderly patients rather than
consider it as a single subject. In the first place diastolic
hypertension and by this is meant diastolic blood pressures of
greater than 100 mmHg are increasingly rare as the patients age and
indeed levels above this in the very old are usually associated with
dissecting aneurysm of the abdominal aorta. Nevertheless in the
patient between 65 - 75, diastolic hypertension will be seen and will
be associated with target-organ damage and if untreated the target-
organ damage will progress and renal failure, cardiac failure or
stroke will result. A series conducted in the Victoria Infirmary
in Glasgow illustrated that the stroke patient under the age of 75
years almost inevitably has diastolic hypertension and has some
evidence of target-organ damage.[3]

Systolic hypertension, and by this I mean those whose diastolic
pressure is usually 95 mmHg or less and whose systolic pressure is
in excess of 160 mmHg, is said to be extremely common among the
elderly. It has been described as being present in up to 30 per cent
of the coloured population in North America and up to 25 per cent of
the white population in North America over the age of 74 years.[5]
This may well be true of a single observation of the blood pressure.
However when attempting to find suitable patients for a double blind
crossover trial for hypertension it is not so easy to find as many as
these earlier papers suggests and often a single reading at hypertensive level is not confirmed on a repeat examination at one week or
two weeks. There is the impression that systolic hypertension,
sustained systolic hypertension is not as prevalent as these earlier
papers would have appeared to indicate. A further observation in
the series of patients examined at the Victoria Geriatric Unit was
that systolic hypertension was just as important as diastolic in the
older patient (over 75 years) with stroke disease and there was often
no associated target-organ damage with the possible exception of
"unfolding" of the thoracic aorta on the x-ray. This is probably
an important early sign which is frequently badly reported because
there are no good measures of what amounts to "unfolding" of the
thoracic aorta.

OBJECTIVES OF TREATMENT

It is also important to determine which kind of hypertension is
being treated because the drugs that are necessary for the treatment
may well be different. In the management of systolic hypertension
recent experience suggests that diuretics, beta-blockers in small
doses and very occasionally a third drug regime either a vasodilator

or perhaps methyldopa may be required.   In the management of diastolic hypertension it may be that larger doses of beta-blockade are
required to reduce the blood pressure than are compatable with an
acceptable heart rate and the capability of exercise tolerance and
in these circumstances a combination of a diuretic and methyldopa
may be more appropriate.   The attending physician should decide which
type of hypertension he is managing and what the objectives of treatment are in each case - reduction of systolic levels, diastolic
levels or both, the amount of reduction desired and the rate at which
it is to be achieved.   In fixing these objectives it is necessary to
appreciate the contraindications to treatment.

     The principal danger is hypotension which may result from
inappropriate treatment or treatment regimes in which drug therapy
has been incremented too rapidly causing the blood pressure to fall
when the patient is in the upright position.   A fall of blood
pressure in excess of 20 mm systolic or 20 mm diastolic in the upright position is labelled as postural hypotension and the significance of this is apparent when the mean arterial flow falls below the
level required to sustain the cerebral circulation. If this occurs
and is not rapidly corrected cerebral infarction will result and the
patient may be left with a hemiplegia.   The clinical importance of
this has been recorded by Jackson et al.[5] Initially this paper was
interpreted as an indication that elderly patients should not be
given hypotensive therapy, but is now interpreted as a good index of
the care needed in the use of hypotensive agents in the elderly and
the importance of differentiating between systolic and diastolic
types of hypertension.   The maintenance of an adequate cerebral
blood flow depends on an adequate input through the carotid and
basilar arteries, the capability of cerebral vasodilatation and an
appropriate level of mean arterial pressure.   Given these factors
the cerebral blood flow has been shown to remain constant over a
wide range of mean arterial pressures.[6] Postural hypotension is an
entity which occurs in some middle age people which chronic neurological disease associated with the basal ganglia but occurs not
uncommonly in elderly people often as a consequence of drug therapy
being prescribed for other problems, but also in association with
cardiac arrhythmias, gastrointestinal hemorrhage, dehydration, sodium
depletion due to diuretic therapy and sometimes for no apparent
reason.   It must be appreciated that the presence of postural hypotension would be a contraindication to prescribing a hypotensive drug
and therefore all patients who attend a hypertensive clinic should
have their blood pressure taken in the erect position as a starting
measure.   Most of the experimental work associated with hypotension
is an acute experiment carried out with a tilting table; whether this
gives a true reflection of what happens to mean arterial pressure
when assuming the erect position or not is doubtful because the tilt
table does not involve all the postural support mechanisms that an
individual uses in the activities related to standing. Similarly
there is a need for caution in extrapolating from the acute experi-

ments carried out in such physiological studies to the effects that will occur in the mean arterial pressure and the baroreceptors if the blood pressure which has been high is reduced slowly over a period of days or weeks or even months. There is good evidence to believe that if blood pressure is reduced slowly the baroreceptor's response is adequate even in fairly severe vascular disease and that cerebral auto-regulation can occur as the blood pressure is lowered. The clinician who is responsible for the initiation of treatment should accept his clinical responsibility and should assess mental function as well as renal function at the outset of treatment and if there is a deterioration in either assume he has reduced the blood pressure either too far or too quickly.

If there are problems associated with the proposition to reduce blood pressure then there are other problems associated with the side effects of the drugs which will be used to lower the blood pressure and all these side effects must be studied very closely since any attempt to lower the blood pressure as a preventive exercise is designed to improve the quality of life for the patient by avoiding stroke and hypertension and should not substitute other risk factors or other symptom complexes which make life less than tolerable.

REGIMES AVAILABLE

There are many regimes currently available for the management of hypertension and very few now use ganglion blocking drugs. Very few even use post-ganglionic blocking drugs. The centrally acting drugs such as reserpine, rauwiloid and clonidine have been shown to cause severe depression when used in elderly people and for this reason are in disfavour. The fact that these reports may be dose-related and these drugs might be used again in different dosages and in different regimes will have to be considered. The biggest impact on the management of hypertension in the elderly has been with the advent of oral diuretic therapy. This not only lowers the blood pressure but by removing the salt and fluid load reduces the risk of cardiac failure and is a very logical method of tackling the problem of hypertension. Unfortunately changes in the ability to handle glucose occur during prolonged therapy with any of the oral diuretic agents and evidence from the European Working Party on Hypertension in the Elderly has shown that although these figures do not reach clinical importance they do achieve statistical significance in their continuing study of these regimes.[7] Similarly uric acid values are also increased and while the significance of this is not known the levels of change have reached statistical significance. It is perhaps because of these two factors that further investigation has been made into the role of oral hypotensive treatment with beta-blockade. In this respect there are both total beta-blockade and cardio-selective blocking drugs available on the market. The importance of the cardio-selective drug is that the risk of inducing broncho-spasm in those with previous pulmonary disease is avoided. The problem with the cardio-selective drug is that it may induce a significant bradycardia before it has achieved the levels of blood pressure control that are desired and further increase in therapy may induce a level of bradycardia that is not acceptable to the patient. Other side effects associated with the beta-blocking drugs are cold and painful tingling of the fingers and toes and in some of the older patients nightmares and bad dreams occur. It is likely that this latter side effect is associated with those drugs which are able to cross the blood brain barrier and this does not apply to all the medications.

Perhaps because both groups of drugs have some limitation in their expected use a combination of the two drugs has seemed a logical step forward in the management of hypertension and fixed dose combinations are now available. Many clinicians prefer to tailor the dose individually to the patient and resist the fixed dose combinations but there is no doubt that they do improve patient compliance. For those patients whose blood pressure still remains above the optimal level of 160/90 without serious side effects and therefore still has the capability of being reduced to 160/90 a third drug is sometimes required and the regime varies from center to center. As experience with vasodilator drugs such as hydralazine, isosorbide, and others increases it may well be that a combination of beta-blockage, vasodilator and diuretic would be the drug regime of choice. However for the patient who has a raised systolic and diastolic pressure at the outset where the combination of diuretic and beta-blockade is not sufficient the addition of a small dose of methyl-dopa such as 125 mgm once daily with gradual increasing of the dosage has been shown to be perfectly adequate and indeed there have been few side effects to methyldopa reported in the European Working Party trial on Hypertension in the Elderly so it should not be discarded and is a useful drug in this group of patients.

SUMMARY

The treatment of hypertensive disease in old people is the management of systolic hypertension and/or diastolic hypertension. It is important that the clinician should know what he is attempting to achieve. It is important that any change in the blood pressure should occur slowly in order to allow appropriate restabilizing of the baroreceptor responses which will be fundamental to maintaining an adequate cerebral perfusion. The regime should be as simple as is compatible with the results desired e.g. a single tablet daily if possible in order to improve compliance. A range of drugs is now available which has the capability of reducing blood pressure. None of these drugs is entirely free from side effects. Small doses of the drugs are often effective without serious side effects and therefore can be tolerated by the patients on maintenance therapy.

The prescribing physician retains the responsibility of ensuring that the level of blood pressure that he wishes to achieve is compatible with good intellectual function and good renal function and these factors should be reviewed constantly during any maintenance therapy. Finally, but of great importance, it should be remembered that the cerebral blood flow is seldom impaired when the patient is recumbent and the danger occurs when the patient is in the standing position and therefore in any treatment regimes that are established the patient should have the blood pressure recorded in the erect position.

REFERENCES

1. W. B. Kannel, Some lessons in cardiovascular epidemiology from Framingham, Am.J.Cardiol. 37:269 (1976).
2. M. A. Colandrea, G. D. Friedman, M. A. Nichman, and C. M. Lynd, Systolic hypertension in the elderly, Circulation 41:239 (1970).
3. J. L. C. Dall, J. P. R. MacFarlane and I. L. Lennox, Proceedings of the Xth Congress of the International Association of Gerontology, Hamburg (1981).
4. Hypertension Detection and Follow up Program Cooperation Group, Blood pressure studies in 14 committees, J.Am.Med.Ass. 237:2385 (1977).
5. B. Jackson, W. Mahon, T. Pierscianowski, and J. Condon, Inappropriate hypertensive therapy in the elderly, Lancet 2: 2317 (1976).
6. S. Stangaard, J. Olesen, E. Skinhoj et al, Auto-regulation of the brain circulation in severe arterial hypertension, Br.Med.J. 1:507 (1973).
7. A. Amery, P. Berthaux, W. Birkenhager et al, Antihypertensive therapy in patients above age 60 years. (Fourth interim report of the E.W.P.H.E.) Clin.Sec.Molec.Med. 55:263 (1978).

TREATMENT OF ORTHOSTATIC HYPOTENSION

William Davison

Consultant Physician in Geriatric Medicine
Addenbrooke's Hospital
Cambridge

A critical component of the aging process is the inability to
maintain homeostasis. Orthostatic (postural) hypotension (OH) is
a good example of homeostatic failure which may be due to aging or
to disease but is not infrequently predominantly iatrogenic. It is
a serious disorder with potentially catastrophic complications
including falls, fractures, infarction of brain and heart, immobility and increasing dependency. Orthostatic hypotension is not a
single nosological entity and the clinical classification, pathology,
biochemistry and treatment remain a challenge.[1]

DEFINITION

The term orthostatic hypotension (OH) means that the systemic
blood pressure falls on standing erect. A small drop in systolic
pressure is acceptable but bigger drops may well cause transient
failure of cerebral function. The symptoms produced range from
lightheadedness or dizziness to complete loss of consciousness.
The definition of orthostatic hypotension should include a fall in
systolic pressure on rising:
>   i)   of 20 mmHg or more
> or ii)   to 80 mmHg or less
> or iii) sufficient to cause symptoms.[2,3]

Normal cerebral blood flow is constant over a wide range of
mean arterial pressure but elderly patients with impaired cerebrovascular autoregulation may be at risk of brain damage from quite
modest falls in blood pressure.[3,4] There is evidence to suggest
that physiological impairment may occur in autonomic neural pathways
with aging.[5] ANS failure might therefore be the cause of both OH

and failed cerebral autoregulation in some elderly patients but it
is unlikely that neuronal degeneration alone will account for most
OH in old people.[3,5,13]   A proper medical assessment is required
and many patients can be helped by appropriate treatment. The
disability must never be dismissed as simply due to 'old age'.

INCIDENCE OF ORTHOSTATIC HYPOTENSION

Orthostatic hypotension is commonly found in sick elderly people.
Thus in USA of 100 mentally disturbed elderly patients living at
home, but attending a psychiatric outpatient clinic, Davie et al[6]
found 34 to have OH.   In many of these cases, psychoactive drugs
were implicated to some extent, viz: phenothiazines, tricyclic antidepressants, benzodiazepines and hypnotics. In an earlier British
study,[7] even in the absence of hypotension-inducing drugs, 17 per cent
of the patients aged 70 years and over admitted to a hospital for
the elderly had OH.   By way of contrast, in a study in New Zealand
of a stratified population sample of 553 subjects aged 65 and over
mostly living in their own homes (only 15 were in hospital), 74
subjects (just over 13 per cent) demonstrated a drop of 20 mmHg or
more of systolic pressure on rising.[8]

MECHANISMS OF ORTHOSTATIC HYPOTENSION

When a fit person stands up from sitting or lying there is a
transient fall in systemic blood pressure (BP) due to a reduction of
venous return to the heart and a consequent drop in cardiac output.
This drop in BP stimulates the baroreceptors in the carotid body,
aortic arch and possibly elsewhere to produce via the autonomic
nervous system (ANS) arterial and venous constriction and an increase
in heart rate and strength of contraction (positive inotropism) to
maintain the mean arterial pressure.[4]* Normally the systolic BP
drops no more than 15 mmHg and the diastolic falls or rises a little
but the MABP is virtually unchanged.   The heart rate rises by an
average of 20 beats/minute but in late life the rise is much less.
The change in heart rate can be calculated from the R - R intervals
on the ECG. In younger folk the shortest interval comes at about the
fifteenth beat and the longest at about the thirtieth beat.
In old people this cardioaccelerator response is impaired whether or
not there is postural hypotension.[5]

Even in healthy young people orthostatic tolerance is often
imperfect and 10 - 15 per cent of fit young men may faint during
head-up-tilt or in prolonged standing.   Exercise and high ambient
temperature increase the likelihood of OH especially in the heat
intolerant subjects.[10]

---

* The mean arterial blood pressure (MABP) may be calculated as the
  diastolic pressure plus one-third of the pulse pressure.[3]

In the common faint, (syncope) or vasovagal attack, seen in young people, there is crescendo autonomic discharge often in association with strong emotion.   There is active dilatation of blood vessels and prominent prodromal cholinergic clinical features, viz: sweating, increased peristalsis, nausea, salivation and brady-cardia, prior to failure of cerebral function.   In OH due to ANS failure alone (referred to as idiopathic orthostatic hypotension, or IOH) there is loss of the normal adrenergic and cholinergic response to the fall in blood pressure so that the prodromal signs of the common faint (vasovagal or vasodepressor attack) are absent and syncope may occur with no warning.[11]

Autonomic nervous dysfunction is not the sole cause of OH. Even with an intact ANS there may be OH due to loss of normal baro-receptor responses following a recent illness (especially viral infection) and prolonged recumbency and there is a well recognised association between OH and varicose veins.[12]

Polinski et al[1] described a form of orthostatic hypotension in younger patients in which pronounced cardioacceleration occurred but the peripheral vasculature failed to respond normally to the intact sympathetic nervous system (sympathotonic orthostatic hypotension (SOH) or orthostatic tachycardia (OT).   Decreased alpha-adrenergic response with normal beta adrenergic response might be the basis of SOH.

Vascular disease in old people may prevent peripheral vascular contractions even with an intact ANS and baroreceptor reflex failure may also be primarily vascular and due to loss of arterial distens-ibility at the baroreceptor sites.[13]   Similarly, the loss of beat-to-beat variation in heart rate in response to deep breathing may be due to a loss of the atrial response to vagal stimuli rather than to a loss of ANS activity.   Additionally, other factors may influence baroreceptor reflexes and indeed Vargas and Lye[14] believe it is somewhat naive to imagine that ANS dysfunction can be readily and separately identified from all the other factors involved in the maintenance of BP in old age.   MacLennan et al[13] found a strong positive correlation between postural drop in BP and the lying systolic pressure.   They suggested that a change in structure of the vascular tree may be more important than ANS dysfunction as a determinant of OH in old people.

ANS disturbance occurs in about a third of severe cases of pernicious anaemia according to White et al[15] who reported a case of neurogenic OH in an elderly male which resolved rapidly with treat-ment of the pernicious anemia.   Most doctors are aware of the possibility of peripheral neuropathy in this disease but may fail to appreciate the ANS damage.

MANAGEMENT OF ORTHOSTATIC HYPOTENSION

Management is often difficult because of lack of full knowledge of the cause of the disorder. As we have seen, there are many possible mechanisms and two or three of these may be operating simultaneously in the one patient. An attempt should be made to identify the likely factors involved in each case (see Table 1). This requires a careful physical examination especially of the nervous and cardiovascular systems. The integrity of the ANS system should be tested using a small battery of tests, despite the difficulties both of execution and interpretation.[5,13,14,18] (Table 2)   Reduced variation in heart rate in response to postural change, reduced vasoconstrictor response to cooling and reduced baroreflex response to lower body negative pressure are regarded as suitable tests in elderly patients by Collins et al[5] but only the first of these three tests will be routinely available to most physicians in geriatric practice. Other clinical evidence of ANS dysfunction should also be sought (Table 3).

Table 1. Etiology of orthostatic hypotension*

Central nervous system

Ischaemic brain	Parkinson's disease
Multiple system atrophy (Shy Drager)	Tumours
Syringomyelia	Tabes dorsalis

Peripheral nerves

Diabetes mellitus	Amyloidosis
Rheumatoid arthritis	Alcoholism
Carcinomatous neuropathy	

Circulatory disorders

Varicose veins	Myocardial insufficiency
Hypovolemia (dehydration)	Vascular rigidity
Electrolyte depletion	Defective venous pump (hemiplegia)

Primary autonomic failure

Familial dysautonomia	Acquired IOH

Drugs - see Table 4

---

* Not an exhaustive list but to indicate the broad spread of possible underlying diseases

Table 2. Tests of autonomic function

(a) Physical.

Heart rate and BP response to standing

Vasomotor response to temperature change

Baroreceptor response to lower body negative pressure

Pressor response to mental and physical stress
    (Arithmetic and handgrip)

(b) Chemical.

Heart rate response to i.v. atropine: ablates vagal inhibition

Plasma NE levels: lying, standing and after tyramine

Heart rate response to i.v. pindolol: differentiates pre and
                                post-ganglionic block

BP response to i.v. phenylephrine: tests for denervation
                                supersensitivity

The patient's drugs should be scrutinised most carefully and
any item likely to cause hypotension (see Table 4) should be
stopped or the dose reduced. Dehydration and electrolyte imbalance
should be corrected. Both sodium[16] and potassium[17] deficiency have
been implicated as causes of OH. Adrenal failure should be excluded.
If possible, plasma levels of norepinephrine (NE) should be measured
[1,18] but this assay is not generally available in the U.K. outside
special research centres. In peripheral ANS failure, low plasma NE
is to be expected whereas in central failure the level should be
normal. Tyramine infusion releases NE from sympathetic nerve endings.
This response will be reduced in peripheral ANS failure. The BP
response to I.V. Tyramine in both central and peripheral failure
will be greater than normal because of 'denervation supersensitivity'[1].
This denervation supersensitivity may contribute to the recumbent
hypertension (sic) which may be found in these cases[19] and which is
liable to be exacerbated by treatment. The heart rate response to
I.V. pindolol will help separate preganglionic from postganglionic
sympathetic failure (v. infra).

Table 3. Autonomic dysfunction

Orthostatic hypotension	Disordered gut motility
Impaired bladder control	Impaired thermoregulation
Disordered sweating	Impotence

Table 4.   Drugs implicated in orthostatic hypotension

Antihypertensives	Vasodilators
Diuretics	Tranquillisers
Betablockers	Antidepressants
Levodopa	Hypnotics

When the general workup is complete, the severity of the OH should be rechecked. The symptoms may already have abated following attention to the drug schedule and fluid and electrolyte balance together with increased physical activity.   A BP drop in the absence of symptoms is probably not worth treating. Ideally the BP should be measured lying and then immediately on standing and each minute thereafter for 5 - 10 minutes and again after exercise depending on the severity of the case.   In severe cases marked symptoms may occur very early so that full testing in this way is neither possible nor necessary.

PHYSICAL TREATMENT

It is generally recommended to try some physical modes of management before going on to drugs.   Patients should be advised to avoid the sudden assumption of the upright position - especially when getting out of bed.   Repeated gradual changes of posture are to be encouraged so as to help reactivate the baroreceptor reflexes.   The patient should sleep well propped up on pillows or have the head of the bed raised on blocks at the maximum tolerated angle.   The object is to reduce renal artery blood pressure and thus stimulate the renin-angiotensin-aldosterone mechanism to retain sodium and so maintain the blood volume and also to allow adaptation of cerebral autoregulation to the chronically low perfusion pressure.

Firm elastic stockings or elastic mesh tights (from metatarsals to waist) are usually recommended and are very helpful if there are gross varicose veins. The object is to reduce venous pooling and so maintain the central blood volume and venous return to the heart. However, Bannister[19] points out that they are best avoided in OH due to chronic autonomic failure because their use might reduce the valuable peripheral vessel myogenic response to stretch. Additionally these garments should be kept on at night if the patient needs to get up for any reason.

DRUG TREATMENT OF ORTHOSTATIC HYPOTENSION

In the absence of a precise diagnosis (e.g. inappropriate medication, primary adrenal failure, pernicious anemia or a positive response to I.V. pindolol (q.v.) drug treatment has to be given on a trial and error basis.

Fludrocortisone acetate

(9 - alpha fluorocortisol) this drug produces an initial expansion of the plasma volume but the effect is thought to be transitory.   The longer term value in OH is thought to be due to an increased sensitivity of the peripheral vessels to NE. This effect is obtained with quite modest doses e.g. 0.1 mg daily, increasing cautiously at weekly intervals while watching for complications, especially supine hypertension and heart failure but also hypokalemia.   Usually 0.2 mg daily is a sufficient maintenance dose.

Monoamine Oxidase Inhibitors (MAOI) and Tyramine

The use of a monoamine oxidase inhibitor such as tranylcypromine (an antidepressant) and cheddar cheese has been advocated.   There must be real doubts about this line of treatment because the lack of close dosage control of the tyramine in the cheese makes it particularly hazardous.   To provide better control, Nanda et al[20] used tyramine in 5 - 10 mg capsules (there is no commercially available pharmaceutical preparation) in addition to tranylcypromine 10 mg thrice daily and increasing to 20 mg thrice daily. The tyramine (which is oxidised in the gut wall) acts indirectly. It causes release of NE from sympathetic nerve endings and so produces a pressor response.   The tyramine is protected from breakdown by the MAOI and dangerous hypertension can occur.

Non-steroidal Anti-inflammatory Drugs

Indomethacin and other non-steroidal anti-inflammatory drugs have been used.   The beneficial effects have been variously ascribed to suppression of prostaglandin synthetase, inhibition of the naturetic hormone cascade or increased sensitivity of the vascular tree to NE (like the longer term effect of fludrocortisone). Flurbiprofen is said to be more potent than indomethacin as a prostaglandin synthetase inhibitor but with fewer adverse effects. Watt et al[21] proposed a stepwise approach to the treatment of idiopathic OH starting with flurbiprofen 50 mg twice daily and increasing; then adding fludrocortisone 0.1 mg twice daily and increasing, finally adding tyramine and phenelzine (a MAOI). The therapeutic escalation can be stopped at any stage if the patient becomes symptom free or if complications develop, especially supine hypertension and heart failure.

Indomethacin increases the absorption of sodium in the proximal tubule and helps the aged nephron to conserve sodium. Neither hypoaldosteronism nor a high level of renal prostaglandin is responsible for the excess naturesis commonly seen in the elderly.[22]

Beta-adrenergic Blocking Agents

These have been used with some success (e.g. propranolol 40 mg
up to 240 mg daily). The rationale being that a decrease in beta
adrenergic activity might reduce active (beta mediated) vasodilation
and thus offset the lack of alpha mediated vasoconstriction. More
recently, pindolol has been shown to be effective in OH due to
peripheral (postganglionic) sympathetic failure[23] but not in OH
due to multiple system atrophy (MSA).[24] This beta blocking agent
has partial agonist activity and the latter effect is enhanced by
the presence of denervation sensitivity. Additionally, pindolol
has appreciable chronotropic and inotropic effects. Given in a
dose of 15 mg daily it increases the heart rate, cardiac output
and arterial pressure and relieves symptoms of OH.[23] The
beneficial effect on the heart in these cases (with postganglionic
sympathetic failure) is in contrast to the lack of effect in cases
with central failure reported by Davies et al.[24] The difference in
response to pindolol between these two types of sympathetic failure
is stressed by Man in't Veld and Schalenkamp who have suggested the
use of this drug as a diagnostic tool.[25] An increase in heart
rate after I.V. pindolol (0.4 - 0.8 mg) points to postganglionic
failure whereas a heart rate which fails to accelerate, or slows,
favours a central or preganglionic block. A positive response to
this test is an indication to try treatment with pindolol. A
negative response suggests that the use of this drug is contraindicated.

Clonidine

This antihypertensive drug with both central and peripheral
sites of action has some alpha adrenoceptor agonist activity and
has also been found to be beneficial in OH with postganglionic
failure but not in OH with a preganglionic lesion.[26]

SOME OTHER TREATMENTS OF ORTHOSTATIC HYPOTENSION

Many other treatments for OH have been recommended included
direct acting pressor amines such as oral phenylephrine, ephedrine
and longacting isoprenaline. These stimulate the supersensitive
alpha receptors to produce vasoconstriction. Combination with
a beta blocker is advised to overcome the problem of supine
hypertension. Lisuride an antagonist of serotonin has been tried
with some success in MSA and ANS failure[27] but with problems due
to psychiatric side effects, especially terrifying visual
hallucinations. Yohimbine an alkaloid with alleged aphrodisiac
properties and an alpha antagonist has been found helpful in OH
induced by tricyclic antidepressants.[28] It has an alpha
adrenergic blocking action of short duration. Dihydroergotamine
(DHE) an ergot alkaloid constricts capacitance vessels and thereby

lessens venous pooling on standing.   It has helped in some cases
of peripheral sympathetic defect provided an adequate dose was
given. Atrial tachypacing was used successfully in a 77 year old
man with an eight year history of progressive and incapacitating
primary OH which was unresponsive to elastic stockings, sympathetic
amines and mineralocorticoids.[30]   The heart rate previously fixed
at 72/minute was boosted to 100/minute and the patient markedly
improved.   However, others have reported lack of success with
tachypacing if the ANS dysfunction is very severe.

Hydrallazine (an arteriorlar smooth muscle relaxant) 75 mg
daily produced dramatic improvement in a 70 year old male with
marked idiopathic OH with ANS dysfunction unresponsive to many of
the usual treatments.[31]   There was a fall in supine and a rise in
the upright BP together with a raised pulse rate. Concomitant
measures used were fludrocortisone, slow release sodium tablets and
elastic stockings.

SUMMARY

The principal factors involved in chronic orthostatic hypotension have been outlined.   Attention is drawn to the need for
accurate diagnosis so that the most appropriate treatment may be
given.   With current knowledge, a specific defect is not always
apparent.   In the absence of a clear indication for a particular
line of treatment it is necessary to conduct a careful empirical
trial of the more likely physical and drug treatments.   Often a
combination of treatment is effective.

REFERENCES

1.   R. J. Polinski, I. J. Kopin, M. H. Ebert, and V. Weise,
        Pharmacologic distinction of different orthostatic hypotension
        syndromes, Neurology 31:1 (1981).
2.   Editorial, Management of orthostatic hypotension, Lancet 2:963
        (1981).
3.   L. Wollner and J. M. K. Spalding, The autonomic nervous system,
        in: "Textbook of Geriatric Medicine and Gerontology",
        J. C. Brocklehurst, ed., Churchill Livingstone, Edinburgh
        (1978).
4.   L. Wollner, S. T. McCarthy, N. D. W. Soper, and D. J. Macy,
        Failure of cerebral autoregulation as a cause of brain
        dysfunction in the elderly, Br.med.J. 1: 1117 (1979).
5.   K. J. Collins, A. N. Exton-Smith, M. H. James,and D. J. Oliver,
        Functional changes in autonomic nervous responses with ageing,
        Age and Ageing, 9:17 (1980).
6.   J. W. Davie, M. D. Blumenthal,and S. Robinson-Hawkins, A model
        of risk of falling for psychogeriatric patients, Arch.gen.
        Psych. 38:463 (1981).

7. R. H. Johnson, A. C. Smith, J. M. K. Spalding, and L. Wollner, Effect of posture on blood pressure in elderly patients, Lancet 1:731 (1965).

8. A. J. Campbell, J. Reinken, B. C. Allen, and G. S. Martinez, Falls in old age: A study of frequency and related clinical factors, Age and Ageing 10:264 (1981)

9. D. J. Ewing, I. W. Campbell, A. Murray, J. M. M. Neilson, and B. F. Clarke, Immediate heart-rate response to standing: simple test for autonomic neuropathy in diabetes, Br.med.J. 1:145 (1978).

10. G. Keren, Y. Epstein, A. Ohri and A. Magazanik, Orthostatic responses in heat tolerant and intolerant subjects compared by three different methods. Aviat.Space Environ.Med. 51:1205 (1980).

11. R. B. Hickler, Orthostatic hypotension and syncope, New Engl. J.Med. 296:336 (1977).

12. F. I. Caird, G. B. Andrews, and R. D. Kennedy, Effect of posture on blood pressure in the elderly. Br.Heart J. 35:527 (1973).

13. W. J. MacLennan, M. R. P. Hall, and J. I. Timothy, Postural hypotension in old age: is it a disorder of the nervous system or the blood vessels? Age and Ageing 9:25 (1980)

14. E. Vargas and M. Lye, The assessment of autonomic function in the elderly, Age and Ageing 9:210 (1980).

15. W. B. White, L. Reik and D. E. Cutlip, Pernicious anemia seen initially as orthostatic hypotension, Arch.int.Med. 141:1543 (1981).

16. W. Fine, Some common factors in the causation of postural hypotension, Geront.clin. 11:206 (1969).

17. J. R. Cox, A. K. Admani, M. L. Agarwal, and P. Abel, Postural hypotension - body fluid compartments and electrolytes. Age and Ageing 2:112 (1973).

18. I. J. Schatz, Current management concepts in orthostatic hypotension, Arch.int.Med. 140:1152 (1980¼

19. R. Bannister, Chronic autonomic failure with postural hypotension, Lancet 2:404 (1979).

20. R. Nanda, R. Johnson, and H. Keogh, Treatment of neurogenic orthostatic hypotension with a monoamine oxidase inhibitor and tyramine, Lancet 2:1164 (1976).

21. S. J. Watt, J. E. Tooke, C. M. Perkins, and M. R. Lee, The treatment of idiopathic orthostatic hypotension: a combined fludrocortisone and flurbiprofen regime, Q.J.Med. 50:205 (1981).

22. J. F. Macias Nunez, C. Garcias Inglesias, J. M. Taberno Romo, J. L. Rodrigues Commes, L. Corbacho Becerra, and J. A. Sanchez Tomero, Renal management of sodium under indomethacin and aldosterone in the elderly, Age and Ageing 9:165 (1980).

23. A. J. Man in't Veld and M. A. D. H. Schalenkamp, Pindolol acts as beta-adrenoceptor agonist in orthostatic hypotension: therapeutic implications, Br.med.J. 282:929 (1981).

24. B. Davies, R. Bannister, C. Mathias, and P. Sever, Pindolol in postural hypotension: the case for caution, Lancet 2:982 (1981).

25. A. J. Man in't Veld and M. A. D. H. Schalenkamp, Pindolol in postural hypotension, Lancet 2:1279 (1981).

26. D. Robertson, M. R. Goldberg, A. S. Hollister, D. Wade.and R. M. Robertson, Clonidine raises blood pressure in idiopathic orthostatic hypotension, Circulation 64:(Supp.V) 10 (1981).

27. A. J. Lees and R. Bannister, The use of lisuride in the treatment of multiple system atrophy with autonomic failure (Shy-Drager syndrome) J.Neurol.Neurosurg.Psychiat. 44:347 (1981).

28. Y. Lecrubier, A. J. Puech and A. des Lauriers, Favourable effects of yohimbine on clomipramine-induced orthostatic hypotension: a double blind study, Br.Clin.Pharmacol. 12:90 (1981).

29. I. N. Olver, G. L. Jennings, A. Bobik,and M. Esler, Low bioavailability as a cause of apparent failure of dihydroergotamine in orthostatic hypotension, Br.med.J. 281:275 (1980).

30. A. J. Moss, W. Glaser, and E. Topol, Atrial tachypacing in the treatment of a patient with primary orthostatic hypotension, N. Engl.J.Med. 302:1456 (1980).

31. D. H. Jones and J. L. Reid, Volume expansion and vasodilators in the treatment of idiopathic postural hypotension. Postgrad.med.J.56:234 (1980).

TREATMENT OF CARDIORESPIRATORY FAILURE IN THE ELDERLY

F. Rengo, D. Bonaduce, N. Ferrara,V. Canonico,
M. Petretta, P. Abete, C. Vigorito and C. Rengo

Department of Gerontology and Geriatrics
Faculty of Medicine and Surgery
University of Napoli, Italy

INTRODUCTION

A rational approach to the treatment of cardio-respiratory insufficiency must take into account a detailed knowledge of the physiological mechanism underlying chronic cor pulmonale and of the clinical conditions most frequently associated with its acute deterioration.

From a nosographic point of view, it has not been still clearly established whether or not the definition of cardiorespiratory insufficiency should be applied to all patients with associated cardiac and respiratory insufficiency, undependently from temporal sequence or clinical relevance of one of these over the other.

The definition of "cardiorespiratory insufficiency" was also proposed by L. Condorelli about 30 years ago[1] as an alternative to "chronic cor pulmonale" (CCP) defined later by the expert Committee of the W.H.O. as "right ventricular hypertrophy provoked by diseases altering lung function or structure with the exception of congenital heart disease and of those pulmonary abnormalities secondary to pathologic events primarily localized in the left heart".[2]

In this present review we will reserve the definition of cardio-respiratory insufficiency to those acute or subacute clinical pictures characterized by significant anatomic and functional lung impairment with severe respiratory failure and secondary development of cardiac insufficiency. With the exception of severe acute infective and thromboembolic processes and of adult respiratory distress syndrome (ARDS) certainly the "reacutization of CCP", gives the largest contribution to this entity.

CCP is a disease with a high degree of social relevance due to its common occurrence and high mortality and morbity rates. The terminal event is usually represented by a sudden exacerbation of the disease, which terminates a natural history characterized by alternating periods of improvement and deterioration particularly present in advanced decades.

Several diseases may lead to CCP, however, particular clinical and social relevance have diseases primarily compromising air transit to and from alveoli. Particularly important is chronic obstructive airways disease (COAD), an entity with clinical spectrum ranging from panacinous pulmonary emphysema to chronic bronchitis, in which a major role is played, particularly in the asthmatic variety, by the bronchospasm.

Social relevance of COAD can be appreciated by the official mortality tables for 100,000 inhabitants published by the WHO (Europe), based on data obtained in the years 1974/75 from Italian Public Social Service.[3]

The gradual clinical course of CCP can change to a dramatic rapid evolution, as for instance when an acute bronchopulmonary infection complicates the natural history of the disease. In this case there is a sudden worsening of respiratory insufficiency, frequently followed by the development of overt cardiocirculatory failure ("reacutization" of CCP).

This condition can lead to death, or, when treated adequately, can regress to the original clinical state; the reversibility of this syndrome indicates the prevalence in this condition of reversible functional over irreversible anatomic factors.

The reacutization of CCP may have two clinical forms that differ in the time required to develop hypercapnic coma, and are influenced by different therapeutic interventions.

THERAPY OF CARDIO-RESPIRATORY FAILURE

The fundamental aims of therapy cardio-respiratory failure are:[4] prompt removal of hypoxia and hypercapnia by improvement of alveolar ventilation; normalization of acid-base balance; improvement of cardiac performance; improvement of respiratory surface; removal of associated infections; reduction of bronchial obstruction and of the alveolar membrane thickness; reduction of polycythaemia and normalization of the blood viscosity.

TREATMENT OF HYPOXIA AND HYPERCAPNIA

The most important therapeutic measures are represented by oxygen administration, mechanically assisted ventilation, and drugs stimulating the respiratory centres. The treatment of hypoxaemia, which should be imperative whenever arterial pO_2 falls below 50 mmHg, reduces hypoxia induced tissue damage which favours interstitial pulmonary oedema with consequent additional impairment of alveolar capillary diffusion of respiratory gases.[5]

Furthermore, it abolishes pulmonary arteriolar vasoconstriction, improves the metabolic function of important organs such as kidney and brain and allows a good clinical response to cardiac glycosides.

a) Oxygen Therapy

While everybody agrees that O_2 inhalation is a fundamental step in the therapy of cardio-respiratory failure, several controversies exist on its modalities of application. It is well known that pCO_2 levels in the blood tend to rise during O_2 therapy, proportionally to the rise of arterial HbO_2 level.

The rise in pCO_2, whose principal clinical manifestation is progressive drowsiness up to hypercapnic coma, is secondary to many factors: 1) worsening of alveolar hypoventilation due to the decrease of hypoxia-induced stimulation of respiratory centres, which are chronically depressed in the hypercapnic state; 2) impairment of ventilation/perfusion ratio due to reduction of pulmonary arterial vasoconstriction in poorly ventilated alveoli; 3) increased CO_2 production from stimulation of aerobic metabolism. For these reasons, O_2 should be administered at low flow (about 1 lt/min). The aim of this approach is to keep arterial pO_2 within values largely lower than normal (about 60 mmHg), thus maintaining the hypoxia induced stimulation on ventilation.[6]

With the nasal cannula advanced into the pharynx, it is possible to reach very high alveolar O_2 percentage concentration (30-50%), even higher than those obtained with oxygen tent. However, the rise of arterial pO_2 beyond 80 mmHg would only increase the risk of alveolar hypoventilation and of hypercapnic coma with only a slight additional improvement of O_2 availability to tissues, due to the peculiarity of the Hb-O dissociation curve. O_2 administration at high flow could also carry the risk of ARDS. Recently there has been introduced in clinical practice "home O_2 concentration device" and "O_2 bottles for continuous low flow O_2 administration".

b) Central Stimulating Drugs

These drugs are mainly indicated when alveolar hypoventilation is secondary at least in part, to a depression of respiratory centres

by drugs or hypercapnia, particularly in the course of O_2 therapy.
Among these drugs, the bulbar analeptics are able to increase the
sensitivity of respiratory centres and, in some cases (i.e. Lobeline)
also of peripheral chemoreceptors, thus increasing the respiratory
work, global and alveolar ventilation, and improving gas exchange.

In this family of agents, the most largely employed are
Prectmamide, Dimefline, Metamivan, Mepixanton, Doxapran etc.
In cardio-respiratory failure one or two of these drugs in association are usually administered by i.v. infusion, diluted in 150-
250 ml of 5-10 per cent physiologic saline solution, starting with
slow infusion rate which is gradually increased until a favourable
clinical response or toxicity is observed. Signs of toxicity are:
systemic hypertension, bradycardia, nausea and vomiting, due to
the stimulating effect of these drugs not only on respiratory
centres, but also on bulbar vasomotor and vagal centres. The
therapeutic index is very low, and doses of these drugs slightly
higher than those necessary to stimulate depressed respiratory
centres are able to induce generalized tonico-clonic convulsions
and hyperpyrexia. At high doses, malignant hyperkinetic
ventricular arrhythmias have also been observed. These drugs
should be avoided when diffuse trachiobronchial obstruction or
severe impairment of thoracic mechanics due to neuro-muscular
diseases prevent the increase of alveolar ventilation; in these
patients the stimulation of respiratory centres would only
unnecessarily increase the respiratory work.

A very useful drug in the treatment of cardio-respiratory
insufficiency, especially in very acute forms, is aminophylline a
soluble ethylendiaminic esther of theophylline.[8]

This drug possesses several modalities of action: a) improves
the rate and depth of ventilation with a better therapeutic index
than other drugs stimulating respiration centres; b) reduces
pulmonary arterial pressure by pulmonary arterial vasodilatation;
c) has a relaxant effect of bronchial smooth muscle, thus increasing
alveolar ventilation.

The frequent disappearance of Cheyne-Stokes respiration with
aminophylline seems rather due to the properties of ethylendeamine.
Aminophylline should be administered by i.v. infusion diluted in
5 per cent glucose or physiologic saline,the starting infusion rate
is usually 6 mg/Kg i.v. in 15 - 20 min. thereafter 1 mg/Kg/h without
exceeding 1,5-2 g in 24/h. Plasma blood levels of the drug should
be kept between 10 and 20 μ/ml.

c) Mechanical Ventilation

When an intense pharmacological intervention with central
stimulating drugs does not succeed in improving alveolar ventilation

and in reducing hypoxaemia, or when hypercapnic coma is impending, it is imperative to start the patient on assisted pulmonary ventilation.[9]

This can be performed with positive or negative pressure, maintaining in either case, continuously or intermittently by a pressure gradient between the pleural cavity and room air. From a practical standpoint, there are two fundamental forms of assisted ventilation, the pressure controlled and the volume controlled ventilation.

In the pressure controlled ventilation, which requires some patient co-operation, the inspiratory flow is triggered by a depression in the system provoked by a spontaneous inspiratory act. This causes a valve opening with consequent respiratory flow at positive pressure, which terminates when pressure in the patient's airways reaches a pre-determined value, usually lower than 30 cm of H_2O.

These systems can allow the patient a spontaneous expiratory phase with or without a negative pressure. Some authors suggest to program, in the course of expiration, progressively increasing resistance (PEEP) or a continuous intrapulmonary positive pressure (CIPP),[10] in order to perform a more complete alveolar distention. With this type of assisted ventilation it is possible to measure the tidal volume and to administer room air or air mixture with 40 per cent O_2. Each course of assisted ventilation should last between 20 and 40 min. and should be repeated several times a day, with frequent determination of arterial blood gases. Pressure controlled ventilation is not indicated in patients with low pulmonary compliance, since in these patients it is very difficult to achieve a satisfactory tidal volume. In these patients, or when the patient is unconscious, it is preferable to adopt a volume controlled assisted ventilation, which allows a pre-established volume of air, or O_2 enriched mixtures, to be administered independently from developed pressure.

In this case, patients should be intubated and one should pre-determine respiratory cycle duration and rate, trying to synchronize the activity of the mechanical device with the patient's spontaneous respiration. When this is difficult, drugs stimulating respiratory centres should be discontinued and a short period of 100 per cent O_2 breathing or sedative drugs (morphine, diazepam, barbiturates, meperidine) to depress patient's spontaneous ventilation must be given. It is also important to program the ratio between inspiratory and expiratory time (I/E ratio). The ideal ratio of 1:2 can be changed to 1:4 in patients with COAD; these patients may also benefit from 1:6 ratio. Since PEEP reduces venous return to the heart, it has been suggested that these patients should be monitored by right heart catheterization with determination of right heart

pressures and cardiac output.   All patients undergoing assisted
mechanical ventilation are at risk of the so called "riventilation
syndrome".[11] This syndrome, whose main clinical expression is cardiocirculatory collapse, is due to the sudden reduction of arterial
pCO_2 with a much slower fall of previously accumulated HCO-, with a
consequent metabolic alkalosis, always associated with hypochloraemia.   The analysis of arterial blood gases will show an elevated
pH, elevated standard HCO_3^- and a total HCO_3^- more or less elevated
than standard HCO- according to the values of pCO_2.   The treatment
of this condition consists of i.v. administration of KCL and NH_4Cl
solutions until normal values of serum Cl- ions (100-105 mEq/1) are
reached.   The preventive administration of small doses of these
solutions has been also suggested by some authors in patients with
ventilatory insufficiency, since they induce a noticeable improvement of ventilation and reduce hypoxaemia and hypercapnia, despite
a transient and mild worsening of acidosis.[12]

d) Bronchodilators

    Before discussing bronchodilator drugs a short comment should
be made on the regulation of bronchial smooth muscle tone.   The
nervous regulation of bronchial smooth muscle cell is predominantly
under parasympathetic drive, while sympathetic fibres are scarcely
and irregularly distributed.   The bronchial smooth muscle cell
possesses however beside stimulating parasympathetic receptors for
acetylcholine, stimulating alpha-adrenergic receptors and inhibiting
beta adrenergic receptors.[13]

    These latter seem to have considerable density or sensitivity
to circulating catecholamines and to adrenergic drugs, and can be
probably identified with the enzyme adenilate-cyclase;[14] the
activation of this enzyme leads to the production of cyclic AMP,
responsible for the relaxation of bronchial smooth muscle cell.
The action of cyclic-AMP is balanced by that of cyclic GMP and vice
versa; both nucleotides are metabolized by the enzyme phosphodiesterase.   Beside cholinergic and alpha adrenergic stimulation,
the contraction of bronchial smooth muslce cells can also be induced
by other substances such as histamine, serotonin, SRSA and leukotrienes released locally or systemically in several patho-physiologic
conditions.   These substances can act directly or through the
synthesis, in the (bronchial) tissues, of arachidonic acid metabolites, such as cyclic endoperoxides, thromboxane A_2 (TXA_2) and prostaglandin PGF_2-α.[15]

    From the above, it is evident that a bronchodilatation can be
obtained, by drugs influencing the neurovegetative regulation of
bronchomotor tone, by drugs inhibiting the liberation of chemical
mediator- of bronchspasm or by drugs which are competitive or
functional antagonists of these mediators.

Among drugs which antagonize the enzyme phospodiesterase, we have already discussed aminophylline.   Ephedrine, another drug in the past widely employed in the treatment of bronchospasm has the advantages of a long duration of action and effectiveness by oral route.   This drug acts through the liberation of catecholamines from storage sites and through the potentiation of intensity and duration of action of endogenous or exogenous catecholamines, although this mechanism is still unclear and quantitatively less important.[16]

The therapy of bronchospasm is mostly based on drugs directly stimulating beta-adrenergic receptors.   The utilization of these drugs, which has received large theoretical recognition, has been limited in clinical practice by adverse secondary cardiovascular effects represented by tachycardia, increase of myocardial work and O_2 consumption and serious cardiac arrhythmias.   Since the demonstration of 2 types of beta-adrenergic receptors, beta   and beta , the former mainly present in the myocardium, the latter in the bronchial smooth muscle, a new class of therapeutic agents has been introduced, the so called beta stimulating drugs[17] with variable but high ratio of $beta_2$/$beta_1$ receptor stimulating activity.   Some of these drugs have not been introduced in clinical practice due to their unfavourable pharmaco-kinetic or toxic properties, while others, such as metaproterenol, fenoterol, trimetochinol, salbutamol, terbutalin, reproterol and carbuterol, are nowadays available commercially and can be administered by oral or parental route, or by aerosol; it is likely that the number of these drugs will increase with the discovery of new similar molecules.   In addition the proper adoption of pressurized aerosols with a pre-established single dose has contributed to a reduction of collateral effects, since it permits a topical administration, thus minimizing systemic effects and reducing the risk of overdosage. The efficacy of treatment is maximized by a correct modality of aerosol administration, which should coincide with a slow and maximal inspiration at the end of a complete expiratory act, and be followed by a brief period of post-inspiratory apnea terminated by a slow expiration.   However, due to the unhomogeneous alveolar ventilation and to the dead space the drug preferentially diffuses in the most ventilated alveoli; in addition the administration of these drugs by aerosol cannot be utilized in the most serious forms of cardio-respiratory failure.   In these patients, broncho-dilator drugs with the best $beta_2$/$beta_1$ activity ratio, such as salbutamol, can be administered by i.m. injection or by slow i.v. infusion.   The clinical usefulness of alpha-adrenergic blocking drugs such as inderamine, phentolamine or timoxamine, is only theoretical, due to the prevalence of cholinergic and beta-adrenergic receptors in bronchial walls.   Experimentally demonstrated favourable effects of these drugs might be explained by the alpha adrenergic mediated inhibition of purinergic bronchodilatation. Alpha blockers might inhibit the alpha stimulated reduction, through activation of the enzyme ATP ase, of the amount of ATP available for purinergic transmission; and also inhibit the release, mediated by calcium-dependent ATPase, of chemical mediator of bronchospasm from the mast-cells.[18]

Anticholinergic drugs have well known disadvantages in the therapy of bronchospasm, because of their untoward secondary effects. Two atropine derivatives active by aerosol, ipratropium, an isopropil substitute of atropine, and ossitropium bromide, seem to have overcome these side effects because of a selective effect on bronchial smooth muscle. These drugs have several mechanisms of action: 1) blockade of muscarinic receptors, with reduced formation of cyclic-GMP, both within bronchial smooth muscle, with secondary bronchodilation, and within mast-cells, with reduced release of chemical mediators of bronchospasm; 2) reduced accumulation of arachidonic acid metabolites, particularly $PGF_{2\alpha}$ and TXA_2. The latter is the most powerful constricting agent of bronchial and vascular smooth muscle presently known.

Glucocorticoid drugs can also be clinically useful in the treatment of cardio-respiratory failure due to their antiphlogistic and bronchodilatating activity.

The antiphlogistic effect is responsible for the reduction of bronchial and interstitial pulmonary oedema with consequent improvement of gas diffusion through the alveolar capillary membranes.

The antiphlogistic effects possessed by glucocorticoids are due to the stabilization of lysosomial membranes, and to the inhibition of leucocytes and mast cells migration. This last effect is, however, potentially unfavourable in patients with infection unless possible complications are prevented bv antibiotic treatment. The glucocorticoid·induced bronchodilatation is mediated by several mechanisms that finally are responible of a potentiation of the catecholamine induced cyclic-AMP production. Glucocorticoids can be administered topically by inhalation, alone or in combination with beta-adrenergic drugs; in the less severe forms of the disease, beclomethasone dipropionate or desamethazone isonicotinate should be preferentially used. Collateral effects with this modality of administration can be represented by dysphonia and oro-pharyngeal candidiasis, which can be treated with topical alkaline solutions containing nystatine. In the most severely compromised patients, i.v. route and steroids with higher antiphlogistic/sodium retentive effect ratio should be preferred, such as prednisone or methyl-prednisolone, since these patients often have a clinically manifest or inapparent retention of salt and water. Other bronchodilator drugs such as disodiocromoglicate, tiaramide, chlorpheniramine, ciproeptadine chlorydrate are preferentially indicated in the chronic stage of COAD rather than in the acute phase of the disease, although they are occasionally useful. Administration by aerosol route of prostaglandins PGE_1 and PGE_2, of prostacycline PGI and derivatives is still experimental, even though their vasodilator pulmonary effect, by improvine alveolar ventilation without impairment of ventilation/perfusion ratio offers favourable clinical results.

e) Fluidificant and Mucolytic Drugs

These drugs are another fundamental aid in the therapy of cardio-respiratory failure. They are useful for removing viscous bronchial secretions, which contribute to the beginning and maintenance of bacterial infections, reduce the airways lumen and favour bronchial phlogistic processes.

Both quantitative and qualitative abnormalities of bronchial secretions are important factors in sustaining reactive bronchospasm and in impairing muco-ciliar clearance.

Acetylcysteine, a direct mucolytic agent, has still a major role in the treatment of bronchial hypersecretion. The main action of this drug is the disruption of disulphur links of mucoproteins, but it also improves muco-ciliary clearance, reduces the viscosity of secretions and has some antibacterial activity probably through competitive inhibition of cysteine utilization. This drug can be administered by aerosol inhalation or endobronchial instillation, by oral or i.m. injection, with a less immediate effect, and is generally well tolerated.

The drugs with an indirect mucolytic action are not mucolytic in vitro, but interfere with the metabolic processes of the mucus producing cells, possibly through metabolites reaching the airways by the blood. Widely used is bromexine, a benzylaminic derivative of vasicin, an alkaloid extracted from Adathoda vasica, which acts through the decomposition of muco-polysaccarides the activation of lysosomial enzymes of bronchial glands and a modification of carbohydrate metabolism, leading to an incomplete or abnormal synthesis of mucopolysaccarides, can be administered orally by aerosol or by the parenteral route and is well tolerated, with only sporadic reports of mild gastric distress when administered orally. Carboxymethylcysteine can only be given by oral administration, is only active on mucus secretion and has a still unclear mechanism of action, however it is effective and devoid of important side effects. Iodine compounds possess some direct in vitro mucolytic effects, particularly on mucopurulent secretions; however, their main mechanism of action is indirect and are often poorly tolerated.

Sobrerol is the most important of the detergent or tensio/ active agents; it has the limitation of being active only on mucus secretions. It can be administered orally, by aerosol, by i.m. injection or rectally, and is usually well tolerated. Other drugs

effective in bronchial hypersecretive states, although not useful
in emergencies, are guaifenesin, guacetisal, ipecacuana, and sodium,
ammonium and potassium salts, these latter usually given in association with other drugs.

It should be stressed that adequate and cautious hydration,
orally or parenterally, is a fundamental part of the mucolytic
therapy.

Postural drainage and other physiokinetic therapies are very
important in the rehabilitation of patients with chronic obstructive
lung disease, but, since these manoeuvres require patient co-operation, they are usually reserved only for patients improving from
the acute phase of cardio-respiratory failure.   Furthermore it
should be remembered that the injudicious use of fluidificant and
mucolytic drugs can inadvertently induce particularly in comatose
patients or when cough depressant drugs are associated, a massive
inundation by the liquefied bronchial secretions, with sudden
development of respiratory failure.   In this event a prompt
endobronchial aspiration is mandatory.

f) The Correction of Acid-base and Electrolyte Abnormalities

This is another fundamental aspect of the treatment of cardio-
respiratory failure.   Therapy should be monitored by frequent
determination arterial blood gases, serum electrolytes total
bicarbonates (TB), standard bicarbonates (SB), and excess of
deficient base values (positive or negative BE).[20]   The most
common abnormality of acid base balance in patients with cardio-
respiratory failure, is respiratory acidosis, compensated or not
compensated.   This condition is characterized by a reduced or
normal pH, elevated arterial pCO_2, elevated TB, SB normal or
increased, according to renal compensation, with TB always higher
than SB.

Therapy in these patients should be addressed toward the
improvement of respiratory failure.   Alkaline solutions should be
reserved for patients with more severe acidosis, progressive hypoxaemia or signs of impending renal failure.

Small amounts of isotonic (1/6 molar) sodium bicarbonate
solution, containing about 170 mEq/l of anions and equivalent sodium
amounts should be administered.   In some patients, a metabolic
acidosis is superimposed on the respiratory acidosis; in these
cases TB are reduced, but always higher than SB because of hypercapnia, and lactic acid levels are increased over the normal
range of 1.2 mEq/l.   In these patients alkaline solutions are
indicated such as 1/6M bicarbonate or lactate, with the aim of keeping TB levels above 18 mEq/l.

From a practical standpoint, the following formula can be used to calculate the amount of bicarbonate to be administered: BE x 0.3 x kg. only 1/2 of the value obtained in this fashion is usually given in the first 24 hours; the remaining amount should be administered according to the results of blood gas and analytic determinations. In these patients the diuretic of choice is ethacrynic acid, which is more effective than other diuretics in patients with acidosis. When a powerful and rapid effect is required in patients with respiratory acidosis, THAM or 8.4 per cent sodium bicarbonate solutions containing 1 mEq of HCO_3/ml should be used. However, careful attention should be given to avoid post-hypercapnic metabolic alkalosis, usually a consequence of an incongruous therapeutic regime and often not recognized by clinicians, who focus their attention on the treatment of acidosis. This condition usually intervenes when arterial pCO_2 levels tend to normalize, but HCO_3 values are still elevated, because of the slow kidney compensatory mechanism and of excessive administration alkaline solutions or alkalinizing diuretics. These latter drugs can favour metabolic alkalosis also by the production of hypo-kalaemia or hypochloraemia.

Post-hypercapnic metabolic alkalosis is characterized by normal or elevated pH, elevated SB, TB more elevated than SB because of hypercapnia, and frequently by hypochloraemia and hypo-kalaemia. The therapy in this condition consists first of all in the removal of the determining causes. In addition, chloride deficiency should be corrected by NH_4 Cl or lysine monochloride administration with the aim of raising chloride plasma levels not above 95 mEq/l.

In the correction of metabolic alkalosis the diuretic of choice is acetazolamide which favours the loss of HCO- at the level of renal tubules, through the inhibition of the enzyme carbonic anhydrase. However, this drug also increases renal elimination of potassium ion, by reducing the availability of hydrogen ion to be exchanged with sodium at the level of the distal tubules. Potassium deficit can be corrected by KCl solutions, but this therapy should be conducted cautiously, because the physiological extracellular concentration of K^+ is very small, and because hypokalaemia can mask high K^+ intracellular concentrations; in this case severe compli-cations may arise when acidosis is corrected and intracellular potassium ion is exchanged with hydrogen ions.

g) Cardiokinetic Drugs

These drugs play a fundamental role in the therapy of cardio-respiratory failure, even though their use has been challenged by some authors in the stable stage of chronic obstructive lung disease because of a possible increase in pulmonary arterial pressure, due to improved right ventricular output against a restricted pulmonary

arterial bed.   A contemporaneous removal of hypoxia is always
necessary, since the hypoxic myocardial fibre is unable to respond
to cardiokinetic drugs.[21]   In these patients the risk of toxicity
is particularly present due to the frequently associated hypokalaemia, which lowers the therapeutic index of digitalis.
Furthermore, since the tachycardia in these patients is mainly
secondary to hypoxaemia and hypercapnia, the dose of cardiokinetic
drugs to be administered cannot be determined by the usual heart
rate criterion.

h) Diuretic Drugs

Ethacrynic acid should be reserved for patients with a tendency
to metabolic acidosis, since this drug, compared to other diuretics,
has more powerful hydrogen, chloride and potassium depleting
properties.   However, ethacrynic acid can precipitate metabolic
alkalosis, with consequent reflex hypoventilation and worsening of
respiratory acidosis up to hypercapnic coma.

Carbonico-anhydrase inhibiting diuretic drugs are only
indicated in the condition of respiratory acidosis associated with
post-hypercapnic metabolic alkalosis.   Usually other potassium
depleting diuretics are used, such as thiazides, xipamide, fenquiz-
one, which can be given only orally, and particularly furosemide
and bumetanide, generally given in association with potassium
sparing drugs such as spironolactone, triamterene, amiloride, or
potassium canreonate an active spironolactone derivative.   Patients
should also be kept on a Na restricted diet (for example, fresh
fruit and boiled rice) until the chloride eliminated with the urine
is less than 2 g/day; thereafter, the Na intake can be raised up to
20 mEq/day until clinical improvement occurs.   The Na depletion
improves cardio-respiratory conditions and the general clinical
state; furthermore, it enhances alveolar ventilation and gas
diffusion through the alveolar capillary membranes by the reduction
of oedema at the level of bronchial mucosa and alveolar capillary
interstitial tissue, respectively.   However, body weight, electrolyte and acid base balance should be constantly monitored, and
dehydration which would preclude the efficacy of mucolytic drugs
should be avoided.

i) Increase of the Respiratory Surface

This is another important aim of therapy of cardio-respiratory
failure, particularly in the reacutization phase of the disease
and is accomplished by modification of pulmonary hemodymics.

Many of the therapeutic interventions discussed increase the
capillary respiratory surface and the alveolar-capillary diffusion
(aminophylline, nicotinic acid).

Vasodilators should be employed in patients with more advanced right ventricular overload. Preference should be given to i.v. or sublingual nitrates, which by reducing venous return, right atrial and pulmonary pressures, decrease right ventricular overload. It is fundamental in these patients to try to improve with vasodilator drugs, the capacity of pulmonary vascular bed, especially when this is severely reduced such as in patients with primary pulmonary hypertension. Unfortunately, the drugs proposed in this condition, such as phentolamine, acetylcholine, tolazoline, hydrallazine, reserpine, isoprenaline, and ganglioplegics even though effective in single patients, especially when infused directly into the pulmonary artery, in the long term lose efficacy in most cases.[22] Despite these limitations, these drugs can be tried in cardiorespiratory failure, particularly in order to reduce the functional pulmonary vasoconstriction; venesection can also be employed when hematocrit is over 60 per cent. This is usually accomplished by withdrawing 250-300 ml of blood every 2 - 3 days until lowering the haematocrit to 48 - 50 per cent, always trying to keep Hb values over 12 g%. An equivalent amount of plasma expanders should always be infused, in order to avoid dangerous reduction of blood volume and of cardiac output.

Venesection, by reducing polycythaemia, increases O_2 disposal to tissues at the same level of O_2 content, thus decreasing tissue hypoxia, and reduces blood viscosity, which contributes to pulmonary arterial hypertension and to the decreased blood flow velocity, and predisposes to thrombo-embolic phenomena. Prevention of thromboembolic episodes should be accomplished either with heparin, oral anticoagulants or platelet anti-aggregating drugs (such as aspirin, dipyridamole, sulphynpirazone) in those patients in whom repeated unrecognized pulmonary microemboli may represent an aetiologic factor in chronic cor pulmonale or of its reacutization. However, thromboembolic phenomena are frequent, although often clinically silent, also in chronic cor pulmonale secondary to chronic obstructive lung disease.

For these reasons, and when absolute contraindications are not present, it is advisable to give anticoagulants in the reacutization of cor pulmonale particularly in patients with severe pulmonary arterial hypertension and polycythaemia reserving thrombolytic therapy only to those in whom an active thromboembolic episode is demonstrated. When acute pulmonary thromboembolism is demonstrated, streptokinase or, preferentially, urokinase should be administered by continuous i.v. infusion or by direct infusion in the pulmonary artery; heparin should also be given in association, followed by oral anticoagulants by the second day. Dosage of these drugs has to be regulated on thrombin time, which should be maintained around values twice normal. Later, with the patients on oral anticoagulants, prothrombin time should be kept between 15 and 35 per cent of normal.

j) Antimicrobic Drugs

They play a major role in the therapy of cardio-respiratory
failure. Acute infections are the most frequent cause of reacutization of CCP patients.

The use of these drugs should be based on the identification
of the aetiologic agent of the infection, its sensitivity to
antibiotics, therapeutic range, pharmacokinetic characteristics,
preferential route of administration and adequate dosages. The
choice of the appropriate antibiotic agent, its dosage and route
of administration should be individualized to the need of the
single patient.

Particular attention should be given to the factors regulating
the diffusion of the drug into the bronchial secretions and exudates.
At the level of bronchopulmonary perfusion, and to the ability of
the drug to pass the haematobronchial barrier, the so called
"pneumotropism" and its modification by bronchopulmonary infection.[23]

In patients with acute bronchopulmonary infection a bacterio-
logic analysis of bronchial secretions should always be undertaken
in order to identify the aetiologic agent or agents and to test
their sensitivity to several antibiotics. This practice is
particularly important, since with increasing frequency bronchopulmonary infections are sustained by so called opportunistic
aetiologic agents, often in multiple association, such as anaerobic
agents resistant to the usual chemio-antibiotics particularly in
hospitalized, cachectic and older patients.[24]

Sputum cultures should be repeated in case of persistence or
recurrence of symptoms since in the course of the infection the
development of bacterial "step resistance" or virulentation of
saprophitic bacteria by antibiotics is frequent.

Considerable controversies have arisen on the modalities of
obtaining samples of bronchopulmonary secretions for diagnostic
purposes. Subglottid sampling is being more widely performed
especially when the choice of the adequate antibiotic is critical.
A useful criterion for differentiating a true superinfection from
the non causal mere presence of germs in the secretions is the
quantitative evaluation of the number of such micro-organisms:
a number over 10^6/ml probably indicates an aetiologic role.

However in ambulatory practice, a differential diagnosis
between bacterial and viral or mycoplasm bronchopneumonia infection
can be easily made on clinical grounds and with simple laboratory
test. In fact the usual aetiologic agents in ambulatory patients
with acute bronchopneumonia complicating CCP are Diplococcus
pneumoniae, Staphylococcus aureus, Streptococcus pyogenes,

Haemophylus influenzae and Klebsiella pneumoniae which with rare exceptions respond to the most wide spectrum bactericidal antibiotics, and particularly to cephalosporin.

In acute cardio-respiratory failure with rapid clinical course the antibiotic of choice should have a wide spectrum and should be given parenterally at high dosages. [25]

In patients with more rapid evolution, it is mandatory to immediately administer parenteral antibiotics, also of different types in association, giving the preference to those with broad spectrum and bactericidal action, without waiting for the results of cultures; these however will be required subsequently in order to control the results of therapy or to readjust the therapy, if unsuccessful.   In such cases hospitalization is also required, and the antibiotics of choice are usually the aminoglycosides and parenteral cephalosporins, (cephamicines, metoximines).

Aminoglycosides are a group of parenterally active drugs similar to streptomycin.   The number of these drugs has progressively increased in the last few years, however their major pharmacological properties are similar, with some difference concerning the activity against myobacteria and Pseudomonas aeruginosa. These antibiotics are active on both gram + and gram - bacteria, such as Enterobacter, Brucella, Haemophilus and sometimes Pyocianeus; they are relatively less active against streptococcus and pneumococcus.   Their side effects are represented by frequent ototoxicity and nephrotoxicity, and by the poor therapeutic index in presence of renal insufficiency.   Caution is particularly required in patients over 60 years, and in those with pre-existing renal disease; in these cases one should avoid treatment over 10 days and maintain adequate hydration in order to preserve renal perfusion.

Cephalosporins are also parenterally active broad spectrum antibiotics. They have the potential disadvantages (particularly cephaloridine) of possible nephrotic acute tubular necrosis with elevated doses; furthermore, they are sensitive to the action of endo-betalactamasis (cephalosporinasis) produced by some gram-bacteria.   However, second and third generation cephalosporin, (cephamycins and metoximins) are resistant to beta-lactamasis. Among cephamicines the only utilized compound is cephoxitin, while cephuroxim and cephotaxime are the most widely used among metoximin group.   These new drugs, resistant to both eso- and endo- beta lactamasis produced by gram + and gram- bacteria, respectively should be reserved only for the most compromised patients, in order to avoid the selection and the diffusion of resistant groups. Psudomonas is, however, resistant to both cephamycine and metoximine. In infections sustained by this agent, carbenicillin should be given by i.v. infusion at doses up to 20 - 30 g/day, in association with gentamicin, tobramicin, amikacin or sisomicin.

Compared to these wide spectrum bactericidal antibiotics, tetracyclins, which are only bacteriostatic are of second choice. However, they are first choice antibiotics in infections sustained by Rickettsiae.

Fosphomicin can also be useful in the most severe cases of cardio-respiratory failure, alone or in association with amino-glycosides; however, large i.v. doses must be given because of the unfavourable pharmacokinetic characteristics of this drug in broncho-pneumonic disease.

Particularly important are infections caused by anaerobic microorganisms, which can be the primary cause of acute deterior-ation of broncho-pulmonary diseases, or rather can superimpose, by development of bacterial resistance, during the hospitalization of cachectic patients. Anaerobic germs form an heterogeneous group (Bacteroides, Fusobacter, Clostridia, Propionibacter, Peptococchi, Peptostreptococchi, Enbacter, Bifidobacter) generally resistant to the usual chemio-antibiotic agents. For these reasons, whenever the presence of fetid sputum and of predisposing factors suggest an aetiological role of anaerobic germs, it is mandatory, before starting antibiotic therapy, to obtain repeated cultures during febrile spells; samples should be taken in absolute anaerobiosis and cultured on specific media.

The antibiotics of choice in these patients are wide spectrum penicillins or an injectable cephalosporin at high doses, in association with gentamicin and chloramphenicol, or tiamphenicol or lincomicin; therapy should be readjusted according to the clinical course and to the results of repeated cultures with anti-biograms.

In the most severe cases, when multiple microrganisms are involved, non specific human immunoglobulins given i.v. at adequate dose are also indicated.

The forms of cardio-respiratory failure with slow clinical evolution, also require prompt antibiotic treatment. preceded by sputum cultures with antibiogram, giving the preference to wide spectrum bactericidal antibiotics and the oral route.

In these patients, the drugs of choice are the oral penicillins similar to Penicillin G, which are however ineffective against most gram -ve bacteria and are inactivated by beta-lactamases, the isossazolil-penicillins, which are resistant to eso-beta-lactamases but have a spectrum of action limited to gram + bacteria, or ampicillin and derivated, which have a wider spectrum but are inactivated by eso-beta-lactamases. Particularly indicated in these patients are the orally active cephalosporins, which have the advantage of being resistant to eso-beta-lactamases, and of being

active against both gram + and - bacteria; they are however
inactive against Pseudomonas and, at least the classic cephalosporins, are inactivated by endo-betalactamases.

The most widely used oral cephalosporins are cephalexin,
cephadrin and cefadroxil. The oral administration of these drugs
yields effective blood concentrations equivalent to those obtained
by parenteral cephalosporins.

Tetracyclins possess the above indicated limitations in most
seriously ill patients; in addition, generally it is preferable to
avoid the association of a bactericidal with a bacteriostatic
antibiotic; although this rule has several exceptions and favourable
results have been obtained with the combination tetracycline-
amoxicillin.

Rifampicin, a well known antibiotic active against mycobacter-
ium tuberculosis, can be employed in association with other drugs
for its capacity for reaching high concentration in the secretions;
however its indiscriminate use should be avoided for the risk of
selecting resistant strains of mycobacteria. In addition, the drug
is not without serious side effects.

Extremely important and of difficult solution is the problem of
prophylaxis of infective recurrences in chronic cor pulmonale. The
administration of short cycles of antibiotics in patients with non
specific chronic bronchopneumonic disease has been widely used.
However, several controlled studies have shed considerable doubt on
the validity of this chemioprophylaxis, particularly for the risk
of favouring bacterial resistance to the most widely used antibiotics.
Parenteral or local (by aerosol) administration of antimicrobial
vaccines although empirically used, does not usually give favourable
results.

Chemio-antibiotic prophylaxis in the course of influenza virus
infections is advantageous in patients with chronic cor pulmonale,
due to the potential risk of bacterial infections in these patients,
compared to the relatively minor risk of selecting resistant
bacterial strains.

However, these situations should be prevented by the wide
application, particularly in older patients, of the anti-influenzal
vaccines. An alternative to such prophylaxis can be represented by
amantadine in short courses (100 mg b.i.d. by mouth). Whenever
the typical symptoms of influenza occur a trial with isoprinosin
4 g/day p.o. in divided doses, can also be attempted.

In order to prevent infection recurrences, in most recent
years courses of therapy with mucolytic or antisecretory drugs have
often been used, considering the lack of unfavourable side effects
as is the case with cycles of chemio-antibiotics.

Rehabilitation is another important aspect of prevention, but this is outside the scope of the present review. However, cardio-respiratory rehabilitation has to be reserved for qualified centres, provided with adequate instrumentation, and with intensive care units. These may be necessary for possible deterioration of cardio-pulmonary status during the course of the rehabilitation program in those patients with more severe anatomo-functional impairment.

In these centres adequage hygienic rules, such as elimination of tobacco smoke, and also cycles of physiokinetic and inhalation therapy can be applied, together with attempts at respiratory reduction with periods of intermitent positive pressure respiration.

REFERENCES

1. L. Condorelli, Il Cuore polmonare cronico, Rif.Med. 73:1393 (1979).
2. Organition Mondial de la Sante': Le coeur pulmonaire chronique. Rapport d'un Comite d'esperts. Series des rappotrs technique. Geneve, 213 (1961).
3. A. Blasi, D. Olivieri, Secrezione bronchiale ed aspetti fisiopatologici dell'ipersecrezione. Il Pensiero Scientifico, Ed. Roma, 107 (1980).
4. M. Condorelli, F. Rengo, D. Bonaduce, Fisiopatologia e clinica del cuore polmonare cronico, in: "Medicina Clinica-Diagnostica e terapia," A. Beretta Anguissola, ed., Ed. Medico Scientif-iche, Torino (1979).
5. P. J. Cohen, The metabolic function of oxygen and biochemical lesion of hypoxia, Anesthesiology 37:148 (1972).
6. E. J. Campbell, The management of acute respiratory failure in chronic bronchitis and emphysema, in: "The J. Burns Amberson Lecture" Am.Rev.Resp.Dis. 96:626 (1967)
7. S. C. Wang, J. W. Ward, Analeptics, Pharmacol.Ther. 3:123 (1977).
8. D. C. Webb, J. L. Andrews, Bronchodilator Therapy, N.Engl.J.Med. 297:476 (1977).
9. J. F. Nunn, Applied/Respiratory Physiology, 2nd ed. Butterworths, London (1977).
10. R. R. Kirby, Intermittent mandatory ventilation, Curr.Probl. Anesth.Crit.Care Med. 5:18 (1977).
11. M. Pasargiklian, G. Fumagalli, A. Ferrara, La terapia del disequilibrio acidobase nel trattamento intensivo della patologia respiratoria, Min.Med. 70:1 (1979).
12. G. Giuntini, A. Santolicandro, Insufficient respiratoria, in: "Terapia Medica Moderna," Vol. 1. L. Baschieri, ed., Ed. Ambrosiana, Torino, (1979).
13. J. B. Richardson, C. C. Ferguson, Neuromuscular structure and function in the airways, Fed.Proc.38:202 (1979).

14. G. A. Robinson, R. W. Butcher, E. W. Sutherland, Cyclic AMP, Academic Press, Inc. New York (1971).

15. B. Samuelsson, G. Folco, E. Granstrom, H. Kindhal, C. Malmsten, Prostaglandins and thromboxanes: biochemical and physiological considerations. Adv.Prostaglandins and Thromboxanes 4:1 (1978).

16. D. M. Aviado Jr.,"Sympathomimetic drugs," Charles C. Thomas, Springfield, Ill, (1970).

17. A. M. Lands, A. Arnold, J. F. McAuliff, F. P. Luduena, T. G. Brown, Differentiation of receptor systems activated by sympathomimetic amines, Nature 214:597 (1967).

18. C. G. Irvin, R. Boileau, J. Tremlay, R. R. Martin, P. T. Macklem, Bronchodilatation: noncholinergic, nonadrenergic mediation demonstrated in vivo in the cat, Science 207:791 (1980).

19. L. Reid,"Mucus in health and disease," New York Plenum Press, New York (1977).

20. H. W. Davemport, "The ABC of acid-base chemistry," The University of Chicago Press, Chicago, (1970).

21. F. Rengo, M. Condorelli, D. Nonaduce, Mechanism of pulmonary hypertension by hypoxemia. 2nd Europ.Symp.Coagulation platelet fibrinolysis and vascular diseases, Palermo (1980).

22. R. P. Gatewood, P. N. Yu, Primary pulmonary hypertension in: "Progress in Cardiology, P. N. Yu, J. F. Goodwin, eds., Vol. III, Lea and Febiger, Philadelphia (1979).

23. J. E. Pennington, Immunological and infections in the lung, C. H. Kirkpatrick, H. Y. Reynolds, Dekker, New York (1977).

24. W. B. Pratt,"Chemotherapy of Infection," Oxford University Press, New York (1977).

25. G. L. Mandell, R. G. Douglas Jr., J. E. Bennett, "Principles and Practice of Infectious Disease," John Wiley and Sons, New York (1979).

INTENSIVE CARE OF ACUTE ISCHEMIC CARDIOPATHY IN OLD PATIENTS

N. Marchionni, R. Pini, A. Vannucci, M. Calamandrei,
A. Conti, M. di Bari, L. Ferrucci, B. Greppi and
F. M. Antonini

University of Florence
Institute of Gerontology

Ventricular power failure is presently the main therapeutic problem during acute myocardial infarction (AMI), since it accounts for more than 50 per cent of deaths in any age,[1] (see Table 1). Direct measurements of cardiac hemodynamics through the use of right heart balloon flow-directed catheters have proved to be of invaluable help in optimizing the clinical management of patients with AMI, as they allow for selection of the most appropriate therapy, based on individual pathophysiological data.[2,3] In this lecture, we will describe the immediate effects of the drugs which are most commonly used in the correction of the hemodynamic disorders complicating AMI, and we will examine their efficacy in young (age <65 years) and elderly (age ⩾65 years) patients.

Treatment of impaired cardiac function due to AMI is directed toward correcting specifically identified hemodynamic derangements without increasing the magnitude of regional ischemia. Since hemodynamic abnormalities are proportional to the amount of ischemic myocardium, which is in turn determined by the mismatch between myocardial oxygen demand and coronary supply, the following are the goals for therapy:
- reduction of abnormally elevated pulmonary artery pressures, to relieve pulmonary congestion
- improvement of depressed cardiac output, to prevent peripheral hypoperfusion
- achievement of these goals without either increasing myocardial oxygen consumption (MVO_2) or decreasing coronary oxygen delivery.

With this in mind, the proper therapy can be selected for each of 5 hemodynamic subsets (Figure 1), which are defined according to

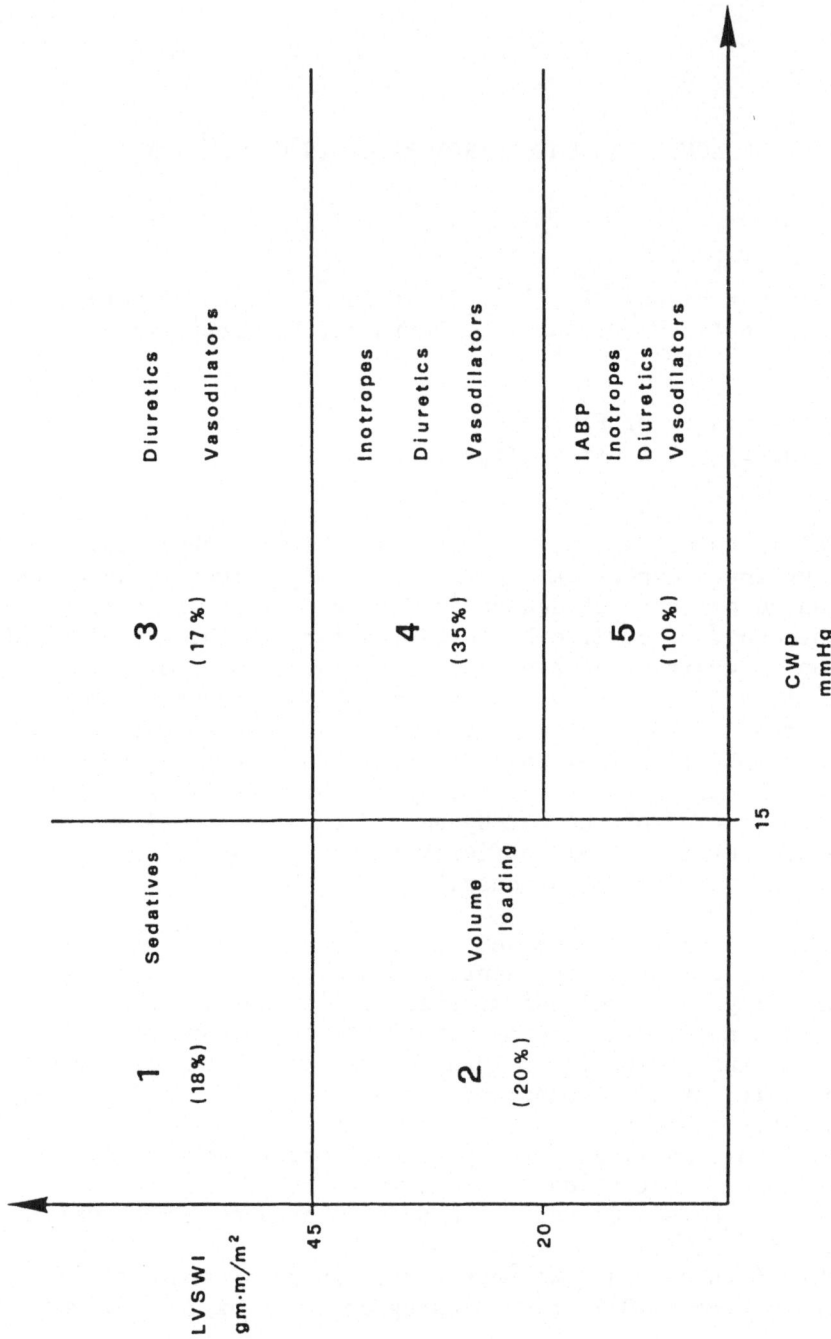

Fig. 1. Hemodynamic subsets based on the relationship between capillary wedge pressure
(CWP) and left ventricular stroke-work index (LVSWI). Numbers between
brackets indicate the incidence rate for each subset.

Table 1. Percentage distribution by age of causes of death in acute myocardial infarction (n = 448)

AGE (YEARS)	LEFT VENTRICULAR FAILURE	CARDIAC RUPTURE	PULMONARY ARTERY THROMBO-EMBOLISM	PRIMARY VENTRICULAR FIBRILLATION	TOTAL DEATHS	
	%	%	%	%	n	%
≤60	50.0	10.0	10.0	30.0	10	5.5
61-70	51.9	40.7	3.7	3.7	27	20.6
71-80	48.5	39.4	12.1	0.0	33	33.3
>80	61.5	30.8	7.7	0.0	13	37.1
TOTAL	51.8	34.9	8.4	4.8	83	18.5

the relationship between the left ventricular filling pressure (measured as capillary wedge pressure, CWP, mmHg) and the left ventricular stroke-work index (LVSWI, gm.m/m$_2$) (Frank-Starling law)[4].

Subset 1 - Normal left ventricular function (CWP ≤15 mmHg; LVSWI >45 gm.m/m^2). 18 per cent of patients with AMI are in this subset, with a fairly good prognosis (average mortality is about 2%). No treatment is needed, except sedation and oxygen.

Subset 2 - Absolute or relative hypovolemia secondary to vomiting, sweating, or to inadequate venoconstrictive reflex (CWP ≤15 mmHg; LVSWI <45 gm.m/m^2). The incidence rate of this subset in AMI is about 20 per cent. The average mortality rate was four times greater than that which we observed in subset 1, probably because of peripheral hypoperfusion. Whether mortality rate can be reduced by increasing the cardiac output, is still unknown. However, left ventricular performance is rapidly improved by volume loading.[5,6]

In 55 patients with transmural AMI (<65 years: n=28; ≥65 years: n=27) we observed significant increases in left ventricular pump performance and mean systemic arterial pressure (MAP, mmHg) (Figure 2) after i.v. infusion of low molecular weight Dextran (5 ml/kg over an average time of 45 minutes). Heart rate (HR, b/min) was unchanged. Hemodynamic responses to plasma volume expansion under and over the age of 65 were analogous.

Subset 3 - Decreased left ventricular compliance (17 per cent of all cases of AMI. CWP >15 mmHg; LVSWI ≥45 gm.m/m^2). Pulmonary congestion is frequently observed, in the absence of peripheral hypoperfusion. The average mortality rate in our series was 14 per cent. Pulmonary hypertension is promptly reversed either by i.v. Furosemide or by nitrates.

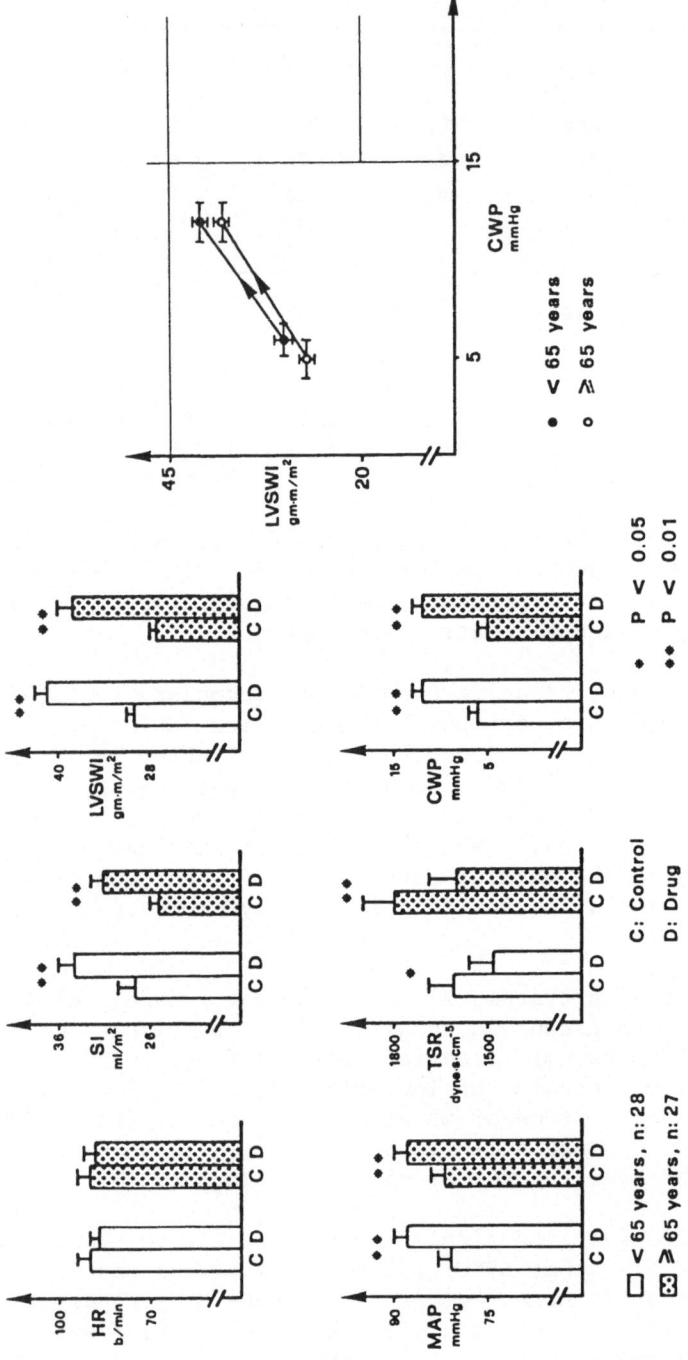

Fig. 2.   Improvement of left ventricular performance after volume loading in subset 2
(hypovolemia).   HR = heart rate; SI = stroke index; LVSWI = left ventricular
stroke-work index; MAP = mean systemic arterial pressure; TS = total systemic
resistance; CWP = capillary wedge pressure.

In our experience, both drugs reduced CWP to the normal range, without significantly affecting other hemodynamic parameters (Figures 3 and 4). The effects were similar in the two age groups.

Isosorbide dinitrate (5 mg s.l.) slightly decreased cardiac output and stroke index (SI, ml/m^2), probably because of its more rapid venodilating action. Occasionally, such a reduction was large enough to induce a dramatic fall in MAP toward critical levels (Figure 5). This side effect, which we observed more frequently in the elderly, as a possible consequence of impaired vasoactive reflexes, is a potential hazard of ischemia in the brain and the myocardium.

If the systemic hypertension is present, the use of anti-hypertensive drugs with rapid action and short half-life is advisable, in order to unload the left ventricle and to prevent a possible enlargement of infarct-size by the protection of the jeopardized myocardium.[7] Intravenous Trimethaphan (average dose 10 mcg/Kg/min) reversed CWP, MAP and total systemic resistance (TSR, dyne.sec.cm^{-5}) to normal levels in a few minutes (Figure 6). SI was unaltered. Reflex tachycardia, which had been observed as a side effect of other vasodilators, such as Phentolamine,[8-11] was absent. Hemodynamic response to Trimethaphan was similar in the two age groups, with the exception of MAP, which was more markedly reduced in the elderly (-29.6 vs -21.6%, P <0.05), despite the lower average dose which was employed (8 vs 13 mcg/Kg/min). We consider this as a further example of the need for a cautious use of vasodilating and antihypertensive drugs in the elderly.

Subset 4 - Moderate left ventricular failure (CWP >15 mmHg; LVSWI 20 - 44 gm.m/m^2). 35 per cent of AMI present with a mild to moderate left ventricular dysfunction. Physical examination most often reveals pulmonary congestion variably combined with peripheral hypoperfusion (cold, clammy skin; mental obtundation; oliguria). a poor prognosis is common (34.3% mortality rate).

Three different therapeutic strategies can be followed, and possibly associated:

- i.v. diuretics (Furosemide)
- i.v. positive inotropic drugs (Digoxin, Dopamine, Dobutamine)
- pre- and/or afterload reducing vasodilators (s.l. Isosorbide dinitrate, i.v. Nitroglycerine or Nitroprusside, oral Hydralazine or Captopril).

Diuretics[12-14] are especially useful in cases with elevated left ventricular filling pressure and only slightly reduced cardiac output. In 49 patients (<65 years: n=19; >65 years: n=30) with these hemodynamic features, CWP substantially decreased, while cardiac output did not change, within 90 minutes from i.v. bolus

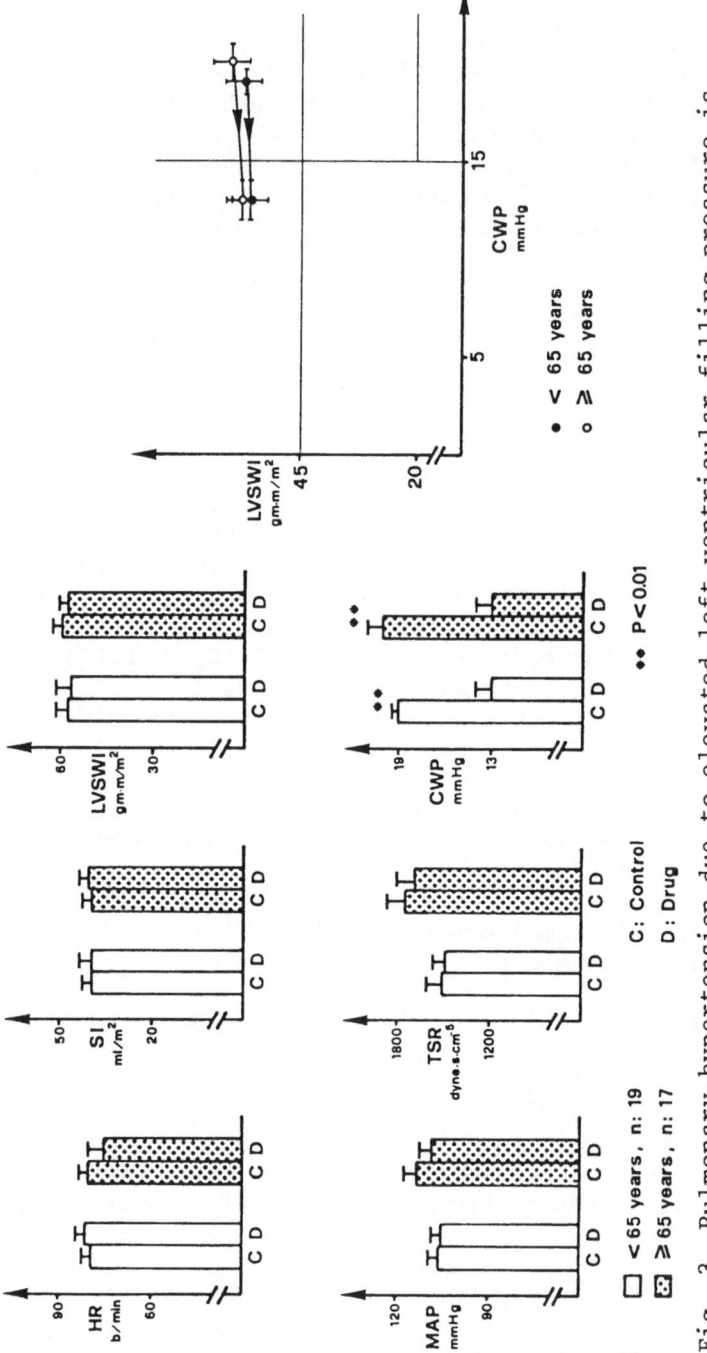

Fig. 3. Pulmonary hypertension due to elevated left ventricular filling pressure is reversed by diuretic therapy in patients with reduced left ventricular compliance. Abbreviations as in Figure 2.

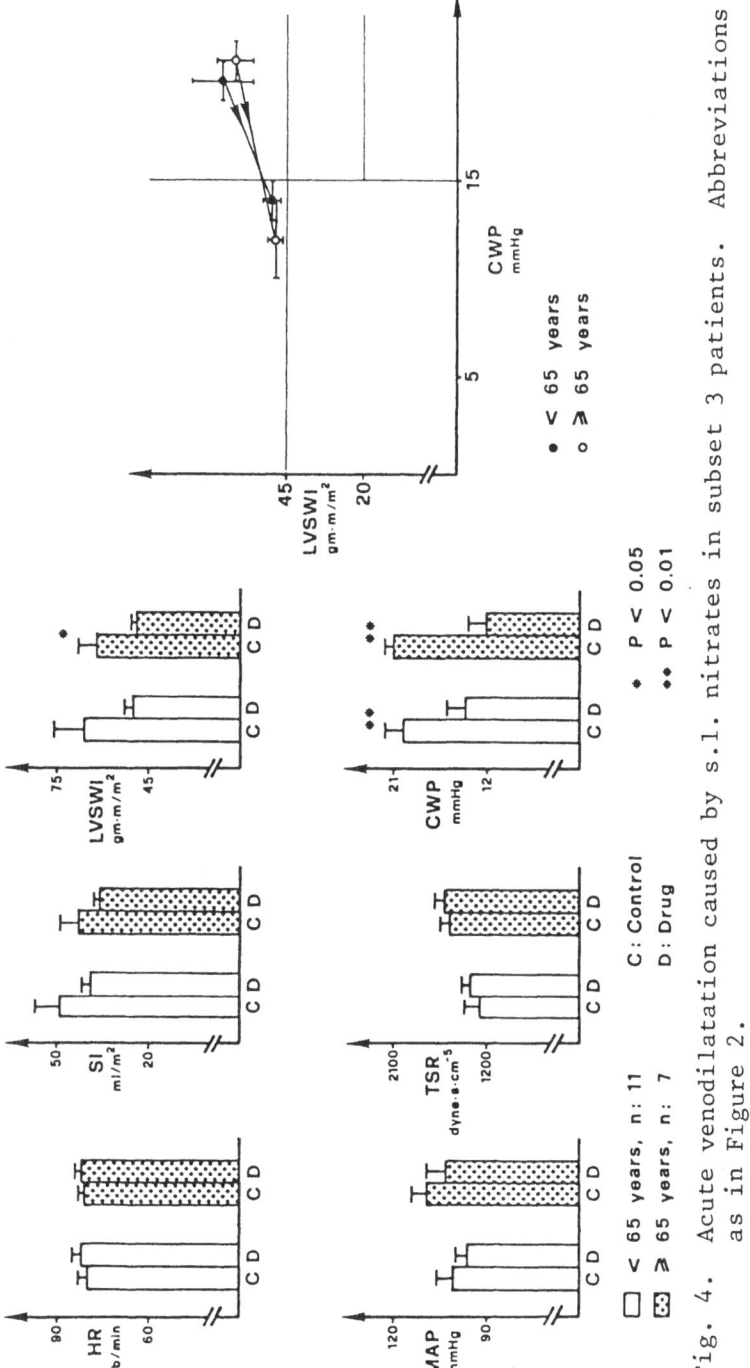

Fig. 4. Acute venodilatation caused by s.l. nitrates in subset 3 patients. Abbreviations as in Figure 2.

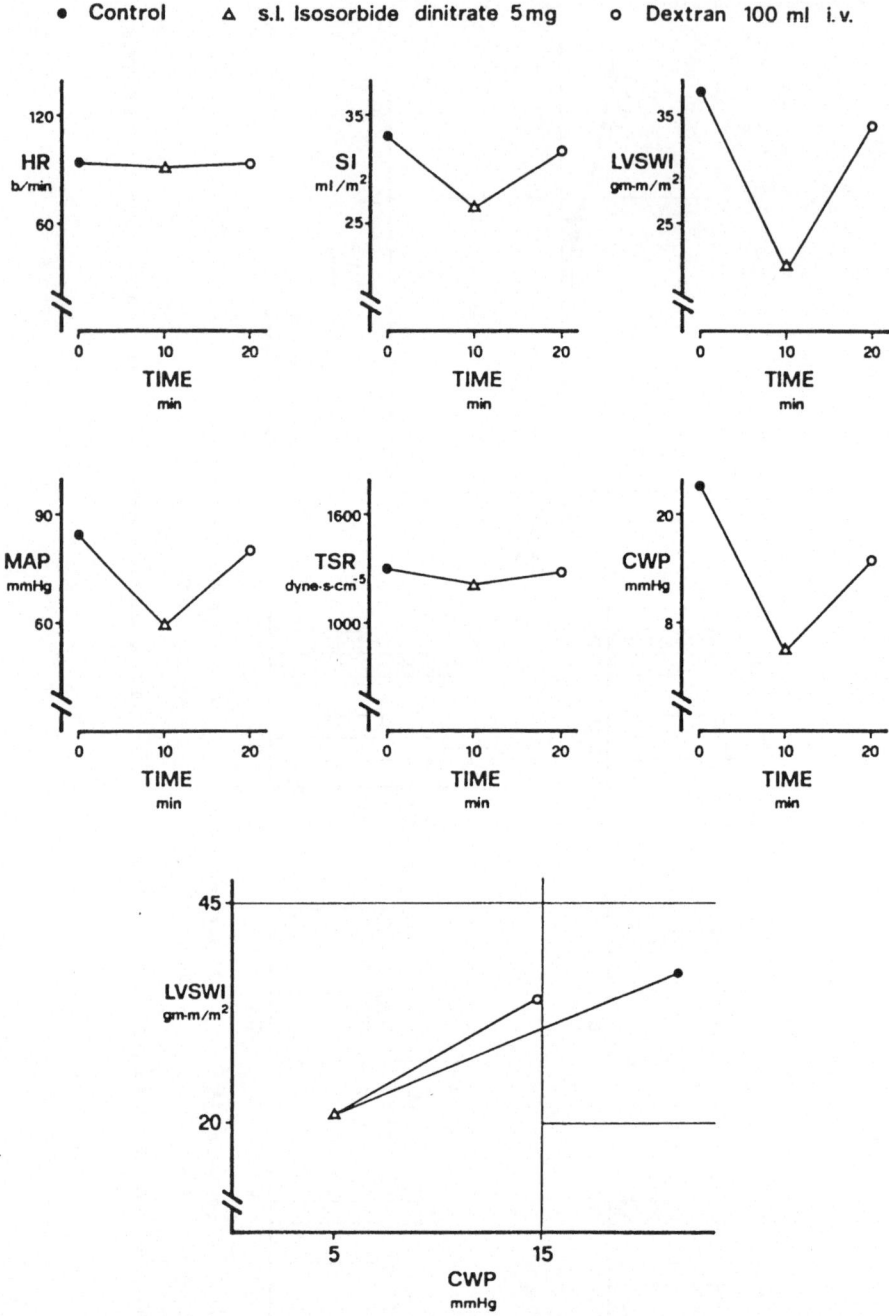

Fig. 5.   Marked hypotension secondary to excessive venodilat-
          ing effect was caused by s.l. Isosorbide dinitrate
          in a subset 4 patient. Normal cardiac output and
          MAP are re-established by rapid volume loading.

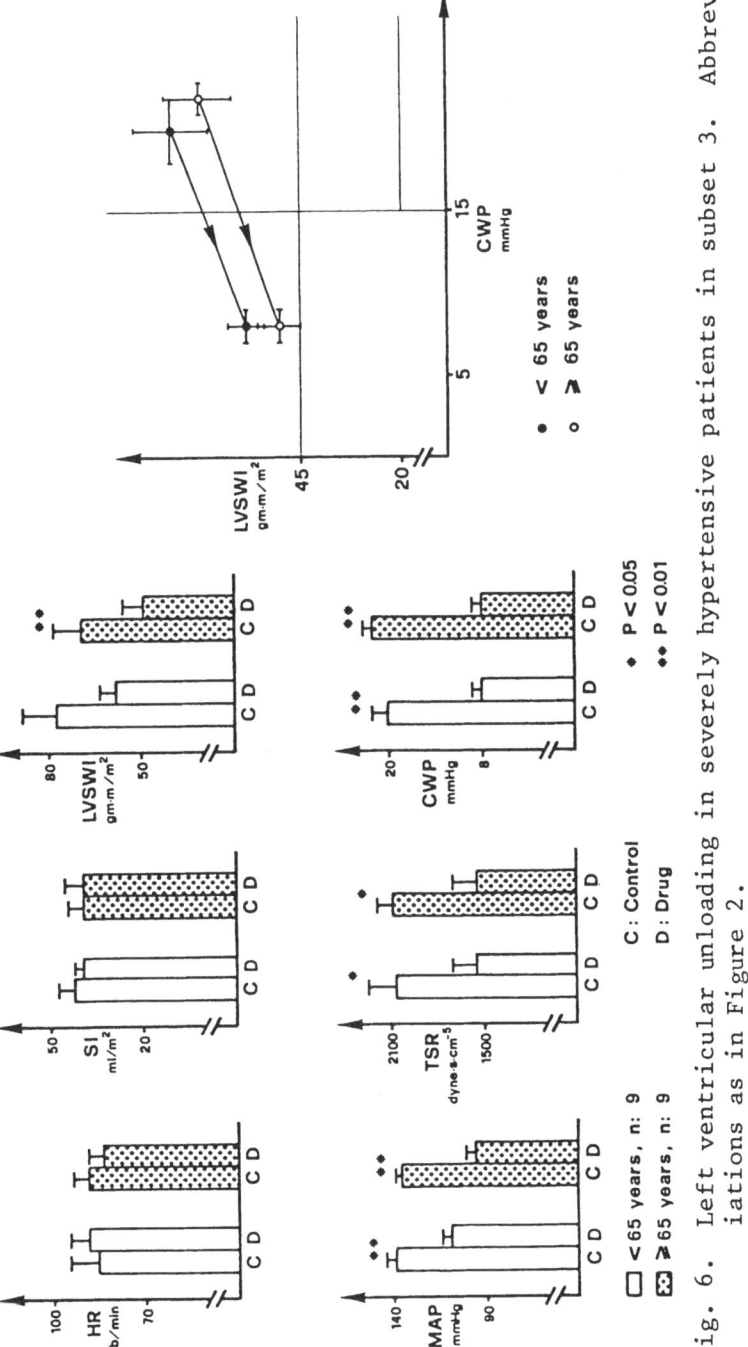

Fig. 6. Left ventricular unloading in severely hypertensive patients in subset 3. Abbreviations as in Figure 2.

injection of Furosemide at an average dose of 0.7 mg/Kg (Figure 7). Moderate improvements of SI and LVSWI were seen after diuretic therapy in the elderly.

The use of inotropic agents in AMI is still controversial, since each of them has been thought to increase myocardial oxygen demand, thereby causing a progressive enlargement of the size of infarction.[15,16]

Two other objections have been raised, concerning Digoxin in particular: cardiac toxicity has been supposed to be increased and hemodynamic efficacy decreased, when the drug is administered in the presence of myocardial ischemia. Nevertheless, several clinical trials[17-19] failed to demonstrate an increased incidence of ventricular arrhythmic complications following digitalis treatment in AMI. On the other hand, Digoxin exerts favourable hemodynamic effects in patients with moderate cardiac failure, even in the early stage of myocardial infarction.[20,21]

Thirty two patients (<65 years: n=10; ≥65 years: n=22) in subset 4 received intravenously 0.50 mg of Digoxin, within 24 hours from the onset of infarction: cardiac output, SI and LVSWI markedly improved, and CWP decreased, as a result of increased inotropic state (Figure 8). Indirect actions of the drug (withdrawal of sympathetic stimulation along with improving cardiac function) might have contributed to the observed reduction in HR and TSR. On the average, maximal effects were reached after 90 minutes. In no case ventricular irritability was induced or electrocardiographic signs of ischemia worsened. Furthermore, the effects of increased contractility on myocardial oxygen demand might have been minimized by the reduction in preload and HR.[22,23] Despite these favourable reports, several limitations to the use of Digoxin in AMI still exist: the drug does not have a rapid onset of action (30 minutes at least); it is less effective as the magnitude of ventricular failure increases; it is scarcely useful in hypotensive states, since generally it has little effect on MAP and it has small hemodynamic effects, as compared with results obtained with newer isotropic agents (Dobutamine).[24]

Dopamine is the agent of choice in cases with depressed cardiac output and systemic hypotension.[25] Its hemodynamic effects are dose-related.[26] Six young patients in our series, who received i.v. Dopamine at an average rate of 4 mcg/Kg/min, showed increases in cardiac output and SI, associated with slightly decreased MAP and TSR due to the prevalence of direct, non-adrenergic vasodilatation in the mesenteric, renal, coronary and cerebral arterial districts, and to the skeletal muscle vasoconstriction that the drug causes by alpha-adrenergic stimulation (Figure 9). In 5 elderly patients, in whom the dose averaged 14 mcg/Kg/min for the presence of marked systemic hypotension, MAP was increased by the

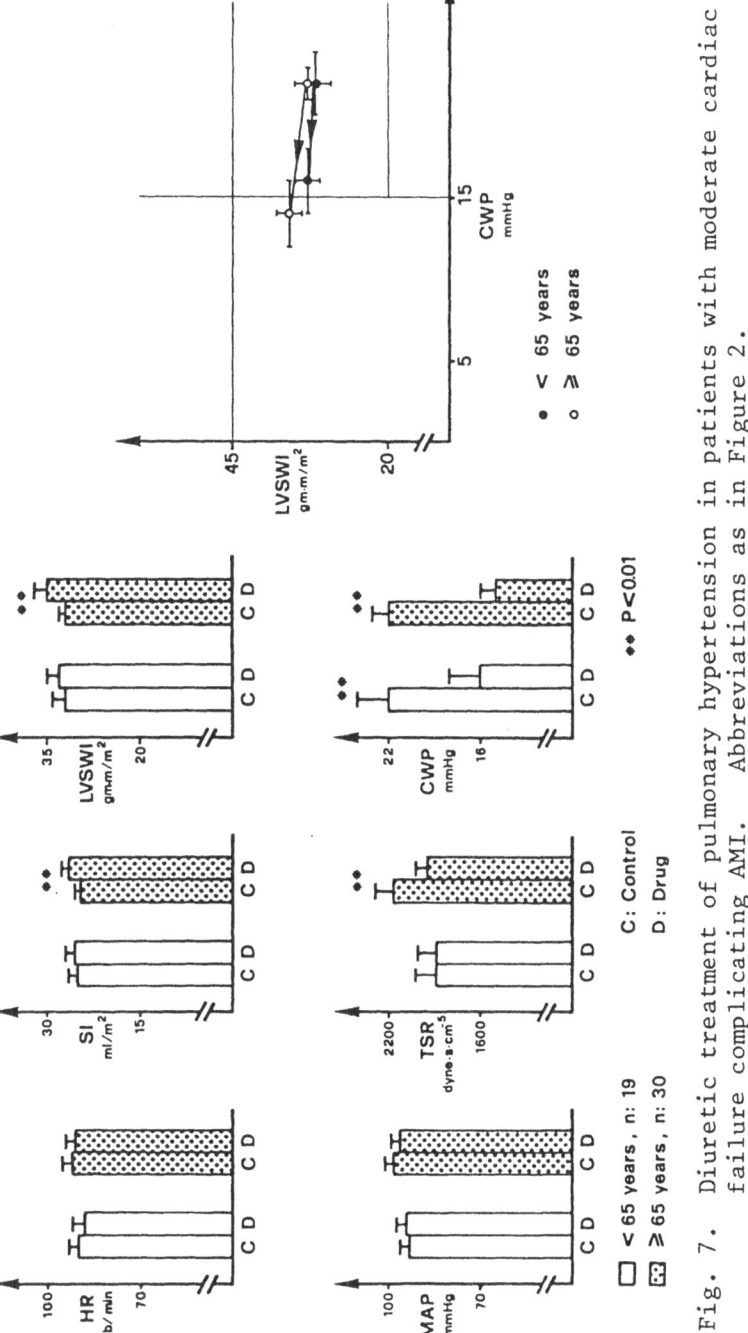

Fig. 7. Diuretic treatment of pulmonary hypertension in patients with moderate cardiac failure complicating AMI. Abbreviations as in Figure 2.

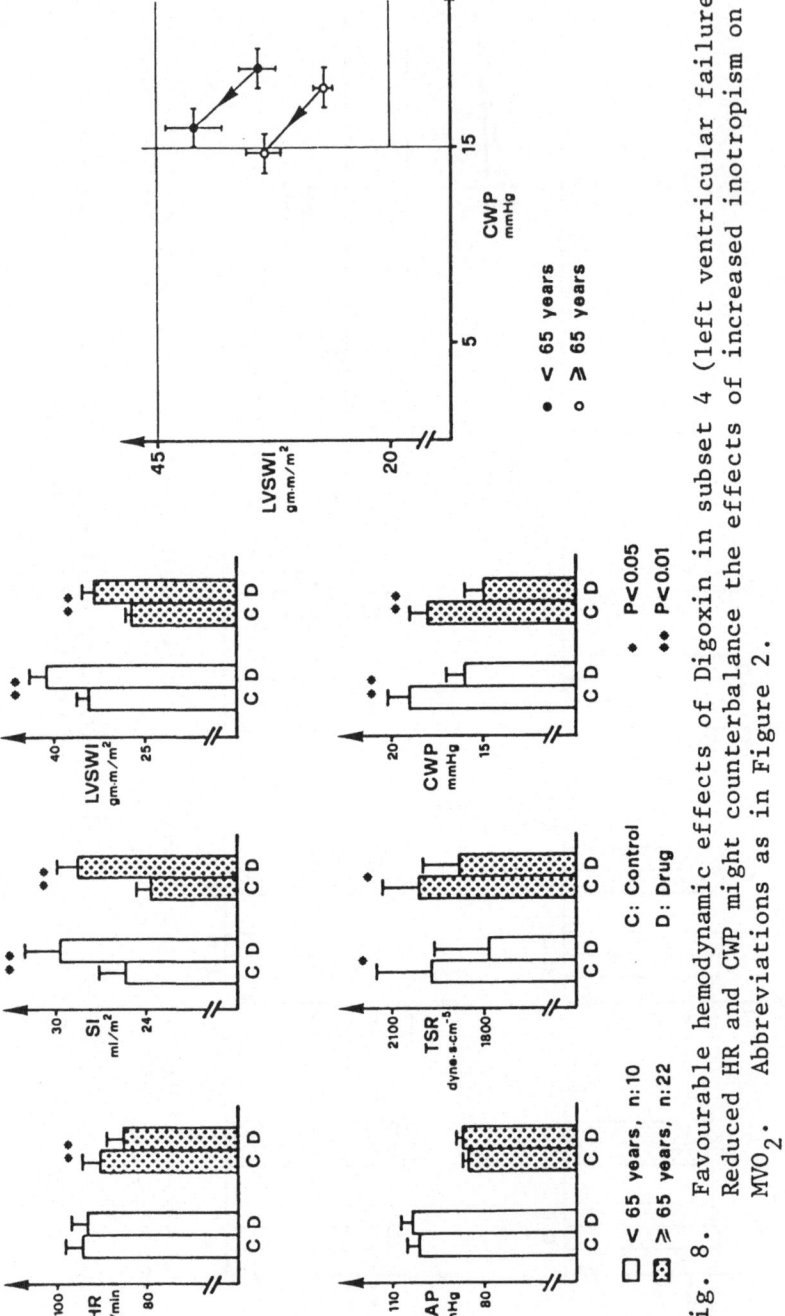

Fig. 8. Favourable hemodynamic effects of Digoxin in subset 4 (left ventricular failure).
Reduced HR and CWP might counterbalance the effects of increased inotropism on
MVO_2. Abbreviations as in Figure 2.

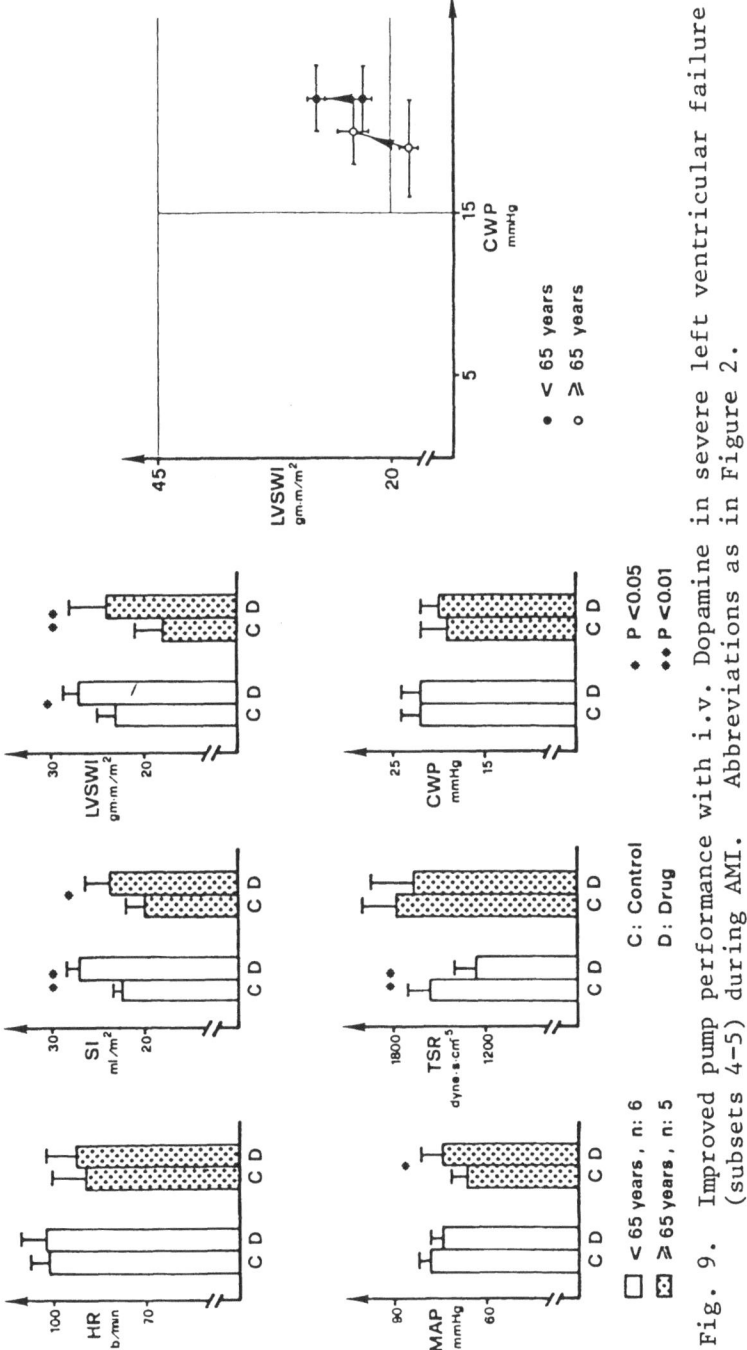

Fig. 9. Improved pump performance with i.v. Dopamine in severe left ventricular failure (subsets 4-5) during AMI. Abbreviations as in Figure 2.

enhanced cardiac output, while TSR remained constant. In no case
did Dopamine substantially reduce CWP, which was even increased in
individual patients; combined diuretic and venodilating therapy is
mandatory in patients with severe pulmonary hypertension.

Sublingual nitrates cause isolated venodilation,[27] therefore
they cannot improve left ventricular ejection fraction and SI.
Hemodynamic effects of s.l. Isosorbide dinitrate (5 mg) are shown
in figure 10: CWP dropped to normal values in 29 patients with
left ventricular failure, but depressed SI and LVSWI were not
changed.

When nitrates are administered intravenously, they possess a
"balanced" vasodilating effect, thereby reducing both pre- and
afterload, with resultant normalization of left ventricular filling
pressure and concomitant improvement in SI due to the lowered
outflow impedence (Figure 11). MAP is relatively unaffected.
The response to i.v. Nitroglycerin was similar in the two age
groups with the same average dose. In no case did we observe
reflex tachycardia or other untoward reactions.

Arterial vasodilators, such as Hydralazine, are useful when
peripheral hypoperfusion (reduced SI and extremely elevated TSR) is
the dominant hemodynamic abnormality, in the absence of marked
pulmonary congestion. These agents, in fact, improve cardiac
output and SI by lowering afterload, but they have little or no
effect on pulmonary artery pressures (Figure 12).[29,30]   Reflex
tachycardia, which was seen as a deleterious side effect mainly in
hypovolemic patients,[31] was not evident in our series. A major
limitation to the use of oral Hydralazine in the intensive care
setting is the slow onset of action (60 minutes).

Subset 5 - Cardiogenic shock (10% of all cases of AMI. CWP >15 mmHg;
LVSWI <20 gm.m/m^2). Whatever the pharmacological treatment might
have been, prognosis in this subset was constantly poor in the past
(mortality rate 98-100%).[4,32-35]   We reported the same mortality
rate also in subset 4 patients who failed to improve their hemodynamic conditions within one hour from the initiation of maximal
pharmacological therapy with combined inotropes, vasodilators and
diuretics (refractory heart failure).

We consider cardiogenic shock (CS) and refractory left ventricular failure (LVF) as absolute indications for mechanical cardiocirculatory assistance with intra-aortic balloon pumping (IABP).[36,37]
IABP is achieved with insertion of a 40 ml balloon, which is
mounted on a catheter, via a common femoral artery into the
thoracic aorta, just distally to the origin of the left subclavian
artery (Figure 13). The balloon is rapidly inflated in the course
of diastole, at the onset of aortic valve closure, thus increasing
diastolic aortic pressure and coronary blood flow. Deflation takes

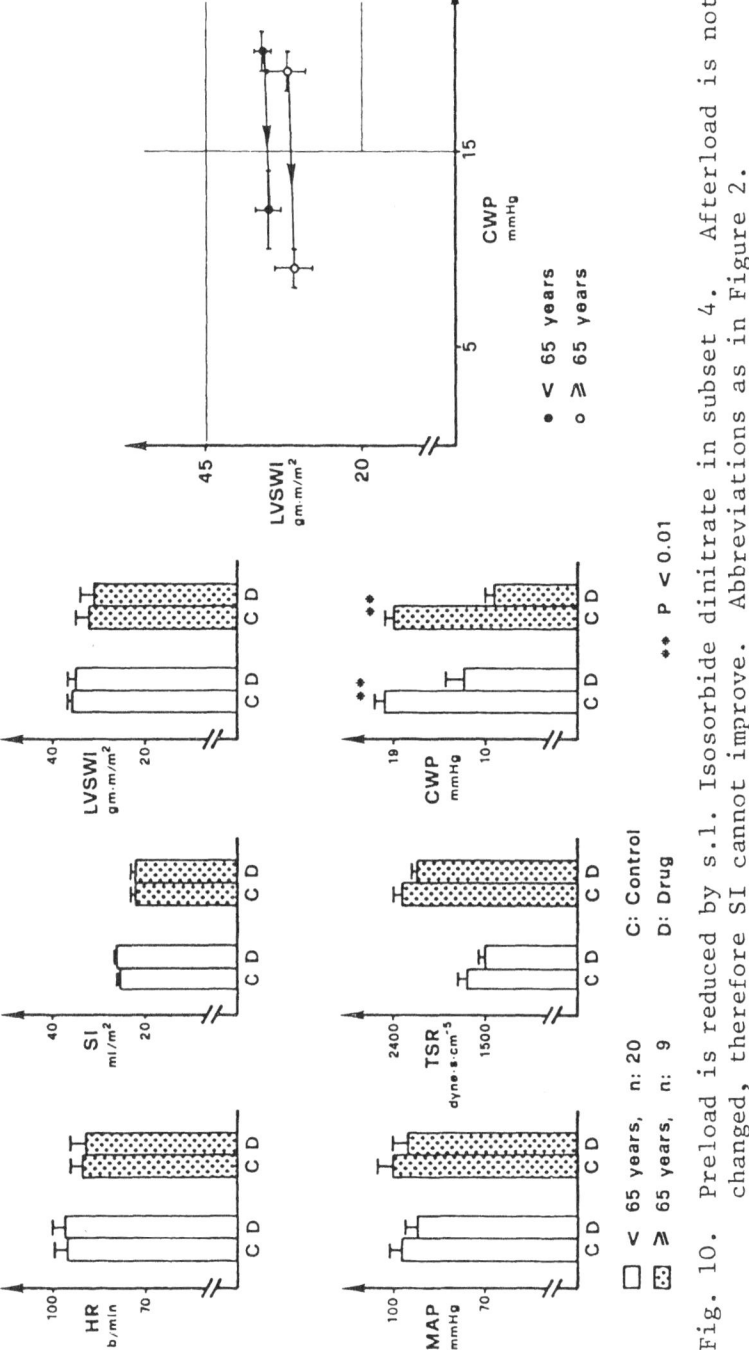

Fig. 10. Preload is reduced by s.l. Isosorbide dinitrate in subset 4. Afterload is not changed, therefore SI cannot improve. Abbreviations as in Figure 2.

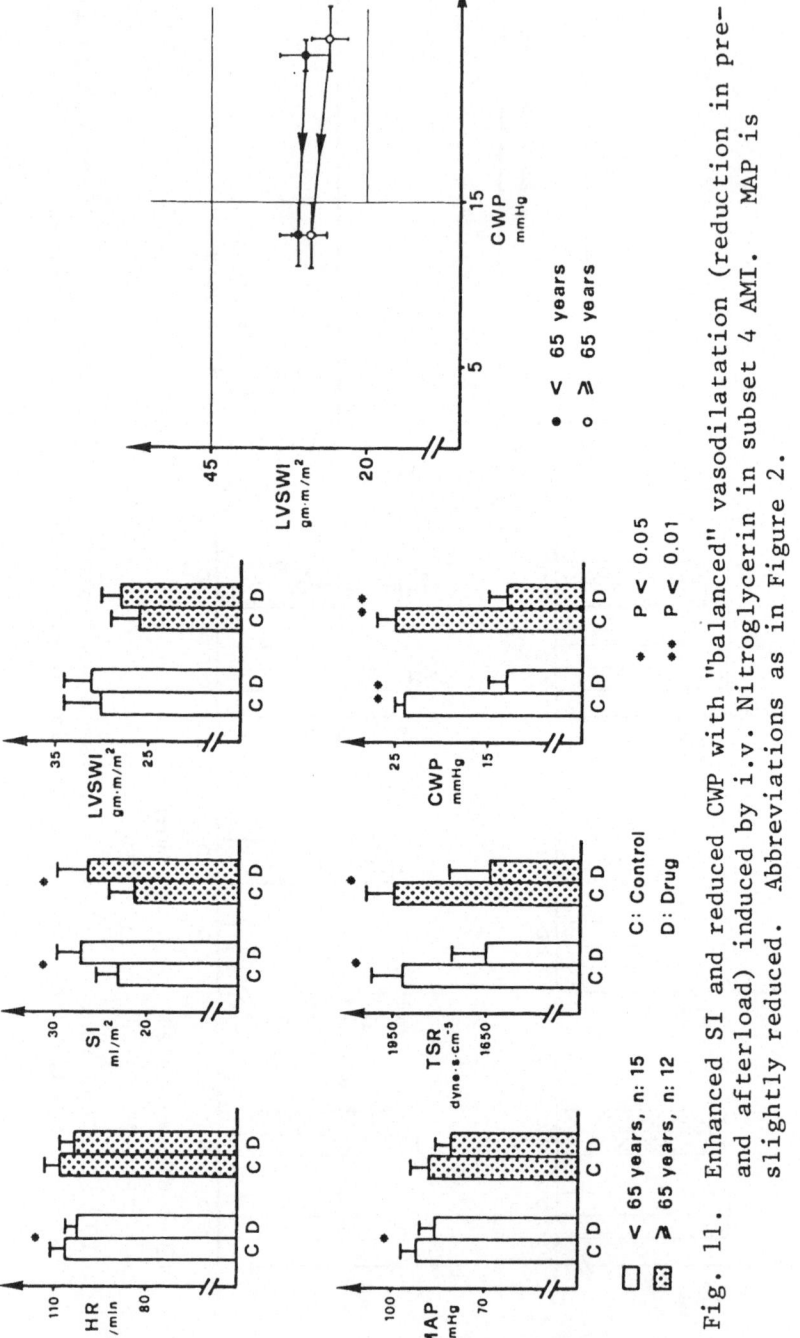

Fig. 11. Enhanced SI and reduced CWP with "balanced" vasodilatation (reduction in pre- and afterload) induced by i.v. Nitroglycerin in subset 4 AMI. MAP is slightly reduced. Abbreviations as in Figure 2.

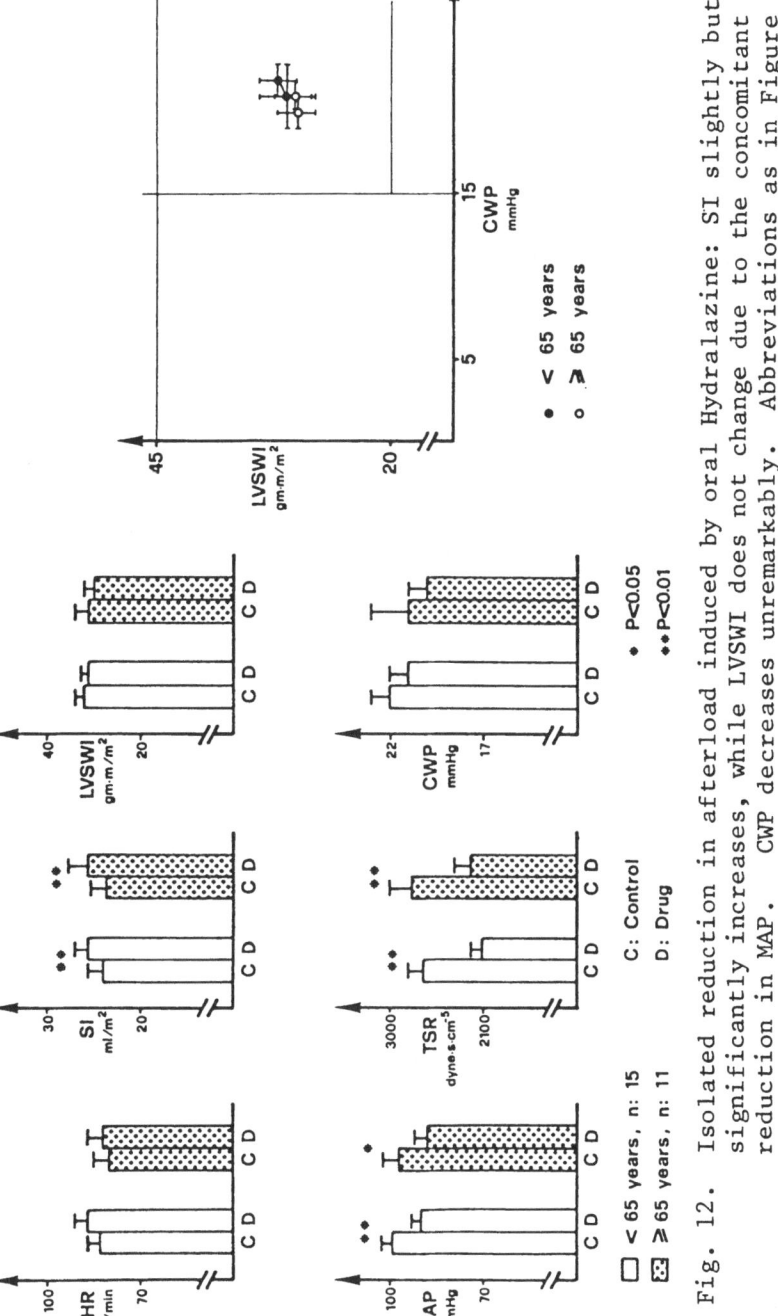

Fig. 12. Isolated reduction in afterload induced by oral Hydralazine: SI slightly but significantly increases, while LVSWI does not change due to the concomitant reduction in MAP. CWP decreases unremarkably. Abbreviations as in Figure 2.

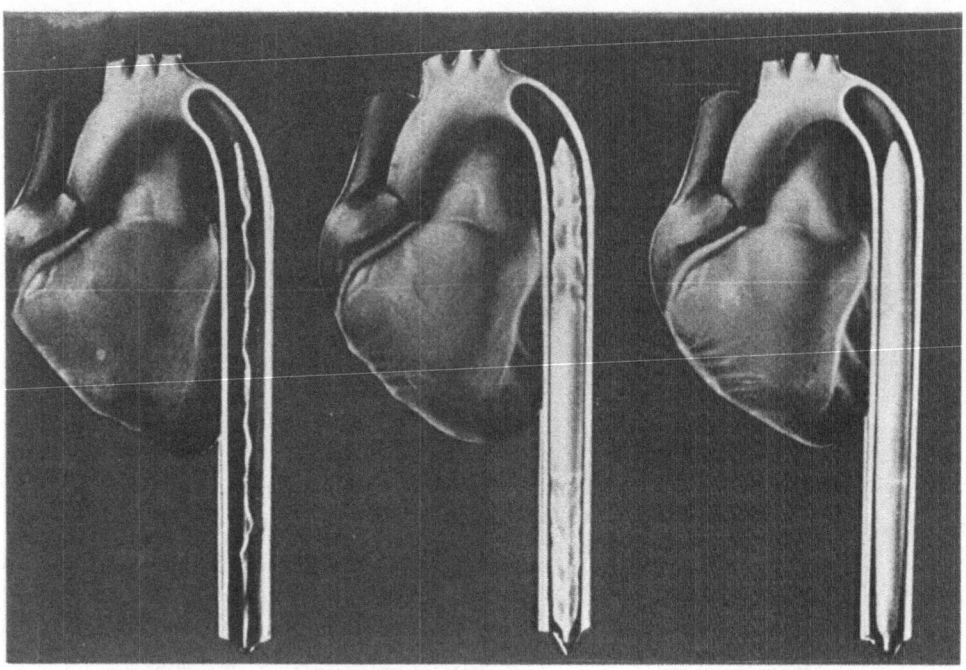

Fig. 13. From left to right: intra-aortic balloon deflated, half
          inflated and fully inflated in the thoracic aorta.

place just prior to left ventricular ejection, thus decreasing pre-
systolic aortic pressure and left ventricular afterload.   Left
ventricular systolic tension is also markedly reduced by balloon
deflation (Figure 14).[38]   The final hemodynamic effects of IABP
consist of an improvement in the supply/demand ratio of oxygen to
the heart, coupled with an increase in ejection fraction and a
decrease in left ventricular filling pressure.   Originally, the
IABP catheter was introduced after surgical exposure of a common
femoral artery and arteriotomy;   in 1980,   a percutaneous
catheter was designed, which represented a considerable advance
for the clinical application of the technique.

     From January 1977 to Deceember 1980 we employed the surgical
IABP catheter to support 38 patients with AMI complicated by CS
(n=28) or refractory LVF (n=10).   Twenty-nine patients were younger
and 9 older than 65 years.   A significant improvement of almost
all hemodynamic parameters was observed from the early phase of
pumping (within 4 hours from onset of assistance) (Figure 15).
Such an improvement was unrelated to age: in particular SI, which
is known to bear prognostic significance,[37,41] was enhanced by the
same extent in young and elderly patients (Figure 16).

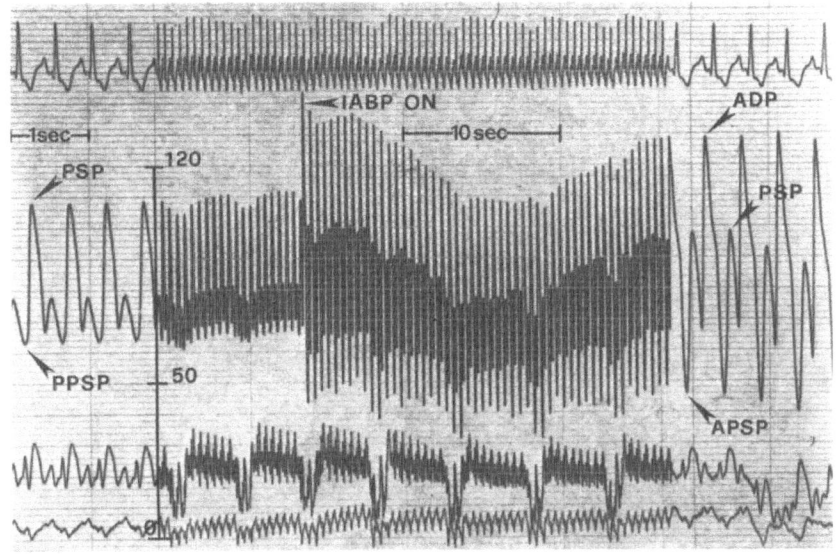

Fig. 14.   Continuous strip-chart recording of (from top
           to bottom) electrocardiogram, systemic arterial
           pressure, pulmonary arterial pressure, right
           arterial pressure, showing the effects of IABP.
           PSP/PPSP= patient's systolic/pre-systolic
           pressure; ADP/APSP= assisted diastolic (aortic
           peak)/pre-systolic pressure.

In accordance with the favourable hemodynamic response,
mortality rate was reduced from the expected 100 per cent to 65.8
per cent, without any significant difference between the two age
groups (<65 years: 65.5% vs ≥65 years: 66.7% N.S.).

The IABP catheter could not be inserted in 10 other patients
(10/48, 20.8%), because of the presence of obstructive atherosclerotic lesions in the femoro-wliac arteries.   The incidence of
failure to introduce the catheter was age-related (Table 2).   This
seemed to be the only factor which could limit the applicability
of IABP to geriatric patients, since the rate of major complications
deriving from IABP did not differ between the younger and the
older group (Table 3).

The difficulties in introducing the IABP catheter over the age
of 65 can be overcome by the use of the percutaneous technique,
which we employed in 28 out of 36 patients in a subsequent series,
from January 1981.   Percutaneous IABP caused a significant decrease
in the rate of failures over the age of 65 (Table 4).

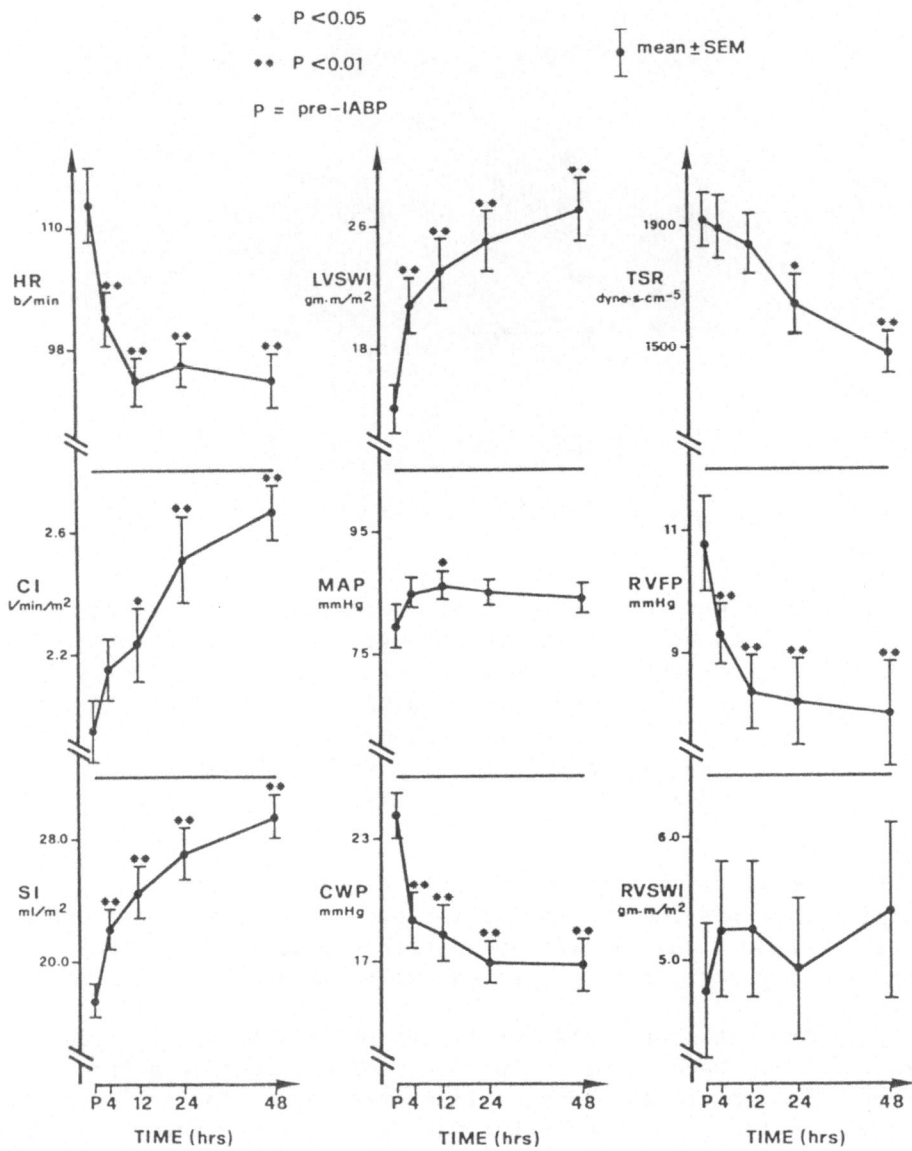

Fig. 15. Hemodynamic effects of IABP assessed after 4, 12, 24
and 48 hours in patients who underwent cardiocirculatory mechanical support for cardiogenic shock or
refractory left ventricular failure during AMI.
CI= cardiac index; RVFP= right atrial mean pressure;
RVSWI= right ventricular stroke-work index.   Other
abbreviations as in Figure 2.

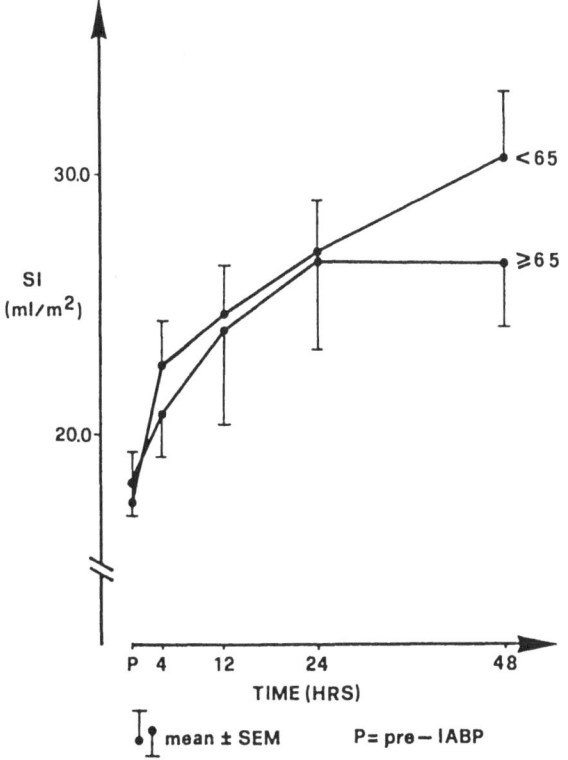

Fig. 16. SI after IABP increases by
the same extent in young and
elderly patients.

Table 2. Failures of insertion of an IABP
catheter in patients under and over
the age of 65

	n	FAILURES	P <
< 65 YEARS	31	2(6.5%)	
			0.01
≥ 65 YEARS	17	8(47.1%)	
TOTAL	48	10(20.8%)	

Table 3. Major complications after IABP insertion

COMPLICATION	AGE< 65 YEARS	AGE≥65 YEARS	P	TOTAL
LEG ISCHEMIA	3	1		4
WOUND BLEEDING	1	—		1
WOUND INFECTION	1	—		1
TOTAL	5 (16.1%)	1 (5.9%)	N S	6 (12.5%)

In conclusion, we have demonstrated that the acute hemodynamic responses to different therapeutic interventions during AMI do not substantially differ between young and elderly patients. In general terms, a more cautious use of antihypertensive drugs and vasodilators is recommended in the elderly, who are more prone to develop hazardous untoward reactions on account of the failure of vasomotor feed-back mechanisms. Usually these iatrogenic derangements can be rapidly and effectively corrected, when continuous hemodynamic surveillance is available, by plasma-volume expanders and vasopressors.

On the basis of hemodynamic assessment, patients at extremely high risk are identified. These patients profit from aggressive, mechanical cardiocirculatory support, whose application proved to be beneficial at any age.

In our experience, the treatment of AMI based on hemodynamic subset classification and monitoring improved significantly the prognosis in the elderly. General mortality, in fact, was reduced after the introduction of hemodynamic facilities in our Intensive Care Unit (Table 5), and the reduction occurred mainly in the group of patients older than 65 years.

Table 4. Improved success rate of IABP insertion
with percutaneous catheter over the age
of 65

	n	FAILURES	P<
SURGICAL	22	9 (40.9%)	
			0.05
PERCUTANEOUS	17	2 (11.8%)	
TOTAL	39	11 (28.2%)	

Table 5. Mortality in acute myocardial infarction before
(1969-1975) and after (1976-1980) the availability of hemodynamic monitoring in the Geriatric
ICU

AGE (YEARS)	1 JAN 69 - 31 DEC 75			1 JAN 76 - 31 MAY 80			
	CASES n	DEATHS n	%	CASES n	DEATHS n	%	P <
< 65	656	132	20.1	475	74	15.6	N.S.
≥ 65	316	149	47.2	229	86	37.6	0.05
TOTAL	972	281	28.9	704	160	22.7	0.01

REFERENCES

1. N. Marchionni, R. Pini, A. Vannucci, Intensive care for the
    elderly with acute myocardial infarction, J.Clin.Exp.Geront.
    3:47 (1981).
2. K. Chatterjee, H. J. C. Swan, Hemodynamic profile of acute
    myocardial infarction, in: "Myocardial Infarction," E. Corday,
    H. J. C. Swan, eds., Williams & Wilkins, Baltimore (1973).
3. J. S. Forrester, G. A. Diamond, Hemodynamic subsets in acute
    infarction: a basis for prognosis and therapy, in: "Clinical
    Strategies in Ischemic Heart Disease," E. Corday, H. J. C.
    Swan, eds., William & Wilkins Co., Baltimore (1979).
4. G. Bertini, N. Marchionni, R. Pini, A. Vannucci, F. M. Antonini,
    Significance of prognostic indexes derived from hemodynamic
    monitoring in acute myocardial infarction, in: "Coronary
    Care Units," A. Maseri, C. Marchesi, S. Chierchia, M. G.
    Trivella, eds., Martinus Nijhoff, The Hague (1981).
5. C. Crexelles, K. Chatterjee, J. S. Forrester, K. Dikshit,
    H. J. C. Swan, Optimal level of filling pressure in the
    left side of the heart in acute myocardial infarction,
    New Engl.J.Med. 289:1263 (1973).
6. H. L. Greene, D. T. Kelly, D. R. Taylor, B. Pitt, Hemodynamic
    effects of plasma volume expansion and prognostic implicat-
    ions in acute myocardial infarction, Circulation 49:106 (1974).
7. W. E. Shell, B. E. Sobel, Protection of jeopardized ischemic
    myocardium by reduction of ventricular afterload, New Engl.
    J.Med. 291:481 (1974).
8. L. Gould, M. Zahir, S. Ettinger, Phentolamine and cardiovascular
    performance, Brit.Heart J. 31:154 (1969).
9. C. Fumagalli, G. Bertini, N. Marchionni, R. Pini, A. Vannucci,
    The use of Phentolamine in the therapy of acute myocardial
    infarction with cardiac failure, Boll.Soc.Ital.Cardiol.
    657:63 (1974).

10. L. Gould, C. V. R. Reddy, T. Kalanithi, L. Espina, R. F.
     Gomprecht, Use of Phentolamine in acute myocardial
     infarction, Am.Heart J. 88:144 (1974).
11. C. Perret, J. P. Gardaz, M. Reynaert, F. Grinbert, J. F.
     Enrico, Phentolamine for vasodilator therapy in left
     ventricular failure complicating acute myocardial
     infarction, Brit.Heart J.37:640 (1975).
12. K. Dikshit, J. K. Vyden, J. S. Forrester, Renal and extra-
     renal hemodynamic effects of Furosemide in congestive
     heart failure after acute myocardial infarction,
     New Eng.J.Med. 288:1087 (1973).
13. J. Kiely, D. T. Kelly, D. R. Taylor, The role of Furosemide in
     the treatment of left ventricular dysfunction associated
     with acute myocardial infarction, Circulation 48:581 (1973).
14. H. Mond, D. Hunt, G. Gloman, Hemodynamic effect of Furosemide
     in patients suspected of having acute myocardial infarction,
     Brit.Heart J. 36:44 (1974).
15. M. Lesch, Inotropic agents and infarct size, Am.J.Cardiol.
     37:508 (1976).
16. Y. Varonkov, W. E. Shell, V. Smirnov, D. Gukovsky, E. Chaxov,
     Augmentation of serum CPK activity by digitalis in
     patients with acute myocardial infarction, Circulation
     55:719 (1977).
17. J. M. Askey, Digitalis in acute myocardial infarction, J.Amer
     Med.Assoc 146:1008 (1951).
18. N. H. Boyer, Digitalis in acute myocardial infarction,
     New Eng.J.Med. 252:536 (1955).
19. R. Bine, The treatment of heart failure and the use of digit-
     alis in acute myocardial infarction, Am.J.Cardiol. 1:250
     (1958).
20. S. H. Rahimtoola, M. Z. Sinno, R. Chuquimia, H. S. Loeb,
     K. M. Rosen, R. M. Gunnar, Effect of Ouabain on impaired
     left ventricular function in acute myocardial infarction,
     New Eng.J.Med. 287:527 (1972).
21. N. Marchionni, A. Vannucci, R. Pini, Hemodynamic effects of
     Digoxin in acute myocardial infarction, Eur.Heart J. 1:319
     (1980).
22. J. W. Covell, E. Braunwald, J. Ross Jr., E. H. Sonnenblick,
     Studies on digitalis, XVI. Effects on mycardial oxygen
     consumption, J.Clin.Invest. 45:1535 (1966).
23. T. Watanabe, J. W. Covell, P. R. Maroko, E. Braunwald,
     J. Ross, Jr., Effects of increased arterial pressure and
     positive inotropic agents on the severity of myocardial
     ischemia in the acutely depressed heart, Am.J.Cardiol.
     30:371 (1972).
24. R. A. Goldstein, E. R. Passamani, R. A. Roberts, A comparison
     of Digoxin and Dobutamine in patients with acute infarction
     and cardiac failure, New Eng.J.Med. 303:846 (1980).

25. L. I. Goldberg, Dopamine. Clinical use of an endogenous cathecolamine, New Eng.J.Med. 291:707 (1974).

26. D. T. Mason, E. A. Amsterdam, A. N. DeMaria, G. Lee, J. A. Joyl, J. M. Foerster, Clinical reappraisal of therapy for congestive heart failure and cardiogenic shock due to acute myocardial infarction, in: "Clinical Strategies in Ischaemic Heart Disease," E. Corday, H.J.C. Swan, eds., Williams and Wilkins Co., Baltimore (1979).

27. R. R. Miller, L. A. Vismara, D. O. Williams, Pharmacological mechanism for left ventricular unloading in clinical congestive heart failure: differential effects of Nitro-prusside, Phentolamine and Nitroglycerin on cardiac function and peripheral circulation, Circ.Res. 39:127 (1976).

28. R. R. Miller, D. O. Williams, A. DeMaria, E. A. Amsterdam, D. T. Mason, Ventricular afterload-reducing agents in congestive heart failure therapy, in: "Congestive Heart Failure," D. T. Mason, ed., Yorks Medical Books, New York (1976).

29. K. Chatterjee, W. W. Parmley, B. Massie, Oral Hydralazine therapy for chronic refractory heart failure, Circulation 54:879 (1976).

30. J. S. Franciosa, G. Pierpont, J. N. Cohn, Hemodynamic improvement after oral Hydralazine in left ventricular failure, Ann.Intern.Med.86:388 (1977).

31. J. Koch-Waser, Hydralazine, New Eng.J.Med. 295:320 (1976).

32. C. Friedberg, Cardiogenic shock in acute myocardial infarction, Circulation 23:325 (1961).

33. S. Scheidt, R. Ascheim, T. Killip, Shock after myocardial infarction. A clinical and hemodynamic profile, Am.J.Cardiol 26:556 (1970).

34. R. A. Ratshin, F. Harrel, D. J. Field, A prognostic index for myocardial infarction from analysis of acute phase hemo-dynamic data, Circulation (Supp.II) 43:214 (1971).

35. K. Chatterjee, H. J. C. Swan, W. S. Kaushik, Effects of vaso-dilator therapy for severe pump failure in acute myocardial infarction on short-term and late prognosis, Circulation 53:797 (1976).

36. N. Marchionni, C. Vassanelli, G. Menegatti, Effectiveness of intraaortic balloon counterpulsation in the elderly, Eur.Heart J. in press (1982).

37. N. Marchionni, C. Vassanelli, G. Menegatti, Management of refractory heart failure complicating acute myocardial infarction in the elderly: effectiveness of intraaortic balloon pumping without cardiac surgery, J.Clin.Exp.Gerontol. in press (1982).

38. H. Bolooki, Physiologic principles of circulatory assist. in: "Clinical Application of Intra-aortic Balloon Pump," H. Bolooki, ed., Futura Publishing Co. New York (1977).

39. H. Bolooki, Methods of insertion of intra-aortic balloon,in: "Clinical Application of Intra-aortic Balloon Pump," H. Bolooki, ed., Futura Publishing Co. New York (1977).

40. D. Bregman, W. J. Casarella, Percutaneous intra-aortic balloon
        pumping: initial clinical experience, Ann.Thor.Surg.
        29:153 (1980)
41. H. Bolooki, Cardiogenic shock, in: "Clinical Application of
        Intra-aortic Balloon Pump," H. Bolooki, ed., Futura
        Publishing Co. New York (1977).

CARDIAC REHABILITATION IN THE ELDERLY

M. Impallomeni

Consultant Physician in Geriatrics and
Honorary Senior Lecturer in Medicine
Royal Postgraduate Medical School
Hammersmith Hospital, London

Heart disease is a major cause of morbidity and mortality in the developed countries and its prevalence progressively increases in old age. The W.H.O. Statistics Report (W.H.O.)[1] shows that worldwide cardiac diseases are the most common cause of death over the age of 65. Kennedy, Andrews and Caird[2] found that in a random sample of people aged 65 and over living at home 40 per cent had evidence of heart disease. This proportion increased to 50 per cent over the age of 75.

Great advances have been made in the treatment of heart diseases over the past two decades, leading to a greater number of patients surviving acute episodes. This has brought about a reversal of attitudes in the medical profession with regard to the management of the long term effects of heart disease, resulting in greater attention being paid to the rehabilitation of these patients.

The W.H.O.[3] has defined cardiac rehabilitation as "the sum total of activities necessary to ensure the best possible physical, mental and social conditions so that patients may by their own efforts regain as normal as possible a place in the community and lead an active and productive life".

Although many types of heart disease may affect the older patient, ischaemic heart disease is by far the most common. Kennedy et al found it in 22 per cent of the random sample of old people living at home.

This paper will mainly deal with the rehabilitation of the elderly patient suffering from this disease, although the same principles of rehabilitation are applicable to most other forms of cardiac diseases.

Ischaemic heart disease (ICD 410/414)[4] is not only more prevalent in old age, but also has a higher mortality as Figure 1 shows. Its mortality approximately doubles every ten years over the age of 65.

Figure 2 shows mortality from ischaemic heart disease as a percentage of all deaths in various age groups in England and Wales in 1980. It is the single most important cause of death in the elderly in the United Kingdom.

A Joint Working Party of the Royal College of Physicians of London and of the British Cardiac Society[5] reported on a survey of cardiac rehabilitation facilities in Britain. They found that formal rehabilitation facilities for heart patients were found in only a small proportion of British hospitals. There is no reason to believe that this situation has improved.

Until 30 years ago there seems to be relatively little need for rehabilitation following a heart attack. The patients were kept on absolute bed rest for six weeks, as pathological studies had shown that this was the time necessary to convert necrotic heart muscle into scar tissue and it was feared that exercise would increase the risk of cardiac rupture and arrhythmias.

In 1952 Levine and Lown[6] showed that sitting in a chair seven days after onset was not dangerous and had physical and psychological advantages. In the last decade it has been accepted that the heart muscle has a remarkable capacity to recover and that even earlier mobilisation is indicated. The dangers of bed rest, such as deep vein thrombosis and pulmonary embolism, postural hypotension, wasting of muscles and physical deconditioning, have been found to be so great that this is now limited to those patients with severe myocardial infarction with evidence of shock, cardiac failure, arrhythmias or persistent pain.

Grading of myocardial infarction into various categories of severity was introduced about 20 years ago. The Peel Coronary Prognostic Index[7] is still widely used. It takes account of age, previous history of heart disease, presence or absence of shock, cardiac failure, arrhythmias and E.C.G. changes. This index is effective for both the short term and long term prognosis.

The Norris Index[8] requires initial radiological assessment and is therefore hospital based. He and his group also used a separate long term coronary prognostic index.[9] These two indices are perhaps more precise than the Peel Index but cannot be applied to the patient at home.

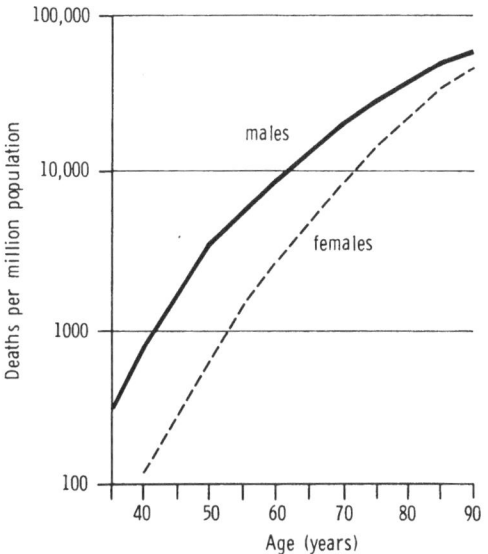

Fig. 1.   Ischaemic heart disease
England & Wales, 1980;
by permission of OHE

A rehabilitation index based not only on the severity of the myocardial infarction, but also on psychological and social factors was introduced by Schiller[10] in Australia in 1972. The success of rehabilitation, however, was measured as return to work, making this index of difficult application in old age.   The Joint Working Party referred to above introduced a simplified grading of severity of myocardial infarction at 7 - 14 days for purposes of rehabilitation (see Table 1).

Several papers have shown that approximately 45 per cent of all survivors from myocardial infarction are of the uncomplicated Grade I type.   Semple and Williams have shown that over the age of 70 this proportion is about 30 per cent, Grade II 50 per cent and Grade III 20 per cent.

STATE OF CARDIAC REHABILITATION IN OTHER COUNTRIES

Questionnaire surveys on methods of cardiac rehabilitation have shown that generally speaking British and American physicians believe that most patients given proper advice and using simple techniques rehabilitate themselves satisfactorily.   Continental Europeans tended to favour more elaborate programmes, often using formal rehabilitation institutes. The above Working Party visited facilities in many countries and found profoundly varying approaches, commenting that enthusiasm for physical conditioning programmes

Table 1. Grading of severity of myocardial
          infarction at 7-14 days for purposes
          of rehabilitation

Grade 1: Mild, uncomplicated; absence of the
         features of grades 2 and 3

Grade 2: Temporarily complicated; absence of
         the features in grade 3 but any of
         the following, even if temporary:

         Sinus tachycardia (over 100/m')
          at rest lasting longer than one
          hour but less than 48 hours
         Dyspnoea during ordinary activity
         Temporary abnormal cardiac impulse
          (dyskinesia)
         Moist sounds persisting after
          coughing, or pulmonary venous
          engorgement on X-ray

Grade 3: Presence at any time of one of the
         following:

         Sinus tachycardia at rest, persisting
          longer than 48 hours
         Arrhythmias still present at time of
          grading
         Dyspnoea at rest
         Alveolar or interstitial pulmonary
          oedema on X-ray
         Third heart sound
         Continuing palpable dyskinesia or
          ventricular aneurysm
         Definite cardiac enlargement
         Persisting heart block, left bundle-
          branch block or bifascicular block

(5) Assistance with social problems.

    Patients with uncomplicated myocardial infarctions can be
mobilised by day 2, allowed to walk on the level by day 7, and
usually allowed home by day 15. Patients with Grade II or
temporarily complicated infarctions need adequate control of their
symptoms before mobilisation can proceed at a slower pace. For
Grade III myocardial infarctions no precise rules can be set, each
patient presenting special problems for the clinician.

    Before discharge from hospital both the patient and his/her
spouse should receive information about ischaemic heart disease.

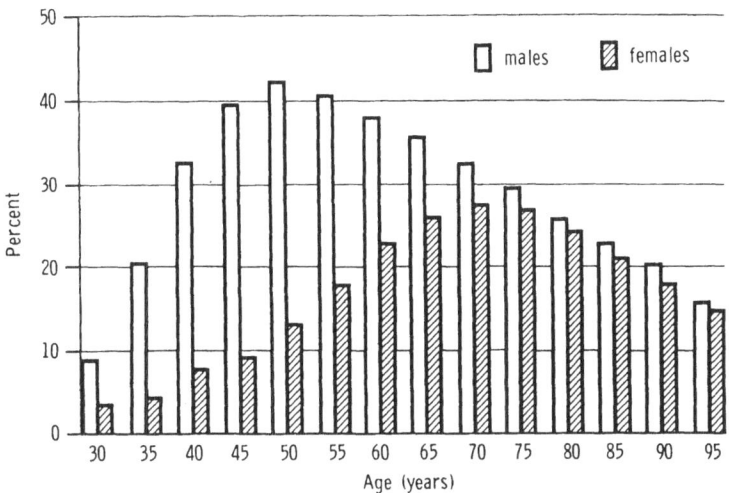

Fig. 2.   Ischaemic heart disease England and Wales,
          1980; by permission of OHE

remains patchy and confined to a minority of doctors and patients.
In particular in West Germany and Eastern Europe Sanatoria and
Spas have been converted to rehabilitation hospitals.   Some are run
on Spa lines including hydrotherapy and skin brushings.   In Israel
special institutes with full time specialists in cardiac rehabilitation have been created.   In Russia there are sanatoria outside
cities where progressive physical rehabilitation lasts one to three
months.

The W.H.O. has tried to set up a multi-centre control study on
the effect of rehabilitation on patients suffering from ischaemic
heart disease without reaching conclusive results.

Rehabilitation of myocardial infarction patients is an essential
part of its management and should start as soon as possible after
the patient's admission to hospital.   It should start together with
the medical treatment of the acute episode and should be comprehensive including all physical, psychological and social problems that
may affect the patient.   The skills of the physician may be
inadequate for such a task, hence the need for a multi-disciplinary
team approach.

Such a rehabilitation programme should include:-

(1)   Adequate treatment of medical problems.
(2)   A programme of graduated increasing physical activity.
(3)   Psychological support to promote adjustment.
(4)   Education about ischaemic heart disease, its course and
      prognosis.

In many hospitals this is done with the aid of booklets and discussion of its contents with patients and their relatives.   Patients
are encouraged to ask questions of the rehabilitation team, and
discuss their problems and worries with doctors, nurses and other
paramedical staff.

Several studies have shown that these steps may reduce the
incidence of late psychological problems complicating myocardial
infarctions.   Psychological problems are common after a heart
attack.   The initial trauma is often a terrifying experience.   In
the first few days anxiety is common, depression occurs a little
later.   If this stage is inadequately managed the patients can
remain maladjusted for a long period of time.   Hellerstein[11]
concluded that anxious doctors produce anxious patients.

The personality of the patient is equally important.   Those
who have coped well with problems during their adult life usually
adjust successfully to the consequences of myocardial infarction.
Cay[12] showed that patients with social problems had a higher
prevalence of anxiety and depression than patients without them.

THE ROLE OF PHYSICAL CONDITIONING IN REHABILITATION AFTER
MYOCARDIAL INFARCTION

A number of studies have shown that cardiac patients may
respond to physical conditioning in the same way and to the same
degree as healthy people of similar age.   Physical conditioning
mainly acts through training the oxygen transport system and its
benefits in cardiac patients are mainly thought to be a consequence
of its effect on the peripheral circulation and muscles rather than
the heart.

The capacity of the oxygen transport system can be studied by
measuring the maximum oxygen uptake (VO2 max) also called maximum
aerobic power, aerobic capacity and physical working capacity.   It
is a reflection of the maximal expenditure of energy of which that
person is capable.   This varies greatly among healthy people.
Its peak is reached at about age 20 and is 30 - 65 mls of O_2 per kg
of body weight per minute for men and 25 - 50 mls of O_2 per kg of
body weight per minute for women.   It is much higher in athletes.
It declines in value after the age of 30 so that by the age of 55
most men and women have lost one-third to one-quarter and by the
age of 80 about half.

Response to physical conditioning depends on the total amount
of work performed, i.e. its intensity, duration and frequency.
The capacity of the body to utilise oxygen maximally may increase
by 7 - 30 per cent through physical training depending on the
initial value.   The physiological stress in sub-maximal work at

Table 2. Regular physical activity
has been shown to:

increase

Maximum oxygen uptake
Work performance
Maximum cardiac output
Mitochondrial enzyme activity
  in skeletal muscle
Joint mobility and skeletal
  muscle strength

and may increase

Safety of vigorous exertion
Coronary collateral circulation
Physical coordination and
  dexterity

any given intensity is reduced and work that was formerly maximal
can now be carried out with relative ease.   The benefits of physical
conditioning go beyond this: improved mobility, co-ordination, muscle
strength, range of joint movement, weight reduction, together with
improvement of mechanical efficiency, contribute to the overall
sense of wellbeing that it brings about.

   Tables 2 and 3 show that proven and possible effects of physical
training.   Despite this, physical conditioning has not received
unanimous approval in the rehabilitation of the cardiac patient.
This is because although several studies have confirmed its beneficial effects on physical performance, especially in patients with
angina, there has often been a very high drop out rate for hospital
based exercise programmes.   Furthermore, no convincing evidence
that it improves life prognosis for the cardiac patient has yet
been produced.   Exercise regimes are probably of greatest value to
those with psychosomatic symptoms, depression and effort angina.

SECONDARY PREVENTION

   Of all the risk factors for coronary heart disease recognised
of importance in middle age, cigarette smoking is the only one which
also applies in old age.   Modification of the other factors has not
been confirmed to have any beneficial effects in the elderly.

Table 3. Regular physical activity
has been shown to:

decrease

Arterial systolic pressure⎤
Heart rate   │   at any
Ventilation   ├   given O_2
Blood lactate   │   intake
Myocardial oxygen uptake ⎦
Obesity
Convalescent depression

and may decrease

Convalescent postural hypotension
Catecholamine production
Serum triglycerides
Vulnerability to arrhythmias
Blood coagulation
Insomnia

CONCLUSION

Rehabilitation procedures are of great importance in the
management of the elderly patient with heart disease. They must
start as early as possible in the course of the disease and must
include psychological rehabilitation, manipulation of the social
environment, as well as medical treatment and the management of
physical problems.

Physical conditioning, although beneficial in some patients
with heart disease, may be difficult to apply in some old patients
because of the occurrence of multiple pathology, and especially of
diseases involving the locomotor system.

Education of the patient and his/her family about the course
and prognosis of heart disease should be given great relevance
especially before discharge from hospital. Rehabilitation involves
so many different aspects of the patient's life that for its
success it must depend on intervention of a multi-disciplinary team
involving doctors, nurses, social workers, physiotherapists,
occupational therapists and other paramedical staff. Good and
successful rehabilitation is good total patient care.

REFERENCES

1. World Health Organisation, World Health Statistics Report,
        27:563 (1974).
2. R. D. Kennedy, G. R. Andrews, F. I. Caird, Ischaemic heart
        disease in the elderly, Brit.med.J. 39:1121 (1977).
3. Report of the W.H.O. Expert Committee on disability prevention
        and rehabilitation. W.H.O. Technical Report Series 668,
        W.H.O. Geneva (1981).
4. International Classification of Diseases, W.H.O. Geneva (1977).
5. Cardiac Rehabilitation. Report of a Joint Working Party of the
        Royal College of Physicians of London and The British Cardiac
        Society on rehabilitation after cardiac illness,
        J.Roy.Coll.Physns.London 9:281 (1975).
6. S. A. Levine, B. Lown,"Armchair" treatment of acute coronary
        thrombosis, J.Amer.Med.Ass. 148:1365 (1952).
7. A. A. F. Peel, T. Semple, I. Wang, W. M. Lancaster, J. L. G. Dall,
        A coronary prognostic index for grading the severity of
        infarction, Brit.Heart J. 24:745 (1962).
8. R. M. Norris, P. W. Brandt, D. E. Caughey, A. J. Lee, P. J. Scott,
        A new coronary prognostic index, Lancet 1:274 (1969).
9. R. M. Norris, D. E. Caughey, L. W. Deeming, C. J. Mercer,
        P. J. Scott, Coronary prognostic index for predicting survival
        after recovery from acute myocardial infarction, Lancet 2:485
        (1970).
10. E. Schiller, Cardiac rehabilitation: its potential in the early
        prevention of disability after myocardial infarction,
        Med.J.Aust. 2:751 (1972).
11. H. G. Hellerstein, Various types of exercise programmes - a
        discussion in exercise testing and exercise training, in:
        "Coronary Heart Disease," J. P. Naughton and H. K. Hellerstein
        eds., Academic Press, New York and London (1973).
12. E. L. Cay, Psychological aspects of cardiac rehabilitation,
        Hospital Update 8:161 (1982).

PATHOPHYSIOLOGY OF URINARY INCONTINENCE

P. W. Overstall

Consultant in Geriatric Medicine
General Hospital
Hereford

The prevalence of urinary incontinence varies according to the population surveyed and the criteria used. Among subjects over the age of 65 the prevalence rates are between 7 per cent and 11 per cent for men, and 12 per cent and 18 per cent for women.[1-3] Prevalence tends to rise with increasing age, although there is no consistent trend through the age groups. Rates are higher among residents of old people's homes and long-stay hospitals; among multiparous compared with nulliparous women, and in subjects with poor mobility or a history of cerebrovascular disease.

ANATOMY AND PHYSIOLOGY

The bladder is under two nervous system controls: the local, sacral reflex arc through parasympathetic cholinergic nerves, (S2-4) and voluntary inhibition of this reflex arc via descending impulses from the frontal cortex. As the bladder fills, the intravesical pressure normally rises to less than 10 cms. H_2O and the first desire to void occurs when the bladder contains approximately 150-200 mls. Voiding can usually be postponed until convenient, and provided the intraurethral pressure exceeds the intravesical pressure, the person remains continent. The bladder is guarded by the proximal sphincter at the bladder neck, and the distal sphincter which consists of both intrinsic urethral striated muscle and peri-urethral striated muscle. The smooth muscle of the male bladder neck is richly supplied with sympathetic, noradrenergic nerve endings, but in the female, the supply is very poor. The proximal and distal urethral sphincters are stronger and more reliable in the male, but in both sexes, continence will be maintained provided that either the bladder neck or the distal urethral sphincter are compet-

ent. The periurethral muscles are incapable of maintaining
continence if both the urethral mechanisms fail.[4]

The normal stable detrusor is relaxed during bladder filling,
and only contracts when voiding is voluntarily initiated. As the
detrusor contracts the bladder neck opens and voiding begins. The
stream can be interrupted by contraction of the periurethral
striated muscle, and at the same time the detrusor contraction is
voluntarily inhibited with the bladder neck closing a few seconds
later.

INVESTIGATION OF BLADDER DYSFUNCTION IN OLD AGE

The history and physical examination will sometimes indicate
the cause of a patient's incontinence, and in women an algorithmic
method of assessment can reduce the need for cystometry by 60 per
cent.[5] A cystometrogram will usually be needed for men with
possible outlet obstruction, women with stress incontinence, and
any patient with atypical clinical presentation, or who is not
responding as expected to treatment. Patients with Parkinson's
disease have a high incidence of acontractile bladders, often not
suspected on clinical grounds, and cystometry is likely to help
management.[6]

DETRUSOR INSTABILITY

Results of cystometrograms in elderly patients show that the
commonest cause of incontinence is an unstable bladder. (Table 1)
Typically this is due to loss of inhibitory control of the sacral
reflex, and may, to a certain extent, be a part of normal aging.
Thus, a person who, when young, was able to postpone micturition
for an hour or so after the sensation of bladder fullness was
experienced, may find in old age that micturition can be delayed
for only a few minutes. Bladder capacity is reduced, and the
bladder fails to empty completely, leaving residual urine. This
will cause frequency and nocturia, but not incontinence, provided
that the person remains mobile and is intellectually unimpaired.
Damage to the frontal cortex by a stroke or Alzheimer's disease will
weaken inhibitory control still further to a stage where bladder
emptying occurs a few seconds after, or even co-incidentally with
the sensation of fullness. These patients will experience urgency
and urge incontinence if they are unable to get to the W. C. in time.

Cystometrogram shows a small capacity bladder, usually less
than 300 ml and sometimes less than 100 ml and much higher pressures
at capacity than normal subjects.[8] Because the bladder neck opens
automatically during an involuntary detrusor contraction, continence
can only be maintained provided that the voluntary distal urethral

Table 1. Causes of Incontinence in the Elderly
(N = 309)

Causes of Incontinence	% of Patients
Unstable bladder	57
Outflow obstruction	13
stable (20%)	
unstable (80%)	
Acontractile (atonic) bladder	7
Pure stress	2
Miscellaneous: e.g. vaginitis, acute infections, drugs	21

(Reproduced with permission of the Editor of the Journal of the
American Geriatrics Society from Overstall et al[7])

sphincter is strong enough to overcome the voiding pressure. This
is possible for some men, provided that their voiding pressure is
not too high, but the weak distal mechanism in women can rarely
maintain continence.

Bladder instability may be seen in elderly patients with no
obvious neurological defect.   Sometimes there is a history of
enuresis as a child, suggesting life-long instability, a pattern
which is surprisingly common, although largely asymptomatic in the
10 per cent of the population affected.[4]   Often incontinence is
the result of treatment with diuretics and sedative drugs, acute
illnesses, or the stress of hospital admission.

Psychological factors clearly have an effect on detrusor
function, as shown by the remarkable results of bladder training on
women with detrusor instability.   Cure rates of 82 per cent have
been described,[9] and are accompanied by the cystometrogram reverting
to normal.[10]   Essentially the method depends on the patient understanding the nature of her bladder disorder, maintaining a
micturition chart, and strictly adhering to a gradually increasing
interval between voidings so that bladder capacity is increased
and urgency reduced.

OUTLET OBSTRUCTION

Elderly men with frequency, urgency and nocturia are frequently misdiagnosed as having "prostatism" when the symptoms are more commonly due to detrusor instability. Furthermore, the same patients may complain of "poor stream" since the small voided volume reduces the flow rate. Thus it is not unusual to find an elderly man with detrusor instability who has an enlarged prostate, but no significant obstruction.[11] The results of prostatectomy in these patients are poor, and a cystometrogram is necessary for accurate assessment. Cystoscopy alone is unhelpful, and will not distinguish between hypertrophic trabeculation due to obstruction, and that simply caused by unstable detrusor exercise.

Although bladder instability appears to be worsened by outlet obstruction, the reason for this is unclear. About 80 per cent of patients with outlet obstruction also have bladder instability (Table 1). Surgery will improve the voiding stream, but only two out of three patients will revert to stability after the relief of the obstruction.[11] Even where there is undoubted obstruction, prostatectomy cannot therefore be guaranteed to relieve unstable symptoms such as frequency and urgency, particularly if there is another cause for the instability such as dementia.

Apart from prostatic hypertrophy, the commonest cause of outlet obstruction is faecal impaction, and clearing of the rectum with enemas may reduce the symptoms of instability. Other less common causes of obstruction include prostatic carcinoma, bladder neck hypertrophy, tumour or stones in the bladder, and pelvic tumours.

STRESS INCONTINENCE

Elderly patients who complain of incontinence associated with coughing, laughing, getting out of a chair, or walking, usually have detrusor instability. If, in addition, they have frequency, nocturia and urgency, then the diagnosis is almost certain. The cystometrogram will usually demonstrate that the rise in intra-abdominal pressure, associated with coughing or standing, provokes a detrusor contraction.

Pure stress incontinence in the presence of a stable bladder is fairly uncommon (Table 1). Both bladder neck and urethral closure mechanisms must be impaired, and this may be associated with anterior vaginal wall prolapse. A Colpo-suspension procedure is probably the operation of choice, and success appears to depend not so much on the anatomical position achieved as on restoring the urethral funtion.[12]

ACONTRACTILE (ATONIC) BLADDER

This is due to loss of bladder sensation, and results in chronic distention with overflow incontinence. In some patients the diagnosis is suspected after finding a large bladder on abdominal examination, but the history and clinical examination is usually a poor guide to diagnosis. Sometimes the condition is suspected when incontinence appears for the first time, or worsens in a patient given anticholinergic drugs (which include tricyclic antidepressants).

Disturbance of bladder efferent nerves may be due to diabetes mellitus, tabes dorsalis, or Vitamin B_{12} deficiency, but it is also commonly part of the autonomic neuropathy seen in Parkinson's disease. In normal aging there may be widespread impairment of autonomic function, and subjects with acontractile bladders show a marked decrease in baroreflex activity.[13]

Cystometrogram shows a bladder capacity greater than 500 ml, and not infrequently, more than 1,000 ml with a normal pressure at capacity.[8] There is a weakness or absence of detrusor voiding contractions.

ATROPHIC VAGINITIS

Estrogen deficiency in post-menopausal women causes both a vaginitis and an accompanying urethritis of the estrogen-sensitive urothelium. Urethral function is impaired, and typically causes incontinence and dysuria. The response to estrogen replacement therapy is excellent.[7]

URINARY TRACT INFECTION

As already discussed, symptoms such as dysuria, urgency and frequency which are a reliable guide to a urinary tract infection in the young, are in the elderly, often due to atrophic vaginitis or detrusor instability. Acute infections may cause incontinence, but chronic ones do not. Significant bacteriuria is present in 20 per cent of women over the age of 65, and in 20 per cent of men over the age of 70.[14] This is largely due to detrusor instability with incomplete bladder emptying, or outlet obstruction, and the resulting residual urine favours bacterial growth.

REFERENCES

1. T. M. Thomas, K. B. Plymat, J. Blannin, and T. W. Meade, Prevalence of urinary incontinence, Brit.med.J 281:1243 (1980).

2. J. W. G. Yarnell and A. S. St. Leger, The prevalence, severity
    and factors associated with urinary incontinence in a random
    sample of the elderly, Age & Ageing 8:81 (1979).

3. N. J. Vetter, D. A. Jones, and C. R. Victor, Urinary incontinence
    in the elderly at home, Lancet ii:1275 (1981).

4. R. T. Warwick, Observations on the function and dysfunction of
    the sphincter and detrusor mechanisms, in: "The Urologic
    Clinics of North America," R. T. Warwick and C. G. Whiteside,
    eds., W. B. Saunders Co., Philadelphia (1979).

5. P. Hilton and S. L. Stanton, Algorithmic method for assessing
    urinary incontinence in elderly women, Brit.med.J. 282:940
    (1981).

6. H. D. H. Eastwood, Urodynamic studies in the management of
    urinary incontinence in the elderly, Age & Ageing 8:41 (1979)

7. P. W. Overstall, K. Rounce, and J. H. Palmer, Experience with
    an incontinence clinic, J.Am.Geriat.Soc. 28:535 (1980).

8. C. M. Castleden, H. M. Duffin and M. J. Asher, Clinical and
    urodynamic studies in 100 elderly incontinent patients,
    Brit.med.J. 282:1103 (1981).

9. W. Frewen, Role of bladder training in the treatment of the
    unstable bladder in the female, in: "The Urologic Clinics of
    North America," R. T. Warwick and C. G. Whiteside, eds.,
    W. B. Saunders Co., Philadelphia (1979).

10. G. J. Jarvis and D. R. Millar, Controlled trial of bladder drill
    for detrusor instability, Brit.med.J. 281:1322 (1980).

11. R. T. Warwick, A urodynamic review of bladder outlet obstruction
    in the male and its clinical implications, in: "The Urologic
    Clinics of North America," R. T. Warwick and C. G. Whiteside,
    eds., W. B. Saunders Co., Philadelphia (1979).

12. R. T. Warwick and A. D. C. Brown, A urodynamic evaluation of
    urinary incontinence in the female and its treatment, in:
    "The Urologic Clinics of North America," R. T. Warwick and
    C. G. Whiteside, eds., W. B. Saunders Co., Philadelphia (1979).

13. K. J. Collins, A. N. Exton-Smith, M. H. James and D. J. Oliver,
    Functional changes in autonomic nervous responses with ageing,
    Age & Ageing 9:17 (1980).

14. J. C. Brocklehurst, J. B. Dillane, L. Griffiths, and J. Fry,
    The prevalence and symptomatology of urinary infection in an
    aged population, Geront.clin. 10:242 (1968).

THE PHARMACOLOGY OF URINARY INCONTINENCE

J. Malone-Lee

Department of Geriatric Medicine
School of Medicine
University College London

The Anatomy

The smooth muscle fibres of the detrusor pass over the body of
the bladder and are continued into the urethra as longitudinal fibres
forming part of the urethral wall. They terminate in the distal
urethra in the female and in the veruomontanum of the male. The
effect of these fibres contracting is a rise in bladder pressure and
shortening of the urethra. The muscle fibres of the ureters are
continued into the trigone which forms a base-plate at the bladder
neck. Fibres continue from the trigone into the urethra being
inserted in a similar manner to the detrusor muscle. When the
trigone contracts the bladder neck funnels. There are some smooth
muscle fibres which arise in the detrusor and are inserted into the
outer surface of the trigone distally. They tend to pull the plates
of the trigone apart thus opening the bladder neck.[1,2]

At the bladder neck there is a concentration of smooth muscle
which is only developed significantly in the male. This is the
internal sphincter. It has little to do with continence but
functions to prevent the retrograde flow of semen during ejaculation.[3]

The urethra is a highly distensible tube. The resistance to
flow is dependent on the pressure gradient down the urethra and the
internal diameter. These factors are governed by the urethral
elastic tissue, the cushioning effect of intra-mural arteriovenous
sinuses, the smooth muscle of the walls, the surface tension of the
urethral mucus, the external sphincter and the periurethral striated
muscle.[3,4]

The external sphincter consists of circularly arranged striated muscle which is slow-twitch in character and capable of continuous contractions. This muscle is quite separate from the periurethral striated muscle which is part of the pelvic sling and fast-twitch in character. Fast-twtich muscle can maintain contractions lasting about ninety seconds. Both these muscles are supplied by somatic efferents from S 2, 3 and 4 which run quite separately from the pudendal nerves.[1],[2],[3],[4]

The prostate gland provides men with a considerable advantage over women in maintaining continence. It consists of a mixture of glandular tissue and smooth muscle. The latter is known to produce a constant resting tone and to influence the process of micturition.[6]

The Neuro-anatomy

The functional neuro-anatomy of the bladder is based as much on pharmacological and physiological experiments as on anatomical investigation. Sometimes the findings of these different disciplines do not tally. This reflects the limitations of our current knowledge. The information in this article must be judged in that light.[5]

The greater bulk of the detrusor muscle is innervated by parasympathetic efferents originating from S 2, 3 and 4. Their influence is excitatory and they are responsible for voiding. Scattered within the detrusor are a limited number of inhibitory $\beta2$ receptors and excitatory $\alpha1$ receptors both supplied by adrenergic neurones from the hypogastric plexus. There is evidence that in certain neurogenic bladders, the alpha receptors play a more important role than usual[5].

The internal sphincter of the bladder is richly supplied with excitatory $\alpha1$ receptors, a number of excitatory cholinergic receptors and a few inhibitory $\beta2$ receptors.[5]

The smooth muscle of the urethra has a rich supply of excitatory $\alpha1$ adreno-receptors. These are also concentrated in the prostatic capsule and adenoma. There are some inhibitory $\beta2$ receptors in the urethral muscle but their significance is ill-understood.[4],[5]

Adrenergic nerve endings have not been satisfactorily demonstrated in the area of the internal sphincter but cholinergic endings certainly are seen. Be that as it may, cholinergic and alpha adrenergic stimulation cause a rise in urethral pressure. It has been suggested that the alpha activity may be located at the pelvic plexus.[3]

The Receptor Anatomy

The existence of "atropine resistant transmission" in both

mammalian and human detrusor has been known for many years and
accounts for between 30 per cent and 40 per cent of human detrusor
activity.   It has been suggested that the responsible receptors
are purinergic and we do know that quinidine will block their
action.[7] [120]

Hindmarsh in 1977 showed that electrically induced, but not
acetylcholine induced, contractions of human bladder muscle were
potentiated by very low doses of serotonin and this potentiating
effect was not altered by methysergide or morphine.[8]

The effects of the prostaglandins F2 alpha and E2 were studied
by Khalaf et al, who found that prostaglandin E2 reduced urethral
pressure and lowered the threshold for induced detrusor contractions.
Prostaglandin F2 alpha in low dose raised urethral pressure but at
higher doses also reduced the detrusor threshold.[9]

It has been known for a while that the female urethelium is
oestrogen sensitive and that it degenerates along with the vagina
as oestrogenisation reduces in old age.   The result is an atrophic
urethra which markedly reduces urethral function.   Recently,
workers have identified large numbers of oestrogen receptors
throughout the urethra and these are most concentrated in the distal
two thirds.   Progesterone receptors have not been detected though
there is a sensitivity of the urethra to progesterone.[10] [11]

With the development of immunological methods, the identification of other transmitters in the central and peripheral nervous
system has become possible.   We know that the neurones innervating
the striated sphincter and those innervating the smooth muscle of
the detrusor are influenced by the somatostatinergic system.[12]

Of greater importance is the recent observation that the
intravenous administration of the opiod antagonist naloxone markedly
aggravates the reactivity of the unstable bladder.   It has also
been demonstrated that the intra-thecal administration of morphine
reduces detrusor instability.[121]

Prostatic Function

I would like to cover prostatic function in some detail.   In
1965 Owman and Sjostrand reported fluorescent microscopic studies
which demonstrated a rich innervation, with short adrenergic
neurones, of the muscular tissue in the prostate glands of various
mammals.[122]   In 1972 Raz and Caine reported in vivo experiments
on the female canine urethra which indicated the presence of
excitatory alpha and inhibitory beta adreno-receptors in the smooth
muscle of the urethra.[123]   At the same time they demonstrated an
alpha mediated excitatory influence of angiotensin.   In vitro
experiments performed by Raz, Zeigler and Caine on the female canine

urethra confirmed these findings as well as demonstrating a certain
amount of excitatory alpha activity in the resting urethral musculature.[124]   The fact that the prostate and urethra are closely
related both anatomically and embriologically led the same workers
to conduct a series of experiments on the rat prostate.[125]
They demonstrated a rich supply of excitatory alpha adreno-receptors,
some excitatory cholinergic receptors and a very few inhibitory
beta adreno-receptors.   In 1975 the same workers reported on in
vitro studies of the human prostate, prostatic capsule and bladder
neck.[126]   The prostatic capsule was found to be very rich in both
excitatory alpha adreno-receptors and excitatory cholinergic
receptors.   The prostatic adenoma was moderately rich in alpha
adreno-receptors, but cholinergic receptors were absent.   Beta
adreno-receptors were absent in the prostatic adenoma, and there was
an equivocal response in less than half the specimens of the
prostatic capsule.   It is extremely important to note that when
studying the effect of noradrenaline on the prostatic capsule they
found that in some cases there was extreme sensitivity to even very
small concentrations of noradrenaline.

     Benign prostatic hypertrophy is a common accompaniment to aging.
Autonomic neuropathy in one form or other is a frequent manifestation of the aging process.   That the two conditions may co-exist
should not be surprising and when this occurs they are likely to
interact.

     A number of workers have observed degenerative changes in
sympathetic and parasympathetic ganglia and in postganglionic nerve
endings in various organs of aging humans and animals.[127]

     Gey, Burkard and Pletscher[12] showed that, due to a reduction
of decarboxylase of aromatic 1-amine acids and dopamine-β-oxidase
activities, the synthesis of noradrenaline is reduced in old age.
Different pathways of catecholamine metabolism change irregularly
in the aged; thus, monoamine oxidase activities increased, catechol-
o-methyltransferase remains unchanged and catecholamine uptake by
sympathetic nerve endings (Uptake 1) is weakened.   Reduction of
noradrenaline synthesis in the adrenergic neurones and their
accompanying destruction result in the weakening of the nervous
influences on tissues.   However, changes of catecholamine
metabolism pathways, weakening of Uptake 1 and an increase in the
number of adreno-receptors, results in an increased sensitivity to
catecholamines as well as persistent responses to stimulation.
There is some point to this, as increased sensitivity of the
innervated cells to transmitters will enable the organism to adapt
to the reduction in nervous control.[49]

     This increased sensitivity of adreno-receptors has been demon-
strated experimentally.   Chebotarev in 1974 demonstrated that
smaller doses of catecholamines induced haemodynamic changes in the

aged.[129]   Shevchuk in 1974 demonstrated an increased sensitivity
of the cardiovascular system of aged animals to blockade with
propranolol, though interestingly the sensitivity to the alpha
blocker rogitine was unchanged. Similar examples of adrenoreceptor supersensitivity have been described in conditions associated with chronic autonomic failure.[131,132,133,134]

There are grounds to suggest that a proportion of men presenting with prostatism are suffering from an autonomic neuropathy
secondary to aging or a specific disease. As a result they have
developed a denervation supersensitivity of the prostatic and
urethral adreno-receptors which over-react to circulating catecholamines, provoking a functional obstruction. This problem may be
aggravated by a relative hypotonia of the detrusor accompanying the
autonomic neuropathy. It is likely that these patients may be
more responsive to pharmacological therapy for their symptoms than
those with true anatomical obstruction. At the same time they
are likely to be more susceptible to the side-effects of the type of
drugs used.

DRUG THERAPIES

The Anti-cholinergics

Anti-cholinergics have proved to be effective for the treatment
of detrusor instability. When administered intravenously most anti-
cholinergics will paralyse the bladder.

Atropine and related drugs are tertiary amines that are well
absorbed from the gastrointestinal tract. They also pass the blood-
brain barrier which may lead to pronounced effects on the central
nervous system and thus limit their clinical use.

The two anticholinergic drugs most frequently used in the treatment of the unstable bladder are propantheline and emepronium
bromide. They are quarternary ammonium compounds which do not
readily penetrate biological membranes and therefore have a very low
and variable absorption leading to an unpredictable clinical response.
Both compounds have marked muscarinic blocking action with some
nicotinic blocking properties. They often produce disturbing
systemic side-effects such as paralysis of accommodation, tachycardia, dryness of the mouth, postural hypotension and impotence.
Emepronium bromide has also been identified as a cause of oral and
oesophageal ulceration as well as oesophageal stricture.

It seems that at the moment Emepronium bromide is the drug most
commonly used and some patients gain considerable benefit, but in
general the responses remain far from satisfactory.[15,16,17,18,19,
20,21,22,23,24,25,26,27]

The Direct Acting Smooth Muscle Relaxants

Flavoxate hydrochloride is a tertiary amine chromone which selectively acts on a muscle receptor producing a muscle relaxant effect without significant anticholinergic action. It is thought to act by influencing one of the steps in the excitation coupling of the smooth muscle of the bladder. This is papaverine-like action and the toxicity of the drug is of the same order as that of papavarine.[28,29,30,31,32,33,34]

Oxybutinin chloride is a newer antispasmodic agent having a direct papavarine-like effect as its most prominent property. It also inhibits the muscarinic actions of acetylcholine on smooth muscles, but its ability to depress spontaneous activity or spasms of certain smooth muscles is due to a strong direct musculotrophic action. The effectiveness of orally administered oxybutinin is good and the duration of action seems satisfactory for therapeutic use. It is also said to have analgesic properties.[35,36,37,38]

Dicyclomine appears to have a non specific direct relaxant action on smooth muscle though a competitive antogonism of acetylcholine at muscarinic sites is also a feature.[39,40]

The publications related to the use of these drugs have been both encouraging and disappointing. Flavoxate has not proved very successful. Dicyclomine has not been used very much and has attracted few publications. Oxybutinin is currently available in the United Kingdom on a named patient basis only. It has been used extensively in the U.S.A. and by all accounts it is a drug very well worth trying and our experience confirms this.[41]

The side-effects of dicyclomine and oxybutinin can be attributed to their anticholinergic action.

The Calcium Ion Antagonists

The release of acetylcholine at the neuromuscular junction is dependent on the presence of calcium ions. During depolarisation of the muscle cell membrane, a slow inward current of calcium ions follows the fast sodium ion current. The separation of the actin and myosin myofibrils during muscle contraction involves a reaction utilising energy from ATP and requiring the presence of calcium ions. It has been shown that irrespective of how isolated human bladder muscle is activated the contractions can be blocked by inhibiting the inflow of calcium through the cell membrane.[42,43,44,45,46,47,48,49,50]

The calcium ion antagonist Flunarizine is the bifluorinated derivative of cinnarizine. Because of its lipophilicity it has a long half-life. It inhibits contractions of smooth muscle induced, in the presence of calcium ions, by depolarising solutions, serotonin, noradrenaline and prostaglandins E2 and F2 alpha. Flunarizine is not a "slow channel entry blocker" like nifedipine or verapamyl and does not interfere with calcium homeostasis under normal conditions.   The initial responses of detrusor instability to flunarizine were in some cases quite remarkable though the effect of this drug on the bladder tends to wear off after eight to twelve weeks.[51,52]

Nifedipine effectively inhibits contractions induced in the isolated human bladder by carbachol, prostaglandins E2 and F2 alpha, potassium and barium chloride. Nifedipine does not affect the resting pressures in the bladder and urethra but has proved effective in the treatment of uninhibited detrusor contractions, though it seems that this action is very limited.[48,49,50]

Terodiline has similar properties to nifedipine but in addition it has an antimuscarinic effect which should give it a considerable advantage.   To date, studies on this drug have been promising and it certainly deserves much further scrutiny.[53,54,55,56]

The actions of verapamyl, prenylamine, perhexiline and lidoflazine on the bladder have not yet been assessed.

The side effects of these drugs include fluid retention, palpitations, angina, headaches, dizziness, vertigo, flushing, postural hypotension and hepatitis.[57]

The Prostaglandin Synthetase Inhibitor

Prostaglandins E1, E2, F1 alpha and F2 alpha may all produce contraction of human bladder muscle.   Prostaglandin E2 and F2 alpha being more stable have been studied directly and F2 alpha would seem to be the most potent of the group.[58,59]

Indomethacin has been shown to relieve the symptoms of daytime and nocturnal frequency in detrusor instability.[60] Flurbiprofen has been found to reduce frequency, urgency and urge incontinence in women.   In another study, flurbiprofen was found to have an effect no different from placebo when used in males.   When successful, flurbiprofen seems to improve daytime and nocturnal frequency most, with less effect on urgency and urge incontinence and no effect on nocturnal enuresis.   Some of the response could be attributed to the antidiuretic action of nonsteroidal anti-inflammatories which by suppressing prostaglandin E2 potentiate the effect of anti-diuretic hormone.[61,62,63]

The side-effects of indomethacin include headaches, sometimes associated with dizziness.   Gastrointestinal effects include dyspepsia and peptic ulcer sometimes with haemorrhage and perforation.   The symptoms and signs of infection may be masked.   Long term use may damage the eye.

Dyspepsia, heartburn and headache are the commonest side effects encountered with flurbiprofen. Occasional skin rashes have also been reported.

The Tricyclic Antidepressants

Both amitriptyline and imipramine have been used to treat detrusor instability. Imipramine is the drug which has been studied most and has been found to be effective in treating adult daytime and nocturnal incontinence, the latter being the most responsive.[66]

Imipramine is a dibenzapine derivative. It is structurally similar to the phenothiazines and has adrenergic, antimuscarinic and antihistaminic properties. It is well absorbed from the gastrointestinal tract and rapidly distributed and metabolised. It is demethylated to an active form desmethylimipramine. Both forms are fat soluble and readily cross the blood-brain barrier. Serum levels of imipramine do not always reflect the oral dose and are highly variable from patient to patient.

Imipramine is anticholinergic but it has been found to be superior to L-hyoscine in its effects on the bladder. It has been found to increase urethral pressure in the dog and this pressure rise was blocked by phenoxybenzamine, suggesting an alpha agonistic action. Other experiments on dogs show that it reduces contractions of the urethra and bladder induced by pelvic nerve stimulation and by the injection of acetylcholine, histamine and serotonin into the iliac artery. Urethral and bladder contractions induced by ATP and noradrenaline are not blocked by imipramine. We also know that imipramine blocks histamine receptors in the brain as well as having a monoamine reuptake inhibitory action, both centrally and peripherally. It also possesses local anaesthetic properties.[64,65,66,67,68,69,70,71,72,73,74,75]

Imipramine certainly changes the sleep profile. It will decrease REM sleep, increase stage II, increase latency to the onset of first REM, decrease dreaming and the number of shifts between sleep stages. It has no effect on stages II and IV. Paradoxically nocturnal enuresis tends to begin during periods of slow wave activity, stages III and IV, and is invariably associated with lightening to stage II or I or wakefulness. It does not seem to be associated with REM sleep.[76,77,78,79,80]

These multiple actions combine to give imipramine a very effective influence on incontinence and it induces extremely good results. Nocturnal enuresis usually responds to a dose of between 10 mg and 25 mg nocte. Some of the patients with daytime incontinence require an extra dose in the morning though the majority respond to a single nocturnal prescription.

Amitriptyline is a less effective alternative which has attracted a small literature.

The side-effects of these preparations largely involve anti-cholinergic reactions.   They also include postural hypotension, tremors, hallucinations, confusion, excitement, precipitation of epilepsy and jaundice and there may be cardiac hazards.

The Beta Receptor Agonists

There are a number of inhibitory β2 receptors in the detrusor. It is possible to suppress premature detrusor contractions with terbutaline, salbutamol and orciprenaline.   The latter combined with emepronium has been found to be superior to emepronium on its own.   In clinical practice, however, it has been found that these drugs do not have a significant role to play.[81,83,86,87]

The side-effects of these drugs include nervousness, tremor, headache, nausea and vomiting, hypotension and hypokalaemia because of a shift of potassium into the cells.

The Dopamine Receptor Agonists

Bromocriptine was originally studies as a potential treatment for benign prostatic hypertrophy.   In experimental animals prolactin increases prostatic size.   As dopamine is a prolactin release inhibitor it was hoped that bromocriptine would cause a reduction in prostatic size.   Urodynamic studies failed to show an improvement in urinary flow with this drug, but there was some evidence that associated detrusor instability was reduced.   Further studies were conducted but it was not found to have a clinical use.[88,89]

Drugs Acting at Polysynaptic Spinal Reflexes

These drugs have been studied in patients suffering from multiple sclerosis or spasticity.   Baclofen, hydramitrazine, and meladrazine have been tried, but there is very little data available supporting their efficacy.[90,91,92]

They do not affect neuronal conduction, neuromuscular transmission or muscle excitability except after nearly lethal doses. They depress spinal polysynaptic reflexes preferentially over mono-synaptic reflexes but this does not account for their therapeutic actions as some other drugs selectively depress these reflexes without exhibiting any muscle relaxant activity.   They are known to prolong synaptic recovery time and to reduce repetitive inter-neurone discharges.

Their side-effects include drowsiness, dizziness, headache,

blurred vision, weakness, lethargy, ataxia, nystagmus, nausea, vomiting and heartburn.

Non Pharmacological Therapy

It should be emphasized that by far the most important principle in managing the unstable bladder is the development of a good relationship with the patient involving confidence and compassionate understanding.   The bladder retraining regimes are an essential part of any treatment schedule.   The aim should be to withdraw from supporting medication.   This requires regular consistent follow-up.   It should be remembered that the term "Detrusor instability" is very broad.   It presents in many different forms,depending on the cause and, in the elderly, is often complicated by another bladder lesion.   Unfortunately much of the work on this condition has been conducted on women with so-called "Idiopathic detrusor instability."   It is naive to attempt to extrapolate findings from this particular group to the whole.   Our understanding of the problem is not likely to develop unless we use more detailed descriptions of the individual patients rather than applying a single label to them all.

The Alpha Blocking Agents

The number of drugs with alpha blocking properties is increasing and they all have differing actions.   They can be used in the management of patients with dyssynergic voiding problems and in some cases of prostatism where orchidectomy, oestrogens, and anti-androgens have not found a role.[93,94,95,96,97,98,99,100,101,102,103,104,105,106]

Phenoxybenzamine is a haloalkylamine which produces a blockade of $\alpha 1$ and $\alpha 2$ adrenoreceptors.[106]   At the outset the binding at the receptor site is competitive and the presence of high concentrations of catecholamines can inhibit the blockade.   However, on binding to the receptor a very stable covalent bond is formed and the binding cannot be reversed by the administration of agonist. Because of this strong bond the blockade roduced by a single dose of phenoxybenzamine can be detected three to four days after administration, despite a plasma half-life of twenty four hours. This means that a long dose interval can be used but because of accumulation, dose increase on initiation must be very slow.

It is interesting to note that phenoxybenzamine also inhibits the uptake of catecholamines into the adrenergic neurone (Uptake 1) and into the extra-neuronal tissues (Uptake 2).   It also inhibits the actions of monamine oxidase and catechol-O-methyltransferase. Couple these actions to the blockade of the $\alpha 2$ receptors and you produce an augmentation of the outflow of noradrenaline from stimulated sympathetic neurones.   As a result phenoxybenzamine

induces an enhanced cardiac chronotropic response to vasodilation
which increases cardiac output.   This may well be a considerable
advantage when this drug is used in patients with reduced autonomic
control of blood pressure responses.

It should be emphasized that the effect of any blocking agent
is highly dependent upon the state of activity of the system on
which it acts.   Thus the responses induced by phenoxybenzamine may
vary markedly in different tissues, depending on the degree of
adrenoreceptor control and the physiological state.

In addition to the blockade of alpha adrenoreceptors phenoxybenzamine can inhibit responses to serotonin, histamine and acetylcholine.

The side-effects of phenoxybenzamine include loss of vasomotor
control, postural hypotension and reflex tachycardia.   Other
effects include miosis, nasal stuffiness, inhibition of ejaculation,
sedation and a general feeling of weakness and tiredness.   In the
studies conducted on phenozybenzamine to treat the symptoms of
benign prostatic hypertrophy side-effects were manifest in about
30 per cent of patients but surprisingly they were not troublesome.
They tended to occur where larger doses were being used and it was
found that they could be greatly improved by reducing the dose
without altering the efficacy.   This could be explained by the fact
that some tissues contain receptors in excess of the number required
for a full response to an agonist.   A considerable proportion can
be inactivated before the tissue is incapable of a maximal response.
If, when compared to the vasculature, prostatic and urethral smooth
muscle had a paucity of adrenoreceptors a low dose response in the
absence of side-effects would be explained.

The Ergot Alkaloids

Ergot is the product of a fungus which grows on rye and other
grains.   It has been described as "a veritable treasure house of
pharmacological constituents" and has certainly had a most astonishing impact on history.   All of the ergot alkaloids are derivatives
of lysergic acid, an indole compound.   The variety of natural
compounds can be extended by chemical manipulation.   One of the
double bonds of lysergic acid can be selectively saturated producing
the dihydrogenated alkaloids.   It is possible to combine lysergic
acid with different amines than those linked to it in nature;
lysergic acid diethylamine is a product of this method.

The alkaloids are classified into three groups according to
their chemical structure and pharmacological properties, which
however greatly overlap.   There are the amino acids alkaloids, the
dihydrogenated amino acid alkaloids and the amine alkaloids.

The properties of these compounds include alpha adrenoreceptor antagonism and agonism, smooth muscle stimulation and a variety of complex effects on the central nervous system.   It is the dihydrogenation of the lysergic nucleus which increases alpha adrenoreceptor antagonism and decreases agonism and smooth muscle stimulating activity.

Dihydroergotamine is a compound with mixed alpha receptor antagonistic and agonistic properties.   Its tonic effect on vascular muscles seems to be most marked in the capacitance vessels with the result that it increases venous tone and prevents excessive venous pooling.   For this reason it has been used to treat orthostatic hypotension and is a prophylactic agent against the development of deep venous thrombosis.   It has also been found effective in the treatment of migraine.

Dihydroergotoxin on the other hand, produces only limited vasoconstriction but as a result of alpha antagonistic properties it tends to induce an overall peripheral vasodilation and fall in arterial pressure.   It would seem that this alkaloid is going to prove of greatest use in the management of the dementias because of its cerebral activating properties.

The potential of producing an ergot alkaloid with mixed propert- ies capable of improving venous return to the heart, blocking some alpha receptors and possibly improving cerebral function is a very tempting idea when planning to treat prostatic symptoms in elderly patients with failing minds.

One ergot alkaloid, nicergoline has been tested in benign prostatic hypertrophy with some promising results.   Very little work has been done in this field but there is a great potential and it demands much further investigation.[135,136]

The side-effects of these drugs range through vascular insuffic- iency and gangrene to flushing and postural hypotension and include nausea, vomiting, abdominal cramps, rashes, nasal stuffiness, and psychological disturbances all of which depend on the individual properties of the compound or mixture of compounds used.

The Alpha Receptor Agonists

Alpha agonists acting at the pelvic plexus or on the urethral smooth muscle will raise the resting urethral pressure.   It is however interesting to note that section of the sympathetic supply to the pelvis does not affect micturition. Ephedrine and nor-ephedrine (phenylpropranolamine) have been found effective in the treatment of detrusor instability and stress incontinence. In the former condition results seem to be better if the alpha agonist is combined with an anticholinergic.[82,84,85]

The use of these drugs is limited by their side-effects which produce alertness, anxiety, insomnia, tremor, nausea and tachyphylaxis.

The Oestrogens

Oestrogen receptors in the urethra have been described and there is an association between the atrophic vagina and urinary incontinence which reflects a deterioration in urethral function. Oestrogen therapy has a beneficial effect on this condition. Oral oestrogen therapy produces more predictable tissue levels but topical oestrogens are rapidly absorbed into the peripheral circulation producing plasma concentrations almost identical to a similar oral dose. The variation in the tissue levels achieved by the different routes of administration may reflect the difficulty of vaginal application. Frequently quite high doses of topical oestrogen are used and these can lead to excessive endometrial hyperplasia and a loss of local effect with prolonged therapy. Recently lower doses of 0.1 mg of topical oestrogen daily have been shown to produce premenopausal vaginal cytology without a tailing off of responsiveness. Though this dose would not control systemic postmenopausal symptoms it will work for local problems.[113,114,115]

Long term oestrogen replacement therapy in postmenopausal women is associated with an increased incidence of gall bladder disease and endometrial carcinoma as well as salt and water retention and thromboembolism. Hypertension, liver disease, migraine, diabetes and fibroids may be made worse.

The Cholinergics

If the efferent supply to the bladder is interrupted it will become hypotonic. This will result in incomplete emptying, a large residual volume and troublesome dribbling overflow incontinence. In the early stages a more forceful detrusor contraction may be encouraged by the use of the anticholinesterase, distigmine, given as 5 mg daily. Its benefit is not sustained and patients with chronic hypotonia usually require management with intermittent catheterisation. Distigmine may produce parasympathetic side-effects such as miosis, salivation, abdominal cramps and bradycardia.   An alternative drug with similar effects is the agonist, carbachol.[116] [117,118,119]

REFERENCES

1. J. C. Brocklehurst, "Textbook of Geriatric Medicine and Gerontology," Churchill Livingstone, London (1978).

2. J. A. Hutch, "Anatomy and Physiology of the Bladder Trigone and Urethra," Butterworth, London (1972).

3. J. A. Gosling, J. S. Dixon, O. D. Hilary and Critchley and Sally Ann Thompson, A comparative study of the human external sphincter and periurethral levator ani muscles, Brit.J.Urol. 53:35 (1981).

4. K. E. Creed and A. S. S. Tulloch, The action of imipramine on the lower urinary tract of the dog, Brit.J.Urol.54:5 (1982).

5. T. F. Fletcher and W. E. Bradley, Neuroanatomy of the bladder, J.Urol.119:153 (1978).

6. M. Caine, S. Ras, and M. Zeigler, Adrenergic and cholinergic receptors in the human prostate, prostatic capsule and bladder neck, Brit.J.Urol.47:193 (1975).

7. A. C. Eaton, A. T. Birmingham, and C. P. Bates, Evidence for the existence of purinergic transmission in the human bladder and a possible new approach to the treatment of detrusor instability, in: "XIth Annual Meeting International Continence Society Lund," Skogs Trelleborg (1981).

8. J. R. Hindmarsh, O. A. Idown, W. K. Yeates and M. A. Zar, Pharmacology of electrically evolved contractions of human bladder, Br.J.Pharmacol. 61:115 (1977).

9. I. M. Khalof, M. A. Ghoneim, and M. M. Elkiloli, The effects of endogenous prostaglandins F2 alpha and E2 and indometh-acin on micturition, Brit.J.Urol.53:21 (1981).

10. P. O. Wilson, G. Barker, A. O. G. Brown, A. Russell, and N. Siddle, Steroid hormone receptors in the female lower urinary tract, in: "XIIth Annual Meeting International Continence Society Lund," Skogs Trelleborg (1981).

11. M. Caine and S. Ras, The effect of progesterone on the adrenergic receptors of the urethra, Brit.J.Urol. 45:131 (1973).

12. Henrik Daa Schrode, Somatostatin in regulation of the bladder and the pelvic sphincters, in: "XIth Annual Meeting International Continence Society Lund," Skogs Trelleborg (1981).

13. G. J. Jarvis and D. R. Millar, Crontrolled trial of bladder drill for detrusor instability, Brit.med.J. 11:1322 (1980).

14. G. J. Jarvis, Bladder drill for the treatment of enuresis in adults, Brit.J.Urol.54:118 (1982).

15. U. Almsten and K. E. Andersson, The effects of emepronium on intravesical and intraurethral pressures in women with urgency incontinence, Scand.J.Urol.Nephrol. ii:103 (1977).

16. J. C. Brocklehurst, J. B. Dillane, J. Fry and P. Armitage, Clinical trial of emepronium bromide in nocturnal frequency of old age, Brit.med.J. ii:216 (1969).

17. J. C. Brocklehurst, P. Armitage and A. J. Jouhar, Emepronium bromide in urinary incontinence, Age & Ageing 1:152 (1972).

18. S. Hebjom and S. Walter, Treatment of female incontinence with emepronium bromide, Urol.Int.33:120 (1978).

19. A. E. S. Ritch, C. M. Castleden, C. F. George and M. R. P. Hall, A second look at emepronium bromide in urinary incontinence, Lancet i:504 (1977).

20. T. Habenshaw and J. R. Bennett, Ulceration of the mouth due to emepronium bromide, Lancet ii:1422 (1972).

21. T. M. Strouthidis, R. G. D. Mankilea and R. E. Irvine, Ulceration of the mouth due to emepronium bromide, Lancet i:72 (1972).

22. S. Kenwright and A.D.C. Norris, Oesophageal ulceration due to emepronium bromide, Lancet i:648 (1977).

23. A. Diokno, C. Hindman, D. Hardy and J. Lapides, Comparison of action of imipramine and propantheline on detrusor contraction, J.Urol. 1:42 (1972).

24. K. S. Jones and R. W. Tibbetto, Pituitary snuff, propantheline and placebos in the treatment of enuresis, J.Ment.Sci. 105: 371 (1959).

25. J. Lapides and A. Dodson, A study of the effects of banthine on the human bladder, J.Urol. 69:96 (1953).

26. D. Leys, Value of propantheline bromide in the treatment of enuresis, Brit.med.J. i:549 (1956).

27. J. C. Brocklehurst and J. B. Dillane, Studies of the female bladder in old age. IV drug effects in urinary incontinence, Geront.Clin.9:182 (1967).

28. K. P. J. Delacre, H. G. E. Michiels, F. M. J. Debruyne and W. A. Moonen, Flavoxate hydrochloride in the treatment of detrusor instability, Urol.Int. 32:377 (1977).

29. F. P. Kohler and P. A. Morales, Cystometric evaluation of flavoxate hydrochloride in normal and neurogenic bladder, J.Urol. 100:729 (1968).

30. R. Gaudenz and A. Weil, A comparison of flavoxate hydrochloride, emepronium bromide and propantheline for the treatment of female urinary incontinence and of bladder instability, in: "VIIth Annual Meeting International Continence Society Manchester," Pergamon Press, (1978),

31. U. Jonas, E. petri and J. Kissel, Effect of flavoxate on hyperactive detrusor mustle, in: "VIIIth Annual Meeting International Continence Society Manchester," Pergamon Press (1978).

32. S. L. Stanton, A comparison of emepronium bromide and flavoxate hydrochloride in the treatment of urinary incontinence, J.Urol. 110:529 (1973).

33. H. H. Meyhoff, T. C. Gerstenberg and J. Nordling, Placebo the drug of choice in female motor urge incontinence, in:"XIth

Annual Meeting International Continence Society Lund,"
Skogs Trelleborg (1981).

34. G. T. Jarvis, A controlled trial of bladder drill and drug
     therapy in the management of detrusor instability, in:
     "XIth Annual Meeting International Continence Society Lund,"
     Skogs Trelleborg (1981).

35. F. G. Anderson and C. M. Fredericks, Characterization of
     oxybutinin in drug-induced spasm in the detrusor,
     Pharmacology 14:31 (1977).

36. A. C. Diokno and J. Lapides, Oxybutinin - a new drug with
     analgesic and anticholinergic properties, J.Urol. 118:
     307 (1972).

37. C. M. Fredericks, R. L. Green and G. F. Anderson, Comparative
     in vitro effects of imipramine, oxybutinin and flavoxate
     on rabbit detrusor, Urology 12:487 (1978).

38. C. U. Moisey, T. P. Stephenson and C. P. Brendler, The urodyn-
     amic and subjective results of treatment of detrusor
     instability with oxybutinin chloride, Brit.J.Urol. 52:472
     (1980).

39. S. A. Awad, S. Bryniak and J. W. Downie, The treatment of the
     uninhibited bladder with dicyclomine, J.Urol. 117:161 (1977).

40. J. W. Downie, A. S. Twiddy and S. A. Awad, Antimuscarinic and
     non-competitive antagonistic properties of dicyclomine
     hydrochloride in isolated human and rabbit bladder muscle,
     J.Pharmacol.Exp.Ther.201:662 (1977).

41. L. D. Cardozo and S. L. Stanton, An objective comparison of
     the effects of parenterally administered drugs in patients
     suffering from detrusor instability, J.Urol. 122:58 (1979).

42. J. C. Ruegg, Smooth muscle tone, Physiol.Rev.31:201 (1971).

43. A. Fleckenstein, Specific pharmacology of calcium in myocard-
     ium cardiac pacemakers and vascular smooth muscle,
     Ann.Rev.Pharmacol.Toxicol.17:149 (1977).

44. M. R. Bennett, G. Burnstock, M. M. Holman and J. W. Walker,
     The effect of $Ca2+$ on plateau type action potentials in
     smooth muscle, J.Physiol. 161:48 (1962).

45. E. Bulbring, H. Kuriyama and B. Twarog, Influence of sodium
     and calcium on spontaneous spike generation in smooth
     muscle, J.Physiol.161:48 (1962).

46. E. Bulbring and T. Tomita, Effects of Ca removal on the smooth
     muscle of the guinea pig toenia coli, J.Physiol.210:217
     (1970).

47. A. Fleckenstein, Pharmacology and electrophysiology of calcium
     antagonists, in: "Proceedings of International Symposium on
     Calcium Antagonism in Cardiovascular Therap," Excerpta
     Medica, Florence (1981).

48. A. Forman, K. E. Anderson, L. Henriksson, T. Rud and U.
     Ulmsten, Effects of nifedipine on the smooth muscle of the
     human urinary tract in vitro and in vivo, Acta Pharmacol.
     et Toxicol. 43:111 (1978),

49. R. L. Vereckson, H. Hendrick and R. Casteds, The influence of
     calcium on the electrical and mechanical activity of the
     guinea pig ureter, Urol.Res. 3:149 (1975).
50. T. Rud, K. Andersson and U. Ulmsten, Effects of calcium
     antagonists in women with unstable bladder, Urol.Int.
     34:421 (1979).
51. J. M. Van Neuten, J. Van Beck and P.A.J. Janssen, Effects of
     flunarazine on calcium induced responses of smooth muscle,
     Arch.Int.Pharmacodyn. 232:43 (1978).
52. J. H. Palmer, P. H. L. Worth and A. N. Exton-Smith, Flunarazine
     - A once daily therapy for urinary incontinence, Lancet
     ii:279 (1981).
53. G. Ekman, K. E. Anderson, T. Rud and U. Ulmsten, A double blind
     crossover investigation on the effects of terodiline in
     women with unstable bladders, Acta.Pharmacol.et Toxicol.
     46 Suppl. 1:39 (1980).
54. T. Rud, K. E. Andersson, N. Boye and U. Ulmsten, Terodiline
     inhibition of human bladder contraction. Effects in vitro
     and in women with unstable bladder, Acta.Pharmacol. et
     Toxicol. 46 Suppl. 1:31 (1980).
55. S. Husted, K. E. Andersson, L. Sommer and J. R. Ostergaard,
     Anticholinergic and calcium antagonistic effects of terod-
     iline in rabbit urinary bladder, Acta.Pharmacol. et Toxicol.
     46 Suppl. 1:20 (1980).
56. T. L. Gerstenberg, P. Klanskor, D. Ramirez and T. Hald, The
     effect of terodiline in women with motor urge incontinence
     in: "XIth Annual Meeting International Continent Society
     Lund," Skogs Trelleborg (1980).
57. J. G. Lewis, Adverse Drug Reaction Bulletin 89:324 (1981).
58. M. J. Bultitude, N. H. Hills and K. E. D. Shuttleworth,
     Clinical and experimental studies on the action of prosta-
     glandins and their synthesis inhibitors on detrusor muscle
     in vitro and in vivo, Brit.J.Urol. 48:631 (1976).
59. P. H. Abrams and R. C. L. Feneley, The action of prostaglandins
     on smooth muscle of the human urinary tract in vitro,
     Brit.J.Urol.47:909 (1976).
60. L. D. Cardozo and S. L. Stanton, A comparison between bromo-
     criptine and indomethacin in the treatment of detrusor
     instability, in: "VIIIth Annual Meeting International
     Continence Society Manchester," Pergamon Press (1978).
61. L. D. Cardozo, S. L. Stanton, H. Robinson and D. Hole,
     Evaluation of flurbiprofen in detrusor instability,
     Brit.med.J. i:281 (1980).
62. P. J. R. Shah, P. H. Abrams and G. Choa, Flurbiprofen and the
     treatment of male detrusor instability, in: "XIth Annual
     Meeting International Continence Society Lund," Skogs
     Trelleborg (1981).
63. J. Nakano and M. C. Ross, Pathophysiologic roles of prosta-
     glandins and the action of aspirin-like drugs, Sth.Med.J.
     66:709 Birmingham (1973).

64. M. Gavanski, Treatment of non-retentive secondary encoprens with imipramine and psychotherapy, CMA Journal 104:46 (1971).

65. A. A. Goldfarb and F. Venutolo, The use of an antidepressant drug in chronically allergic individuals, Annal.Allerg. 21:67 (1963).

66. A. D. Korcyn and I. Kish, The mechanism of imipramine in enuresis nocturnal, clinical and experimental, Pharmacology and Physiology 6:32 (1979).

67. K. E. Peterson, O. O. Andersen and T. Hansen, Mode of action and relative value of imipramine and similar drugs in the treatment of nocturnal enuresis, Europ.J.Clin.Pharmacol. 7:187 (1974).

68. C. M. Castleden, C. F. George, A. G. Renwick and M. J. Asher, Imipramine – a possible alternative to current therapy for urinary incontinence in the elderly, J.Urol. 125:318 (1981).

69. D. T. Mahony, Observations on sphincter-augmenting effect of imipramine in children with urinary incontinence, Urology 1:317 (1973).

70. A. T. Coles and F. A. Fried, Favourable experiences with imipramine in the treatment of neurogenic bladder, J.Urol. 107:44 (1972).

71. L. S. Goodman and A. Gillman, "The Pharmacological Basis of Therapeutics," MacMillan, New York (1975).

72. K. E. Creed and A. S. S. Tulloch, The action of imipramine on the lower urinary tract of the dog, Brit.J.Urol54:5 (1982).

73. O. P. Khanna, D. Heber, G. Elkouss and P. Garrick, Imipramine hydrochloride, pharmacodynamic effects on lower urinary tracts of female docts, Urology 6:48 (1975).

74. P. D. Kanof and P. Greengard, Brain histamine receptors as targets for antidepressant drugs, Nature 272:329 (1978).

75. C. M. Fredericks, R. L. Green and G. F. Anderson, Comparative in vitro effects of imipramine, oxybutinin and flavoxate on rabbit detrusor, Urology 12:487 (1978).

76. T. F. Anders and P. Weinstein, Sleep and its disorders in infants and children, A Review J.Paed. 50:377 (1972).

77. R. J. Broughton, Sleep disorders: Disorders of Arousal? Science 159:10170 (1966).

78. E. R. Ritvo, Arousal and non-arousal enuretic events, Amer.J. Psych. 126:77 (1969).

79. E. R. Ritvo, Effects of imipramine in the sleep-dream cycle: An EEG study in boys, Electroenceph.Clinic Neurophysiol. 22:465 (1967).

80. E. Hartman, Antidepressants and sleep in: "Sleep: Physiology and Pathology," A. Kales, ed., Lippincott, Philadelphia (1969).

81. J. Lapides, Physiology of the urinary sphincter and its relation to operations for incontinence, Brit.J.Urol. 33:284 (1961).

82. R. M. Balyeat and J. M. Rinkel, Urinary retention due to the use of ephedrine, J.Amer.Med.Ass. 98:1545 (1932).

83. O. R. Langworthy, A new approach to the diagnosis and treatment of disorder of micturition in diseases of the nervous system, Int.Clin.3:98 (1936).

84. A. C. Diokino and M. Taub, Ephedrine in the treatment of urinary incontinence, Urology 5:634 (1976).

85. B. H. Stewart, L. H. W. Banowsky and D. K. Montague, Stress incontinence: Conservative therapy with sympathomimetic drugs, J.Urol.115:558 (1976).

86. L. Norlen, T. Sandin and F. Wagstein, Beta adrenoceptor stimulation of the human urinary bladder in vivo, Acta. Pharmacol. et Toxicol. 43 Suppl. 11:26 (1978).

87. A. D. G. Brown, E. P. Arnold and P. J. L. Worth, The unstable bladder: A study of drug treatment in women, Unpublished.

88. D. J. Farrar and J. L. Austin, The use of bromocriptine in the treatment of the unstable bladder, Brit.J.Urol.48:235 (1976).

89. K. P. J. Delaere, F. M. J. Debrayne and W. A. Moorien, Has bromocriptine a place in the treatment of the unstable bladder? Brit.J.Urol.50:167 (1978).

90. E. Pedersen and V. Grynderup, Effect of hydramitrazine in the treatment of neurogenic bladder dysfunction, Brit.med.J. 2:271 (1966).

91. S. Helsjorn, Treatment of detrusor hyperreflexia in multiple sclerosis. A double-blind, crossover, clinical trial comparing methantheline bromide, flavoxate chloride and meladrazine tartrate, Urol.Int.32:209 (1977).

92. M. C. Taylor and C. P. Bates, A double-blind, crossover trial of Baclofen - a new treatment for the unstable bladder syndrome, in: "IXth Annual Meeting International Continence Society Rome," Guido Guidotti, ed. (1979).

93. A. C. Diokino, S. A. Ross and W. Anderson, Combined cystometry and perineal electromyography in the diagnosis and treatment of neurogenic urinary incontinence, J.Urol. 115:161 (1976).

94. A. C. Diokino, S. A. Ross and L. F. Bender, Periurethral striated muscle activity in neurogenic bladder dysfunction, J.Urol. 112:743 (1974).

95. T. Sundin and I. Peterson, Cystometry and simultaneous electromyography from the striated urethral and anal sphincters, Unpublished.

96. A. B. Rossier, A. Bushra, M. Fam Di Beneditto and M. Sarkarati, Urethro-vesical function during spinal shock, Urol.Res. 8:53 (1980).

97. J. Nordling, The effects of alpha blocking agents on the urethral pressure in neurological patients in: "Proceedings of the VIIth Annual Meeting International Continence Society,"

98. S. Raz and R. B. Smith, External sphincter spasticity syndrome in female patients, J.Urol. 115:443 (1976).

99. Koyanagi, Denervation supersensitivity of the urethra to alpha -adrenergics in chronic neurogenic bladder, Urol.Res. 6:89 (1978).

100. H. Hedlund, K. E. Andersson, A. Ek, Effects of Prazosin on
     the human bladder and urethra, Unpublished (1979).
101. J. J. Kaufman and W. E. Goodwin, Hormonal management of the
     benign obstructing prostate: Use of combined androgen-
     oestrogen therapy, J.Urol.81:165 (1959).
102. J. Geller, C. G. Nelson, J. D. Albert and C. Pratt, Effects
     of megestrol acetate on uroflow rates in patients with
     benign prostatic hypertrophy, Urology Vol.XIV N.5 (1979).
103. D. J. Fanar and J. S. Prior, The effect of bromocriptine in
     patients with benign prostatic hypertrophy, Brit.J.Urol.
     48:73 (1976).
104. J. E. Castro, H. J. L. Griffiths and D. E. Edwards, A double-
     blind, controlled, clinical trial of spironolactone for
     benign prostatic hypertrophy, Brit.J.Surg. 58:485 (1971).
105. M. Caine, S. Ras and M. Zeigler, Adrenergic and cholinergic
     receptors in the human prostate, prostatic capsule and
     bladder neck, Brit.J.Urol. 47:193 (1975).
106. M. Caine, A. Pfau and S. Perlberg, The use of adrenergic
     blockers in benign prostatic obstruction, Brit.J.Urol.
     48:255 (1976).
107. P. F. Boreham, P. Braithwaite, P. Milewski, H. Pearson,
     Alpha adrenergic blockers in prostatism, Brit.J.Surg.
     64:756 (1977).
108. M. Caine, S. Perlberg, S. Meretyk, A placebo controlled
     double blind study of the effect of phenobenzamine in
     benign prostatic obstruction, Brit.J.Urol. 50:551 (1978).
109. M. Caine, Can benign hypertrophy of the prostate be treated
     medically? Ann.Urol. 13:159 (1979).
112. J. Osborne, Post menopausal changes in micturition habits and
     urine flow and urethral pressure studies, in: "Management
     of Menopause and Post Menopausal Years," S. Campbell,ed.,
     M.T.P. Press Ltd., Lancaster (1976).
113. M. I. Whitehead, J. Minardi, Y. Kitchen and M. J. Sharples,
     Systemic absorption of oestrogen from premanin vaginal
     cream, in: "The Role of Oestrogen/Progesterone in Manage-
     ment of the Menopause," I. D. Cook, ed., M.T.P. Press Ltd.,
     Lancaster (1978).
114. O. Widhohn and E. Vartianen, The absorption of sodium
     oestrone sulphate from the vagina, Ann.Chir.Gynaecol.
     63:186 (1974).
115. G. I. Dyer, O. Young, P. T. Townsend, W. P. Collins, M. I.
     Whitehead and J. Jelowitz, Dose related changes in vaginal
     cytology after topical conjugated equine oestrogens, OC27,
     88B, Brit.med.J. 1:789 (1982).
116. J. Condo, Clinical evaluation of the effects of Ubretid on
     neurogenic bladder dysfunction and high serum beta-
     glucuronidase activity, Acta.Urol.Japan 14:158 (1968).
117. R. Chiba, K. Imabayashi, M. Igarachi, S. Saside, S. Matsumara
     and J. Satake, Experimental studies on the use of Ubretid
     in the field of urology especially in the neurogenic
     bladder, New Remedies and Clinics Japan 16:1368 (1967).

118. J. Njima and M. Asano, Clinical appraisal of Ubretid a new anti-cholinesterase inhibitor in urology, Acta Urol.Japan 13:423 (1967).

119. J. Lapides, Urecholine regime for resolving the atonic bladder, J.Urol. 19:658 (1964).

120. C. Burnstock, T. Cocks, R. Crow, L. Kasakov, Purinergic innovation of the guinea pig urinary bladder, Br.J.Pharmcol. 63:125 (1978).

121. F. Furuyas, Kumamotoy Yoko, Yamae Tsakumotot, Suppressive effect of intrathecal morphine in detrusor hyper-reflexia, Proceedings of the XIIth Annual Meeting of the ICS, Leiden, September (1982).

122. C. Owman, and Sjöstrand, Short, adrenergic neurones and catecholamines containing cells in vas deferens and accessory male genital glands of different mammals, Zeitschrift Für Zelforschung, 66 300 (1965).

123. S. Ras and M. Caine, Adrenergic receptors in the female canine urethra investigative urology, 9:319 (1972).

124. S. Ras, M. Zeigler, M. Caine, Isometric studies on canine urethral musculature, Urology 9:443 (1972).

125. S. Ras, M. Zeigler, M. Caine, Pharmacological receptors in the prostate, Br.J.Urol. 45:663 (1973).

126. M. Caine S. Ras M. Zeigler, Adrenergic and cholinergic receptors in the human prostate, prostatic capsule and bladder neck, Br.J.Urol. 47:193 (1975).

127 V. V. Frolkis, The autonomic nervous system in the ageing organism, Triangle 8:322 (1968).

128. K. F. Gey, W. P. Burkard, A. Pletscher, Variation of norepinephrine metabolism of the rat heart with age, Gerontologia 11:1 (1965).

129. D. F. Chebotref, Principien der medickamentozen therapie in alter, in: "Gerontology," B. Steinman, ed., Bern Haus Huber (1973).

130. V. G. Shevchulz, The effect of stimulation and blocking of adreno receptors upon haemodynamic indices in animals of different age, Pharmakol.Tokiskue 37:322 (1974).

131. U. Trendelenberg, Supersensitivity and subsensitivity to sympathomimetic amines, Pharmacol.Review 15:225 (1972).

132. U. Trendelenberg, Factors influencing the concentration of catecholamines at the receptors, Handbook der Experimentellen Pharmakologie, 33:726 (1972).

133. D. M. Cambrough and H. C. Hartzell, Acetylcholine receptors: Number and distribution at neuromuscular junctions in rat diaphram, Science N.Y. 176:189 (1972).

134. R. Bannister, B. Davies, E. Holly, T. Rosenthal and P. Sever, Defective cardiovascular reflexes and supersensitivity to sympathomimetic drugs in autonomic failure, Brain 102:163 (1979).

135. Scientific data supplied by Farmitalia Carlo Erba Ltd. (1982).

136. F. Rochi, A. Margonato, R. Ceccardi, P. Rigatti and B. M.
      Rossini, Symptomatic treatment of benign prostatic
      obstruction with nicergoline; A placebo controlled clinical
      study and urodynamic evaluation, in print (1983).

PERSPECTIVES ON THE PROPHYLAXIS AND THERAPY OF

THROMBOEMBOLISM IN THE AGED

Pietro de Nicola

Institute of Gerontology and Geriatrics
University of Pavia
Pavia, Italy

In a meeting on the pharmacotherapy of the aged, a session on thromboembolism forms an essential part, due to the frequency, to the severity and to the consequences of this condition;namely there is a steep increase of the occurrence of thromboembolism in the presenile age, i.e. after 45, up to the actual senile age. The severity is due to the fact that vital organs are affected such as the brain and the heart. The consequences are represented by disability, thus requiring rehabilitation measures, which are in effect the quintessence of modern geriatrics (Table 1).

The increased incidence of thromboembolism in the aged may be explained on the basis of the modifications of blood coagulation and haemostasis during aging. In the presenile age there is an increase in blood coagulability and in platelet aggregation, and a decrease in fibrinolytic activity. These modifications are most evident between 50 and 60 and may be observed also during the subsequent few years. After the age of 70 a reversal of these findings may occur, and even a reduction of blood coagulability may be observed, which may account for the onset of haemorrhagic episodes, also in relation to the increased capillary fragility of the actual senile age (Table 2).

AETIOPATHOGENETIC FUNDAMENTALS

We cannot speak of prophylaxis and therapy of thromboembolism without considering the aetiopathogenetic fundamentals and the new developments in this respect. More than one century ago Virchow had proposed his well known triad, with parietal, haematic and haemodynamic factors (Table 3). They are still valid today, with some enrichments.

441

Table 1. Thrombophilia = tendency toward intravascular
          formation of thrombi (= thrombosis)

Medical importance:

formation of thrombi in vital
organs (heart, brain)

Social importance:

invalidity states
reduction of the working capacity
increased needs of medical, social,
economic care

Haemodynamic factors become haemorheology, parietal factors
are related to the modern developments in arteriosclerosis, and
haematic factors are concerned with the tremendous, explosive
evolution of knowledge on blood coagulation and haemostasis during
the last fifty years.

Blood coagulation and fibrinolysis

What's new and important in blood coagulation and haemostasis
with respect to thromboembolism? Around the middle of the century
attention was almost only paid to blood coagulation, trying to
establish a correlation between any of the newly discovered blood
coagulation factors and the tendency to thrombosis, i.e. thrombophilia.

These attempts were destined to wreck and to be replaced by the
most recent knowledge. The discovery of the molecular structure
of prothrombin, factor X and other plasmatic factors of blood
coagulation made it possible to identify the actual trigger mechanism in the activation of prothrombin, to establish which factors
are involved in thrombophilia and to explain how anticoagulants act.

Table 2. Modifications of haemostasis in the aged

Presenile age (45 to 60 years)

Tendency toward thrombophilia with

- increased coagulability and
  platelet aggregation
- reduction of fibrinolytic activity

Actual senescence (over 72 years)

- normal or decreased coagulability
- increased capillary fragility
- tendency to haemorrhages

Table 3. Thrombophilia

Virchow's Triad

Parietal factors: atherosclerosis

Haemodynamic factors: rheology
        (macro- and microcirculation)

Haematic factors: coagulation
                   fibrinolysis
                   aggregation
                   viscosity

The old scheme of the intrinsic and extrinsic mechanism of blood coagulation (Figure 1) has been modified by the introduction of the cascade of enzymatic reactions, like in a fountain, and above all by the central, dominant role of factor X activation in the activation of prothrombin. In the cascade of enzymatic reactions the primum movens or trigger mechanism is still the so-called contact factor, or factor XII, which is activated by the contact with foreign surfaces and is closely related to fibrinolysis, complement and kinins. There should be no better possibility for evaluating a thrombophilic tendency than the quantitative assay of factor XII, just on the basis of its function at the starting point of blood coagulation and its correlation with other important systems, but unfortunately, this is not yet possible.

By contrast, much is known about the activation of factor X and prothrombin through a proteolytic process (Figure 2). The molecules of these two factors are surprisingly similar also in the splitting mechanism, resulting in the formation of activated factor X (or factor Xa) and of activated prothrombin, i.e. thrombin. Factor V, phospholipids and calcium are required for these reactions, which become slow if any of these factors is missing or not in the optimal concentration. This concept had been anticipated by Armand J. Quick using simple experiments, and was confirmed during the last few years as a result of studies on the molecular structure and reactions of these factors.

In these studies the presence of gamma carboxyglutamic acid residues was identified in the molecules of prothrombin, factor X and other plasmatic factors. During treatment with indirect anticoagulants i.e. antivitamins K, no residues of gamma-carboxy-glutamic acid were found, due to the formation of abnormal proteins. Vitamin K is essential for the formation of this acid, which cannot be synthesized when vitamin K is inhibited or lacking. The abnormal proteins which are thus formed, such as prothrombin and factor X, act as inhibitors of activated factor Xa, through a mechanism of competitive inhibition due to their structural analogy, (as for folic

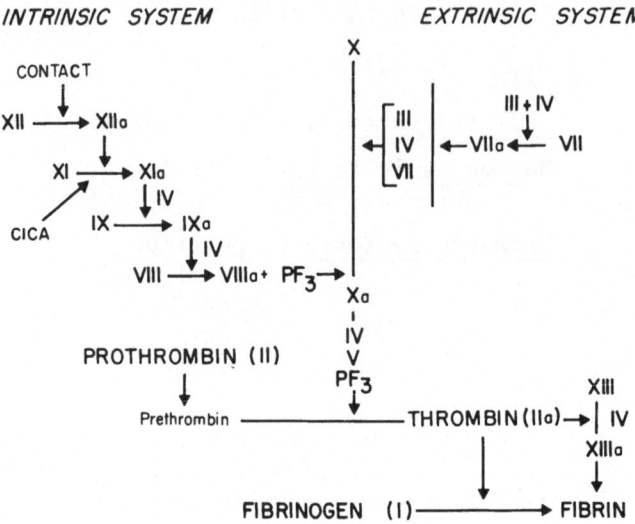

Fig. 1.   Perspectives on the prophylaxis and
           therapy of thromboembolism in the
           aged.

acid and antifolics, ADP and its inhibiting analogues, sulphonamides
and para-aminobenzoic acid).   Inhibition of factor Xa represents
the most important effect of antithrombin III as well as of heparin.
Thus, indirect anticoagulants act like antithrombin III, i.e. in
inhibiting factor Xa (Table 4).

     This represents the most impressive, almost paradoxical
consequence of these studies on the molecular structure of coagulation factors, establishing an unbelievable analogy between indirect
and direct anticoagulants in their mechanisms of action.   Furthermore, the recent developments in antithrombin III have shown that
there is a definite reduction of its activity in thrombophilic
states, including those observed in women given contraceptive drugs.
This finding justifies the use of drugs enhancing the activity of
antithrombin III, such as anticoagulants, both direct, like heparin,
and indirect.

     Besides factor Xa as related to antithrombin III, other
plasmatic factors have been considered with respect to thrombophilia,
particularly factor VIII (coagulant fraction) and fibrinogen (factor
I).   For factor VIII, in addition to the coagulant fraction, which
is related to the pathogenesis of haemophilia, and is increased in
thrombophilic states, also the other fraction, the so-called von
Willebrand factor, (Table 5), has been found to be increased in
these conditions, including diabetes and arteriosclerosis.

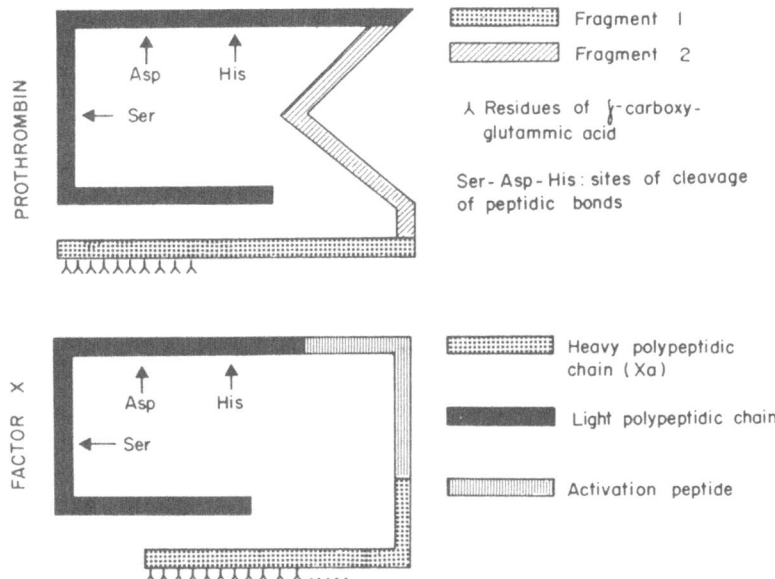

Fig. 2.   Perspectives on the prophylaxis and therapy
          of thromboembolism in the aged.

    As far as factor I is concerned, in addition to the actual
increase in its level, there is an increased fibrinogen turnover,
with a reduction of its half-life time.   Such finding has been
interpreted as a sign of a latent disseminated intravascular
coagulation syndrome, without haemorrhagic manifestations, but with
a thrombophilic tendency (Table 6).   The administration of anticoagulants may counteract the increased fibrinogen turnover (Figure
3), thus representing a further justification for their use in the

Table 4.

INDIRECT ANTICOAGULANTS

Diminution of factors II, VII, IX, X

but also

Formation of endogenous inhibitors
(PIVKA)   inhibition of factor Xa

increase of antithrombin III

DIRECT ANTICOAGULANTS

Increase of antithrombin III
(= inhibition of factor Xa)

THROMBOPHILIC STATES

Diminution of antithrombin III

Table 5. Fractions of Factor VIII

F VIII R:C   Coagulant
F VIII R:vW   Platelet adhesion
                    to collagen
                    (ristocetin)
F VIII R:AG   Antigenic
                    (synthesized in
                    endothelial cells)

prophylaxis of thromboembolism.

In studies on the pathogenesis of thrombophilia and of thromboembolism it was soon recognized, however, that blood coagulation
did not represent the whole story. The second front in these
investigations was opened by fibrinolysis, during the 1950s, on the
basis of the decreased fibrinolytic activity in these diseases.
Furthermore there was the ambitious idea of finding out a counterpart to anticoagulants for prophylactic purposes by means of drugs
activating fibrinolysis, and above all to discover an actual
thrombolytic therapy. This in fact took place in the course of
these studies.

A reduction of the fibrinolytic activity or increase of the
fibrinolysis inhibitors (Table 7) can be observed in cases of
thrombophilia. These findings may be useful also for diagnostic
purposes.

These trends were supported also by the newly developed theory
of the haemostatic balance: namely, it was postulated that under
normal conditions there should be a balance between coagulation and
fibrinolysis, thus ensuring a normal haemostasis (Figure 4). The
onset of haemorrhages should be due to a deficiency of the coagulation mechanism or to an excess of fibrinolytic activity, and the
opposite should take place in case of thrombosis, with a reduction
of fibrinolysis and an excess of coagulation or hypercoagulability
(Table 8).

Table 6. Disseminated intravascular coagulation
          syndrome = thrombophilia

Malignancies   with haemorrhagic
Haemoblastosis   syndrome
Liver cirrhosis
Arteriosclerosis   heparin therapy

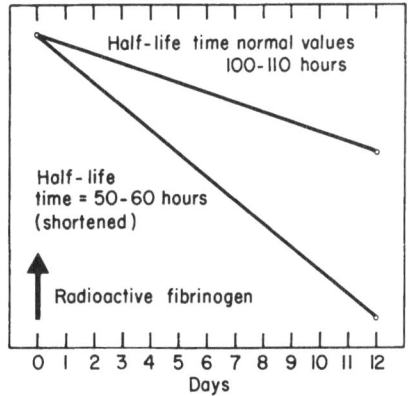

Fig. 3.   Perspective on the prophylaxis and therapy
          of thromboembolism in the aged.

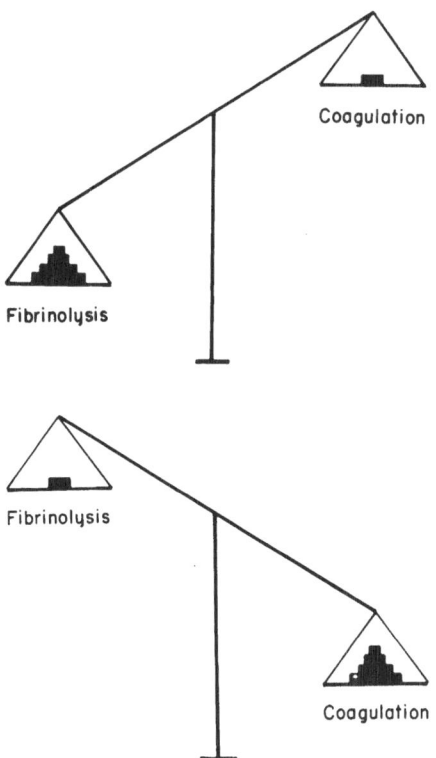

Fig. 4.   Perspectives on the prophylaxis and therapy
          of thromboembolism in the aged.

Table 7. Plasmin inhibitors

Alpha-2-plasmin inhibitor
Alpha-2-macroglobulin
Alpha-1-antitrypsin
Inter-alpha antitrypsin
C-1-esterase inhibitor

Platelets

A third front in studies on thrombophilia was opened by platelets, during the 1960s, even though the role of platelet adhesion and aggregation had been already identified one century ago by the Italian Bizzozero (Figure 5). The increase of platelet aggregation in thrombophilic states is still of utmost importance.

Today there are, however, other parameters to be considered in this respect. First of all, the role of thromboxanes and prostacyclin as derivatives of arachidonic acid was clearly established in the mechanism of platelet aggregation. A fundamental step in the formation of these substances, all belonging to

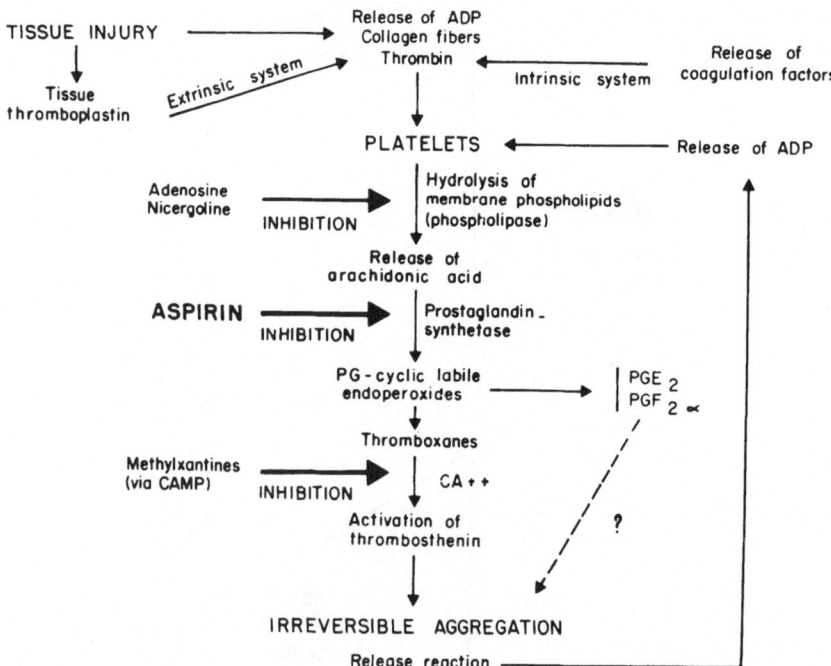

Fig. 5. Perspectives on the prophylaxis and therapy of thromboembolism in the aged.

Table 8. Haemostatic balance

Diminuation of   )
    coagulability   )   Tendency
)   to
Increase of   )   haemorrhages
    fibrinolytic activity)
)

Increase of   )
    coagulability   )   Tendency
)   to
Decrease of   )   thrombosis
    fibrinolytic activity)
)

the group of prostaglandins, is represented by the formation of
cyclic endoperoxides from arachidonic acid in the presence of
cyclo-oxygenase.   This phase is particularly important insofar as
cyclical endoperoxides are the precursors of both thromboxanes, in
the presence of thromboxane-synthetase, and of prostacyclin
(Table 9).   The synthesis of thromboxanes takes place in platelets,
while the synthesis of prostacyclin is due to the vessel wall,
i.e. up to 90 per cent in the intima, and 10 per cent in the media
and adventitia.   Thromboxanes are potent aggregating agents,
while prostacyclin has a marked antaiaggregating action.

It is quite understandable how these two opposite actions are
important also from the diagnostic and therapeutic viewpoints.
In effect it is possible to evaluate quantitatively thromboxanes
and prostacyclin in the blood, which are the most potent aggregating and antiaggregating agents, respectively.   The ratio between
these two substances should be the real key for establishing the
role of platelet aggregation in the pathogenesis of thrombophilia.
At least thromboxanes may be now currently evaluated, and probably
also for prostacyclin there will be a dependable method of assay
in the coming future.

During the release reaction of platelets in the course of
aggregation, particular importance was assigned to a substance
derived from platelets, beta-thromboglobulin, which may be increased
in thrombophilic states, in arteriosclerosis and also in diabetes.

Microcirculation

Besides blood coagulation, fibrinolysis and platelets, which
represent the haematic factors in Virchow's triad, haemodynamic
factors are considered today within the newly developed haemorrhagic
studies and, in general, microcirculation.   The term microcirculation actually represents the counterpart of macrocirculation; as
the latter deals with the large vessels and their alterations, e.g.

Table 9.

Arachidonic acid

Cyclo-oxygenase

Cyclic endoperoxides
Thromboxane-
synthetase

Thromboxane A_2

Thromboxane B_2

Prostacyclin

in cardiovascular diseases, the former is concerned with the small
vessels, thus including morphologic and haemodynamic aspects, which
are both important also in the aetiopathogenesis of thromboembolism.

What is the actual meaning of microcirculation? and when was
the term microcirculation introduced in current use? Just considering the historical development of microcirculation (Table 10)
one should say that the first observations on the small vessels
are due to the Italian Malphighi, in the seventeenth century, but
the actual commencement of these studies is relatively recent, after
the introduction of the capillaroscopic techniques for the identification of the small vessels in vivo in humans, as well as in
experimental animals. This trend was particularly applied during
the first decades of this century, and became stronger and stronger
during the forties, especially due to the fundamental work of Bloch
and Knisely, who established the following principles: the modifications of the small vessels in one region of the organism are
largely superimposable to those which are observed also in other
regions.

This statement is particularly important, because it does not
only apply to animals, but also to humans, in whom only a few
regions may be examined in vivo, e.g. the conjunctiva and the nailbed. Furthermore, there are also other implications in this
statement, insofar as the modifications of the small vessels may be
due to the administration of drugs, and general and not limited
conclusions may be therefore drawn from the findings in the
conjunctiva and in the nailbed.

This was the actual beginning of studies on microcirculation
in Europe and all over the world, and this term was selected in
contradistinction to macrocirculation, i.e. to the study of the
large vessels, which is the aim of cardiovascular research sensu
strictiori.

Table 10. History of microcirculation

1652 Marcello Malpighi: anastomoses
between arteries and veins

1674 Leewenhoech: red cells

1768 Spallanzani: white cells

1823 Purkinje: skin capillaroscopy

1850 Coccius: conjunctival capillaroscopy

1882 Bizzozero: platelets

1922 O. Müller: fundamental studies
on capillaroscopy

1946 Knisely and Bloch: fundamental
concepts on microcirculation

But, just when the term microcirculation was introduced, some
other trends within the studies on microcirculation were followed,
insofar as also the haemodynamic aspects were considered, thus
developing the studies on haemorheology or bio-haemorheology, i.e.
on the characteristics of blood flow in the vessels. Today,
haemorheology actually represents a subdivision of microcirculation
or even an independent branch although the two terms are often
considered almost synonims (Table 11).

In haemorheology, particularly blood viscosity and red cell
deformability are studied, and there are today experts of these
very particular aspects of blood flow, with important physiopatho-
logic and pharmacologic implications.

A tendency toward thromboembolism can be identified through
the evaluation of some microcirculation and haemorheologic parameters. For instance, an increase of blood viscosity and a
reduction of erythrocyte deformability may be observed in patients
who are predisposed to thromboembolism, and certain drugs, which
are used for prophylactic purposes, may counteract these alterations.
From the morphologic viewpoint, some capillaroscopic findings are
typical of the states, in which thromboembolism is likely to occur,
as for instance dilatations and tortuosities, as well as stasis
and congestion in the small vessels, intravasal erythrocyte
aggregation, reduction of speed and homogeneity in the blood stream,
etc. All these data have added new criteria of evaluation of
thrombophilic states, and represent an important aspect in the
study of microcirculation.

Table 11.   Definition of microcirculation

study of the small vessels
morphologic as well as

haemodynamic

biorheology

haemorheology

viscosity and
red cell deformability

Lipidic metabolism

Today, parietal factors in Virchow's triad include all the
studies on arteriosclerosis in general, as well as the alterations
of lipidic metabolism, which are involved thereby.

Without entering into details about the pathogenesis of atherosclerosis and the interrelationships between atherosclerosis and
dyslipidemia, it has been already well known for many years that
lipids may enhance blood coagulation and platelet aggregation and
inhibit fibrinolysis, thus predisposing to thrombosis.   Drugs
which reduce blood lipids are also able to counteract these modifications and, therefore, also the tendency to thrombosis.   In our
studies on this subject several drugs were used both in animals
and humans, and the validity of these assumptions was widely
confirmed (Table 12).

It may be interesting to mention the fact that lipids cause
the onset of intravascular red cell aggregation, thus further
contributing to the thrombotic tendency.   If alcohol is given at
the same time as lipids in experimental subjects, the effect of
lipids is considerably reduced also in this respect.

DIAGNOSTIC PROBLEMS

From a practical viewpoint, all these data on the role of
haemostasis, microcirculation and arteriosclerosis in the pathogenesis of thromboembolism, have also had consequences for diagnostic purposes.   As far as blood coagulation is concerned,
antithrombin III, and, if possible, factor VIII (coagulant fraction)
should be currently evaluated, in addition to the old and still
valid thrombelastographic method (Table 13).   For the detection of
a latent disseminated intravascular coagulation syndrome, there are
specific methods, which make it possible to identify the fibrinogen
degradation products due to intravascular coagulation, as well as
those due to the consequent reactive fibrinolysis.   However, they

Table 12. Experimental and clinical data
on lipids and thrombophilia

1. enhancement of blood coagulation and
   platelet aggregation;

2. inhibition of fibrinolysis;

3. intravascular red cell aggregation.

cannot be used currently and the simplest, even though rather
empiric method for this purpose, is represented by the evaluation
of the fibrin monomere, which is the very demonstration of an
initial intravascular coagulation (Table 14).

Platelet aggregation, and if possible, also thromboxanes and
beta-thromboglobulin in the blood should be evaluated. No value
should be assigned to the study of the vitamin K factors, nor to
prothrombin time itself.

Studies on the microcirculation and on lipids may give further
information on their role in the pathogenesis of thromboembolism.

PROPHYLAXIS

The complexity of the pathogenetic mechanism in thrombophilic
states represents the background for a variety of prophylactic
measures, thus preventing the onset of thrombo-embolic disease.
On the basis of the pathogenetic importance of athero- and arterio-

Table 13. Diagnosis of thrombophilia

Always: Thrombelastography
        Platelet aggregation
        Antithrombin III

If possible: fibrinogen turnover

Limited value: Factor VIII
               Heparin tolerance
               Clotting time

No value: Prothrombin time
           Factors II, V, VII, X

IMPORTANT: identification of
predisposing factors: obesity,
diabetes, hypertension, hyper-
dyslipidaemia, etc.

Table 14.   Demonstration of the FDP due to the
            action of thrombin (coagulation)

1. Protamine sulphate: precipitation of fibrin
   monomers
   not sensitive enough

2. Thrombin (or coagulase)/time not specific

3. Diminution of factors I, II, V, VIII,
   platelets   not sensitive enough

4. Fibrin monomer: specific

sclerotic changes in the arterial wall, one of the most important
treatments should be represented by drugs, which are able to reduce
the extension and the entity of already present lesions and to
prevent the occurrence of further lesions.   Furthermore, other
predisposing factors and diseases should be properly treated, in
order to remove or at least reduce them (Table 15).   This is
particularly true for hypertension, diabetes, obesity, and others.
As soon as such a goal is reached, other treatments against thrombophilia may be discontinued.

Prolonged treatment with anticoagulants for prophylactic
purposes has been used for a long time and is still widely discussed
and criticized.   For long-term treatment, oral anticoagulants have
been mostly employed, such as mono- and dicoumarinic compounds and
indanedione derivatives.   These drugs are only able to reduce blood
coagulability, thus influencing only one of the pathogenetic factors,
but just in a superficial, symptomatic, and transitory way, without
modifying the conditions, which caused the increase of blood coagulability.   Prothrombin (factor I) and factors VII, IX, and X, i.e.
the so-called vitamin-k-dependent factors, are reduced under the
action of oral anticoagulants.   The effect takes place after a
latency period of around 24 h, and, unless a tendency to accumulation is present, the effect is no longer demonstrable within 24-48
h after the last therapeutic dose.

Oral anticoagulants require regular and reliable controls by
means of the prothrombin time, reflecting the variations of
prothrombin (factor II), factor VII and X, among the factors
affected by oral anticoagulants, or by means of the thrombotest,
reflecting the fluctuations of all these factors (prothrombin
(factor II), and factors VII, IX, and X).

Controls are necessary in order to avoid an excessive anticoagulant effect, with risk of haemorrhage, or an insufficient
therapeutic effect, thus predisposing to relapses of thrombosis or
embolism.

Table 15. Thrombophilia

Aetiology

Hereditary, constitutional

Acquired: atherosclerosis
          dyslipidaemia
          obesity
          diabetes
          hypertension
          malignancies

A therapeutic level has been suggested on the basis of the evaluation of per cent values of prothrombin time, i.e. around 20 - 25 per cent. Therefore, values of 40 - 50 per cent are insufficient, especially if we also consider the fact that normal values range between 70 and 130 per cent. Even when laboratory data are accurate and therapeutic levels are regularly obtained, the risk of haemorrhage is not completely avoided, insofar as other factors besides blood coagulation, may be involved in the onset of haemorrhages, above all capillary fragility, which is increased in aged persons.

The effect of oral anticoagulants may be increased or reduced by concomitant diseases, by drugs, by stress, and by other factors, which should be accurately valued during a long-term anticoagulants in cachectic patients, in malnutrition, liver and renal diseases, and in congestive heart failure with liver enlargement, in which there is an increased sensitivity to oral anticoagulants, thus requiring more frequent and accurate controls.

Cortisone and its derivatives reduce the effect of oral anti-coagulants, while barbiturates, MAO-inhibitors, and other drugs potentiate it (Table 16).

The true value of a prolonged treatment with oral anticoagulants is still being debated. After the initial enthusiastic statements about the prophylactic effect of anticoagulants in coronary heart disease, a more cautious re-evaluation took place, and, without denying any favourable effect of oral anticoagulants, the present attitude is to administer them when a marked thrombophilic state is demonstrable on the basis of clinical and laboratory data. This particularly applies to obese, atherosclerotic, hyperdyslipemic, diabetic patients, in whom there is also a history of previous thrombo-embolic accidents. Such treatment may be continued as long as these predisposing factors are present, and a good policy is to start as soon as possible a correct treatment of these factors, as already mentioned, so that the anticoagulant treatment may be discontinued as soon as conditions permit.

Table 16.   Effect of oral anticoagulants

Increased or reduced by drugs,
    concomitant diseases, stress

Examples:

- malnutrition, cachectic patients,
  liver and renal diseases, congestive
  heart failure   increased sensitivity;

- cortisone and derivatives
        reduced effect

- barbiturates, MAO-inhibitors
        increased effect

No anticoagulant treatment should be prescribed if no reliable laboratory controls are possible and if the attending physician has no experience in the correct way of conducting the treatment. In the aged, anticoagulants may be used with more caution than in other ages and if the general contraindications are not present.

Heparin and heparinoid substances have not been currently used for long-term anticoagulant treatment, insofar as the anticoagulant effect may be reached only by using high doses and parenteral routes, which are not always feasible for a long time, and predispose to the onset of haemorrhages even more than with the use of oral anticoagulants.   Heparinoids may be used for prophylactic purposes at low doses, which do not significantly affect blood coagulation such as, for instance, 5,000 units subcutaneously twice a day, while the high intravenous or intramuscular (retard) doses are indicated in the early phases of thrombosis and embolism, by properly evaluating the possible contraindications. For instance, heparin treatment is in general less suitable for cerebral accidents than for peripheral thrombosis (Table 17).

Today the most recommended trend is represented by the administration of drugs which exert a complex effect on various pathogenetic factors of thrombophilia, including fibrinolysis, platelet aggregation and vascular factors.   One of the first trials used in this respect was nicotinic acid, which proved to enhance fibrinolysis, in addition to its already known cardiovascular action and to the antihyperlipidemic effect (Table 18).

A prolonged activation of fibrinolysis of slight degree (i.e. not to be compared with the actual thrombolytic drugs) may be justified as is the prolonged anticoagulant treatment, insofar as they act on either one of the facets in the haemostatic balance. However, drugs acting on fibrinolysis are also often able to influence other pathogenetic factors of thrombophilia, such as the increasing platelet aggregation and hyperlipidaemia.

Table 17. Heparin and heparinoid substances

Not for long term anticoagulant
    treatment (high doses, i.v.;
    risk of haemorrhage)

Yes for prophylactic purposes at low doses
    (e.g. 5,000 units subcutaneously twice a
    day)

Several drugs are chiefly concerned with the inhibition of
platelet aggregation.   After the first trials with adenosine
derivatives, which are able to inhibit platelet aggregation through
a mechanism of competitive antagonism due to structural analogy
(as for sulphonamides and PABA), but are not always well tolerated,
consistent effects were obtained by means of dipyridamole, sulphinpyrazone, and acetylsalicylic acid, which are at the present time
the most recommended drugs for the inhibition of platelet aggregation (Table 19).

In many other drugs this effect was observed, with or without
a concomitant activation of fibrinolysis.   For instance, anti-
inflammatory drugs often inhibit platelet aggregation.   The mechanism of action at the different phases of platelet aggregation has

Table 18.   Indirect anticoagulants

Negative aspects:

- danger of haemorrhage
- periodical controls
- limited action: only
  blood coagulation

Antiaggregants
Fibrinolytic agents

— no dangers, no controls
- action on aggregation and
  fibrinolysis, but also:

    anti-hyperdyslipidaemic
    vasodilating (nicotinics)
    coronary (dipyridamol)
    anticoagulant (moderate)
            for heparinoids

Table 19. Inhibition of platelet aggregation

Dipyridamole   )
Sulphinpyrazone   )   most
Acetyl-salicylic acid[+]   )   used

+ Attention: haemorrhagic complications
(GI, cutaneous, nose, urinary)

been identified for several drugs. The administration of acetyl-
salicylic acid may be accompanied by haemorrhagic complications,
such as epistaxis, gastro-intestinal haemorrhages, cutaneous
haemorrhages, etc. which occur particularly in aged persons.

The site of gastro-intestinal bleeding cannot be easily
identified, also because the small erosions of the mucosa due to
aging are often the very cause of the haemorrhages. Gastro-
duodenoscopic observation may locate them. With other drugs
there is no risk of haemorrhages, e.g. with dipyridamole, which
also acts favourably on the coronary circulation.

The new developments in platelet aggregation had some
consequences on the choice of antiaggregating drugs. It is well
known that aspirin, counteracting both prostacyclin and thromboxanes, cannot be recommended in high doses, which are able to
inhibit also the favourable action of prostacyclin in relaxing
smooth muscles in the vessel wall. Small doses of aspirin are
supposed not to present this negative aspect, but the ideal drug
in this respect is still to be found.

Aspirin and indomethacin chiefly inhibit cyclo-oxygenase, thus
acting on cyclic endoperoxides and influencing both thromboxanes
and prostacyclin. Drugs acting only on thromboxane-synthetase
are to be preferred, for instance benzidamine-imidazole, but so far
there are no conclusive data on their use.

All these data on the prophylaxis of thromboembolism by means
of anticoagulants, fibrinolytics and antiaggregants certainly
represent fundamental advances in fighting thrombosis, but unfortunately their prophylactic action in vivo is not yet definitely
demonstrated in spite of the well established action in vitro and
the extremely detailed studies on their mechanism of action.

Even though several multicentre controlled trials were started
and concluded, we are not yet conclusively convinced about the
actual reduction of thromboembolism and above all about the actual
reduction in mortality after these treatments.

Table 20. Thrombolytic therapy
(streptokinase, urokinase)

Indications:

in recent cases of
- pulmonary infarction
- renal infarction
- peripheral thrombosis
  (femoral, popliteal, etc)

Myocardial infarction?

Cerebral thrombosis: caution!

Spectacular effects (not confirmed):
chronic aortic, iliac thrombosis

THERAPY

This is one of the reasons why other trends have been proposed in order to achieve a real cure of thrombosis, besides the prophylactic trials. One of them is represented by thrombolytic therapy (Table 20), which will be discussed later in detail by Professor Schmutzler.

Another possibility was offered by intravital defibrination, even though the data available are still relatively scarce. Intravital therapeutic defibrination by means of purified extracts of viper venom (ancrod and batroxobin) has been suggested on the basis of occasional observations in a woman after a snake bite, and preliminary favourable observations in myocardial infarction and other cases of thrombosis. The indications for intravital defibrination must still be clearly established and some perspectives have also been opened by the possible use of these agents for prolonged prophylaxis.

THROMBOLYTIC THERAPY IN THE ELDERLY

R. Schmutzler

Medical Clinic "Bergisch-Land"
Wuppertal, Germany

Suffering from the obliterative arterial disease is considered to be typical of the elderly. Therefore those practising geriatric medicine are interested in all the possibilities of treatment in this field.   This lecture is based on our experience of thrombolytic treatment in general, especially in chronic arterial occlusions and stenoses.[1,2,3]

Anticoagulant therapy with heparin, oral anticoagulants or inhibitors of platelet aggregation is primarily prophylactic, and when a thrombus has already formed can only prevent its growth, and reduce the frequency of thromboembolic complications.

Kwaan and Astrup[4] have shown, that as a result of local ischemia - stenosis, thrombosis, embolism - a fibrinolytic activator is released from the intima of the vessel wall.   The resultant spontaneous fibrinolytic activity is usually small, and though it may be enough to clear smaller occlusions, angiography has shown that only about 18 per cent of minor embolic arterial occlusions and about 1 per cent of arterial thromboses are subject to spontaneous lysis.[5]   Experience so far has shown that for the lysis of more extensive thrombi it is necessary to administer a powerful activator, such as high purified streptokinase obtained from hemolytic streptococci and/or the non-antigenic urokinase obtained from human urine or in the recent years from kidney cultures.

PRINCIPLES OF THERAPY

After streptokinase has overcome antistreptokinase, its immunological inhibitor in the plasma, it activates plasminogen

indirectly by way of a pro-activator-activator system to form the
active proteolytic enzyme plasmin (Figure 1). Plasmin is able to
destroy fibrin and particularly other coagulation factors. To
induce fibrinolytic activity the level of streptokinase must always
be higher than that of antistreptokinase. If the resulting
production of plasmin is large and rapidly achieved the binding
capacity of the anti-plasmin may be temporarily exceeded, causing
hyperplasminemia. A continued strong hyperplasminemia, in addition
to the desired fibrinolysis, brings about partial fibrinogenolysis
and reduction in other plasma proteins important for coagulation
such as Factors V and VIII, so that split products appear. This
results in a temporary coagulation defect. With the continued
activity of the streptokinase the greater part of the plasminogen
is finally converted into plasmin after 2 - 8 hours, and so the
hyperplasminemia gradually abates. The thrombus plasminogen
remains the appropriate substrate for the streptokinase (Figure 2).
Apparently the amounts of plasminogen within the thrombus are
sufficient to achieve the lysis of the thrombus fibrin when they
have been activated into plasmin. Since the plasma plasmin, which
is reversibly bound by α_2-antiplasmin and α_2-macroglobulin, is
possibly released in the presence of the thrombus surface fibrin, it
can contribute to exogenous thrombolysis. In this way the further
diffusion of streptokinase within the thrombus is assisted. It
seems correct to regard the mechanism of local thrombolysis as a
combined exogenous-endogenous form of lysis.[6]

CLINICAL RESULTS

     The fibrous organization of an intraluminar thrombus normally
lasts about 8 to 14 days, but is retarded by arteriosclerotic
changes of the vessel wall. Therefore, in stenoses or occlusions
of arteriosclerotic vessels produced through thrombus formation,
the fibrin remains accessible for fibrinolysis for a long period of
time.[3] (Figure 3).

     In 7 German medical centres (between 1968 and 1972) fibrinolytic
therapy with streptokinase (SK) was used in 708 patients with subacute or chronic peripheral arterial stenoses or occlusions.[7]
In all cases, thorough clinical angiological tests, mechanized
oscillography as well as arteriography were conducted in order to
establish the diagnosis and to control the results of treatment.
The improvement in walking distance was determined by using a
"conveyor-belt" ergometer. Seventy-five per cent of the patients
were within the ages of 50 to 70 years. The analysis of coagulation and fibrinolysis included the streptokinase resistance test to
determine the individual initial dose; the Quick test, the thrombin
time, and the level of fibrogen. The loading streptokinase dose
was either a schematic 250,000 units dose or the equivalent of the
individually titrated antistreptokinase content. The maintenance

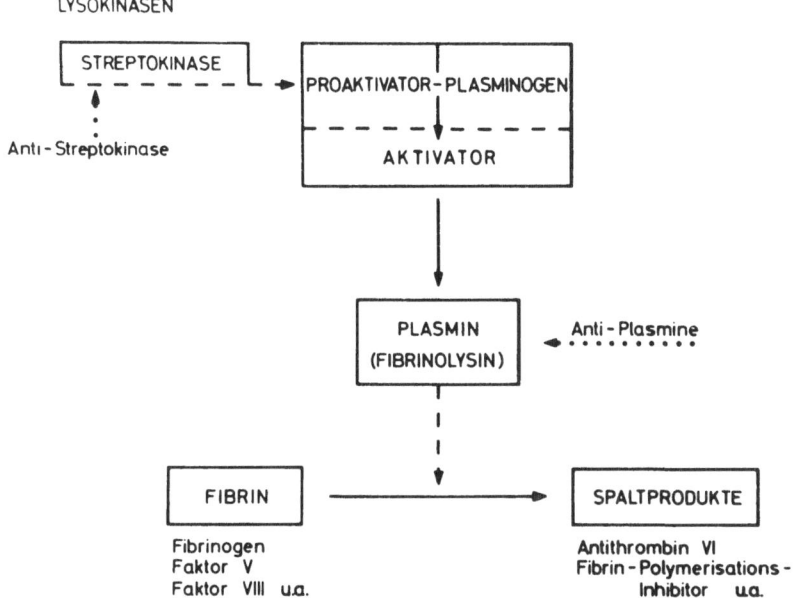

Fig. 1. The activation of plasminogen

dose applied in most of the cases was 100,000 units SK/hr. in a continuous drip infusion. The duration of therapy lasted between 3 to 4 days and was followed by anticoagulant therapy, phenprocumon, to prevent rethrombosis.[8]

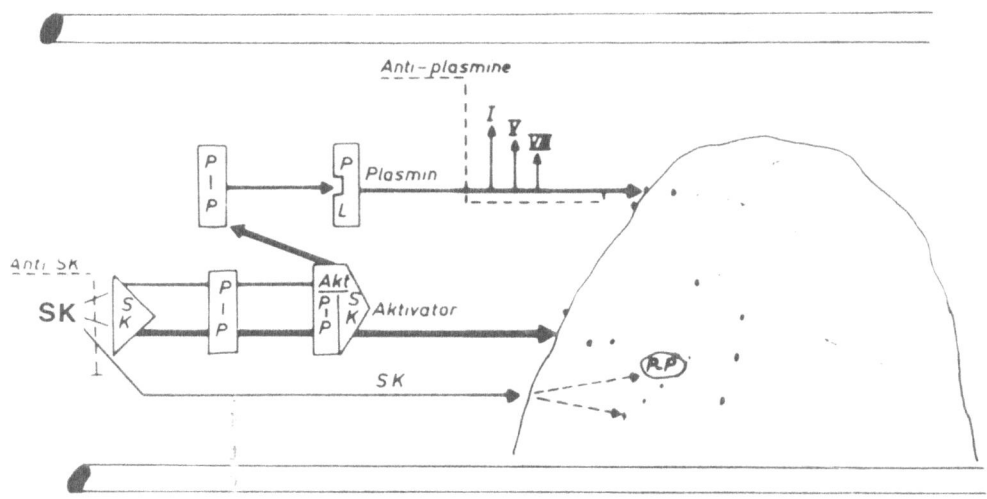

Fig. 2. The mechanism of thrombolysis

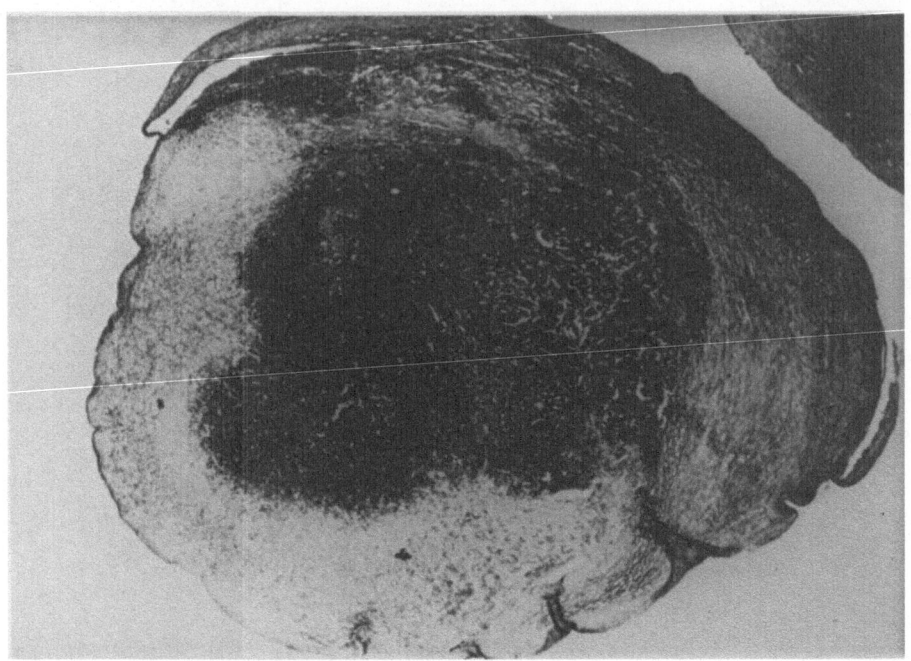

Fig. 3.   Thrombus organization within an arteriosclerotic
          changed vessel.
          At the edge: an arteriosclerotic plaque - plenty of
          thrombotic material is bordering.
          On the side: organized connective tissue.

        Detailed opinion would be reached if one regards stenoses
separately from occlusions, if one takes special localization of
blood vessel changes as well as the period of claudication into
consideration.   After the indications had been narrowed down in
time, the limits for the ability to lyse were set: in the case of
femoropopliteal vessels to 8 weeks, occlusions of pelvic arteries
to 1 year, and for stenoses in this region to 2 years.   No time
limits were made in the terminal aorta.   When taking these limits
into consideration, it can be shown that the success rate increased
in the case of vascular stenoses and in occlusions.

Some Clinical Examples

        Case report. (60 years old man), first claudication of the
right calf after walking 100 m. 6 weeks ago.   Angiography:
occlusions of the right femoral artery.   After 3 days SK-infusion,
the artery could be reopened.   No claudication.

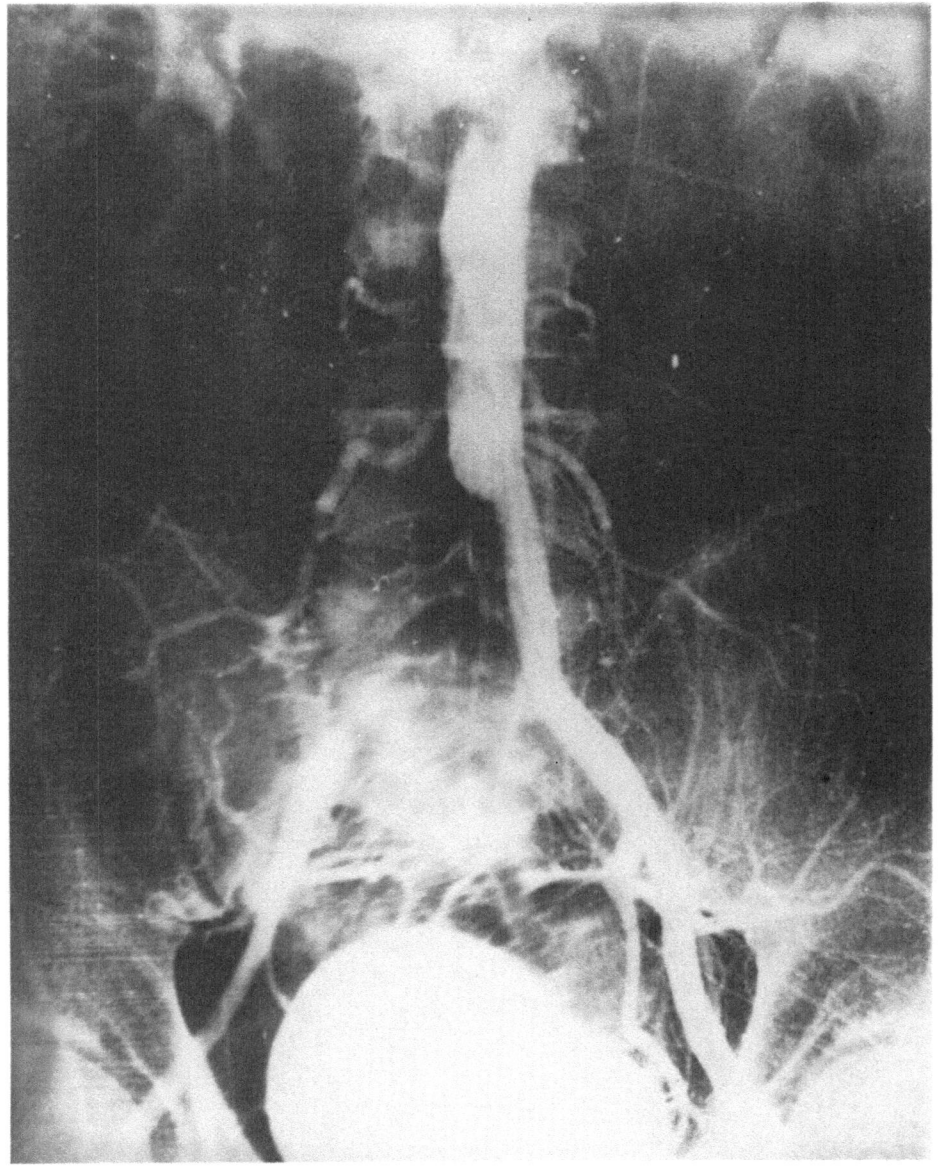

Fig. 4. Chronic arterial occlusion: arteriography before SK
         therapy.

     Case report. 62 years, male, for 6 months claudication after
100 m. Angiography: occlusion of the right common iliac artery.
(Figure 4)   After 4 days of thrombolysis, the reopening of the

iliac artery was observed, walking without pain more than 1000 m.
(Figure 5).

    Case report. 72 years, female, the beginning of claudication
after 10 m. 1 year ago. Aortography: a segmental, single occlusion

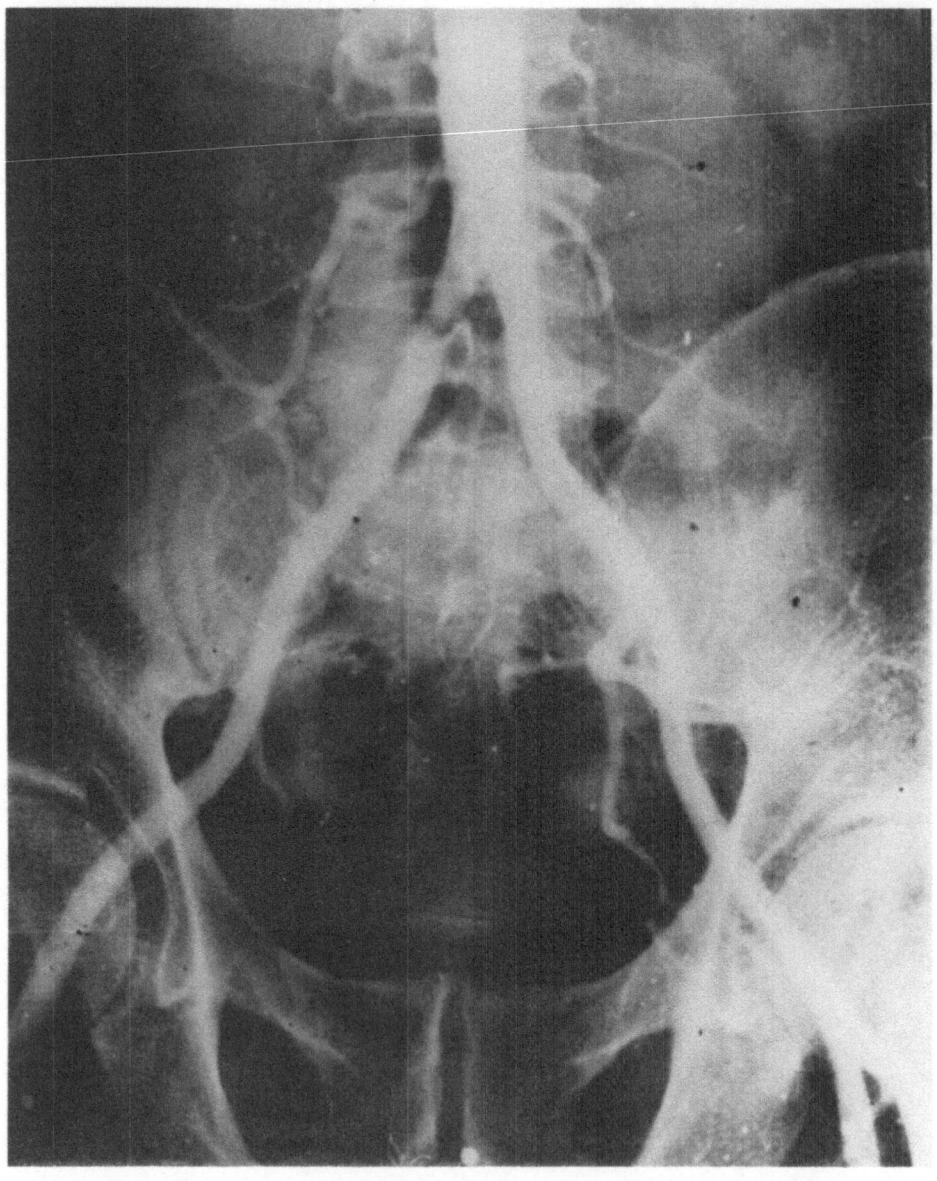

Fig. 5 Thrombolysis after SK-therapy

of the abdominal aorta. (Figure 6)   After 4 days of thrombolytic
therapy the oscillogram normalized, pain disappeared.   The control
angiography showed the continued flow in the aortic vessel. (Figure 7)

Case report. 69 years, female, partial occlusion of the right
femoral artery, 4 weeks ago.   Thrombolysis completely after 48
hours; the arteriosclerotic changes of the vessel wall are impressive.

Case report.   75 years, male, total occlusion of the right
femoral artery, 4 weeks ago.   Thrombolysis and reopening totally of
the arteriosclerotic changed vessel after 30 hours.

Case report. 86 years, female, the oldest patient, occlusion
of the distal right femoral artery, 4 weeks ago.   Thrombolysis
after 48 hours, resting with a small stenosis; amputation could be
prevented.*

The relation between start of fibrinolysis and the clinical
stage of the chronic arterial occlusive disease according to
Fontaine (stages II, III and IV) showed that patients in stages II
and III succeeded in 50 per cent, but in stage IV by only 34 per
cent.   The reason may be that in stage IV the incidence of old
occlusions was higher than in stage III and the indications of SK-
treatment was an ultimo ratio, because the surgeon was often
reluctant to perform the desobliteration.[9]

Our material of 708 patients was composed of single and multiple
occlusions, altogether 1041 obliterations – 613 occlusions (58.9%)
and 428 stenoses (41.1%).

In 613 arterial occlusions 160 (21.1%) could be reopened.   In
relation to the localization, the occlusion of the common femoral
artery showed in 43 per cent the highest success rate and occlusion
of the calf vessels in 20 per cent the lowest success rate.   The
success rate of the other localized occlusions was between 22.6 per
cent and 29 per cent.

In 428 arterial stenoses 216 (50.5%) could be widened.
The stenoses of the popliteal artery in 14.3 per cent and the calf
vessels in 26.7 per cent showed the lowest success rate in comparison
to the good results of therapy in the aorta, the common and external
iliac artery and the common femoral artery with 58.4 per cent to
65.4 per cent.   The strict indication increased the total results
of success: 39 per cent in occlusions, 66 per cent in stenoses.

No correlation between lytic success rate and patient's age
was seen.   Figures 8 and 9 show in the different localizations of

---

* I thank Dr. Martin for kindly lending the figures of the last 3
  cases.

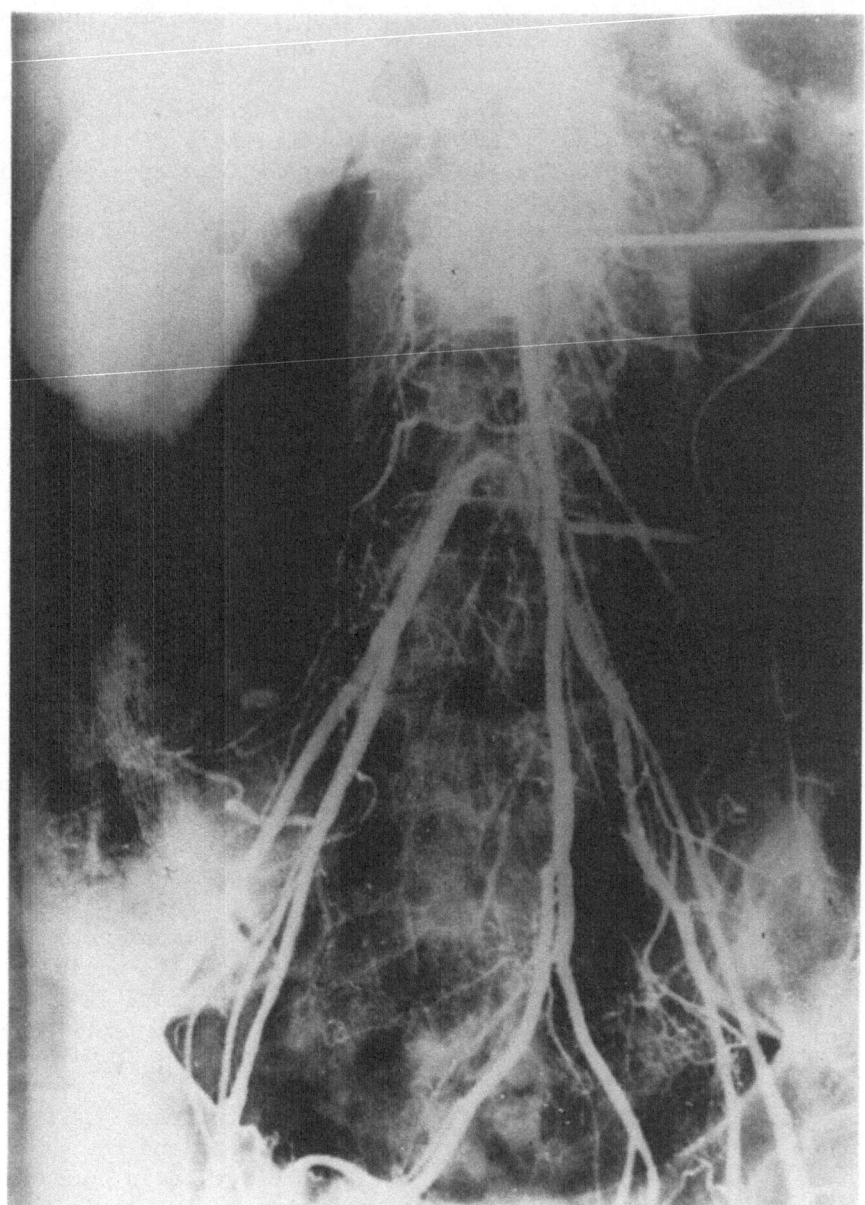

Fig. 6. Chronic arterial acclusion: Aortography before SK
         therapy

arterial occlusions and stenoses the line of regression of the
success rate in relation to the time interval between onset of
symptoms and starting treatment.

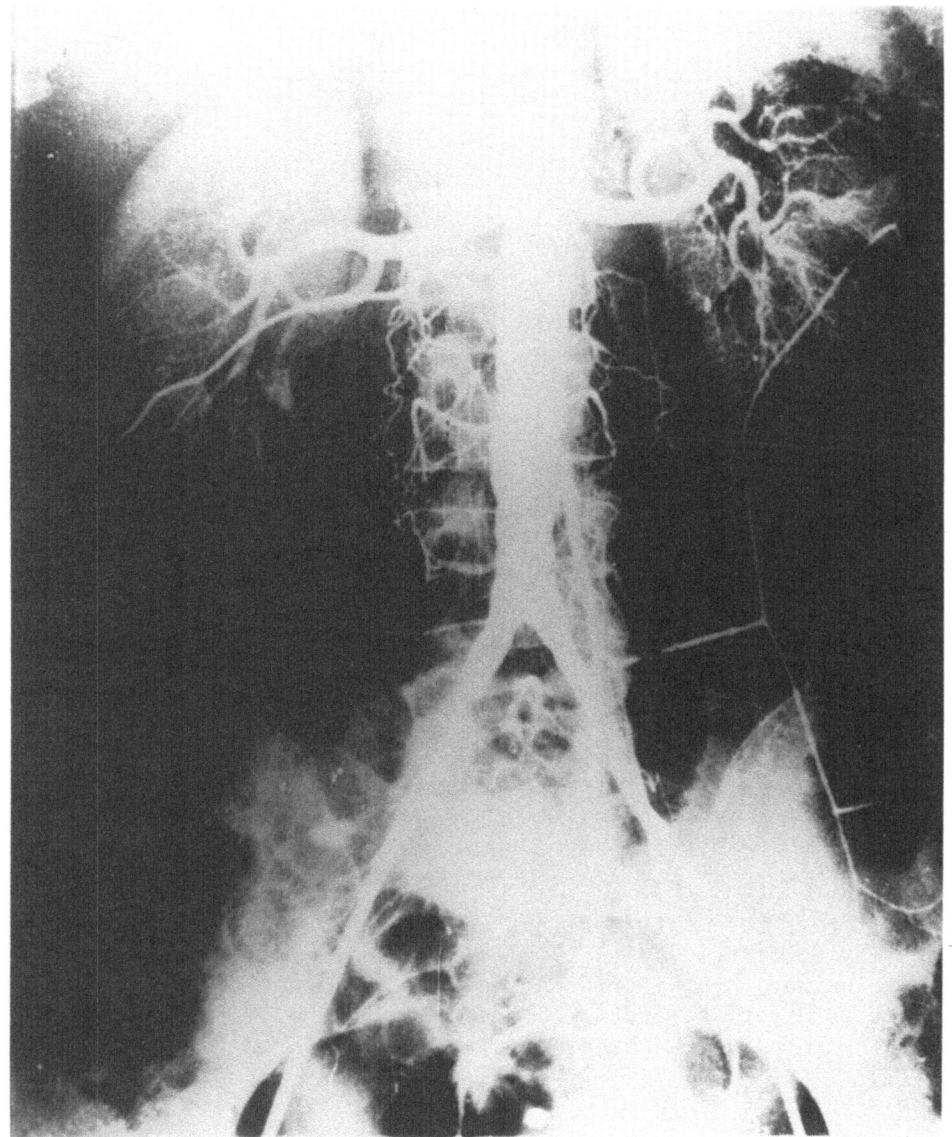

Fig. 7   Thrombolysis after SK-therapy

Occlusions

The common femoral artery has - if the time interval between symptoms and therapy is short - the highest success rate but also the highest rate of regression in success.   On the other hand the

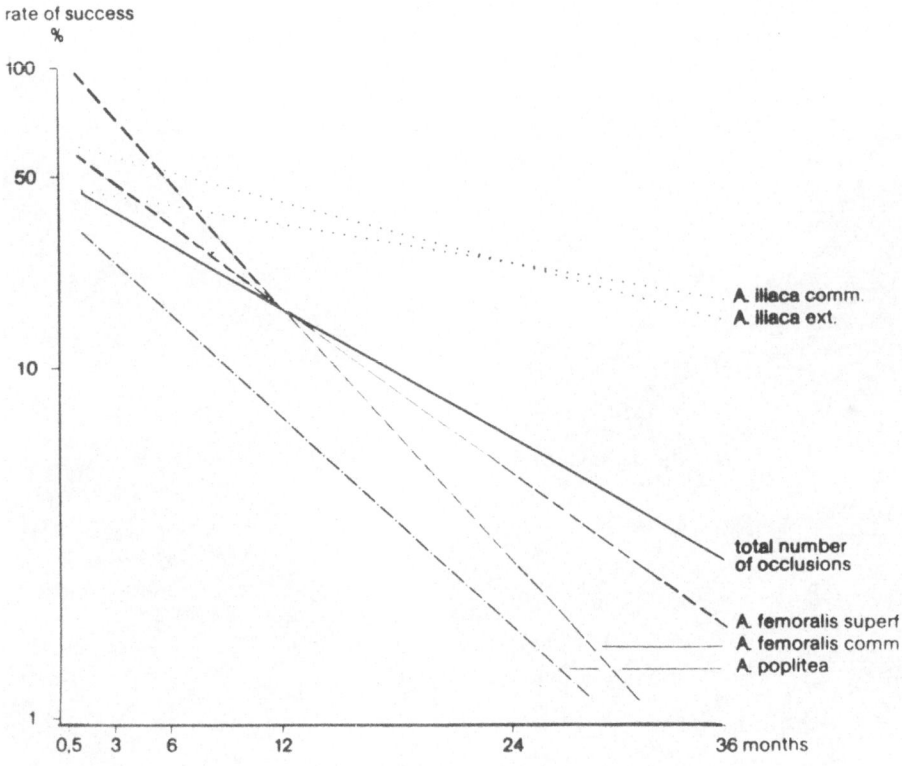

Fig. 8   Rate of success and case history with <u>occlusions</u>

iliac arteries have only a slight regression rate of success.   The
reason may be that in the pelvic region the fibrous organization
of intraluminar thrombi originating from the arterial wall is more
retarded by arteriosclorotic changes than in the distal regions;
thus there is a better chance to get lysis by SK.

<u>Stenoses</u>

     The principle of the regression rate is similar to the
occlusion but on a smaller scale.   The total success rate is
higher than in occlusions.   The reason may be that SK is able to
act simultaneously from the side of blood stream and the vessel wall.

SIDE EFFECTS OF STREPTOKINASE TREATMENT

     These may be subdivided into early reactions that are connected
with the application of the initial dose, and reactions that may

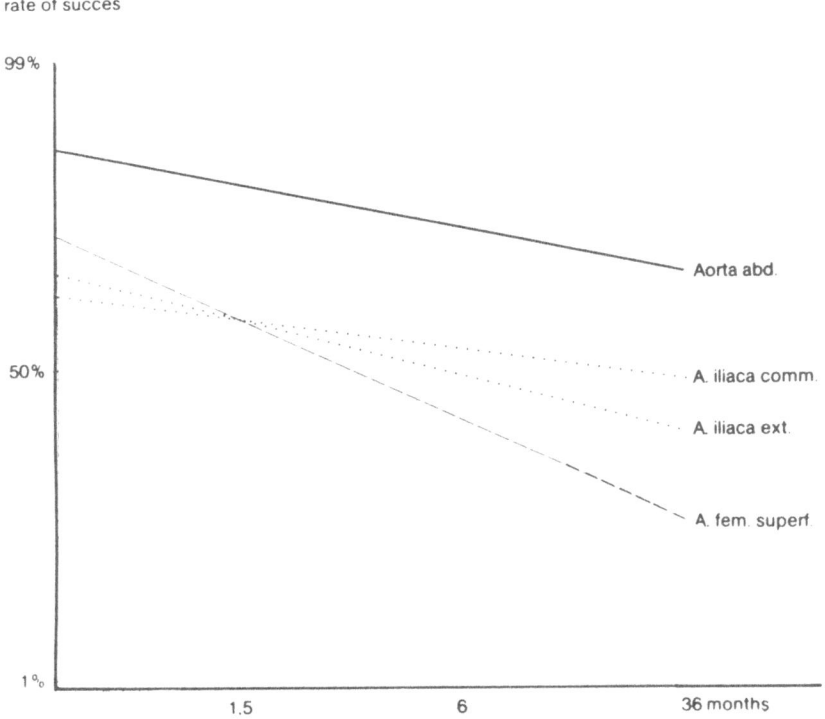

rate of succes

99%

Aorta abd.

50%

A. iliaca comm.

A. iliaca ext.

A. fem. superf.

1%

1.5   6   36 months

Fig. 9. Rate of success and case history with <u>stenoses</u>

appear in less than one day to six days after streptokinase therapy.
Reactions may also be distinguished according to mild side effects
and severe reactions that either lead to disrupting therapy or take
a lethal course. [10]

It may happen that fever occurs during the course of therapy.
This appeared in 57.3 per cent of the patients, primarily from the
2nd to the 4th day. A temperature rise over 38.5°C, which some
patients subjectively found unpleasant, was registered in a maximum
of 17.4 per cent of the cases. The numbers correspond also to the
findings of other authors. Shivering reactions, in contrast to
beginning stages of the streptokinase era, are today no longer
observed. When using urokinase side reactions such as anaphylaxis
or fever have been markedly decreased.

Hemorrhages have the greatest significance of all side effects,
for they not only possibly compel an early discontinuance of
therapy, but in individual cases they may also be linked with lethal
consequences. In the multicentre study, hemorrhaging occurred in
209 (29.5%) cases, 156 (22%) of which could be considered mild.
They were primarily hemorrhages from puncture sites following

drawing of blood samples, and were promptly halted with compression bandages.   Also needing mention are microscopic hematuria and bleeding hemorrhoids and gums.

In 53 patients (7.5%) bleeding was severe enough to warrant stopping treatment.   In 10 cases (1.4%) cerebral hemorrhage occurred, 5 of which (0.7%) led to death.   In all, there were 8 cases developing lethal complications, giving a mortality rate of 1.12 per cent.   Strict adherence to the contraindications is the first commandment for elimination of complications.

Thrombolysis is contraindicated in cases of generalized hemorrhagic diatheses, suspected or proved peptic ulcer, fixed hypertension, (that is, systolic pressure above 200 mm Hg) because of the danger of hemorrhagic cerebral stroke.   A high titre of antistreptokinase, requires the use of urokinase owing to the risk of allergic side-reactions.   High age of the patient (above 80 years) is a relative contraindication of systemically induced thrombolysis because of the increased permeability of the vessels. Local thrombolysis is discussed later.

The allergic reactions and febrile side effects may be reduced or diminished through medicinal prophylaxis in the form of steroids (25 mg daily) and antipyretics.   This applies especially for patients with higher antistreptokinase titres.

Of the hemorrhages, the intracranial is without a doubt particularly impressive, for while it appeared primarily in older patients it was still individually unforseeable.   Martin in his study saw cerebral accidents occurring during fibrinolytic treatment amounting to 1.42 per cent in the age group under 65 years and to 5.26 per cent in the group above 65 years. [11]

Nevertheless, its mortality rate of 0.7 per cent in the multi-centre study may be accepted.   An overall mortality rate for streptokinase therapy, which according to the latest publications ranges between 1.2 per cent and 1.4 per cent, is probably justifiable for a causal, different therapy, in which in individual cases the risks outweigh each other.

Of special interest is the question of the reocclusion rate, after removal of chronic arterial occlusions by streptokinase treatment.   Follow-up studies were made by Martin et al in a 6 year retrospective study including 48 patients of our material.   The reocclusion rate showed a steady rise up to 21 per cent during the first three years following streptokinase treatment, and fluctuated around that value up to the end of the sixth year.   The iliac artery permitted the best results among the vascular segments cleared.   The reocclusion rate was very low at 0 to 14 per cent throughout the 6-year observation period.   The femoral artery reached a reocclusion rate of 50 per cent after 4 and 5 years. The aortic reocclusion figure was between the figures ascertained for the iliac and femoral arteries.[9]

Regular and therapeutically effective oral treatment with anticoagulants produced a significant drop in the reocclusion rate. These results are in striking accordance with the data from follow-up studies of surgical reconstructive vascular treatment reviewed by Heberer et al, 1966.[11] The reocclusion rate in the femoral region after endarteriectomy or by-pass is between 30 per cent and 70 per cent within a one to ten-year observation period. The reocclusion rate in the iliac and aortic region on the other hand showed only 0 to 30 per cent. The mortality rate of patients undergoing surgery is 1 per cent and quite similar to the rate in thrombolytic surgery.

LOCAL THROMBOLYTIC THERAPY

In contrast to systemic induced thrombolytic treatment local thrombolytic therapy in arterial occlusion had been originally performed by Dotter[13] and Zeitler,[14] but so far as known it is no longer used clinically. In the last 3 years Hess et al[15] from Munich have obtained the most clinical experience. An intra-arterial thrombus can be dissolved by direct infiltration of only a few thousand units of streptokinase or urokinase, whereas systemic thrombolysis needs millions of units to achieve the same. Low-dose local thrombolytic therapy eventually in combination with catheter dilatation is the treatment of choice in cases with femoral-popliteal obliteration including the trifurcation.

Cases and Statistics

Case report. 73 years, male occlusion of the right popliteal artery, 3 days old. Dilatation therapy combined with local infiltration of 67,000 units of streptokinase; 2 hours later the artery could be reopened. (Figure 10)

Case report. 71 years, male claudication after 200 m, 3 weeks ago; occlusion of the right femoral artery and stenosis of the left femoral artery. Local thrombolysis at the right and catheter dilatation at the left side; reopening afte 4 hours. The total success was shown by angiography 2 weeks later.* (Figure 11)

The investigated 100 patients, 59 male and 41 female, were in the age range 14 to 89 years; most of the patients were older than 70 years. The table summarizes localization of the occlusions and

---

* I thank Dr. Hess for kindly lending the figures of the 2 cases.

Table 1. Clinical results in local thrombolysis

Localisation	n	recanalization	early reocclusion	open 2 weeks	later reocclusion
femoral/popliteal artery, trifurcation open	64	52	14	38	6
femoral/popliteal artery including trifurcation	29	15	4	11	1
aorta-iliac area	5	2		2	
renal artery	2	1		1	
total	100	70	17	52	7

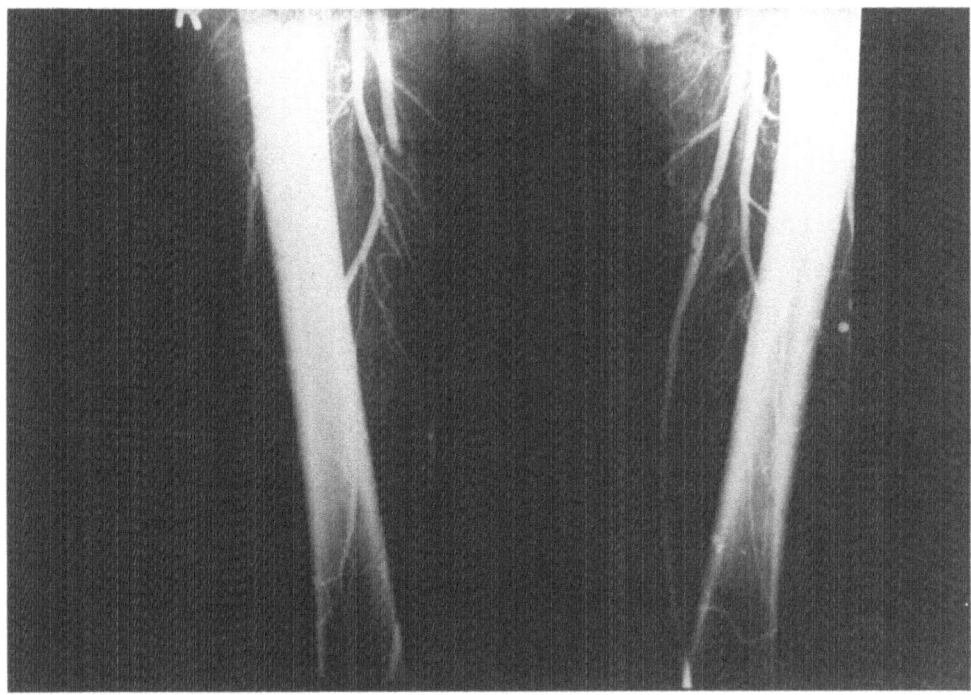

Fig. 10   Local thrombolysis and dilatation in chronic arterial
          occlusion: Arteriography before SK-therapy

the results obtained.   Clinical improvement was evident in those
patients in whom recanalization was permanently achieved.   The
change for the better was most impressive in cases with femoral-
popliteal obliterations including trifurcation. In 4 patients the
planned amputation was avoided.   The risk of general bleeding is
small, and therefore the contraindications, in contrast to
systemic thrombolysis, are few and so the technique is especially
suitable for the elderly.   But there is a risk of severe local
bleeding and of deterioration of the local arterial circulation
after treatment.   Therefore the indication must be made as
carefully as the decision for vascular surgery.   If thrombolytic
treatment is unsuccessful surgical intervention can always follow
without disadvantage.   The indications for the therapeutic
measures should always be made conjointly by the angiologist and
vascular surgeon.   Both procedures being complimentary to each
other.

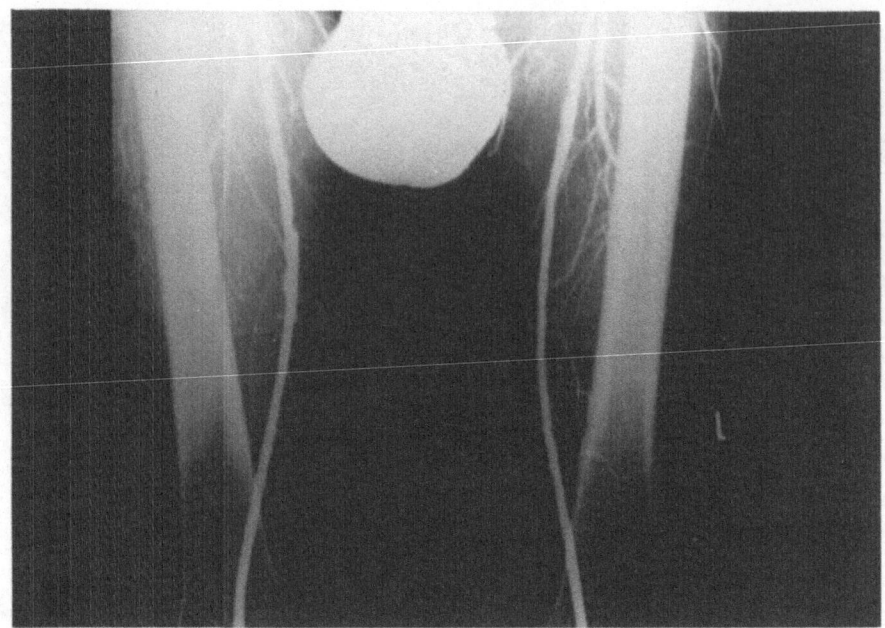

Fig. 11   Arteriography after SK-therapy and dilatation

REFERENCES

1. R. Schmutzler, G. Beneke, F. Eisenreich, and F. Heinrich,
    Thrombolytische Therapie des chronischen thrombolytischen
    Arterienverschlusses, in: "Thrombose und Embolie," R. Marx
    und H. A. Thies, Hrsg., Schattauer, Stuttgart-New York (1970).
2. R. Schmutzler, G. Beneke, and F. Heinrich, New therapeutic
    aspects of fibrinolysis: Basis studies of fibrinolytic
    treatment in chronic arterial stenoses and occlusions,
    in: "Current Concepts of Coagulation and Hemostasis,"
    R. Losito, and B. Longpre, ad., Schattauer, Stuttgart-
    New York (1971).
3. R. Schmutzler, and F. Heinrich, Critical review of one indicat-
    ion for fibrinolytic therapy. Recent experiences in 100
    cases with chronic arterial stenoses and occlusions, in:
    "Present Status of Thrombosis," R. Losito, ed., Schattauer,
    Stuttgart-New York (1973).
4. H. C. Kwaan, and T. Astrup, Fibrinolytic activity in thrombosed
    veins, Circ.Res. 17:477 (1965).
5. V. Hiemeyer, Thrombolytische Therapie bei akuten Gefäßversch-
    lüssen, Dtsch.med.Wschr. 92:955 (1967).
6. R. Schmutzler, Fibrinolytic agents, Medicine Today 8:32 (1969).
7. F. Heinrich (Hrsg.), "Streptokinase-Therapie bei chronischer
    arterieller Verschlußkrankheit. Ergebnisse einer multizentri-
    schen Studie. "Med.Verlagsgesellsch. Marburg (1975).

8. R. Schmutzler, Dosierung der thrombolytischen Therapie, in:
    "Streptokinase-Therapie bei chronischer arterieller Verschluß-
    krankheit," F. Heinrich, Hrsg. Med. Verlagsgesellsch. Marburg
    (1975).

9. M. Martin, U. Martin und H. Auel, Reokklusionsraten nach erfolg-
    reicher Streptokinase-Behandlung arterieller Verschlüsse,
    Klin.Wschr. 55:489 (1977).

10. V. Tilsner, Nebenwirkungen, in: "Streptokinase-Therapie bei
    chronischer arterieller Verschlußkrankheit," F. Heinrich,
    Hrsg., Med. Verlagsgesellsch. Marburg (1975).

11. M. Martin, Die Bedeutung des Lebensalters im Rahmen der thrombo-
    lytischen Therapie mit Streptokinase, Akt.gerontol. 8:169
    (1978).

12. G. Heberer, G. Rau, H. H. Löhr, "Aorta und große Arterien,"
    Springer, Berlin-Heidelberg-New York (1966).

13. C. T. Dotter, and M. F. Judkins, Transluminal treatment of
    arteriosclerotic obstruction. Description of a new technique
    and preliminary report of its application. Circulation
    30:654 (1964).

14. E. Zeitler, and J. C. Demski, Infusion into thrombus material
    via catheter, in: "New Concepts in Streptokinase dosimetry,"
    M. Martin, W. Schoop, J. Hirsh, ed., Huber, Bern-Stuttgart-
    Wien.

15. H. Hess, personal communication (1982).

ANTIPLATELET AGGREGATING AGENTS

M. R. P. Hall

Department of Geriatric Medicine
Southampton General Hospital, Southhampton

INTRODUCTION

Arterial or venous occlusion by blood clot are common occurrences in old age and the incidence rises as age progresses. They both give rise to considerable morbidity as well as mortality. The prevention of thrombosis and its consequences would inevitably promote and maintain health in old people. To attempt to do this is a worthy cause and it is not surprising that many people have set out along this road with much enthusiasm, but often little planning.

Venous thrombosis is a common complication of illness in all age groups and particularly so in the elderly. Wright and his colleagues[1] in Exeter reported a series of 300 consecutive autopsy examinations which showed an 18 per cent incidence of pulmonary embolism. Presentation was often atypical and one third of the subjects presented with what might be termed "a failure to thrive syndrome." Other series have reported an even higher incidence of pulmonary embolism and have compared treatment with a low dose heparin regime to a physiotherapeutic regime, (de Brito-Paiva et al)[2] If drugs are to be used there seems little doubt that low dose subcutaneous heparin is the treatment of choice (Pareti and Mannucci)[3] for a short period and anticoagulants for longer or permanent therapy.

The ability to prevent arterial thromboembolism with drugs is, however, much less clear. Perhaps the first question therefore to be asked is what is to be gained by prevention? The answer is obviously much, for arterial thromboembolism commonly affects the vital organs. Brain and heart as well as the limbs and other organs such as the kidney, gut etc. As thromboembolism is the basis for

much of the underlying pathology affecting these organs then it
should be possible to test the efficacy of treatment aimed at
preventing the occurrence of such pathology.   Yet reviews of
therapy, (Pareti and Mannucci,[3]Weiss,[4] and de Gaetano et al[5])all
stress the doubtful value of antiplatelet aggregating drugs in
treating the condition.   The reasons for this are worth reviewing.

PLATELET AGGREGATION

     Platelets play a major role in the development of a thrombus
and therefore in thromboembolism.   However, it is still by no
means certain if platelet aggregation initiates thrombus or whether
platelets just become involved in the clotting cascade which has
been triggered by some other mechanism.   As Mitchell[6] has put it,
is thrombosis clotting in the wrong place, or haemostasis in the
wrong place?   If it is the latter then we need to understand the
mechanism of platelet aggregation.   Initially two phases are
involved in the formation of a platelet thrombus.

     1.   Adhesion of platelets to a non-endothial surface.
     2.   Platelet aggregation, i.e. adhesion to other platelets.

If either process is prevented then the thrombus will not form.
Consequently, there are two potential processes with which drugs can
interfere.

     Unfortunately the exact chemical process or mechanism involved
in platelet adhesion or aggregation is not clearly understood.
There seems little doubt that it is energy dependent and is mediated
by adenosine diphosphate (ADP) perhaps coming initially from red
blood cells and then being released from the platelets themselves
in the 'release reaction' which is mediated by cyclic adenosine
monophosphate (cAMP).   Adhesion can be induced by collagen which
may be present below the endothelial surface, and also by thrombin
which interacts to develop the thrombus.   It is generally agreed
that haemostatic balance is maintained as a result, amongst other
things, of homeostasis between platelet thromboxane and vascular
prostacyclin (PGI_2). PGI_2 is an unstable potent inhibitor of
platelet aggregation generated by vessel walls and in the lungs.
It acts by potentiating the action of adenylate cyclases, enzymes
which catalyse the breakdown of adenosine triphosphate (ATP) to
cAMP.   The increase in cAMP blocking platelet function (Steer and
Salzman).[7] PGI^2 production by cells however, may be limited by a
negative feedback system which switches off PGI_2 production.
Potentiation of the PGI_2 effect is therefore one possible solution
to the problem of preventing aggregation.

     On the other hand thromboxane A_2 (TXA_2) is a powerful aggregat-
ing agent.   It would seem that TXA_2 production in platelets is

associated with the transformation of arachidonic acid into cyclic endoperoxide PGG_2, the reaction being catalysed by cyclo-oxygenase (Moncada and Korbut).[8] PGG_2 is then converted into TXA_2 by Thromboxane synthetase.   Inhibition of cyclo-oxygenase e.g. by aspirin and thromboxane synthetase inhibition, e.g. by imidazoles, would appear to be anti-thrombotic strategies, (Vermylen et al).[9] However, this does not prevent platelet aggregation in all subjects and this may be due to activation of thromboxane receptors by endoperoxides, (Hornby and Skidmore).[10] The use of thromboxane receptor antagonists may therefore be a better strategy, (Bertele, et al).[11]

If the foregoing hypothesis relating to platelet aggregating and disaggregating mechanisms is true, and this is by no means certain, (de Gaetano et al)[5] then drugs could affect aggregation in various ways by altering the PGI_2/TXA_2 balance in favour of disaggregation (Figure 1).

DRUGS WITH A PLATELET DISAGGREGATING EFFECT

Drugs which may have an anti-aggregating effect are shown in Table 1.   The list is considerable but only the first three and perhaps the fourth have been used to any extent.   Reviews by Weiss[4] and Pareti and Mannucci[3] have outlined the way in which they may produce their effect and since many of these other drugs are commonly used in the management of other illnesses in the elderly it is perhaps briefly worth considering their action.

Heparin prevents thrombin induced platelet aggregation and its greatest value is in the prevention of deep vein thrombosis which so frequently occurs as a complication of illness in old age. Dextran has also been similarly used.   It decreases platelet adhesiveness and interferes with fibrin formation rendering clots more susceptible to lysis.   Both however, have to be given via a needle and dextran can cause circulatory overload and renal tubular blockage with renal failure.   Clofibrate may have an anti-platelet effect but the mechanism is unclear.   Sympathetically induced aggregation is an α-adrenergic effect and can be prevented by alpha blocking drugs such as ergot derivatives and β-blockers may act indirectly by blocking release of free fatty acids.   Anti-serotonin drugs may inhibit platelet function.   Imipramine is thought to prevent uptake at the plasma membrane.   How the others work is uncertain but perhaps through the serotonin and adrenaline pathways.   This then leaves the first three in the table, each of which is worth separate consideration.

Sulphinpyrazone

There is evidence (Cerletti et al)[12] that sulphinpyrazone inhibits cyclo-oxygenase by binding to the same site as the enzyme.

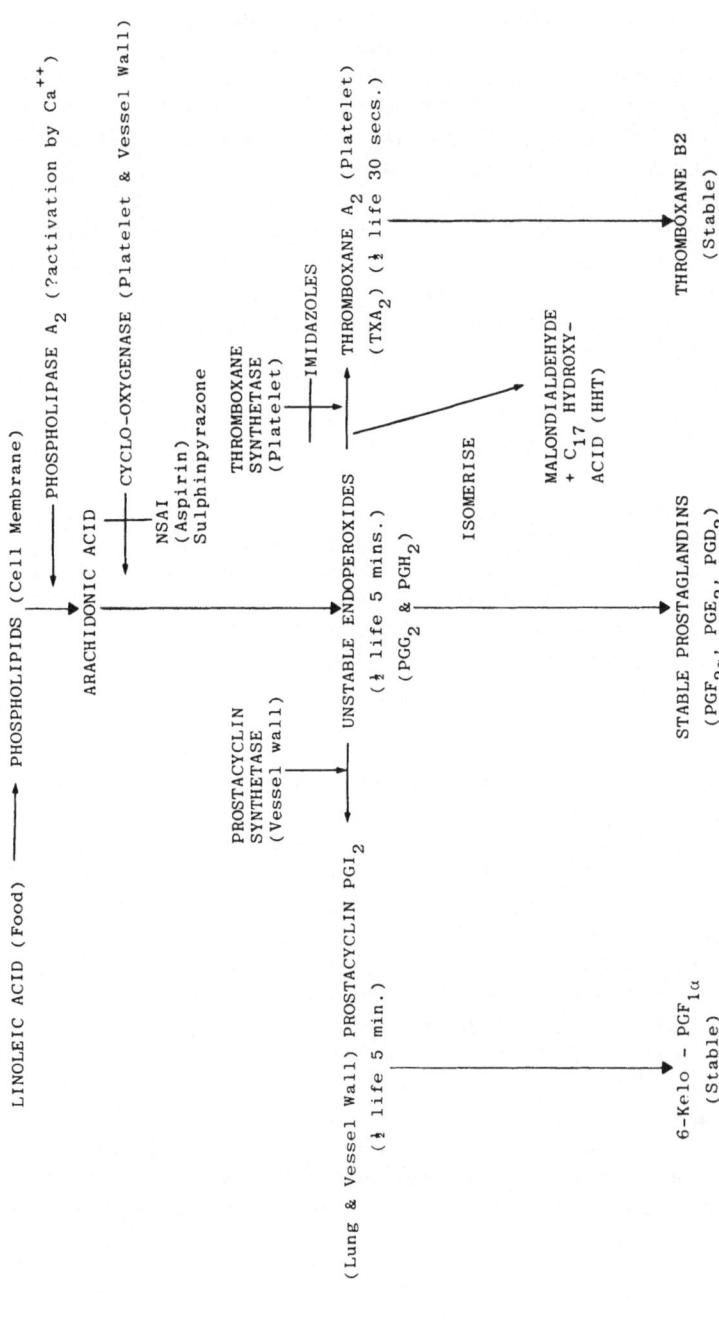

Fig. 1. The development of arachidonic acid and its metabolites in the maintenance of haemostatic balance as a result of homeostasis between prostacyclin and thromboxane A$_2$.

Strategies for anti-platelet aggregation therapy lie in potentiation of prostacyclin (PGI$_2$) (dipyridamole) or preventing formation of thromboxane A$_2$ (TXA$_2$) by cyclo oxygenase inhibition (aspirin, NSAI's, sulphinpyrazone), thromboxane synthetase inhibition (imidazoles), thromboxane receptor blocking agents (not yet available)

Table 1. Drugs which may have an anti-aggregating
effect

DIPYRIDAMOLE (Persantin)
ASPIRIN (and other NSAI's e.g. Indomethacin)
SULPHINPYRAZONE (Anturan)
CLOFIBRATE
ALPHA-ADRENERGIC BLOCKERS (Hydergine)
BETA-BLOCKERS
DEXTRAN
ANTISEROTONIN DRUGS (Cyproheptadine - Imipramine)
HEPARIN
OTHERS (MAOI's, Phenothiazines, Ethanol, etc.)

It also reduces platelet consumption and increases platelet
survival. It was one of the first drugs to be shown to inhibit
platelet function (Smythe et al) and has been widely used in trials
designed to protect against the effects of thrombo-embolism.

Dipyridamole

This drug acts primarily as a phosphodiesterase inhibitor.
How it acts is not clear but Moncada and Korbut[8] have proposed, as
a result of work on rabbits, that it acts by potentiating the action
of PGI_2. However, de Gaetano and his colleagues[5] have argued that
this hypothesis is non-proven.

Aspirin

The effect of aspirin on platelets was described by O'Brien[14]
in 1968. However, the measured effects seem variable and to vary
with the dose and to a certain extent in the biochemical system
being measured, e.g. is the effect in the platelet or the vascular
cell. De Gaetano and his colleagues[5] call this the "aspirin
dilemma", perhaps enigma would be a better word. It would seem,
however, that small doses of asprin inhibit platelet cyclo-oxygenase
thereby preventing production of cyclic endoperoxides and therefore
TXA_2. In small dosage arterial wall cyclo-oxygenase may not be
affected to the same degree and consequently formation of prostacyclin (PGI_2) is not inhibited. However, in larger doses PGI_2 is
inhibited, and may be inhibited in vein walls by doses as low as
81 mg (Hanley et al)[15] However, this effect may wear off quite
quickly, and the action of other substances such as Dipyridamole
potentiated for up to 72 hours. Indeed Moncada and Korbut's work[8]
suggests that Dipyridamole also potentiates the effect of small dose
aspirin.

THE CLINICAL USE OF ANTI-PLATELET AGGREGATING DRUGS

As has been suggested, aspirin, sulphinpyrazone and Dipyridamole are the drugs which have been most used and studied both alone and in combination.   Most of the studies however, have been in younger subjects and have specifically looked at coronary artery disease with a view to attempting to prevent reinfarction. As a result we have had the report of Anturan Reinfarction Trial (ART) research group,[16] the Anturan Reinfarction Italian Study (ARIS)[17] and the Persantin-Aspirin Reinfarction Study Research Group (PARIS).[18]   While there is much to be learned from these studies they are not of great interest to the physician in Geriatric Medicine.   'Stroke' prevention is surely his prime goal.

Consequently, the Canadian Co-operative study group's report is of more importance.   This studied the effect of aspirin alone and in combination with sulphinpyrazone.   No positive effect was shown for sulphinpyrazone but aspirin was effective in men.   In contrast, Blakely and Gent[19] have shown that sulphinpyrazone was associated with a reduction of cardiovascular mortality in elderly males with a history of stroke or myocardial infarction or both in a prospective double blind trial.   Acheson et al[20] however, found no evidence for dipyridamole in reducing transient ischaemic attacks (TIAs).   Nevertheless aspirin in low dosage and dipyridamole 100 mg - 600 mg daily are considered more or less standard therapy after a cerebral ischaemic episode in the U.K.   Cynics however, might say that the aspirin is only given to relieve the headache caused by the dipyridamole.

A recent Danish study (Hansen et al)[21] has, however, reported a more rational approach to the problem in a clinical study of platelet aggregation in focal cerebral ischaemia.   In this aggregation studies were performed and increased aggregation found in 74 per cent of TIAs 4 weeks after the TIA and in 37 per cent of completed strokes.   Abnormal aggregation was treated with aspirin or if this was ineffective with aspirin and dipyridamole, treatment being continued for 3 months.   Results were compared to subjects with normal aggregation or no treatment and to others being treated with anticoagulants.   There was no difference between groups with regard to TIA's but the stroke cases showed a trend in favour of the antiaggregating drugs though this did not reach statistical significance.

Despite its negative result this paper does, I believe give a lead to future studies.   More attention should be paid to measurement of platelet aggregation and trials should be undertaken on a double blind controlled basis to assess the value of drugs which have an anti-platelet aggregating effect.   These should be given in a dose sufficient to correct the abnormal aggregation.   It is suggested that Dipyridamole in a dose of 3 mg/kg/day should be

tried to which aspirin 5 mg/kg could be added, given on alternate
days initially.   Once a treatment regime has been established it
should be continued for at least 1 year and preferably 2 years.
However, such a study would require careful planning and the
employment of research workers.   It would not be cheap.   It would,
however, meet the criticism implied in the last sentence of
de Gaetano's article for as Mitchell[6] has pointed out, oral anticoagulants are effective in preventing myocardial infarction if
properly prescribed and used.   Antiaggregating drugs may be
equally successful in preventing cerebral infarction in old people.

REFERENCES

1. W. B. Wright, Geriatric diagnosis - a survey, Geront.Clin.
    11:75 (1969).
2. E. de Brito Paiva, J. Borgykowski, and J. P. Junod, Physio-
    therapeutic prophylaxis of pulmonary embolism in the elderly
    high risk patient: Indications and limitations, J.Clin.
    Exp.Geront. 3:245 (1981).
3. F. I. Pareti, P. M. Mannucci, Drugs affecting platelet behaviour,
    J.Roy.Coll.Phycns. 10:194 (1976).
4. H. J. Weiss, Antiplatelet drugs - a new pharmacologic approach
    to the prevention of thrombosis, Am.Heart J. 92:86 (1976).
5. G. de Gaetano, C. Cerletti, and V. Bertele, Pharmacology of
    antiplatelet drugs and clinical trials on thrombosis
    prevention: a difficult link, Lancet 2:974 (1982).
6. J. R. A. Mitchell, Anticoagulants in coronary heart disease -
    retrospect and prospect, Lancet 1:257 (1981).
7. M. L. Steer, and E. W. Salzman, Cyclic nucleotides in hemostasis
    and thrombosis, in: "Advances in Cyclic Nucleotide Research,
    P. Hamer and H. Sands, eds., Vol. 12, Raven Press, New York
    (1980).
8. S. Moncada, and R. Korbut, Dipyridamole and other phosphodi-
    esterase inhibitors act as antithrombotic agents by potent-
    iating endogenous prostacyclin, Lancet 1:1286 (1978).
9. J. Vermylen, G. Defreyn, L. O. Carreras, S. J. Machin, J. Van
    Schaeren, and M. Verstraete, Thromboxane synthetase
    inhibition as antithrombotic strategy, Lancet 1:1073 (1981).
10. E. J. Hornby, and I. F. Skidmore, Evidence that prostaglandin
    endoperoxides can induce platelet aggregation in the absence
    of thromboxane A2 production, Biochem.Pharm. 31:1158 (1982).
11. V. Bertele, C. Cerletti, A. Schieppati, G. Di Minno, G. de
    Gaetano, Inhibition of thromboxane synthetase does not
    necessarily prevent platelet aggregation, Lancet 1:1057 (1981).
12. C. Cerletti, M. Livio, and G. de Gaetano, Non-steroidal anti-
    inflammatory drugs react with two sites on platelet cyclo-
    oxygenase.   Evidence from in vivo drug interaction studies
    in rats, Biochem.Biophys.Acta. 714:122 (1982).

13. H. A. Smythe, M. A. Ogryzlo, E. A. Murphy, and J. F. Mustard,
     The effect of sulphinpyrazone (Anturan) on platelet
     economy and blood coagulation in man, Can.Med.Assoc.J.
     92:818 (1965).
14. J. R. O'Brien, Effects of salicylates on human platelets,
     Lancet 1:779 (1968).
15. S. P. Hanley, J. Bevan, S. R. Cockbill, and S. Heptinstall,
     Differential inhibition by low-dose aspirin of human venous
     prostacyclin synthesis and platelet thromboxane synthesis,
     Lancet 1:969 (1981).
16. Anturan Reinfarction Trial Research Group, Sulphinpyrazone in
     the prevention of sudden death after myocardial infarction,
     N.Eng.J.Med. 302:250 (1980).
17. Anturan Reinfarction Italian Study, Sulphinpyrazone in post-
     myocardial infarction, Lancet 1:237 (1982).
18. The Persantin-Aspirin Reinfarction Study Research Group.
     Persantin and aspirin in coronary heart disease,
     Circulation 62:449 (1980)
19. J. A. Blakely, and M. Gent, Platelets drugs and longevity in
     a geriatric population, in: "Platelets Drugs and Thrombosis,"
     J. Hirsh, J. F. Cade, A. S. Galius, eds., S. Karger, Basel
     (1975).
20. J. Acheson, G. Danta, and E. C. Hutchinson, Controlled trial
     of dipyridamole in cerebral vascular disease, Brit.med.J.
     1:614 (1969).
21. P. E. Hansen, J. H. Hansen, and S. Stenbjerg, Platelet
     aggregation in focal cerebral ischaemia: a clinical study.
     Acta Neurol.Scand. 65:212 (1982).

ENDOGENOUS OPIATES REGULATING PAIN PERCEPTION IN THE AGED RAT

S. Govoni, G. Pasinetti, C. Missale, P. F. Spano
and M. Trabucchi

University of Brescia
Brescia, Italy

The brain is a unique structure mediating the relationship between the environment and the "self"; the investigation of age related changes in cerebral functions is a major goal in the study of the aging process.

Experiments on laboratory animals have shown the presence of widespread alteration of neurotransmitter synthesis, catabolism and receptor activity within the aging CNS.[1-6]

The present paper will be focussed on the age-related changes in the transmission of nociceptive signals. Clinical observations suggest that pain sensitivity is decreased in old people.[7] In addition, aged patients show an increased analgesic response to morphine that seems to be unrelated to changes in drug metabolism or disposition, since the incidence of side effects for a given dose is similar to that observed in young people.[7] On the other hand experimental data show that old rats have an increased tail-flick latency.[8]

The processing of nociceptive information is complex and involves both peripheral and central nervous structures as well as a number of neurotransmitters, among which are the more recently discovered polypeptides such as met-enkephalin, β-endorphin, substance P and others.

The response to noxious stimuli mediated through nervous circuits, such as the response to infectious agents mediated through the immune system, is a key event in the recognition of the "self" as different from the environment and in the expression of a number of behaviours or activities having as a final goal the preservation of life or well being.   These functions decline with age.

The purpose of the present study is to determine whether the age-related changes in the efficiency of neuronal circuits carrying nociceptive information is paralleled by biochemical changes in the neuropeptides involved in this process.

MATERIALS AND METHODS

Young (3 month-old) and aged (25 month-old) male srague Dawley rats were used.   The animals were exposed to a light cycle of 12 hours (from 06.00 to 18.00 h) and had free access to food and water. Senescent rats with pathological affections (such as gross tumours, pituitary tumours, lung infections) were excluded from the study. The rats were decapitated, the brain and the spinal cord dissected and frozen on dry ice.

Met-enkephalin immunoreactive material (ME-IR) was radi-immuno-assayed according to Yang et al.[9] ^{125}I-Met-enkephalin was prepared according to Miller et al.[10]   Antiserum against β-endorphin (BE) and ^{125}I-β-endorphin were prepared as described by Fratta et al.[11]   Aliquots of neutralized 0.1 M acetic acid tissue extracts were incubated at 4°C for 16 hours (ME assay) or 48 hours (BE assay) with the appropriate antibody and labelled peptide in 0.5 ml of 0.2 M Tris-HCl buffer, pH 7.4, containing 0.1 per cent bovine serum albumin and 0.15 per cent dextran.   The labelled peptide bound to the antibody was separated by the addition of 0.2 ml of 1.5 per cent charcoal slurry containing 0.15 per cent dextran and 0.9 per cent NaCl.   After centrifugation the supernatant was decanted into a scintillation vial.

Opiate receptors were analyzed using (^{3}H)-d-Ala$_2$-met-enkephalin-amide (20 Ci/mmol., Radiochemical Centre, Amersham) as ligand. Tissues were homogenized in 30 vol. (w/v) of 0.05 M Tris-HCl buffer (pH 7.7) and centrifuged at 49,000 x g for 15 min.   The pellet was resuspended in the original volume of buffer and incubated at 37°C for 30 min.   After centrifugation the pellet was finally resuspended in 45 volumes of the same buffer.   Aliquots of the homogenate were incubated at 25°C for 30 minutes with the tracer in the presence or absence of 1 μM D-Ala$_2$-met-enkephalinamide.

The reaction was stopped by filtration on GF/B filters.   The specific binding was defined as the difference between the total binding and the binding in the presence of 1 μM D-Ala$_2$ -met-enkeph-alinamide.

Protein content was measured according to Lowry et al.[12]

Table 1. ME-IR content in the spinal cord of young (3 months) and
          aged (25 months) rats

Age	ME-IR (ng/mg of protein)			
	Cervical		Thoracic	
	Dorsal	Ventral	Dorsal	Ventral
3 months	3.9_0.40	1.3_0.20	3.5_0.71	1.1_0.16
25 months	1.8_0.18^o	1.2_0.14	1.9_0.19^o	1.4_0.21

Each value is the mean $\pm$ SD of at least ten rats

^{o}P < 0.001 respect to 3 month-old rats, two tailed student's test.

Results and Discussion

     Immunohistochemical and anatomical studies indicate that
analgesia is related to the activation of spinal enkephalinergic
interneurons forming synapses with pain afferents (the neurotransmitter of which is substance P) in the superficial laminae of
the dorsal horn.[13][14]   The aging process seems to interfere with
this regulatory mechanism.   Behaviourally, an increase of tail-
flick latency (4.5 $\pm$ 0.31 sec. for 3 month-old rats, 6.7 $\pm$ 0.39 sec.
for 25 month-old rats) was observed in old rats.

     Biochemically, in the same animals ME-IR content is decreased
in thoracic and cervical portions of the spinal cord.   The decrease
of ME-IR at spinal level is confined to the dorsal half, while no
changes are observed in the ventral one (Table 1).

     These data although describing an age-related decrease of ME-
IR content do not provide information on neurotransmitter dynamics.
In order to get insight, at least indirectly, on the functional
activity of enkephalinergic neurons we studied the D-Ala$_2$-met-
enkephalinamide binding to crude synaptic membrane preparations from
old rat spinal cord.   The number of binding sites for this opiate
receptor agonist is markedly increased in 25 month-old rats as
compared with 3 month-old rats (Figure 1).

     The receptor binding data strongly suggest that the decreased
ME-IR content reflects either decreased functioning or number of
ME interneurons, followed by receptor supersensitivity.   Possibly,
these biochemical changes are related to the altered pain sensitivity and morphine response.   However the spinal cord is not the
only structure in which endogenous opiates are involved in the
processing of nociceptive information.   In this light we explored

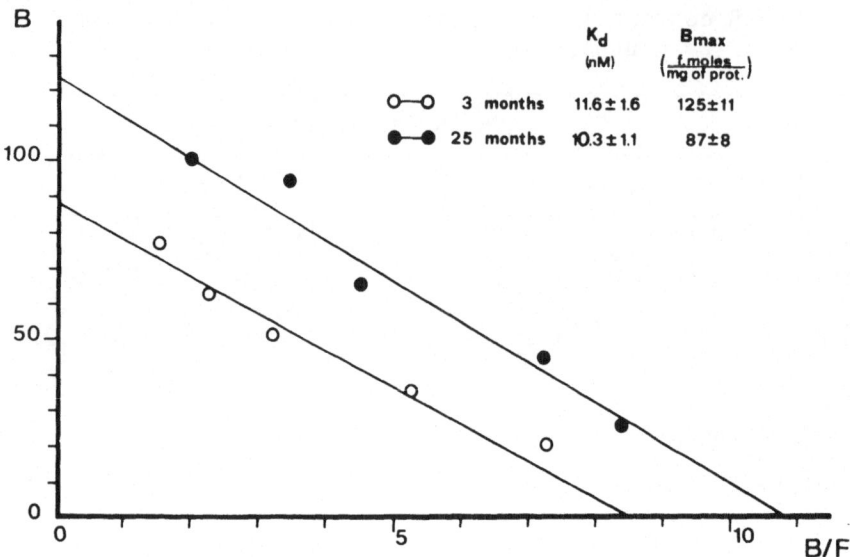

Fig. 1.   Scatchard analysis of (^3H)-D-Ala_2-met-enkephalinamide
binding to crude synaptic membrane preparations from
thoracic spinal cord of young (3 months) and aged
(25 months) rats

the possibility of age-related changes of the β-endorphin (BE-IR)
content in various brain areas and in the pituitary.

The results show opposite trends in the pituitary compared to
the brain (Table 2).

BE-IR content is greater in the pituitary, while it is reduced
in the hypothalamus and periaqueductal gray of old rats.

The increase of pituitary BE-IR content seems to be of functional importance since the in vitro release from the gland[8],[15]
as well as the plasmatic concentrations[16] of BE-IR are also
increased.   Experimental data indicate that the pituitary is the
main source of the circulating peptide,[17],[18] although the physiological regulation of in vivo endorphin secretion is not known in
detail.   Stressful situations are able to promote β-endorphin
secretion together with ACTH and -LPH.[18],[19]   β-endorphin,
peripherally injected, is a potent analgesic agent; there is some
doubt whether the plasmatic concentrations of the endogenous
peptide observed in vivo, even after stress, are sufficient to
induce analgesia.

Nevertheless, the increased circulating levels observed in aged
rats are consistent with a decrease of pain sensitivity.   Since
old animals show an impaired ability to cope with stress,[20] it

Table 2. BE-IR content in pituitary, hypothalamus and periaqueductal gray of rats at different ages

Age	BE-IR		
	Pituitary (ng/gland)	Hypothalamus (ng/mg of protein)	Periaqueductal gray (ng/mg of protein)
3 months	1872 ± 163	6.1 ± 0.35	4.5 ± 0.21
25 months	2941 ± 257°	2.2 ± 0.39^o	2.7 ± 0.35^o

Each value is the mean ± SD of at least 10 rats

^{o}P <0.001 compared to 3 month-old rats.

would also be interesting to know whether or not old rats are able to increase further circulating BE-IR concentrations during stress.

The decrease of BE-IR in the periaqueductal gray indicates that the brain circuits mediating analgesia are also affected by aging, although it is difficult to reconcile this modification with the decreased pain sensitivity observed at the functional level. However, this is just a description of changes in content without considering neurotransmitter turnover.

Our data showing age-related changes of endogenous opiates in brain, pituitary and spinal cord provide a biochemical approach to the understanding of age-related changes in pain sensitivity.

Pain perception is mediated at different neuronal levels and each of the observed modifications should be considered as a possible contribution to determine an altered noxious stimuli transmission. Hopefully the studies will provide a rationale to a correct management of pain therapy in the elderly.

REFERENCES

1. C. E. Finch, Catecholamine metabolism in the brain of ageing male mice, Brain Res. 52:261 (1973).
2. K. R. Brizzle, J. M. Ordy, J. Hanscke, and B. Kaack, Quantitative assessment of changes in neuron and glia cell packing density and lipofucsin accumulation with age in the cerebral cortex of nonhuman primate (Macaca Mulatta), Aging 3:229 (1976).
3. S. Govoni, P. Loddo, P. F. Spano, and M. Trabucchi, Dopamine receptor sensitivity in brain and retina of rats during aging, Brain Res. 138:565 (1977).

4. S. Govoni, P. F. Spano, and M. Trabucchi (H)-Haloperidol and (H)-Spiroperidol binding in rat striatum during ageing, J.Pharm.& Pharmacol. 30:448 (1978).

5. S. Govoni, M. Memo, L. Saiani, P. F. Spano and M. Trabucchi, Impairment of brain neurotransmitter receptors in aged rats, Mechan.Ageing & Develop. 12:39 (1980).

6. S. N. Pradhan, Minireview: central neurotransmitters and aging, Life Sciences 26:1643 (1980).

7. J. W. Belville, W. H. Forrest, E. Miller, B. W. Brown, Influence of age on pain relief from analgesics, J.Amer.Med Ass. 217:1835 (1971).

8. C. Missale, S. Govoni, R. Rozzini, P. F. Spano, and M. Trabucchi,"Altered pain sensitivity in the aged rat, a biochemical hypothesis," Proceedings of the tenth Aharon Katrir Katchalsky Conference on the aging of the brain (Mantua, Italy, March 1982), Raven Press, New York, in press (1983).

9. H. Y. T. Yang, J. S. Hong, and E. Costa, Regional distribution of Leu- and Met-enkaphalin in rat brain, Neurpharmacology 16:303 (1977).

10. R. J. Miller, K. J. Chang, J. Leighton and P. Cuatrecasas, Interaction of iodinated enkaphalin analogues with opiate receptors, Life Sciences 22:379 (1978).

11. W. Fratta, H. Y. T. Yang, B. Majane, and E. Costa,Distribution of β-Endorphin and related peptides in the hypothalamus and pituitary, Neuroscience 4:1903 (1979).

12. O. H. Lowry, H. J. Rosenbrough, A. L. Farr, and R. J. Randall, Protein measurement with the Folin phenol reagent, J.Biol. Chem. 193:265 (1951).

13. T. Hükfelt, A. Lijungdahl, L. Terenius, R. Elde, and G. Nilsson, Immunohistochemical analysis of peptide pathways possibly related to pain and analgesia: enkephalin and substance P, Proc.Natl.Acad.Sci.USA 74:3081 (1977).

14. M. L. Fields, P. C. Emson, B. K. Leigh, R. F. T. Gilber, and L. L. Iversen, Multiple opiate receptors sites on primary afferent fibers, Nature 284:351 (1980).

15. C. Missale, S. Govoni, A. Bosio, L. Croce, P. F. Spano, and M. Trabucchi, Changes of β-Endorphin and met-enkephalin content in the hypothalamus-pituitary axis induced by aging, J. Neurochem. 40:20 (1983).

16. L. J. Forman, W. E. Sonntag, D. A. Van Vugt, and J. Meites, Immunoreactive -Endorphin in the plasma, pituitary and hypothalamus of young and old male rats, Neurobiol.Ageing 2:281 (1981).

17. H. Akil, S. J. Watson, J. D. Barchas, and C. H. Li, β-Endorphin immunoreactivity in rat and human blood: Radioimmunoassay, comparative levels and physiological alterations, Life Sciences 24:1659 (1979).

18. K. Nakao, Y. Nakai, S. Oki, K. Horii, and H. Imura, Presence of immunoreactive β-Endorphin in normal human plasma, J.Clin. Invest. 62:1395 (1978).

19. D. T. Krieger, A. Liotta, and C. H. Li, Human plasma immuno-
     reactive β-Lipotropin: Correlation with basal and
     stimulated plasma ACTH concentrations, Life Sciences
     21:1771 (1977).
20. S. Ritter, N. L. Pelzer, Magnitude of stress-induced brain
     norepinephrine depletion varies with age, Brain Res.
     152:170 (1978).

PATHOPHYSIOLOGY OF PROSTAGLANDINS, WITH PARTICULAR REFERENCE

TO THE ELDERLY

C. Galli

Institute of Pharmacology and Pharmacognosy
University of Milan, Italy

Prostaglandins (PG) are formed in mammals, in practically all types of cells with the exception of erythrocytes, in response to a variety of often unspecific stimuli. The long chain 20 carbon polyunsaturated fatty acid (PUFA) which is the comon PG precursor, i.e. arachidonic acid (AA), is mainly formed from the short chain 18 carbon essential fatty acid (EFA), linoleic acid (LA), supplied by the diet. AA is ubiquitously present in structural phospholipids (PL) located in cellular and subcellular membranes, whereas the enzyme system responsible for the conversion of LA to AA through desaturation and elongation, and the machinery for the oxidative conversion of AA to PG and related products, are located in the endoplasmic reticulum. Utilization of AA for PG synthesis is dependent upon its release from the PL stores through activation of lipolytic processes (phospholipase acting on certain PL), which have been investigated in detail in various biological systems.

The complete sequence of factors involved in the formation of PG, including the 20 carbon PUFA precursor(s) and the enzymes sequentially converting the precursor(s), after its release from PL, through oxydative steps (cyclooxygenase and lipooxygenase) to the intermediate unstable products (e.g. the cyclic endoperoxides) and finally to the end products (primary PG, prostacyclin, thromboxane, leukotrienes) has been defined "the eicosanoid" system. A schematic representation of the eicosanoid system is shown in Figure 1. PG precursor(s) and synthetic machinery appear thus to be ubiquitously distributed, and the formed products are subsequently responsible for the modulation of a variety of biological responses in different systems.

For this reason physiological and pharmacological factors involved in the modulation of eicosanoids can be considered and

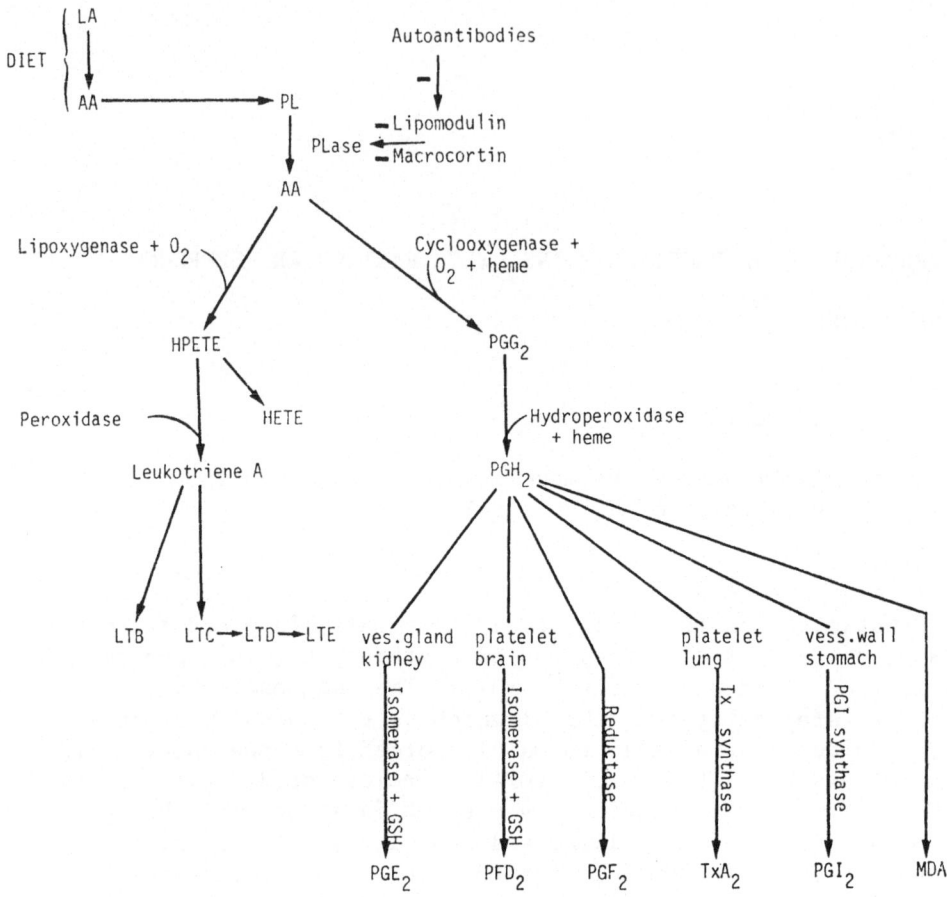

Fig. 1. The eicosanoid system

discussed only in a very general way, whereas participation of
eicosanoids in physiological and pathological processes are more
conveniently analyzed in distinct systems and districts, also in
relation to functional changes associated with the aging process.

    This paper will briefly review some of the information on the
factors known to modulate PG formation in tissue and, subsequently,
consider the eicosanoid cascade in the cardiovascular system,
where the eicosanoid products are known to affect various functional
parameters, under the influence of physiological conditions (e.g.
nutritional state) or in the course of pathological states (e.g.
the atherosclerotic process or diabetes mellitus, which are
generally associated with aging).

FACTORS RESPONSIBLE OF THE MODULATION OF EICOSANOIDS

PG Precursor FA

The formation of biologically active products of lipid nature from PUFA, which ultimately derive from EFA, makes the eicosanoid system dependent upon dietary influences.   Dietary induced changes of tissue levels of PG precursor FA, may, in fact, affect the formation of the biologically active products.

Eicosanoids may theoretically derive from various 20 carbon PUFA accumulated in tissues either through conversion from short chain precursor(s) or through exogenous supply of the preformed compound.   AA (20:4, n-6, Δ5,8,11,14) is the precursor of the PG_2 series-which appears to be physiologically the most important-since it is the predominant 20 carbon PUFA present in tissues under normal nutritional conditions.   PG of the 1 and 2 series are formed from 20:3, n-6, Δ8,11,14 (di-homo-j-linolenic acid, DHGLA) and from 20:5, n-3, Δ5,8,11,14,17 (eicosapentaenoic acid, EPA) respectively.[1] These PG precursors are, however, virtually absent in tissues, since AA is practically the only 20 carbon PUFA of the n-6 series accumulated in tissues after dietary supply of the precursor LA, and linolenic acid, which is present in very low concentration in the diet, is converted to the 22 carbon, highly unsaturated derivative (22:6, n-3, Δ4,7,10,13,16,19, docosahexaenoic acid) rather than to the 20 carbon compound, EPA. DHGLA and EPA are incorporated in tissues after administration of the preformed compounds and EPA is present in appreciable concentration in lipids of edible marine animals.[2]

The sequential conversion of 18:2, n-6, and 18:3, n-3, to the PG precursor FA are as follows:

$$
\begin{array}{ccccccc}
 & PG_1 & PG_2 & & & & \\
 & \uparrow & \uparrow & & & & \\
\text{n-6:} \quad 18:2 & \to 18:3 & \to 20:3 & \to 20:4 & \to 22:4 & \to 22:5 \\
\end{array}
$$

LA   γ-linolenic   DHGLA   AA

$$
\begin{array}{cccccc}
 & & PG_3 & & & \\
 & & \uparrow & & & \\
\text{n-3:} \quad 18:2 & \to 18:4 & \to 20:4 & \to 20:5 & \to 22:5 & \to 22:6 \\
\end{array}
$$

α-linolenic   EPA

Of general importance are the following observations:
a) Metabolic conversions have been studied mainly in liver endoplasmic reticulum preparations, and limited information is available on the actual rates of conversion and on the overall metabolic capacities of other tissues and organs. This is

relevant when considering the effects of changing levels of
dietary EFA on the FA-composition of various cells, tissues and
organs.

b)   Species differences in PUFA metabolism have been described and
it appears that felines are unable to convert LA to AA,[3] which
ingested preformed by carnivores since it is present in
appreciable concentrations in PL of animal tissues (e.g. muscle
and liver).

Detailed studies on the conversion of 18:2, n-6, and 18:3,
n-3, to their respective longer chain PUFA derivatives, have shown
that reciprocal competitive inhibition between the two FA series
occurs.[4],[5] In tissues where high levels of very long chain PUFA
appear to modulate important physico-chemical parameters of structural PL (e.g. the C.N.S.) the total level of unsaturation of
specific PL is retained, in spite of appreciable changes of the
relative dietary supply of n-6 and n-3 FA.[6] This indicates that
"in vivo", desaturation and elongation of the two EFA series are
modulated in a complex and co-ordinated manner.

Studies on the "in vitro" conversion of LA to AA in liver
endoplasmic reticulum revealed that desaturases - $\Delta 6$ involved in
the conversion of LA (18:2 $\Delta 9,12$) to γ-linolenic (18:3 $\Delta 6,9,12$)
acid and $\Delta 5$ responsible of the conversion of 20:3 8,11,14 to AA
(20:4 $\Delta 5,8,11,14$)-are the most important limiting steps and are
modulated by various hormonal and nutritional factors.[7] Hormones
which are released during fasting, e.g. adrenaline, glucocorticoids
and glucagon, inhibit $\Delta 6$ desaturase, possibly through stimulation
of hepatic adenylate cyclase,[9] whereas metabolic and hormonal
changes following feeding (increased glycemia and insulin levels,
decrease of adrenaline, glucagon and glucocorticoids) exert an
opposite effect.

The positive effect of insulin on $\Delta 6$ desaturase,[10] explains
the reduced desaturation of LA by diabetic rats.[11] Reduction of
LA desaturation is also presumably responsible for the reported
reduction of AA levels in tissue PL during the aging process in
animals[12] and also in humans.[13] It is not know whether changes
in tissue levels of PG precursor FA, occurring in pathological
states or in the aging process are responsible for altered PG
production. However, factors others than changes in PG-precursor
levels appear to be prevalent in the modulation of PG formation.

Changes of tissue levels of PG-precursor FA are obtained also
through modifications of dietary FA. Administration of diets
rich in either LA, or DHGLA, or EPA results in changes of tissue
levels of AA and, generally, in the accumulation of the administered
FA in tissue PL. This would result in a modified production of
PGs of the 2 series and in the production of PG of the 1 and 3
series respectively. Predictions of the effects induced by changes

of dietary FA on the production of PG in tissues and organ systems,
and on functional parameters are, however, difficult, since a)
the affinities of the various 20 Carbon PUFA for the cyclo-oxygenase
are very different and inhibition of the enzyme system, rather than
stimulation, may occur,[1] and b) productions of eicosanoids with
opposite biological activities (e.g. thromboxane and prostacyclin)
are usually both affected.[14]

Release of PG Precursor FA

Stimulation of eicosanoid production is obtained mainly through
activation of lipolytic enzymes releasing the FA precursor(s) from
the PL stores. Detailed studies in simplified systems, such as
blood platelets, have shown that the initial step activated in the
release of AA after platelet stimulation with various agents is the
activation of a phospholipase C specifically acting on phosphatidyl
inositol (PI), which is converted to diacylglycerol (DG).[15]
A diglyceride lipase subsequently releases AA from DG.[16] This
mechanism may be responsible of receptor-mediated stimulation of
PG formation also in other systems,[16] but massive and unspecific
release of PG precursor FA and of PG, such as that occurring during
inflammation, may be obtained through stimulation of various types
of lipase and phospholipases.

The recent discovery of endogenous protein factors inhibiting
the release of AA in leukocytes (lipomodulin)[17] and lung tissue
(macrocortin)[18] appears of importance for the understanding of
altered PG formation in pathological states. Antibodies against
lipomodulin have been detected e.g. in patients with rheumatic
disease.[19] Corticosteroids appear to decrease PG formation in
tissues[20] by stimulating the formation of the phospholipase
inhibiting proteins.[18]

Conversion of PG Precursor FA

The 20 carbon PUFA, mainly AA, released from PL, are converted
through various oxidative steps to intermediate and end products
of the eicosanoid family. The major pathways so far explored
are the cyclooxygenase and lipogenases. The cyclooxygenase leads
to the initial generation of unstable intermediates (the cyclic
endoperoxides) followed by the formation of various end-products
(primary PG_s, thromboxane A_2 and prostacyclin) through the activity
of different synthases and isomerases. The balance in the production of the various compounds differs from one tissue to another.

Lipooxygenases are responsible of the formation of a family
of compounds, some of which have potent biological activities.
The 5-lipoxygenase is specifically involved in the pathway leading
to the production of SRS-A substances, such as the recently
described leukotrienes,[21] which are formed by cells participating

in inflammatory and immunoreactive processes (e.g. leukocytes and mast cells), and appear to be important mediators of these pathological states.

EICOSANOIDS IN THE CARDIOVASCULAR SYSTEM

A vast literature has been produced on the formation of eicosanoids in different tissues and organ systems and on the activities of the various products on different biological preparations. In the cardiovascular system, the most important compounds appear to be prostacyclin (PGI_2), which is produced by vessel walls and exerts potent anti-aggregatory and vasodilating activities,[23] and thromboxane A_2 (Tx) generated by platelets after stimulation[24] and exerting potent pro-aggregatory activity and vasoconstricting effects on the coronary circulation.

Also considerable information is available on the pharmacological control of the cyclooxygenase and of the Tx synthase.

EICOSANOIDS IN THE VASCULAR DISTRICT: CHANGES IN PATHOLOGICAL AND AGE-RELATED STATES

The role of eicosanoids in controlling important functional parameters in the cardiovascular system has promoted investigations on their possible involvement in pathological states of this district, such as the atherosclerotic and thromboembolic disease and conditions which may favour their onset and progression (e.g. diabetes). A greater body of information is available on the Tx generating system, also in humans, since platelets and their response to various stimulating agents are easy to handle in "in vitro" systems, whereas limited and often contradictory data have been obtained on the PGI_2 generating system, due to lack of adequate experimental models. Even more limited is the knowledge of changes of responsiveness to eicosanoids in tissues of the vascular district in the course of pathological states.

The complexity of the biological machinery, based on a number of still unexplored endogenous factors, for the control of PG generation, makes it difficult to investigate changes in tissue PG formation. In fact, studies based on determinations of conversion rates of labelled PG precursor FA to the various products in various tissues do not provide information on the biologically controlled process of PG formation from endogenous precursor(s), which, however, is more difficult to investigate.

Platelet function and the thromboxane generating system have been extensively studied in vascular pathologies, where impairments of the haemostatic balance appear to play a role in the progression

of the disease, such as the atherosclerotic process.   Increased
platelet aggregation[25,26] associated with enhanced platelet Tx
formation in response to stimulation[27,28] have been reported in
human hyperlipaemia.   However, the responsiveness of the Tx
forming system to different stimulating agents appears to be
differentially affected, at least in experimental hypercholesterolaemia.[29]   Increased platelet Tx production has been reported also
in other vascular pathologies, such as coronary heart diseases,[30]
and arterial and venous thrombosis.[31] The well known vasculopathy
associated with the diabetic condition may also be partly based on
altered haemostatic balance and impaired eicosanoid function, since
increased platelet Tx formation from AA has been reported in
diabetic patients.[32]   Most of the data of enhanced Tx production in
pathological conditions have been obtained by using exogenous
substrate.   However, since the response of the Tx generating
system in platelets after stimulation with aggregating agents may
be instead affected (for instance in experimental hyperlipaemia),[29]
the actual significance of the reported data in the disease
processes should be critically evaluated.

More limited information is available on the PGI_2 system in
pathological conditions.   Circulating PGI_2 levels, measured as the
stable metabolite 6 keto $PGF_1\alpha$ do not appear to be representative
of the endogenous rates of formation, and urinary metabolites may
represent renal production[35] more than systemic levels.   Involvement of PGI_2 has been claimed in essential hypertension, since
urinary PGI_1 metabolites are decreased in hypertensive patients.[34]
Studies in experimental animals revealed that alterations of PGI_2
production occur also in hyperlipaemia, although both decreased[35]
and increased[29] formation by isolated arterial walls of hyperlipaemic rabbits have been reported.

Evaluations of isolated parameters, such as thromboxane
formation in platelets and of PGI_2 production in the vascular
district, in pathological states, do not provide, however, adequate
information on the overall haemostatic balance, since responsiveness
of cells to eicosanoids should also be taken into account.   In
fact PGI_2 production and platelet sensitivity to PGI_2 appear to be
inversely correlated as shown in Table 1.   Decreased sensitivity
to PGI_2 has also been reported for platelets of diabetic subjects.[36]

From the reported data it appears that formation of, and
responsiveness to eicosanoids are indeed altered in various pathological states of the vascular district, but it is still unclear
whether the observed abnormalities play a role in the disease states
or are just secondary processes.

Since the incidence of the most common disease of the vascular
system increases with age, age-related changes of eicosanoid
production could influence the onset of pathological processes.

Table 1.   Aortic PGI_2 production and platelet sensitivity to PGI_2
           in hypercholesterolaemic rabbits

	6 keto $PGF_1\alpha$ pg/μl	PGI_2 M.I.C. M/l
N (10)	0.82$\pm$ 0.1	12.0x10^{-9}$\pm$2.1
HC (10)	1.86$\pm$ 0.3	20.5x10^{-9}$\pm$3.3

PGI_2 production by perfused (0.3 ml/min) isolated aortas was
measured as 6 keto PGF levels (pg/μl) in the perfusate and
sensitivity of platelets to the antiaggregatory activity of PGI_2
was measured as the Minimal Concentration (M/l) Inhibiting (M.I.C.)
arachidonic acid (0.4 mM) induced PRP aggregation, in normal (N)
and hypercholesterolaemic (HC) rabbits.   Values are the mean $\pm$
S.E.M. of the number of determinations in brackets.   Differences
N vs HC are significant at the p <0.01 level.

Information on the eicosanoid system in the vascular district in
relation to aging, is, however, rather limited.   Data in
experimental animals are also difficult to extrapolate to humans,
since life spans and aging characteristics are obviously different
in different animal species.

    Levels of 6 keto $PGF_1\alpha$ in human urines are high in the neonate,
rapidly decline after one year and remain unchanged until the adult
age.[37]   The prostacyclin biosynthetic activity from exogenous
substrate in rat aortic smooth muscle cells[38] and vascular tissue[39]
are decreased with aging, although the total cyclooxygenase
activity[38] and basal prostacyclin production[39] are not changed.
The reduced maximal capacity to produce prostacyclin, associated
with unchanged basal formation, would suggest that during aging,
the ability of the vascular tissue to increase prostacyclin
production to cope with increased demand, is impaired.   Since it
has also been shown that formation of PGE_2 - a modestly prothrom-
botic agent in the rat - is increased in the vascular system of
aged rats,[38] this may further contribute to promote the pathogenesis of arterial thrombosis during aging.

    Furthermore, PGI_2 biosynthesis is strongly inhibited by
hydroperoxyacids derived from PUFA.[40]   Since lipid peroxidation
preceded by free radical formation occurs during aging,[41] this
may lead to decreased prostacyclin biosynthesis and thus contribute
to the onset and progression of atherosclerosis.

    In conclusion, several reports of modified eicosanoid
production, in aging and age-related pathological conditions have

appeared.   Further research is however required in order to
elucidate the mechanisms responsible for these alterations and
the functional correlations between PG abnormalities and pathological states.

REFERENCES

1. P. Needleman, M. O. Whitaker, A. Wyche, K. Watters, H. Sprecher
   and A. Raz, Manipulation of platelet aggregation by prostaglandins and their fatty acid precursors: pharmacological
   basis for a therapeutic approach, Prostaglandins, 19:165
   (1980).
2. J. Dyerberg and H. O. Bang, Haemostatic function and platelet
   polyunsaturated fatty acids in Eskimos, Lancet 2:433 (1979).
3. J. P. W. Rivers, A. J. Sinclaire and M. A. Crawford, Inability
   of the cat to desaturate essential fatty acids, Nature
   258:171 (1975).
4. R. R. Brenner, The desaturation step in the animal biosynthesis
   of polyunsaturated fatty acids, Lipids 6:567 (1971).
5. R. T. Holman, Nutritional and metabolic interrelationships
   between fatty acids, Fedn.Proc. 23:1062 (1964).
6. C. Galli, H. I. Trzeciak and R. Paoletti, Effects of dietary
   fatty acids on the fatty acid composition of brain ethanolamine phosphoglyceride: reciprocal replacement of n-6 and
   n-3 polyunsaturated fatty acids, Biochem.Biophys.Acta
   248:449 (1971).
7. R. R. Brenner, R. O. Peluffo, O. Mercuri and M. A. Restelli,
   Effect of arachidonic acid in the alloxan-diabetic
   rat, J.Physiol. 251:63 (1968).
8. R. R. Brenner, Nutritional and hormonal factors influencing
   desaturation of essential fatty acids, in: "Progress in
   Lipid Research," vol. 20, R. T. Holman, ed., Pergamon
   Press (1982).
9. I. N. T. De Gomez-Dumm, M. J. T. De Alaniz and R. R. Bremer,
   Comparative effect of glucagon, dibutyril cyclic AMP and
   epinephrine on the desaturation and elongation of linoleic
   acid by rat liver microsomes, Lipids 11:833 (1976).
10. O. Mercuri, R. O. Peluffo and R. R. Brenner, Depression of
    microsomal desaturation of linoleic to α-linolenic acid in
    the alloxan-diabetic rat, Biochem.Biophys.Acta 116:409
11. R. O. Peluffo, S. Ayala and R. R. Brenner, Metabolism of
    fatty acids of the linoleic acid series in testicles of
    diabetic rats, Am.J.Physiol. 218:669 (1970).
12. A. Keranen, P. Kankare, and M. Hallman, Changes of fatty acid
    composition of phospholipids in liver mitochondria and
    microsomes of the rat during growth, Lipids 17:155 (1982).
13. H. B. White, Jr., C. Galli and R. Paoletti, Ethanolamine
    phosphoglyceride fatty acids in aging human brain,
    J. Neurochem. 18:1337 (1971).

14. G. Hornstra, E. Christ-Hazelhof, E. Haddeman, F. ten Hoor and
    D. H. Nugteren, Fish oil feeding lowers thromboxane and
    prostacyclin production by rat platelets and aorta and does
    not result in the formation of prostaglandins I₃,
    Prostaglandins 21:727 (1981).
15. S. Rittenhouse-Simmons, Production of diglyceride from phosphat-
    idylinositol in activated human platelets, J.Clin.Invest.
    63:580 (1979).
16. R. L. Bell, D. A. Kennerly, N. Stanford and P. W. Majerus,
    Diglyceride lipase: A pathway for arachidonate release from
    human platelets, Proc.Natl.Acad.Sci. USA 76:3238 (1979).
17. F. Hirata, E. Schiffmann, K. Venkata-Subramanian, D. Salomon
    and J. Axelrod, A phospholipase A₂ inhibitory protein in
    rabbit neutrophyls induced by glucocorticoids, Proc.Natl.
    Acad.Sci.USA. 77:2533 (1980).
18. G. J. Blackwell, R. Carnuccio, M. Di Rosa, R. J. Flower,
    L. Parente, and P. Persico, Macrocortin: a polypeptide
    causing the anti-phospholipase effect of glucocorticoids,
    Nature 287:147 (1980).
19. F. Hirata, R. Del Carmine, C. A. Nelson, J. Axelrod, E.
    Schiffmann, A. Warabi, A. L. De Blas, M. Nirenberg, U.
    Manganiello, M. Vaughan, S. Kumagai, I. Green, J. L. Decker,
    and A. D. Steinberg, Present of antiantibody for phospholipase
    inhibitory protein, lipomodulin, in patients with rheumatic
    disease, Proc.Natl.Acad.Sci.USA 78:7190 (1981).
20. A. Danon and G. Assouline, Inhibition of prostaglandin bio-
    synthesis by corticosteroids requires RNA and protein
    synthesis Nature 273:552 (1978).
21. P. Borgeat and B. Samuelsson, Metabolism of arachidonic acid
    in polymorphonuclear leukocytes. Structural analysis of
    novel hydroxylated compounds, J.Biol.Chem. 254:7865 (1979).
22. R. A. Lewis and K. F. Austen, Mediation of local homeostasis
    and inflammation by leukotrienes and other mastcell dependent
    compounds, Nature, 293:103 (1981).
23. S. Bunting, R. Gryglenski, S. Moncada, and J. R. Vane, Arterial
    walls generate from prostaglandin endoperoxides a substance
    (prostaglandin X) which relaxes strips of mesenteric and
    coeliac arteries and inhibits platelet aggregation,
    Prostaglandins 12:897 (1976).
24. M. Hamberg, J. Svensson and B. Samuelsson, Thromboxanes: A new
    group of biologically active compounds derived from prosta-
    glandin endoperoxides, Proc.Natl.Acad.Sci.USA 72:2994 (1975).
25. A. Carvalho, R. W. Colman, and R. S. Lees, Platelets function
    in hyperlipoproteinemia, New Engl.J.Med. 290:434 (1974).
26. A. Nordoy and J. M. Rodset, Platelet function and platelet
    phospholipids in patients with hyperbetalipoproteinemia,
    Acta Med.Scand. 189:385 (1971).
27. E. Tremoli, P. Maderna, M. Sirtori and C. R. Sirtori, Platelet
    aggregation and malon dialdehyde formation in type II a
    hypercholesterolemic patients, Haemost. 8:47 (1979).

28. E. Tremoli, G. C. Folco, E. Agradi and C. Galli, Platelet thromboxanes and serum-cholesterol, Lancet 1:107 (1979).

29. E. Tremoli, A. Socini, A. Petroni and C. Galli, Increased platelet aggregability is associated with increased prostacyclin production by vessel walls in hypercholesterol-emic rabbits, Prostaglandins 241:397 (1982).

30. P. D. Hirsch, L. D. Hillis, W. B. Campbell, B. G. Firth and J. T. Wilderson, Release of prostaglandins and thromboxane into the coronary circulation in patients with ischemic heart disease, New Engl.J.Med. 304:685 (1981).

31. M. Lagarde and M. Dechavanne, Increase of platelet prosta-glandin cyclic endoperoxides in thrombosis, Lancet 1:88 (1977).

32. P. V. Halushka, R. C. Rogers, C. B. Loadholt and J. A. Colwell, Increased platelet thromboxane synthesis in diabetes mellitus, J.Lab.Clin.Med. 97:87 (1981).

33. F. F. Sun and B. M. Taylor, Metabolism of prostacyclin in rat, Biochemistry 17:4096 (1978).

34. J. H. Grose, M. Lebel and F. M. Gbeassor, Diminished urinary prostacyclin metabolite in essential hypertension, Clin.Sci. 59:1215 (1980).

35. A. Zmuda, A. Dembinska-Kiec, A. Chytowski and R. J. Gryglewski, Experimental atherosclerosis in rabbits: platelet aggregat-ion, thromboxane A_2 generation and antiaggregatory potency of prostacyclin, Prostaglandins 14:1035 (1977).

36. M. Lagarde, P. Berciaud, M. Burton and M. Dechavanne, Refract-oriness of diabetic platelets to inhibitory prostaglandins, Prostaglandins Med. 7:341 (1981).

37. B. Scherer, S. Rischer, W. Siess and P. C. Weber, Analysis of 6 keto prostaglandin $F_1\alpha$ in human urine: age-specific differences, Prostaglandins 23:41 (1982).

38. W. C. Chang, S. I. Murota, J. Nakao, and H. Orimo, Age-related decrease in prostacyclin biosynthetic activity in rat aortic smooth muscle cells, Biochem.Biophys.Acta 620:159 (1980).

39. R. S. Kent, B. B. Kitchell, D. G. Shand and A. R. Whorton, The ability of vascular tissue to produce prostacyclin decreases with age, Prostaglandins 21:483 (1981).

40. S. Moncada, R. J. Gryglewski, S. Bunting and J. R. Vane, A lipid peroxide inhibits the enzyme in blood vessel micro-somes that generates from prostaglandin endoperoxides the substance (prostaglandin x) which prevents platelet aggreg-ation, Prostaglandins 12:715 (1976).

41. T. F. Slater, "Free Radical Mechanisms in Tissue Injury," London (1972).

PAIN TREATMENT IN OLD PATIENTS

G. P. Vecchi, M. Neri, R. Lugli
*L. A. Pini and **L. Calza

Dept. of Gerontology, *Dept. of Clinical Pharmacology
**Dept. of Physiology
University of Modena, Italy

When taking into account the problem of pain in the old, it must be realised that the measure of pain is based on interpersonal relationship and is therefore conditioned by the subjective capacity to explain the symptom. Every personality, from the behavioural point of view, is also formed on the basis of the learning process; this is tied to the understanding of pain, which in time is tied to the nervous pathways of reward and punishment. Reports on pleasure neurochemistry, have to be added to the many on pain neurotransmitters.[1]

On the theme of neuropeptides in development and aging, research on the effects of morphine and naloxone on learning, memorization and nociception in young and old rats, must be remembered; among the various results obtained, it was noted that the number of opiaceous receptors were reduced in various sites of the central nervous system (striatum and anterior cortex). In the frontal lobes one can also note an increase of the apparent affinity with morphine, a variation which, according to McGaugh and his colleagues[2] could compensate for the decrease of the receptorial sites of the maximum binding capacity.

During the development, from birth to death, this patrimony is enriched and the organism will establish a behavioural pattern which may include disregarding painful stimuli which are of a high physical intensity but of little consequence and vice-verse. The structure and functions which condition the biological and psychological behaviours will also determine the way in which pain is manifested. Among the mechanisms which could deteriorate with aging, new important data are now available.[3,4,5]

In recent research on synaptic transmission, the concept of one nervous termination to one neuro-transmitter, has been widely re-examined; it has been demonstrated that at a receptorial level, several neuro-transmitters are coexistent and the utilization of which is conveniently modulated.[6]   The concepts of the coexistence of transmitters and receptor-receptor interactions have increased our understanding of the integrative processes regulating synaptic homeostasis.[7]   One way of studying the homeostatic mechanism is by causing a controlled disturbance in the system and then to evaluate the changes which are brought about by the system in order to minimize the disturbance.

Thus, dopaminergic receptor supersensitivity may be the result of a lesion of the ascending dopaminergic pathways.   This receptor supersensitivity is associated with an increase in the number of 3H-spiperone (naloxone analogous) binding sites in the striatum. Normal and supersensitive receptors of very old rats in the striatum are less sensitive to other neuro-transmitters (or analogues) than those of younger animals.[8]   In the same way, the binding characteristics of 3H-spiperone binding sites are no longer modulated. Therefore, the plasticity of dopaminergic receptors may be reduced with age leading to the deficiencies in the integrative processes existing in the local circuit of the striatum.[9,10,11]

It is unlikely that the role of the different parts of the central nervous system in the pain mechanism will be adequately understood in terms of independent motor/sensory or autonomic functions, or only in relation to some aspects of pain such as experience.   It seems reasonable to assume that the above mentioned synaptic processes take part in the control of synaptic homeostasis and synaptic plasticity and that the synaptic regulation is involved in the integrative functions of the central nervous system. Furthermore, changes in the synaptic homeostasis are important phenomena in aging.[10,12]

CLINICAL CASES

Our results have been found from 46 subjects who were over 68 years of age.   These patients were all affected by neoplastic diseases of various origin, nature and location.

Before reporting on the terms of the therapeutical schedule, we must point out that in every case all necessary therapeutical procedures were adopted.   Only the pain was treated following a schedule and it is possible that this influenced the progression of the cases.   These cases were homogeneous also from this point of view, as they were with regard to age; and probably with regard to the neoplastic etiology of pain.

The subjects treated were 24 females and 22 males. The primary site of the tumour was: the breast (9 women), ovaries and uterus (5 women), lungs (6 men and 2 women), the brain (2 men and 2 women), stomach (3 men and 2 women), oesophagus (1 man and 1 woman), the prostate (6 men), bladder (2 men), kidney (1 woman), the salivary glands (2 women): one case of costal localization of a myeloma (1 woman) must be added.

THERAPEUTIC SCHEDULE

Our method for the treatment of pain was codified on the basis of a hierarchy of analgesic drugs with central and peripheral action and, if possible or necessary, with surgery. This permitted us to carry out our research while preserving the principle of giving the patient the greatest benefit in terms of symptoms and quality of life.

The procedure was as follows:

1) Placebo in vein. In this period we determined the basic conditions and psychological situation of the patient and his availability to collaborate. Such a period was prolonged in cases of favourable effects. Such effects were eventually favoured by physiotherapy and by electrical transcutaneous stimulation. The placebo treatment was known to the medical and paramedical staff.

2) Drugs with peripheral action. Diflunisal: 750 mg administered in 3 doses per day for an expected effectiveness of 8 - 12 h/dose.[13] This drug has been demonstrated to be effective and well tolerated in old people.[14] Its efficacy is also high in the case of neoplastic pain.[13]

3) Drugs with central action.

   3a - pentazocine oral: 50 mg
   3b - methadone oral: 20 mg
   3c - pentazocine i.m.: 30 mg

The treatment was not carried out on patients who could benefit from analgesic surgery at regional nervous level. The major neuro-surgery was reserved for cases which did not benefit from any of the 3 types of treatment.

Every successive treatment consisted of a period of experimental observation lasting 24 hours, and lasted as long as possible on the basis of its subjective effectiveness (Huskisson's scale).[15] In this way an evaluation of the analgesic activity of different drugs based on the number of days in which the effect was positive could be made.

EVALUATION OF RESULTS

The following parameters have been evaluated previous analgesic
treatments and their effectiveness (yes,no), the location of pain
(visceral, osseous, and soft tissues, head-neck, face, chest,
lumbar regions, abdomen, pelvis, arms and legs) and its characteristics (continuous, discontinuous, irregular, localised, radiated,
diffuse, burning, dull and 'stitch').   The subjective evaluation
based on Huskisson's scale was carried out 4 times per day.   The
collateral effects (anorexia, sleep disturbances, etc.), and the
therapeutical variations induced by symptomatic treatments also in
the case of emergency were noted.

An important review debating pain evaluation methods regarding
geriatrics, concludes by affirming that the actual methods used are
not satisfactory.[16]   Our evaluation was based on a test derived
from our previous experience[17] on the administration of currents of
different amps to old people.   We have used an electrical stimulation technique on the skin, consisting of administering variable
amp. impulses on the volar surface of the forearm by annular
concentric electrodes, with trains of impulses (50 cycles/sec) at
10 ms intervals.   The intensity varied from 0 to 0.5 mA and was
progressively increased up to the minimum amperage able to give a
sensation of burning and unbearable pain.[18,19]

The Ss were instructed to give three verbal responses to
indicate: detection threshold (DT) (first sensation of touch or
tickling), pain threshold (PT) and pain tolerance/threshold (PTT)
(upper stimulus intensity tolerated by Ss).

In 100 healthy subjects (50 over 60 years and 50 under) we have
adopted the same method and have studied the differences between
the stimulation of the right and left limbs.   When testing
responses to painful stimuli, studies of man's behaviour suggesting
that right and left central hemispheres may have a different ability
in evaluating emotions and perhaps pain, should be taken into
consideration.[20]   Many authors state that with aging, the patterns
of cognitive performances are impaired, and some suggest that
functions of the right hemisphere deteriorate more than those of
the left, verbal ability being preserved the most.[21]   Furthermore,
a slowing down of complex reaction time is also described.[22]

This research[23] has been carried out in order to ascertain
whether:

1 - aging can provoke modifications of the psycho-physical para-
    meters of experimental pain;
2 - there is a difference between the left and right limb;
3 - none or some of these parameters show significant correlations
    with the concomitant affective state.

   The Hamilton scale for anxiety was administered, physical and
psychic items were separately scored.[24,25]   A manual dexterity test
was used to exclude left handed Ss.   The data characterizing the
two groups were grouped as follows:

a) If we consider the results of psychophysical measurements (DP,PT,
   PTT) we can observe that

1 - regardless as to which limb was examined the scores of the
    younger people were lower
2 - right scores were always lower than the left ones;

b) The younger subject group scores were always lower than the
   corresponding older subject group scores.

   When a 2-way analysis of variance is applied, psychophysical
scores, namely the left arm pain threshold and perception threshold
in both arms, showed significant differences only in respect to age.
The groups behaved in the same way with respect to affective scores.
In the younger group no significant correlation between psychophysical and affective scores could be observed, but in the older group
psychic and psychophysical scores correlated positively (p <0.01).

   It has been demonstrated that particularly in motor tasks,
intrahemispheric pathways are faster than interhemispheric ones
owing to the time lost in transcallosal transmission.[22]   It can be
presumed that in the same way this phenomenon may play an important
role in determining the common trend occurring in our results.
It must be stressed that the subject's task was a 'verbal labelling'
of the perceptual experience, a specific ability of the left cerebral
hemisphere.   Furthermore, the significant right/left difference
in PT present in the younger group may account for a controlateral
inhibiting effect.   In fact, only a binary choice (yes/no) was
requested for perception and tolerance thresholds but a quality
choice was requested for pain.   In this case, the stimulus had to
be deeply processed and then labelled.   The maintenance of a
specific cognitive hemispheric specialisation presupposed the
controlateral inhibition of potentially rival systems in the opposite
side of the brain.[22]   In this case with the left arm stimulated the
right and left hemispheres were operating and controlaterally
inhibiting.   The aging process emphasizes the right/left imbalance
so that the difference between right/left values becomes constantly
significant in the older group.   Our results do not explain how
and where the impairment took place.   It may be a slowing down of
the central processing, transcallosal information crossing, or an
enhanced sensitivity of the right hemisphere causing controlateral
inhibition.   A slowing down of complex reaction time may account
for the difference between the groups for perception threshold in
both arms.   This phenomenon appears to be negligible for pain
tolerance values.   That is, old people free from peripheral

impairments, have the same ability as young people to produce rapid
and effective behaviour when the stimulus becomes highly unpleasant.
Furthermore, the evidence shows that old people are as sensitive as
the young to noxious stimuli, when they reach a critical level.
For this reason, in order to test the effectiveness of the therapeutic approach (physical, pharmacological, psychological) pain
threshold must be taken into consideration.

Our recordings were taken at 6 a.m., 12 a.m., 6 p.m. and
midnight in basal conditions (placebo therapy) and during the cycles
of therapy with the analgesics. Figure 3 shows the behaviour of the
variables (DT, PT, PGG) in the patients and in control cases.

The results were compared to those already obtained.

We could obtain two kinds of quantitative data:
a) The number of days during which each treatment was effective
   expressed in terms of the degree of subjective improvement.
b) The modification of the parameters related to DT, PT, PTT.

a) Figure 1 refers to the drugs we used in successive periods
of 24 hours varying according to the effectiveness of the drug.
Contrary to the opinion that old people react positively to placebo
therapy, in our subjects the period of treatment with placebo was
very short (on the average about $2\frac{1}{2}$ days). The treatment with
diflunisal lasted for an average of 21 days, and more days in some
cases. Pentazocine taken orally and parenterally, induced a
considerable pain reduction but it also provoked very remarkable
torpor and confusion, in some cases associated with urinary and
fecal retention and in others with incontinence. The effect of
methadone taken orally was very positive, well tolerated and lasted
for an average of 25 days; the morphine parenteral period of
improvement lasted 19 days ($\pm$ 10).

It must be pointed out that, with time, the conditions of our
patients were obviously deteriorating. The therapeutic schedule
was stopped for the following reasons: in 21 cases the patients
died, 12 cases were treated by analgesic surgery and in 13 cases
the quantity of morphine requested for pain relief exceeded our
schedule.

With the above mentioned schedule we could obtain a good
subjective improvement and avoid the side effects of antiphlogistic
analgesic drugs (gastric disturbances, anemia, renal damage, pH
derangements, neuro-toxicity).[14] Among the peripheral analgesics,
diflunisal proved to be the most appropriate drug in the treatment
of neoplastic pain owing to its effectiveness, prolonged action and
minimum toxicity.[13]

Orally taken, methadone is also very useful in neoplastic pain

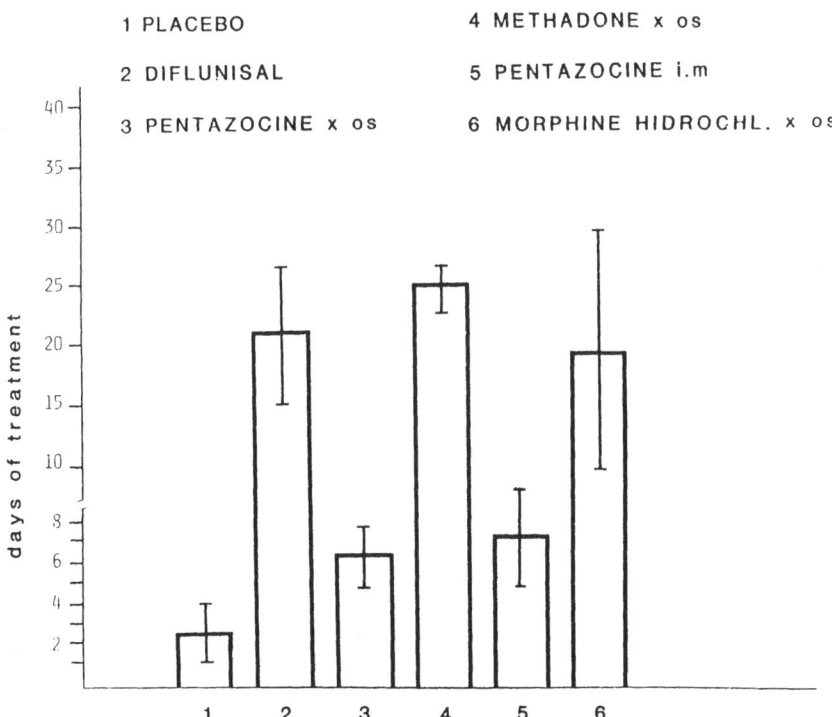

Fig. 1. Drugs used in successive periods of 24 hours varying according to the effectiveness of the drug.

therapy.   Figure 2 shows the relationship between the euphorizing and analgesic activity of morphine taken orally and i.m., and of methadone taken orally.   The data were collected on the basis of Huskisson's scale up to 60 per cent.   Orally taken methadone is absorbed with a peak after 4 hours.   The plasm half-life is 13 to 46 hours with an analgesic activity of 4 to 8 hours.   After oral administration, euphoria is rare and only due to high doses; dependence and tolerance may derive from orally taken methadone as from other narcotics, but are infrequent when used therapeutically and with biochemical monitoring of blood concentration.   As the effect of methadone is long lasting it is used in long-term oral administration, with 10 mg administered 2 - 3 times per day, satisfactory analgesia may be obtained for the whole day.

When taken orally morphine induces an analgesic action which lasts 4 to 5 hours with variable doses of 30 - 90 mg (Figure 2). After 10 mg given parenterally the analgesic effect is reached very quickly, but at the same time, euphoria may also be reached, especially after repeated administration, which may derive from the

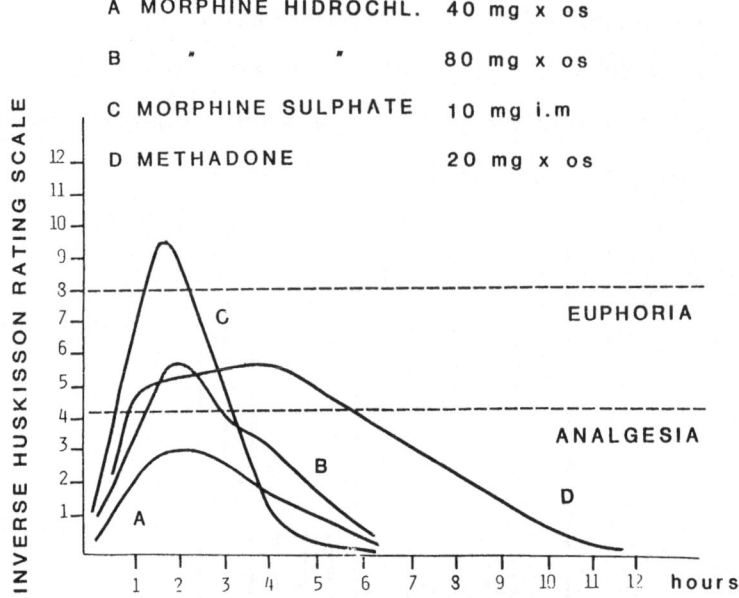

Fig. 2.   Relationship between the euphorizing and
          analgesic activity of morphine taken orally
          and i.m., and of methadone taken orally.

establishment of a dependence.   The importance of selecting not
only the daily morphine dose, but also the interval between doses
and the method of administration is clear; the aim is to maintain
the analgesic but not the euphorising effect and to avoid the
anxious expectation of pain.   With a 10 mg dose, parenteral
morphine induces orthostatic hypotension and myosis, with a 20 mg
dose drowsiness, disphoria and a dangerous respiratory depression
with the phenomenon of sleep-induced apnea may appear.   One may also
observe urine retention especially in subjects with an enlarged
prostate.[26]

Morphine and other narcotics induce tolerance, an effect which
should not be forgotten in the programming of an analgesic therapy.

b) In normal subjects and in cases with neoplastic pain we
have evaluated the threshold of experimental pain, before and during
therapy.   In Figure 3 our data on the detection of the stimulus
(DT) the appearance of burning pain (PT) and the tolerance threshold
(PTT) are noted.   DT was manifested starting at 0.11 mA in older
controls and at 0.09 mA in younger controls; whereas in patients
with neoplastic pain in basal conditions (placebo treatment) it was
much higher: in fact the administration of 0.5 mA did not give any
sensation.   During analgesic treatment DT adjusted to a normal
value (0.12 mA with diflunisal and 0.13 mA with methadone).   PT and

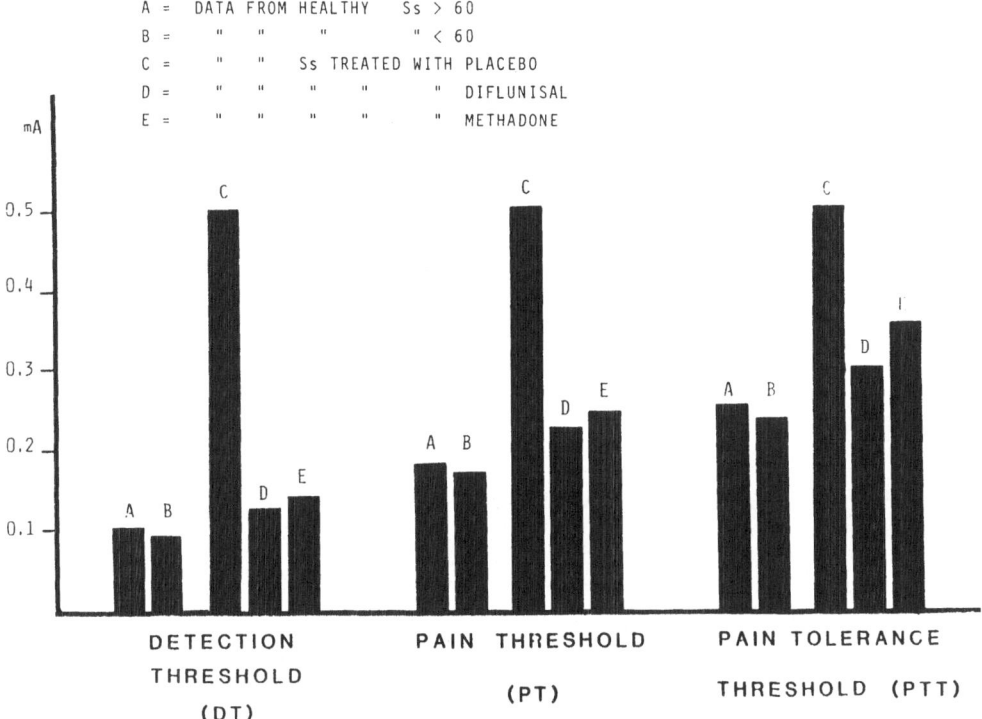

Fig. 3. Pain threshold from electrical stimulus

PTT data showed the same behaviour at the correspondingly higher amperages.

The evaluation of 3 parameters in the controls was related to the Hamilton Scale for anxiety and the Pfeiffer Scale for the Short Psychiatric Evaluation Schedule to ascertain whether:

1) there is a modification with aging
2) there is a correlation between the pain threshold and the presence of anxiety and/or depression symptoms.

Preliminary data suggest that:

1) PT and PTT are negatively correlated with aging
2) PTT and the scores of somatic symptoms in the Pfeiffer Scale are correlated. This result is difficult to explain; one may suggest that old people tend to be an 'augmenter' at least in experimental conditions.[27]

Conclusively our results suggest that the experimental procedure adopted is fairly effective.

SUMMARY

Experimental and clinical research on the pain phenomenon in old people is interlocutory.   This depends, above all, on the difficulty of removing the bias due to personal experience and social-cultural, existential and somatic components.

Worth noting are the results we have obtained with a programmed pharmaceutical schedule, based on peripheral and central analgesics selected on the basis of their effectiveness. length of action and tolerability.   During our studies we have also collected some data about the evaluation of subjective pain symptoms.

REFERENCES

1. L. Calza, Dolore, piacere, neurochimica, Giorn Geront. 29:383 (1981).
2. J. L. McGaugh, R. B Messing, R. A. Jensen, B. J. Vasquez, J. L. Martinez, and V. L. Spiehler, Memory, opiate receptors and aging, in: "Neuropeptides, development and aging," Nat.Inst.Aging, Rochester N.Y. (1980).
3. F. Agnati, K. Fuchs, M. Ferri, F. Benfenati and S. O. Ogren, A new hypnosis of memory.   A possible role on local circuit in the formation of the memory trace, Med.Byol. 59:224 (1981).
4. J. E. Birren, A. M. Woods and M. V. Williams, Speed of behaviour as indicator of age changes and the integrity of the nervous system, in: "Brain function in old age," F. Hoffmeister and C. Muller, eds., Springer Verlag, Berlin (1979).
5. D. Kliss, Neuropsychological evaluation in older persons, in: "The clinical psychology of aging," M. Storandt, I. C. Siegler and M. F. Elias, eds., Plenum Press, New York (1978).
6. T. Hokfelt, J. M. Lundberg, M. Schultzber, O. Johansson, A. Ljungdahl and J. Rehfeld, Coexistence of peptides and putative transmitters in neurons, in: "Neural peptides and neuronal communication," E. Costa and M. Trabucchi, eds., Raven Press, New York (1980).
7. G. Burnstock, T. Hwkfelt, M. D. Gerhson, L. Iverseni, H. W. Kosterlitz and J. H. Szurszewski, non adrenergic, non cholinergic automatic neurotransmission mechanisms, N R P Bulletin 17:3 (1979).
8. L. F. Agnati, K. Fuxe, V. Locatelli, F. Benfenati, A. E. Panerai, M. K. El Eltreby, T. Hokfelt and I. Zini, Neuroanatomical methods for the quantitative evaluation of coexistence of transmitters in nerve cells.   Analysis of the ACTH and endorphin immunoreactive nerve cell bodies of the mediobasal hypothalamus of the rat, J.Neurosci.Meth. 5:203 (1982).
9. L. F. Agnati, K. Fuxe, T. Hokfelt, F. Benfenati, L. Calza, O. Johansson and J. De Mey, Morphometric characteristics of transmitter-identified nerve cell groups: analysis of mesencephalic 5-HT nerve cell bodies, Br.Res.Bull.in press (1982).

10. L. F. Agnati, K. Fuxe, L. Calza, T. Hokfelt, O. Johansson, F. Benfenati and M. Goldstein, A morphometric analysis of transmitter-identified dentrites and nerve terminals, Br.Res.Bull. in press (1982).

11. K. Fuxe, L. F. Agnati, K. Andersson, L. Calza, F. Benfenati, T. Hokfelt, L. Cavicchioli and G. Galli, Morphological, biochemical and functional aspects of the modulatory role played by cholecystokinin and substance P peptides in the regulation of monoaminergic mechanisms, The Second European Winter Conference on Brain Research, Chamonix March 7-13 on neuropeptides by Vanderhaegen (1982).

12. M. Trabucchi, S. Govoni, K. Missale, A. Bosio and P. F. Spano, Brain neurotransmitter receptor changes during aging, Abs.Int.Symp: "Aging brain and ergot alkaloids," 14 (1981).

13. L. Carratelli, Nuovi analgesici sistemici nel trattamento del dolare da cancro, Giorn Geront 29:371 (1981).

14. G. Barbagallo Sangiorgi, A. Di Sciacca, G. Frada Jr., R. Malta and C. Botindari, Principi farmacologici della terapia del dolore nell'anziano, Giorn Geront 29:391 (1981).

15. J. Scott and E. C. Huskisson, Graphic representation of pain, Pain 2:175 (1976).

16. M. Neri and R. Grassi, La valutazione del dolore, Giorn Geront 29:339 (1981).

17. G. P. Vecchi, M. Neri and R. Lugii, La sensibilita al dolore nella vecchiaia, Giorn Geront 29:327 (1981).

18. S. L. Notermans, Measurement of the pain threshold determined by electrical stimulation and its clinical application, Neurology 19:1071 (1966).

19. B. B. Wolff, Measurement of human pain, in: "Pain," J. J. Bonica, ed., Raven Press, New York (1980).

20. E. Lavadas, C. Umilta and P. E. Ricci-Bitti, Evidence for sex differences in right hemisphere dominance for emotions, Neuropsychologia 18:361 (1980).

21. P. Flor-Henry and Z.J. Koles, EEG studies in depression, mania and normal: evidence for partial shifts of laterality in the affective psychoses, in: "Clinical neurophysiological aspects of psychopatological conditions, "C. Perris & L. von Knorring, eds., Karger, Basel (1980).

22. A. Malfara and B. Jones, Hemispheric asymmetries in motor control of guided reaching with and without optic displacement, Neuropsychologia 19:483 (1981).

23. G. P. Vecchi, L. Calza, M. Neri, Age and pain threshold, in: "Pain Therapy." M. Visentin and R. Rizzi, eds., Elsevier/North Holland Biomedical Press, Amsterdam (in press).

24. M. Hamilton, The assessment of anxiety states by rating, Br.J.Med.Psychol. 32:50 (1959).

25. R. C. Oldfield, The assessment and analysis of handedness. The Edinburgh inventory, Neuropsychologia 9:97 (1971).

26. L. A. Pini, Il trattamento del dolore conico grave, Giorn Geront 29:445 (1981).

27. L. von Knorring, F. Johansson and B. Almay, Augmenting/
     reducing response in visual evoked potentials in patients
     with chronic pain syndromes, Adv.Biol.Psychiat. 4:55 (1980).

INDEX